CELL and MUSCLE MOTILITY

Volume 5

The Cytoskeleton

Cell and Muscle Motility

A Continuation Order Plan is available for this series. A continuation order will bring delivery of each new volume immediately upon publication. Volumes are billed only upon actual shipment. For further information please contact the publisher.

CELL and MUSCLE MOTILITY

Volume 5

The Cytoskeleton

Edited by

Jerry W. Shay

University of Texas Health Science Center
Dallas, Texas

PLENUM PRESS • NEW YORK AND LONDON

The Library of Congress cataloged the first volume of this title as follows:

Main entry under title:
Cell and muscle motility.

Includes bibliographies and indexes.
1. Muscle contraction. 2. Cells—Motility. I. Dowben, Robert M. II. Shay, Jerry W. [DNLM: 1. Cytology—Periodical. 2. Muscles—Periodical. 3. Movement—Periodical. W1 CE127].
QP321.C365 599.01′852 81-13827

ISBN 0-306-41443-0

A Division of Plenum Publishing Corporation
233 Spring Street, New York, N.Y. 10013

Printed in the United States of America

Contributors

David F. Albertini, Department of Anatomy and Laboratory of Human Reproduction and Reproductive Biology, Harvard Medical School, Boston, Massachusetts, 02115

Zafrira Avnur, Department of Chemical Immunology, Weizmann Institute of Science, Rehovot 76100, Israel

H. Randolph Byers, Department of Anatomy, Harvard Medical School, Boston, Massachusetts 02115

Fernando Cabral, Departments of Medicine and of Biochemistry and Molecular Biology, University of Texas Medical School at Houston, Houston, Texas 77025

David J. DeRosier, Rosenstiel Basic Medical Sciences Research Center, Brandeis University, Waltham, Massachusetts 02254

Keigi Fujiwara, Department of Anatomy, Harvard Medical School, Boston, Massachusetts 02115

Benjamin Geiger, Department of Chemical Immunology, Weizmann Institute of Science, Rehovot 76100, Israel

Allen M. Gown, Department of Pathology SM-30, University of Washington, Seattle, Washington 98195

Brian Herman, Department of Anatomy and Laboratory of Human Reproduction and Reproductive Biology, Harvard Medical School, Boston, Massachusetts 02115

Yasuo Ichikawa, Department of Cytochemistry, Chest Disease Research Institute, Kyoto University, Kyoto, Japan

Thomas E. Kreis, Department of Clinical Immunology, Weizmann Institute of Science, Rehovot 76100, Israel

Jon C. Lewis, Department of Pathology, Bowman Gray School of Medicine, Wake Forest University, Winston-Salem, North Carolina 27103

Timothy W. McKeithan, Department of Biology, Yale University, New Haven, Connecticut 06511 *Present address:* Department of Pathology, University of Chicago, Chicago, Illinois 60637

Kazuhiro Nagata, Department of Cytochemistry, Chest Disease Research Institute, Kyoto University, Kyoto, Japan

Joel L. Rosenbaum, Department of Biology, Yale University, New Haven, Connecticut 06511

Joseph Schlessinger, Department of Chemical Immunology, Weizmann Institute of Science, Rehovot 76100, Israel

Manfred Schliwa, Department of Zoology, University of California at Berkeley, Berkeley, California 94720

Lewis G. Tilney, Department of Biology, University of Pennsylvania, Philadelphia, Pennsylvania 19104

Richard B. Vallee, Cell Biology Group, Worcester Foundation for Experimental Biology, Shrewsbury, Massachusetts 01545

Arthur M. Vogel, Department of Pathology SM-30, University of Washington, Seattle, Washington 98195

Glenn E. White, Department of Anatomy, Harvard Medical School, Boston, Massachusetts 02115

Preface

The preceding volumes of *Cell and Muscle Motility* have focused on various aspects of motile systems in both muscle and nonmuscle cells. These essays have been critical reviews on topics of current interest and, hopefully, have provided a base from which future investigations may develop. During the past decade, however, much attention in the fields of biochemistry and cell biology has focused on motile systems in nonmuscle cells. Our current understanding of the three-dimensional organization of the cytoplasm involve three major fibrous proteins which are collectively known as the cytoskeletal system. These polymorphic cytoskeletal proteins are microtubules (25-nm diameter), microfilaments (6-nm diameter), and intermediate filaments (10-nm diameter). Microtubules consist of tubulin and several well-characterized microtubule associated proteins (MAPs) including MAP_1, MAP_2, tau, and others. Microfilaments consist of actin and associate with actin-binding proteins including α-actinin, filamin, myosin, tropomyosin, vinculin, and others. Intermediate filaments (10-nm filaments) consist of at least five different tissue-specific classes, including desmin or skeletin (muscle), prekeratin (epithelial), vimentin (mesenchymal), neurofilament (nerve), and glial acidic fibrillary protein (astrocytes). These major fibrous proteins apparently interact with each other as well as other cytoplasmic components and appear to be intimately associated with such biological processes as cell shape changes, growth, motility, secretion, cell division, and uptake of materials from the exterior of the cell. Thus, the term cytoskeleton may be misleading since it is becoming clear that the function of these various filaments is not merely "skeletal."

Volume 5 of *Cell and Muscle Motility (The Cytoskeleton)* is distinctive from the preceding volumes in that it focuses exclusively on these cytoskeletal elements. These comprehensive reviews are directed not only at the structural organization and biochemical properties of the cytoskeleton but also at their function in the living cell. Even though many excellent reviews and books have appeared in recent years on the cytoskeletal system, it is my hope that the reviews in the present volume will foster an interchange of concepts among various researchers and serve as a reference for students and others who wish to familiarize themselves with the most current progress in this field.

I wish to thank Larry Goldes, Kirk Jensen, and their colleagues at Plenum Publishing Corporation for making my task much easier and the distinguished authors who have contributed chapters to this volume.

Jerry W. Shay

Contents

Chapter 4

Changes in Actin during Cell Differentiation

Kazuhiro Nagata and Yasuo Ichikawa

Chapter 5

The Dynamics of Cytoskeletal Organization in Areas of Cell Contact

Benjamin Geiger, Zafrira Avnur, Thomas E. Kreis, and Joseph Schlessinger

Chapter 7

The Biochemistry of Microtubules: A Review

Timothy W. McKeithan and Joel L. Rosenbaum

Chapter 8

MAP_2 (Microtubule Associated Protein 2)

Richard B. Vallee

Chapter 9

Genetic Dissection of the Assembly of Microtubules and Their Role in Mitosis

Fernando Cabral

Chapter 11
Monoclonal Antibodies to Intermediate Filament Proteins: Use in Diagnostic Surgical Pathology
Arthur M. Vogel and Allen M. Gown

1

Mechanisms of Intracellular Organelle Transport

Manfred Schliwa

1. Introduction

All eukaryotic cells exhibit some form of intracellular motility. The spectrum of motile activities observed inside living cells is extraordinarily broad, ranging from processes barely detectable even in time-lapse recordings, to the breathtaking bulk transport of cytoplasm best observed by slow-motion analysis. On the basis of their phenomenology, motile activities of cytoplasmic constituents may be subdivided into three broad categories:

1. Bulk movement of cytoplasm in association with, or as a consequence of, cell deformation. Good examples include the extension and retraction of cell processes (pseudopodia) in amoeboid cells, an activity accompanied by flow of cytoplasm into or out of these cell extensions. Cellular inclusions are carried along in the cytoplasm in a seemingly passive manner. Organelle translocation ceases as the protrusive or retractive activity of the cell comes to a standstill.

2. Uniform, continuous transport of organelles and cytoplasm along more or less defined pathways in the absence of cell deformation. Prime examples include the rotational or vectorial "streaming" of endoplasm in many plant cells, and shuttle streaming in slime molds.

3. Discontinuous, erratic, "saltatory" movements of particles and organelles. This form of transport is observed in a wide variety of eukaryotic cells and includes such diverse phenomena as organelle movements in protists and fast transport of materials in neurons.

Manfred Schliwa • Department of Zoology, University of California at Berkeley, Berkeley, California 94720.

It is the objective of this review to consider several examples of intracellular transport in some detail and to examine possible mechanisms by which the forces responsible for movement may be generated. Although the borders between the three classes of transport listed above are ill-defined, a widely held belief has developed that they differ with respect to the cytoskeletal components with which they are associated and on which they depend for force generation: the first two classes of cytoplasmic transport have been associated with microfilaments and the third with microtubules. Further, by comparison with the two well-studied paradigms for microfilament- and microtubule-based motility—muscle contraction and ciliary movement—it has often been suggested that microfilament-associated cytoplasmic transport is an actomyosin-based contractile process, while microtubule-associated saltatory transport involves a dynein-like ATPase. This review specifically examines the evidence for (and against) the view that in fact two different molecular mechanisms have developed that are responsible for the different categories of intracellular transport.

Since space considerations make it impossible to cover the entire spectrum of motile events in one review, the discussion here will be limited to a specific subset of intracellular transport, namely, that involving the movement of organelles (often referred to as "particles" because of their unknown submicroscopic identity) and/or supramolecular organelle complexes. I will not deal with the first class of intracellular movement (passive bulk flow), nor will I discuss transport of metabolites, or proteins. I will also ignore chromosome transport during mitosis and meiosis since many excellent reviews already exist. Finally, the discussion of cytoplasmic streaming will be restricted to plant cells. Throughout this review, emphasis is placed on findings during the past decade or so, without restating the basic properties of intracellular transport so excellently summarized in Rebhun's (1972) classical overview.

Many other good review papers with a more narrow scope exist. Hyams and Stebbings (1979), for instance, have written a useful summation of microtubule-associated transport. Reviews of intracellular movement in certain cell types will be mentioned in Section 2 when I discuss selected model systems. Compared to other fields of cell motility, however, the number of overview articles is modest, owing to the fact that progress in our knowledge of the mechanisms of intracellular transport is relatively slow. Despite a wealth of detailed information, suggestions as to the biochemical nature of the "motor" are based mainly on circumstantial evidence. The main reason for this deficiency lies in the lack of applicability of some biochemical approaches which, in other areas of research, have led to a deeper understanding of the properties of cell components involved in motility. There are indications, however, that this situation may change in the near future. A summary of what has been achieved so far therefore seems appropriate.

This paper begins by outlining the basic properties of selected cellular model systems frequently used for the study of organelle movements. The physiology of movement will then be discussed in the following paragraphs.

2. *Some Cell Systems Used in the Study of Intracellular Particle Transport*

2.1. Neurons

The neuron is a cell of remarkable anisotropy. Its cell body extends one or more axonal or dendritic cell processes which, in some mammals, may reach a length of several meters. Frequently, the total mass of the extensions may exceed that of the cell body several hundred times. In addition to this morphological asymmetry, there is a functional asymmetry since the protein-synthesizing capacity essentially resides in the cell body; all materials destined for the cell periphery are manufactured there and have to be transported along the axon or the dendritic processes. The mechanism of this transport which for reasons of simplicity has been studied predominantly in axons has been the focus of considerable scientific attention. A voluminous literature has accumulated over the past few decades dealing with various aspects of intra-axonal transport. Some selected topics have been the subject of recent reviews (Samson, 1976; Hanson and Edström, 1978; Kristensson, 1978; Rambourg and Droz, 1980; Wilson and Stone, 1979). More comprehensive surveys of the literature have been provided recently by Lubinska (1975), Heslop (1975), Schwartz (1979), and Grafstein and Forman (1980); workshop proceedings were edited by Thoenen and Kreutzberg (1981) and Weiss (1982).

Initially, studies of anterograde transport revealed basically two rates of transport, one constant slow movement of materials along the axon at about 1–2 mm/day (Weiss and Hiscoe, 1948), and a much faster (more than 100 times) wave of transport (Miani, 1960; Lubinska, 1964) which could be correlated with rapid movements of microscopically visible organelles (for reviews of the earlier literature, see Pomerat *et al.*, 1967; Dahlström, 1971b; Ochs, 1972b). As a result of more detailed studies, at least five different transport groups are now distinguished (Willard *et al.*, 1974; Lasek and Hoffmann, 1976; Black and Lasek, 1980; Ochs, 1972a; Stone *et al.*, 1978; for reviews, see Grafstein and Forman, 1980; Wilson and Stone, 1979).

Group I. Materials moving at a rate of several hundred (usually 200–400) mm/day. This group has been designated the fast component (FC) and seems to consist almost exclusively of membrane-bounded organelles. Fast transport of materials is not only directed away from the cell body (anterograde transport), but also occurs prominently in the opposite direction (retrograde transport).

Group II. Materials transported at about 20–60 mm/day depending on the cell system used (Karlsson and Sjöstrand, 1971a; Lorenz and Willard, 1978). This transport group, like FC, probably consists of membrane-bounded materials.

Group III. Transport rate about 10 mm/day. Relatively few components have thus far been resolved in group III; one of the proteins transported at this rate probably corresponds to a myosin-like protein (Willard, 1977).

The material transported at the intermediate rates II and III constitutes only a minor proportion relative to that transported in group I and the two slow components of groups IV and V.

Group IV. Transport in this group, also termed slow component b (SCb) of axonal transport, proceeds at a rate of 2–4 mm/day. Two-dimensional polyacrylamide gel electrophoresis of radioactively labeled proteins discloses this wave of transported proteins to consist of more than 100 polypeptides. Only some of these proteins have so far been identified, including actin (Black and Lasek, 1979), clathrin (Garner and Lasek, 1981), calmodulin (Brady *et al.*, 1981), and two enzymes of intermediary metabolism, enolase and creatine phosphokinase (Brady and Lasek, 1981).

Group V. This group of proteins designated slow components a (SCa) by Lasek and Hoffman (1976), moves at a rate of 0.2–1 mm/day and is much simpler than that of SCb. It consists primarily of the subunit polypeptides of neurofilaments (the so-called neurofilament triplet: polypeptides of 68, 145 and 200 kd) and microtubules (Hoffmann and Lasek, 1975; Black and Lasek, 1980).

Although for the remainder of this review I will deal primarily with organelle transport (i.e., the fast component of axonal transport), the relationship between the different components of axonal transport, notably the FC and the two groups of SC, should briefly be considered here.

It is clear from the analysis of the different transport groups by gel electrophoresis that FC, SCa, and SCb have a distinctly different composition with virtually no overlap in their constituting polypeptides (Tytell *et al.*, 1981). The different components are able to move past one another with little interference or "mixing" of constitutents: Actin and what appears to be proteins of the cytoplasmic matrix (SCb) move as a coherent complex past the slower-moving microtubule–neurofilament network (SCa), while both are continuously passed by rapidly moving membrane-bounded organelles (FC). Lasek and collaborators, whose work has significantly contributed to our knowledge of this aspect of transport in neurons, have developed a structural hypothesis of axonal transport which holds that the three rate components represent the movement not of individual proteins, but supramolecular cytological structures (see, e.g., Tytell *et al.*, 1981). They suggest that the proteins comprising these classes cotranslate either as an integral part of, or in longterm association with, one of these structures. One interesting aspect of this hypothesis concerns the behavior of cytoplasmic constituents normally thought to be soluble or freely diffusible; many of these constituents are transported in SCb in a coherent fashion and thus appear to be strongly associated with the actin-based cytoplasmic matrix in this component. This observation may have far-reaching consequences for our understanding of the organization of the cytoplasm.

The structure of most axons, myelinated or unmyelinated, vertebrate or invertebrate, is generally rather simple. It comprises a cylindrical cell extension whose cytoskeleton consists mainly of longitudinally arranged microtubules and neurofilaments (Weiss and Mayr, 1971a,b; Wuerker and Kirkpatrick, 1972). The relative proportion of these two structural components depends on species, cell type, size of the axon, and developmental stage of the neuron. Some invertebrate nerve cells do not contain any intermediate filaments, e.g., crayfish neurons, whereas others may be packed with intermediate filaments but are virtually free of microtubules, e.g., giant axons of *Myxicola.* In transverse sections of most mammalian axons, microtubules are loosely clustered locally (Wuerker and Kirkpatrick, 1972; Tsukita and Ishikawa, 1981; Papasozomenos *et al.,* 1981). The continuity of these two cytoskeletal structures along the length of the axon has been a matter of some controversy. Whereas Weiss and Mayr (1971b) originally suggested that microtubules might be continuous along most of the axon, later studies could not support this notion (Zenker and Hohberg, 1973; Nadelhaft, 1974). On the contrary, the first study that reconstructed the course of microtubules in a specialized neuron of a nematode by serial thin sections suggested that most microtubules are very short (5–25 μm) compared to the length of the axon (~500 μm; Chalfie and Thomson, 1979). Subsequent serial section analyses of mammalian axons (Tsukita and Ishikawa, 1981; Bray and Bunge, 1981) support this conclusion, although in these studies the average length of a microtubule (assuming that all microtubules are of equal length) was calculated to around 500 and 100 μm, respectively. The number of neurofilaments decreases dramatically at the node of Ranvier, whereas microtubules pass this region of the axon without a decrease in number (Fig. 1). Thus, most neurofilaments appear to be discontinuous at the nodal regions.

The organization of the third major cytoskeletal element, actin microfilaments, is less well understood. Metuzals and Tasaki (1978) demonstrated the presence of a subaxolemmal network of filaments in squid giant axons, some of which have been identified as actin by the HMM-binding technique. Using the same method, LeBeux and Willemot (1975) have shown an open network of actin filaments in glycerinated tissues derived from the rat caudate nucleus and substantia nigra. Fluorescently labeled HMM-S1 will stain chick dorsal root ganglion cells in culture (Kuczmarski and Rosenbaum, 1979b), but the resolution of the light microscope does not allow to distinguish between a peripheral or central localization within the neurites. Filaments resembling F-actin have, however, been identified in growth cones of cultured ganglion cells (Kuzcmarski and Rosenbaum, 1979b). Actin filaments are a prominent component of cultured neuroblastoma cells where they occur primarily in the cell cortex or as parallel bundles in the microspikes (Burton and Kirkland, 1972; Chang and Goldman, 1973; Ross *et al.,* 1975; Kuczmarski and Rosenbaum, 1979b). Myosin has been isolated from brain (Puszkin *et al.,* 1968; Berl *et al.,* 1973; Kuczmarski and Rosenbaum, 1979a) and has been shown to be present in synaptosomes by subcellular fractionation (Puszkin *et al.,* 1972) and in neurites of cultured dorsal root ganglion cells by immunofluorescence microscopy (Kuczmarski and Rosenbaum, 1979b).

In addition to these major cytoskeletal components, a number of accessory proteins has been identified in nerve cells (for a recent review, see Bray and Gilbert, 1981). These include: (1) α-actinin, the Z-line protein of muscle fibers

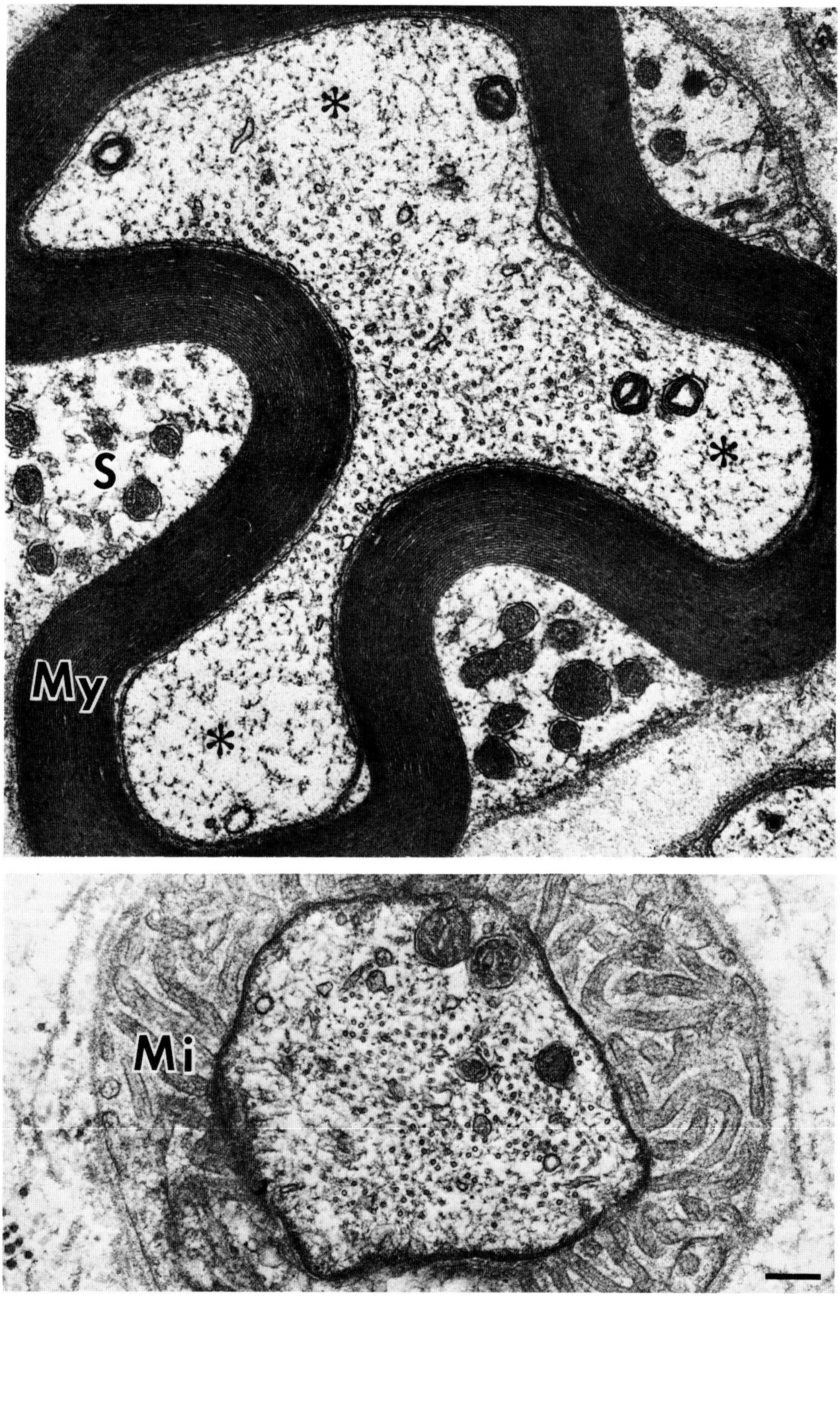

S
My
Mi

(Jokusch *et al.*, 1979), (2) the nonmuscle form of tropomyosin (Bretscher and Weber, 1978), (3) an actin-binding protein similar to, but not identical with, filamin (Walach *et al.*, 1978); (4) profilin, a low molecular weight protein that binds monomeric actin (Carlsson *et al.*, 1977), (5) a pair of two high-molecular-weight proteins termed fodrin (Levine and Willard, 1981; Glenney *et al.*, 1982) that seem to be localized in the cortex of axons, and (6) the high-molecular-weight microtubule-associated proteins MAP_1 (about 300 kd) and MAP_2 (about 280 kd) and the low-molecular-weight MAPs designed tau proteins (for reviews of the voluminous literature on these two classes of proteins, see Stephens and Edds, 1976; Kirschner, 1978; Scheele and Borisy, 1979). The precise function of any of these components in axonal transport (or neuronal integrity for that matter) is only poorly understood.

Microtubules and microfilaments are interconnected by filamentous projections that make contact with the surfaces of both of these structures (Metuzals, 1969; Wuerker and Kirkpatrick, 1972; Burton and Fernandez, 1973; Fernandez *et al.*, 1971; Ellisman and Porter, 1980). These filamentous elements integrate microtubules, microfilaments, and other organelles into a three-dimensional superstructure (Ellisman and Porter, 1980). Many of these links are stable enough to resist extraction in detergent-permeabilized axons (Hirokawa, 1982; Fig. 2). The biochemical composition of these projections is not known. They have been thought to have a motile function and to be composed, at least in part, of the HMW MAPs, one of which (MAP_2) will decorate isolated microtubules with slender projections reminiscent of those seen in axons (Dentler *et al.*, 1975; Murphy and Borisy, 1975; Kim *et al.*, 1979). Later it was shown that MAP_2 is confined to the dendritic portions of neurons (Matus *et al.*, 1981) and is not present in axons, challenging the notion of an involvement of MAP_2 in axonal transport. The filaments of this matrix structure have an affinity for lanthanum ions (Burton and Fernandez, 1973), bismuth, and ruthenium red (Tani and Ametani, 1970; Burton and Hinkley, 1974), indicating a strongly polyanionic character and/or the presence of mucopolysaccharides moieties. Tytell *et al.* (1981) suspect that some of its components might represent the structural assemblies of proteins that travel as a coherent wave in slow component b.

Aside from the cytoskeletal and matrix components, basically three different classes of membraneous organelles have been described in the axon: mitochondria, an anastomozing system of axonal SER, and various types of vesicular structures such as dense core granules, multivesicular and multilamellar bodies, and clear vesicles (Grafstein and Forman, 1980). That transport in FC is associated with one or more of these membraneous organelles

Figure 1. Two transverse sections from a section series through a myelinated axon of the mouse saphenous nerve. (Top) Section at a distance of about 5 μm from the node of Ranvier. Microtubules are packed in the central portion of the axon, whereas the peripheral portion is occupied by neurofilaments (asterisks). (Bottom) At the level of the node of Ranvier, the number of microtubules is almost the same as in the top section, whereas the number of neurofilaments has decreased dramatically. S, Schwann cell; My, myelin; Mi, microvilli of the Schwann cell. Bar = 0.25 μm. Reprinted with permission from Tsukita and Ishikawa (1981).

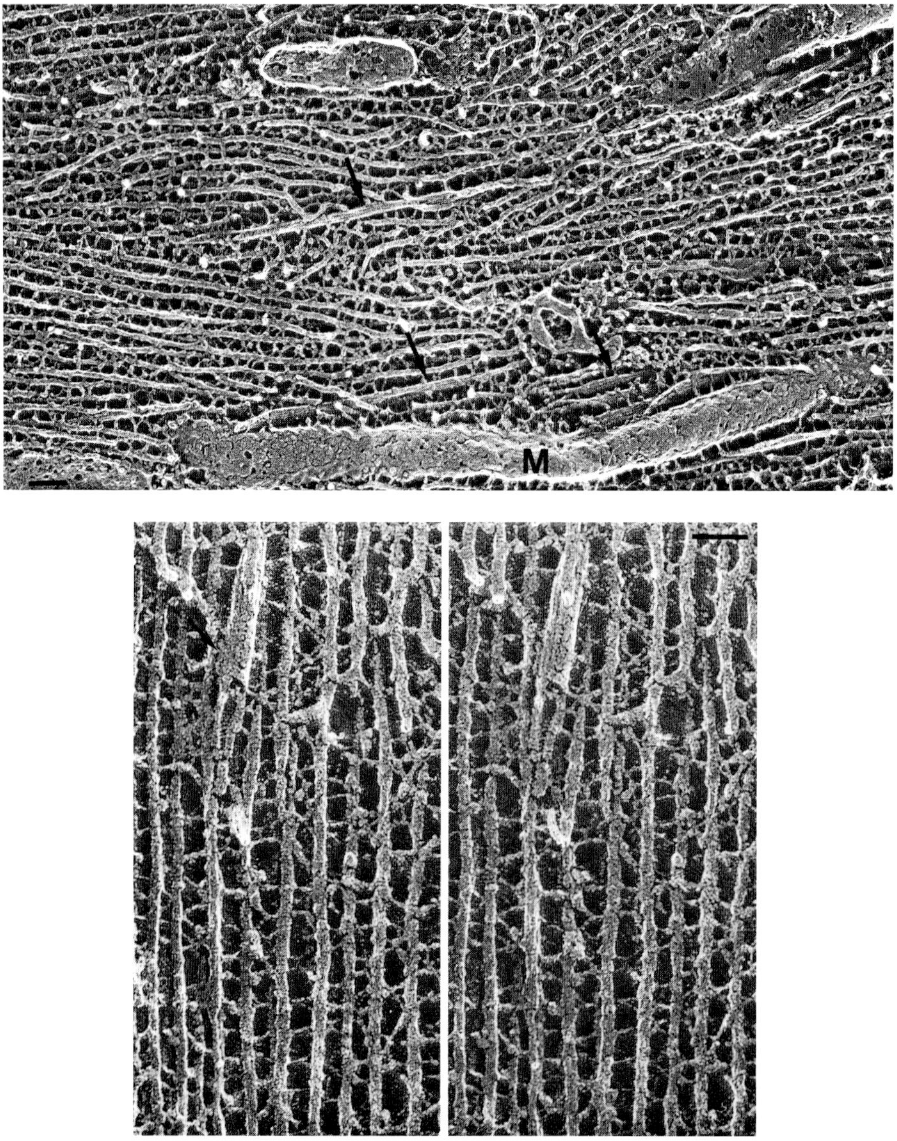

Figure 2. (Top) Low magnification overview of a saponin-treated frog axon prepared by the quick-freezing, deep-etching, rotary-shadowing technique. Saponin treatment did not significantly affect the mitochondrial membrane (M). The axoplasm is filled with numerous neurofilaments aligned parallel to the long axis of the axon. Several microtubules (arrows) are interspersed between the neurofilaments. An extensive system of saponin-resistant cross-linkers interconnects the filaments and the microtubules. Bar = 0.1 μm. (Bottom) Stereo micrograph of part of the axoplasm, showing the arrangement of cross-linkers between the neurofilaments. The cross-linkers are 4–6 nm in diameter and up to 30 nm in length. Cross-linkers also extend between a microtubule (arrow) and neighboring neurofilaments. Bar = 0.05 μm. Reprinted with permission from Hirokawa (1982).

became established more than a decade ago, mainly on the basis of three lines of evidence: tracer studies, accumulation experiments, and direct microscopic observation.

2.1.1. *Tracer Studies*

Radioactively labeled glycoproteins (Edström and Mattsson, 1972b; Elam and Agranoff, 1971; Elam *et al.*, 1970; Forman *et al.*, 1971, 1972; Karlsson and Sjöstrand, 1971b; Zatz and Barondes, 1971) and lipids (Miani, 1963; Abe *et al.*, 1973; Grafstein *et al.*, 1975), both of which are believed to be components of membraneous organelles, are transported in FC along with a certain fraction of labeled proteins (Barondes, 1967; Grafstein, 1967; Lasek, 1968; Ochs *et al.*, 1967).

2.1.2. *Accumulation Experiments*

Membraneous materials, including neurotransmitter storage granules, accumulate at the site of a lesion or block (Dahlström and Häggendal, 1966, 1967; Jeffrey and Austin, 1973; Tsukita and Ishikawa, 1980; Smith, 1980). There is a distinct difference in the morphological appearance of materials accumulating at either side of the block (Figs. 3 and 4). The organelles piled up proximally consist of small vesicles or tubes and dense core granules (the tubovesicular system; Hendrickson, 1972; Schonbach *et al.*, 1971; Tsukita and Ishikawa, 1980), while the retrograde fraction of transported organelles consists mainly of large membraneous structures such as multivesicular or multilamellar bodies (Tsukita and Ishikawa, 1980). The materials of the latter are believed to be destined for degradation and possibly recycling. In addition, there is retrograde transport of materials that have been taken up by the nerve terminals, such as horseradish peroxidase, nerve growth factor, tetanus toxin, and wheat germ agglutinin (Hendry *et al.*, 1974; La Vail and La Vail, 1974; Schwab, 1977; Schwab and Thoenen, 1978; Dumas *et al.*, 1979; Broadwell and Brightman, 1979; Bunt and Haschke, 1978). Some authors have invoked tubular elements of the SER in the retrograde transport of these materials (Nauta *et al.*, 1975; Sotelo and Riche, 1974), while others have presented evidence for transport in vesicular organelles (Schwab, 1977; Schwab *et al.*, 1979; La Vail *et al.*, 1980; Broadwell and Brightman, 1979; Tsukita and Ishikawa, 1980).

2.1.3. *Direct Observation*

Appropriate light microscopic techniques directly visualize the movement of "particles" in both the anterograde and retrograde direction (Berlinrood *et al.*, 1972; Forman *et al.*, 1977a,b; Leestma and Freeman, 1977; Smith, 1972; Smith and Koles, 1976). As a rule, movement in the retrograde direction is much more prominent (Cooper and Smith, 1974; Forman *et al.*, 1977a). The identity of these particles is uncertain, but many of them probably correspond to the large membraneous vesicles identified by electron microscopy as

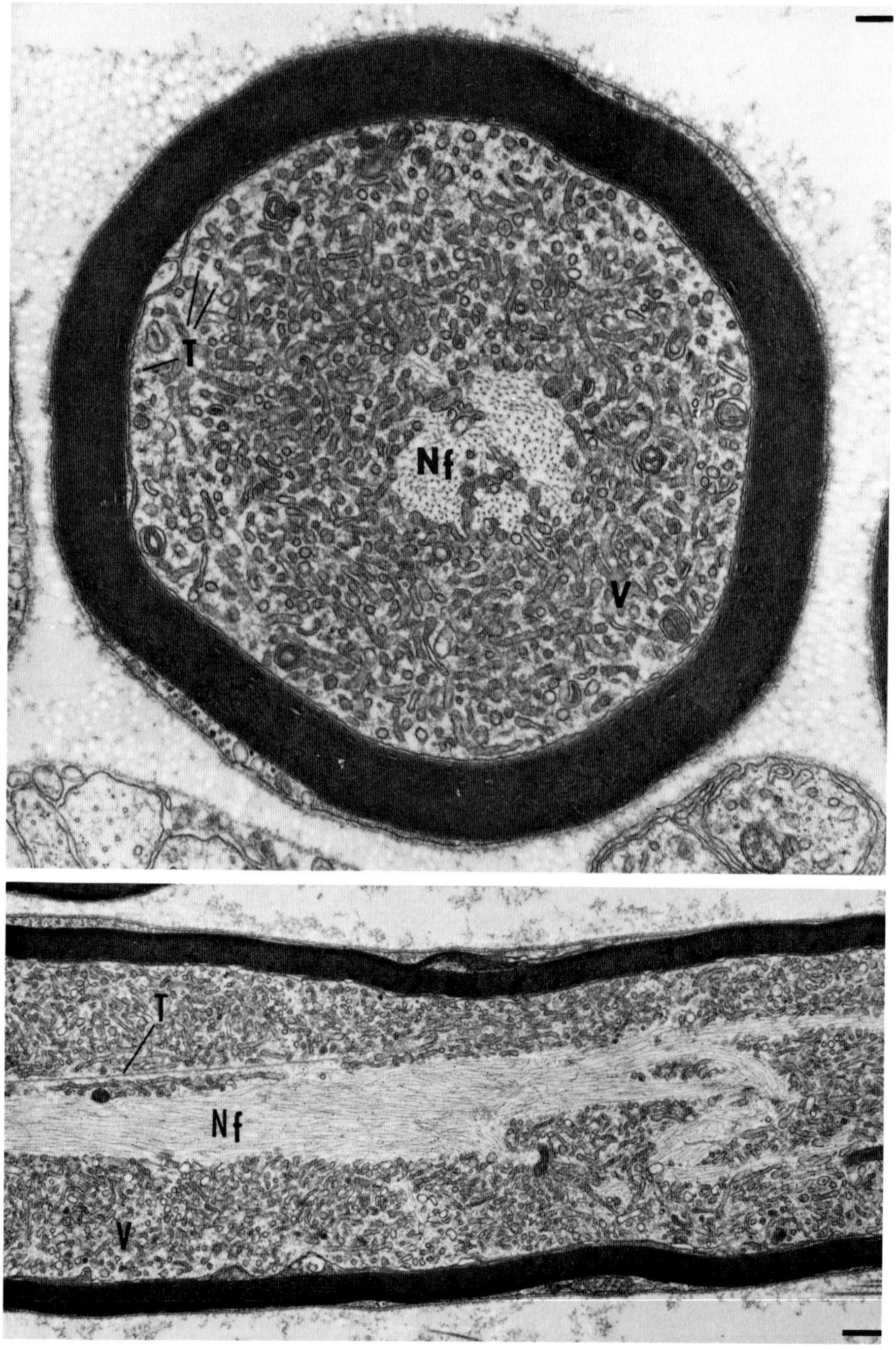

Figure 3. Internodal region of the mouse saphenous nerve just proximal to the site where a cold block had been applied. The material that has accumulated proximally by anterograde axonal transport consists mainly of vesiculo-tubular membraneous structures (V). Neurofilaments (NF) are clustered in the center of the axon, some microtubules (T) are present in a more peripheral location. (Top) Transverse section; bar = 0.25μm. (Bottom) Longitudinal section; bar = 0.5 μm. Reprinted with permission from Tsukita and Ishikawa (1980).

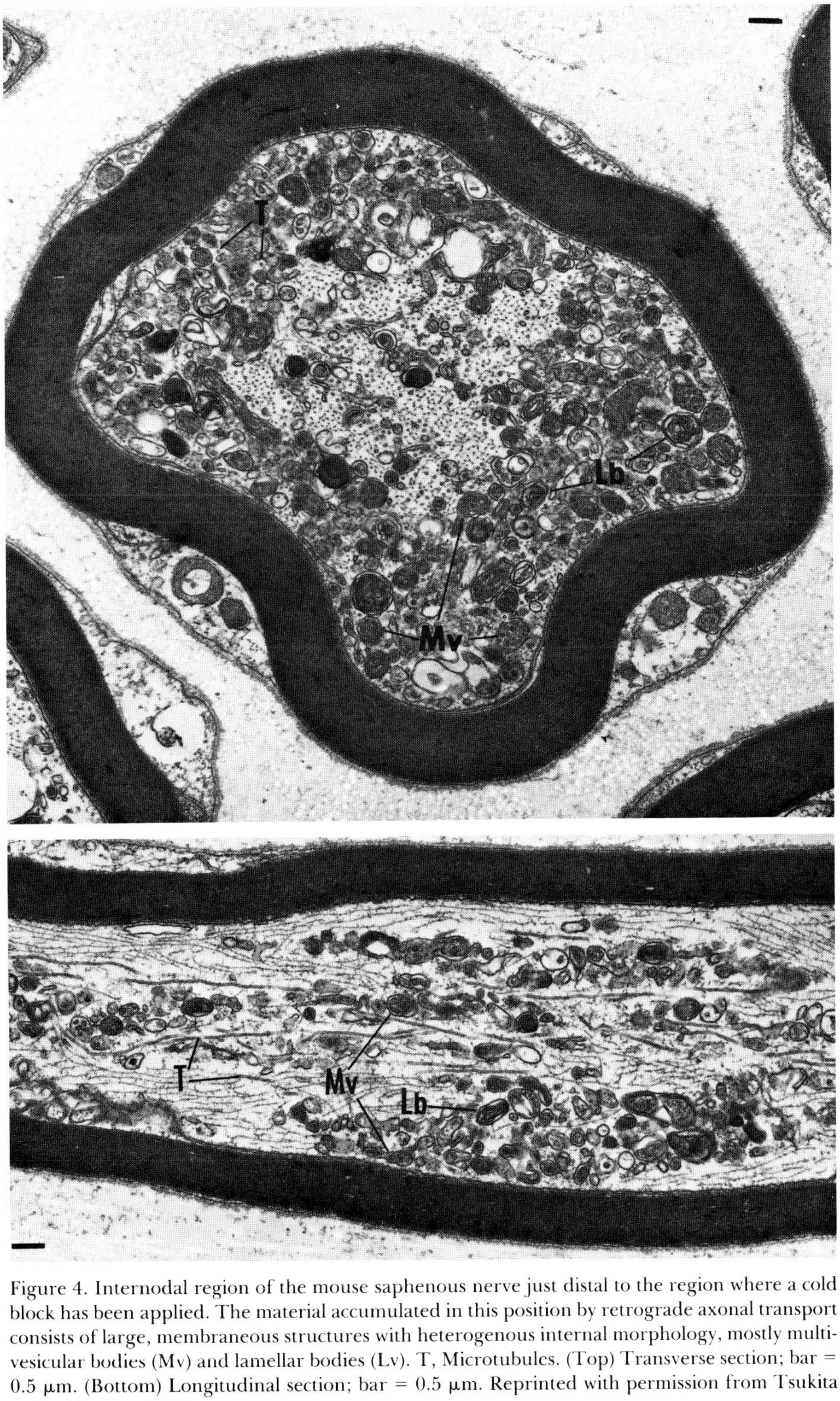

Figure 4. Internodal region of the mouse saphenous nerve just distal to the region where a cold block has been applied. The material accumulated in this position by retrograde axonal transport consists of large, membraneous structures with heterogenous internal morphology, mostly multivesicular bodies (Mv) and lamellar bodies (Lv). T, Microtubules. (Top) Transverse section; bar = 0.5 μm. (Bottom) Longitudinal section; bar = 0.5 μm. Reprinted with permission from Tsukita and Ishikawa (1980).

multivesicular bodies (Breuer *et al.*, 1975). Because of the limits of the resolution of the light microscope it was not possible to observe smaller organelles such as synaptic, neurosecretory, coated, or other vesicles. This situation has changed recently with the development of powerful video techniques that enhance visibility of "submicroscopic" organelles dramatically (Allen *et al.*, 1981a,b; Inoue, 1981). Using this new technique, particles similar in size to synaptic vesicles have been seen to move at rates compatible with the FC of axonal transport (Allen *et al.*, 1982, Brady *et al.*, 1982). Particle velocities are in the range of 1–5 μm/sec (Forman *et al.*, 1977a,b; Leestma and Freeman, 1977). A different class of axoplasmic organelles frequently observed to move in both the retrograde and anterograde direction are rod-shaped or thread-like structures which most probably correspond to mitochondria. Both particles and mitochondria undergo a saltatory type of motility characterized by frequent starts and stops or occasional reversals in the direction of movement (Fig. 5). There is, however, a noted difference in the behavior of "particles" and mitochondria. Whereas many of the latter remain stationary for long periods of time, particles may move continuously for relatively long distances. In fact, particle movement sometimes appears more as constant, smooth streaming rather than erratic, discontinuous jumping. Particles have a preferred direction of movement, i.e., a retrogradely traveling particle does not suddenly reverse its direction of movement and continues to move in the anterograde direction, except for very brief periods of time. It seems as if particles have a destination and "know" which way to go. Particles traveling in opposite directions may pass each other in close proximity to one another. Preformed tracks or channels for either the anterograde or retrograde direction of movement do not seem to exist. The velocities of movements in either direction are approximately the same (Cooper and Smith, 1974; Leestma and Freeman, 1977).

Several electron microscopic studies have noted the existence of cross-bridges between membrane-bounded organelles and axonal microtubules (Jarlfors and Smith, 1969; Smith, 1972; La Vail and LaVail, 1974; Smith *et al.*, 1977; Raine *et al.*, 1971; Hirokawa, 1982; see Fig. 6). That these links do in fact represent the morphological equivalent of a microtubule-associated, polarized force-generating mechanism has been proposed by Cooper and Smith (1974). It is equally plausible, however, that these links represent stable connections between a microtubule and an organelle, the latter in this case being stationary (Schnapp and Reese, 1982).

2.2. *Pigment Cells*

Many invertebrates and lower vertebrates possess stellate or discoid pigment cells (chromatophores) in their integument that chiefly function in the chromatic adaption of the animal. According to the nature of the pigment they contain, four major groups of chromatophores are generally distinguished: (1) melanophores bearing brown to black melanin granules, (2) erythrophores containing pteridine pigments and/or carotenoid droplets of a

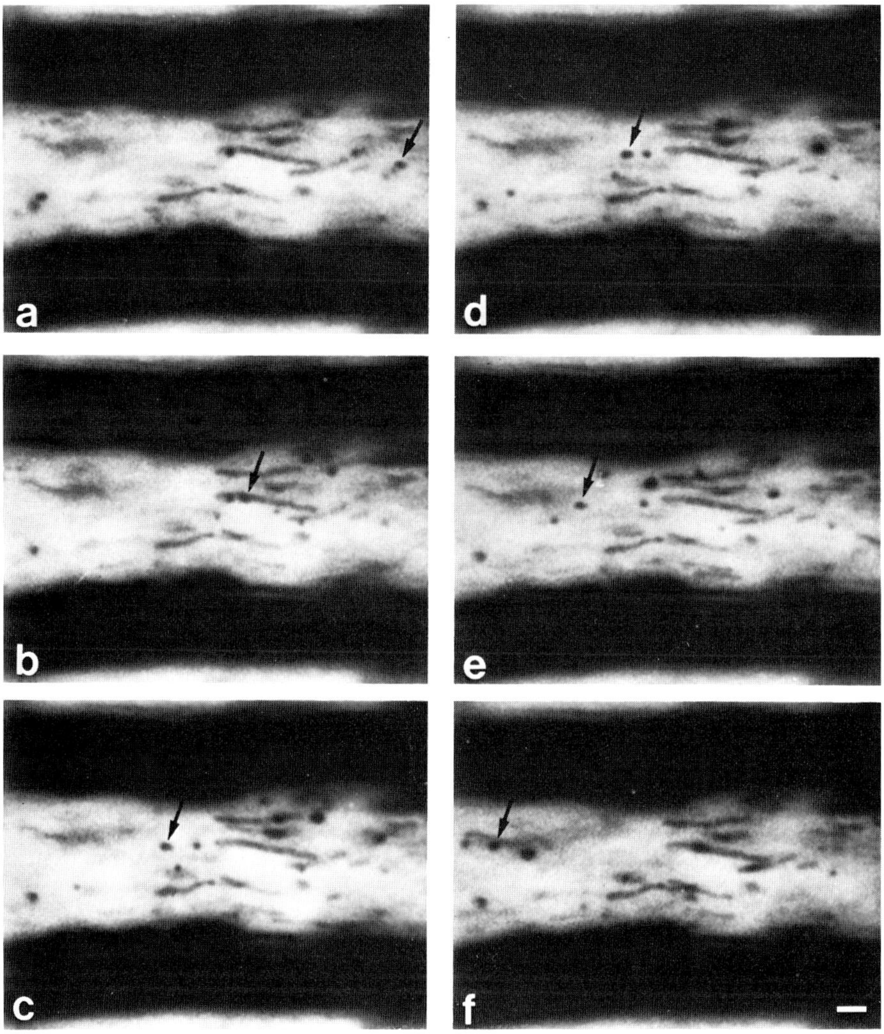

Figure 5. Six frames from a 16 mm movie made by Forman, Padjen, and Siggins. The film shows the movement of particles in a myelinated frog axon in dark field illumination. The frames shown here are printed in reverse contrast. A particle (arrow) traveling in the retrograde direction (a,b) stops transiently (c), and reverses its direction of movement (d) before continuing its movement towards the cell body. The elongated, moderately dense structures are mitochondria. Bar = 2 μm. By permission of Dr. David Forman.

reddish color, (3) xanthophores with yellowish carotenoid droplets, and (4) iridophores containing light-reflecting platelets. Occasionally, a pigment cell may carry two or more types of granules or pigments; some crustacean chromatophores, for instance, contain up to four different types of pigment inclusions (Robison and Charlton, 1973).

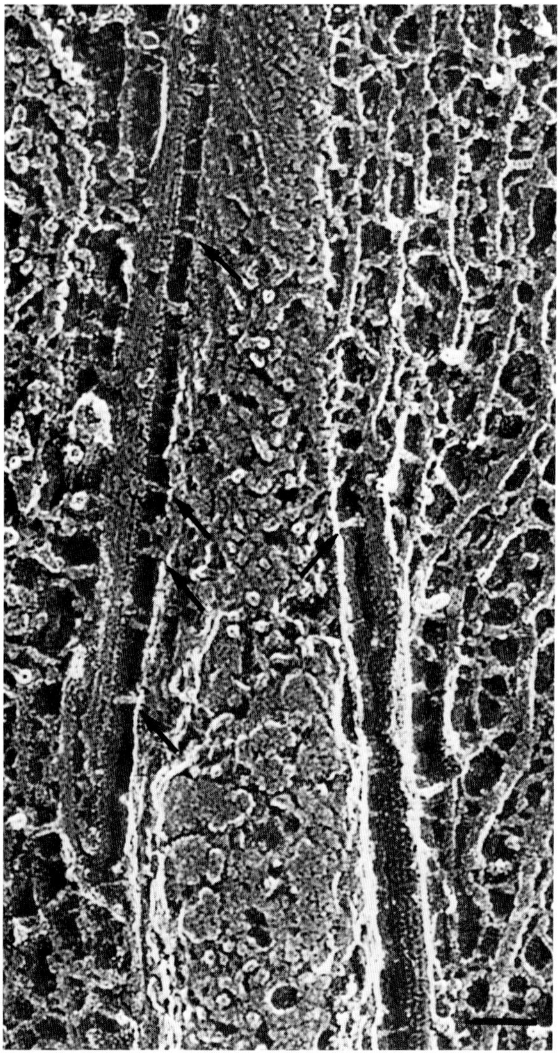

Figure 6. Association between microtubules and a mitochondrion in a saponin-treated axon prepared by quick-freezing. Thin cross-bridges (arrows) extend between the microtubule wall and the outer mitochondrial membrane. Bar = 0.05 μm. Reprinted with permission from Hirokawa (1982).

Since Brücke (1852) first discovered that color change in animals is based on the intracellular redistribution of pigment-containing organelles, the movement of pigments in chromatophores has been the subject of numerous investigations (for recent reviews, see Fingerman, 1965; Fujii, 1969; Novales and Davies, 1969; Bagnara and Hadley, 1973; Schliwa, 1981, 1982b; Luby-Phelps and Schliwa, 1982). In adaptation to light intensity or background coloration, pigment cells may redistribute the bulk of pigmentary organelles, the two extreme states being complete dispersion where granules are more or less evenly distributed throughout the cell, and aggregation where granules are piled up in the central cell region (Fig. 7). In general, the outline of the cell remains unchanged, i.e., in the aggregated state the cell still possesses cell extensions (arms or processes) which now are devoid of pigment granules. Exceptions to this rule are known, however (Weber and Dambach, 1972; Gras and Weber, 1977; Kopenec, 1949). In many chromatophores, the shape of the cell as seen in transverse section may change dramatically (Fig. 8).

The range and speed of pigment granule redistribution is remarkable. Probably the fastest are the erythrophores of the squirrel fish *Holocentrus* studied by Porter and his associates. In these cells, the velocity of pigment granule movement may reach 20 μm/sec and aggregation is accomplished within 3–5 sec. In the much larger melanophores of the angelfish, aggrega-

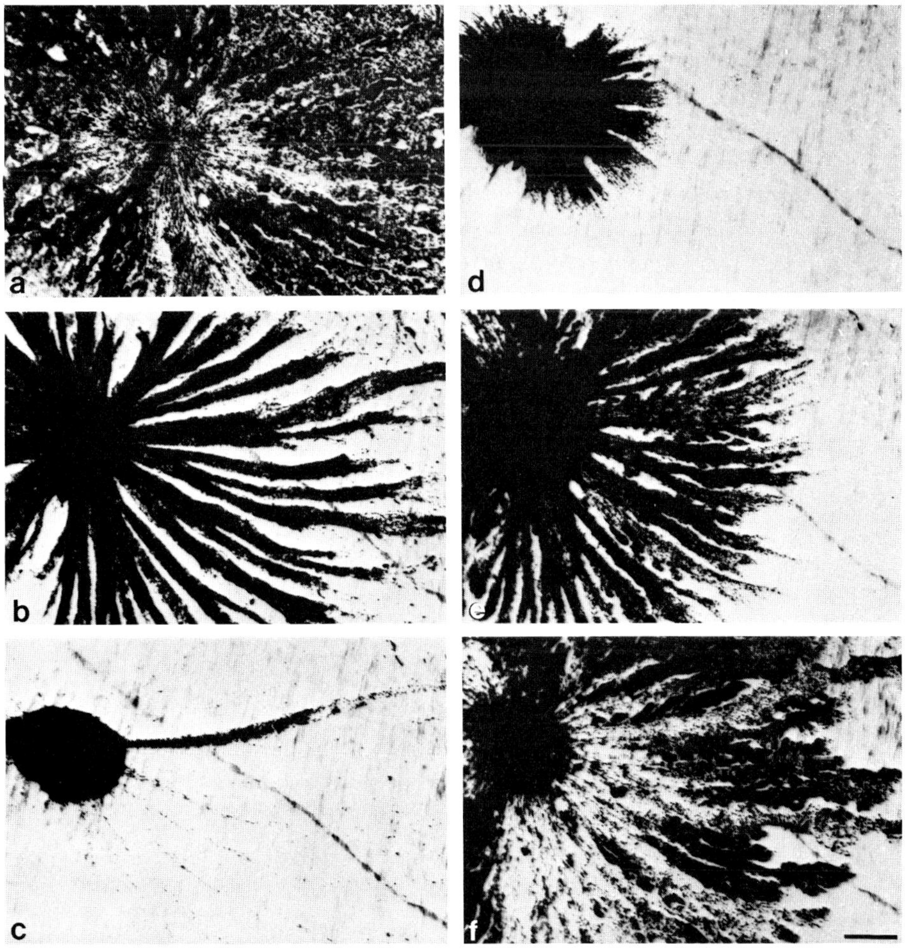

Figure 7. Pigment aggregation and dispersion in a melanophore of the black tetra, *Gymnocorymbus ternetzii*. For this series of photographs, a scale was excised and viewed in a brightfield light microscope. (a) Dispersed state, (b) 30 sec after initiation of aggregation with 10^{-5} M adrenaline, (c) almost completely aggregated state 2 min after adrenaline stimulation, (d) 20 sec after initiation of dispersion with 10^{-5} M atropine, (e) 2 min after atropine stimulation, and (f) almost completely dispersed state 4 min after addition of atropine. Bar = 5 μm.

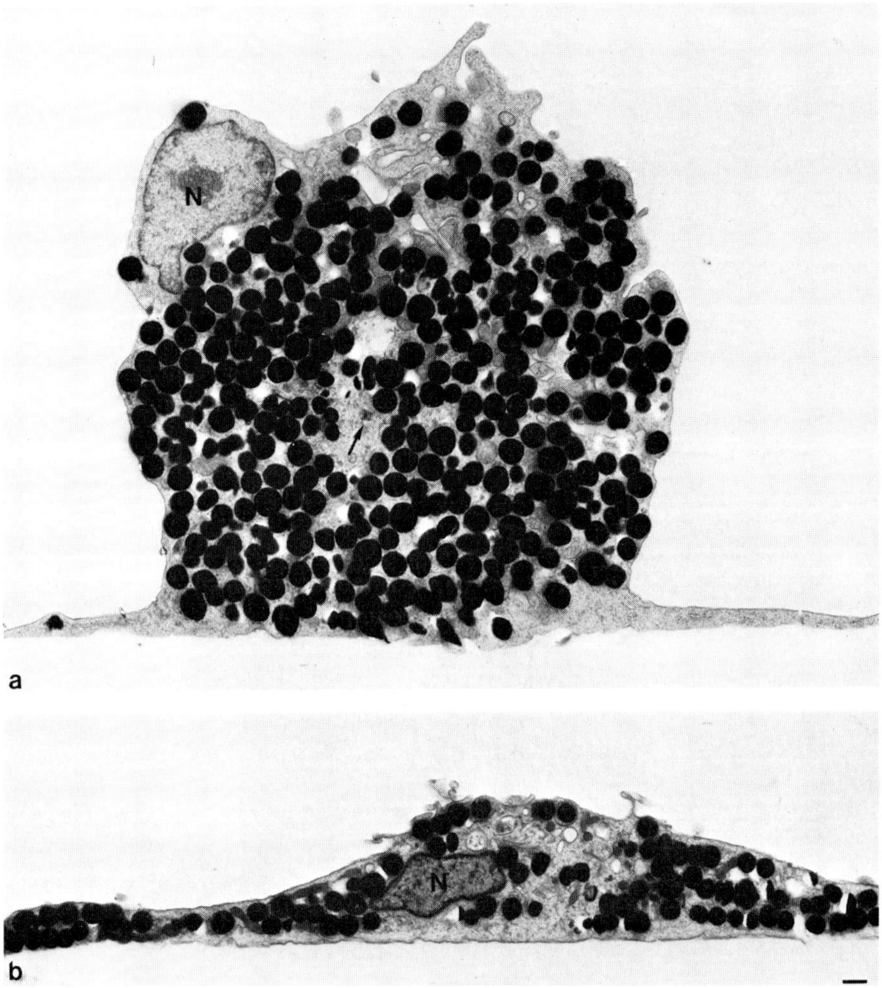

Figure 8. Transverse sections of isolated melanophores of the angelfish spread on a coverslip. (a) Aggregated state, pigment granules are piled up in the cell center, while the cell periphery is collapsed and devoid of pigment granules. The nucleus (N) is displaced towards the periphery of the pigment mass. The arrow indicates a centriole. (b) Dispersed state, pigment granules are evenly distributed throughout the cell. The average thickness of the cell is 1.5 μm. Bar = 0.5 μm. Reprinted with permission from Schliwa and Euteneuer (1978a).

tion requires approximately 10–20 sec for completion (Egner, 1971; Schliwa and Bereiter-Hahn, 1973a). Most other chromatophores described complete aggregation within a matter of minutes, with granules travelling at a rate much slower than 1 μm/sec.

Early in the study of chromatophore motility, it was noted that granules tend to move in linear paths arranged radially around the cytocentrum (Heidenhain, 1907). These pathways are in part defined by the shape of the cell processes, some of which are too narrow to allow the passage of more than two or three rows of pigment granules. However, other cell types, such as the *Holocentrus* erythrophore are perfectly disk-shaped without cell extensions; nevertheless, pigment granules are arranged in a striking radial array. As early as 1914 this observation led to the proposal that the protoplasm of chromatophores is organized into *canaliculi* in which pigments move (Ballowitz, 1914). Electron microscopic studies of the past 15 years strongly suggest that the radial alignment of granules is strictly correlated with the presence of microtubules of a similarly precise arrangement. This remarkable correlation alone has led some investigators to believe that microtubules play an important role in granule movements and possibly even deliver the motive force (see Section 3.1.1).

Light microscopic observations of living chromatophores, notably of fishes, show that even during maintenance of the dispersed state, pigment granules are anything but stationary. There is a constant shuttling of granules back and forth over distances of a few micrometers. The vector sum of all these centrifugal and centripetal short-range movements is zero, so the cell remains in the dispersed state. During mass aggregation, this balance is eliminated and centripetal movement takes over, with all granules moving at essentially the same velocity in a continuous, resolute fashion without shuttling. Dispersion usually takes two or three times longer than aggregation and has been described as "irresolute" by Porter (1973) since granules frequently stop or transiently reverse direction, and might also move at a slightly reduced speed. This difference in aggregation and dispersion rates is not apparently observed in amphibian chromatophores where both processes are much slower than in fish and require roughly the same time (Malawista, 1971b; Novales and Novales, 1972). The behavior of individual pigment granules has not been studied in most invertebrate chromatophores; rather, movements have been analyzed by measuring changes in light transmission or reflectance.

In search for clues to the mechanochemical basis of granule movements, workers have been intensely interested in the organization of the cytoplasm of motile pigment cells. From the first ultrastructural study of a pigment cell (Bikle *et al.*, 1966), microtubules have been reported to be prominent components of chromatophores; in cells with rapid granule movements they are especially abundant (see Schliwa, 1981, for an overview). Their arrangement parallels the radial direction of granule movement. If stellate or discoid chromatophores are viewed by immunofluorescence microscopy with tubulin antibodies, this radial arrangement gives rise to strikingly esthetic images (Schliwa *et al.*, 1978). In both amphibian (Gartz, 1970) and fish chromatophores (Schliwa and Bereiter-Hahn, 1973a; Porter, 1973; Schliwa, 1978; McNiven and Porter, 1980), microtubules emanate from a region in the cell center characterized by an extensive and complex array of fibrillar or amorphous electron-dense material and the presence of centrioles (Fig. 9). This complex, variably termed cytocentrum, motility center, or central apparatus, is the equivalent of the centrosome of animal cells. It seems to be endowed with the capacity to nucleate and, to some extent, orient the growth of microtubules

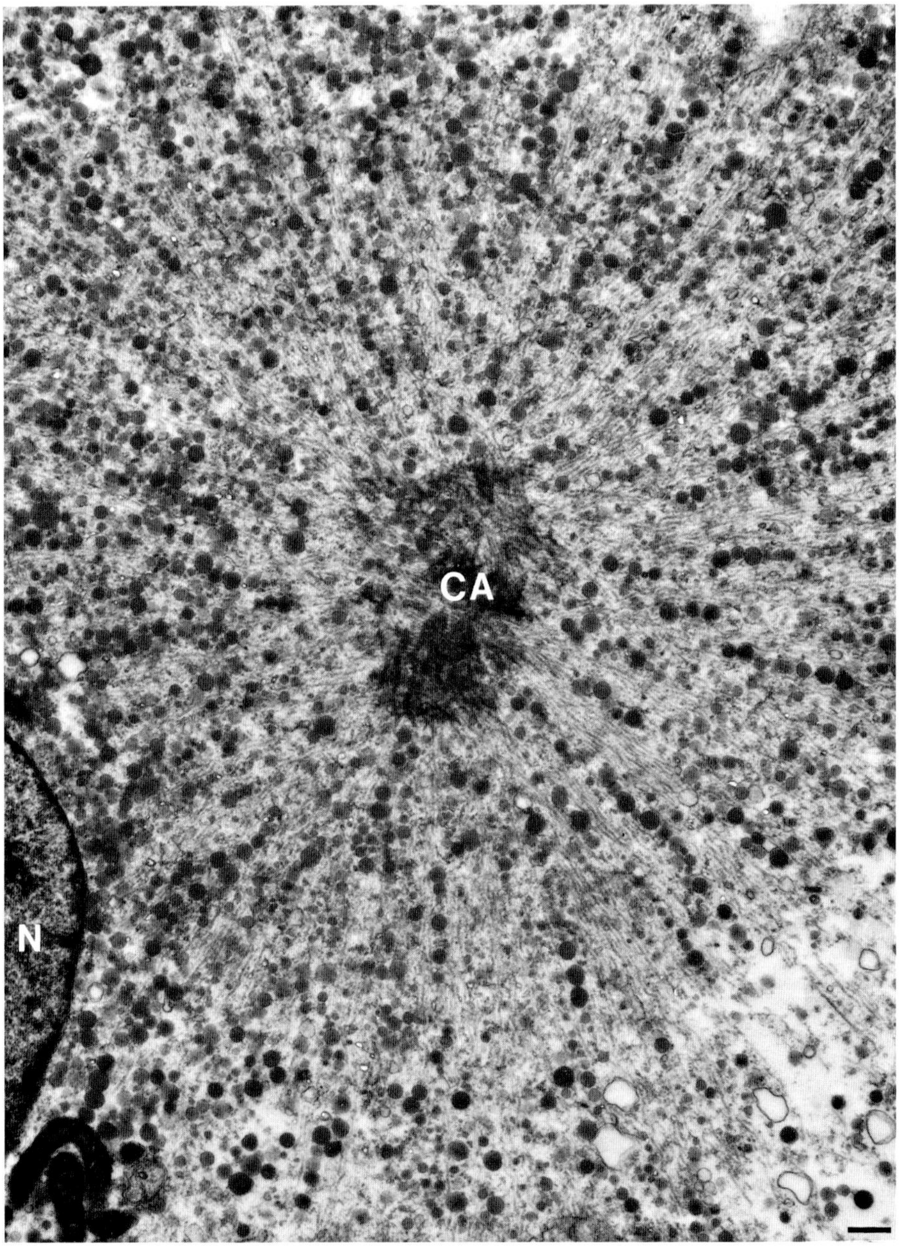

Figure 9. Horizontal thin section of an isolated erythrophore of *Holocentrus ascensionis* spread on a glass coverslip, showing the microtubule-organizing center or central apparatus (CA) with hundreds of microtubules radiating from it. The nucleus (N) is displaced towards the cell periphery. Bar = 0.5 μm. Micrograph kindly provided by Mark McNiven.

(Schliwa *et al.*, 1979a). The morphological appearance of the central apparatus changes more or less strikingly with the state of pigment distribution (Gartz, 1970; Porter, 1973; Schliwa *et al.*, 1979a; Porter and McNiven, 1982), and so does its capacity to initiate the growth of microtubules from exogenous brain tubulin (Schliwa *et al.*, 1979a). However, whether its presence is absolutely required for granule movements is unclear since cell fragments produced by vigorous shearing apparently transport granules in a perfectly coordinated fashion (Fig. 10). Most microtubules initiated at or near the central apparatus probably extend uninterrupted along the entire length of the cell processes, as shown by a serial section analysis (Schliwa, 1978) and by high voltage electron microscopy of whole mount preparations (Byers and Porter, 1977). At some distance from the centrosome, near the bases of the cell processes, many microtubules assume a cortical position with remarkably regular center-to-center spacings. In angelfish melanophores, more than 90% of

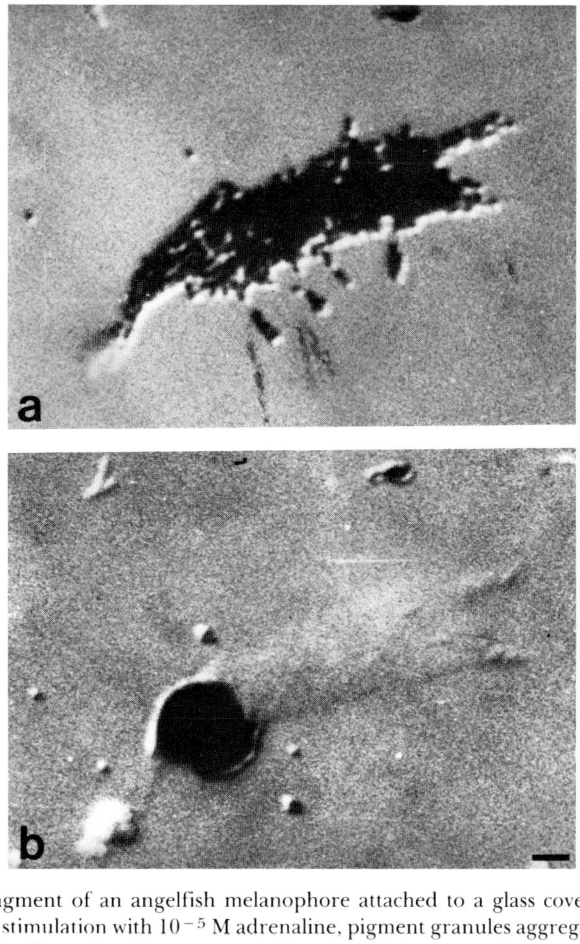

Figure 10. Fragment of an angelfish melanophore attached to a glass coverslip. (a) Dispersed state. (b) After stimulation with 10^{-5} M adrenaline, pigment granules aggregate towards one end of the fragment. Bar = 2 μm.

the microtubules in a cell process reside within 100-nm distance from the limiting membrane, giving the appearance of a microtubule palisade lining the tubular cell processes (Schliwa, 1978). *Holocentrus* erythrophores, too, have a cortical population of microtubules (Porter, 1973; Murphy and Tilney, 1974), but an equally well-developed complement of microtubules resides in the center of the cell extensions. It is not clear whether the arrangement of microtubules is of particular importance for the transport of pigment. Figure 11 shows a diagrammatic representation of three types of microtubule organization in stellate pigment cells and two bipolar or elongated cells. Despite the contrasting patterns of microtubule organization, all these cell types actively translocate pigment granules, though the rates at which transport occurs may vary.

It has not been possible to demonstrate the existence of either direct contacts between pigment granules and microtubules or associations mediated by cross-bridges analogous to those between mitochondria and microtubules in some neurons (Smith, 1971). One study, however, suggests an affinity of microtubules and pigment granules for one another on the basis of a near-neighbor analysis (Murphy and Tilney, 1974).

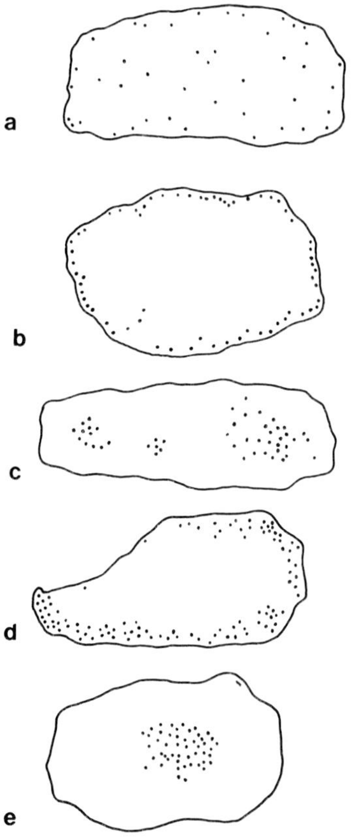

Figure 11. Schematic representation of microtubule organization as seen in transverse views in five different cell types showing prominent pigment granule movements. Each dot represents a microtubule in cross section. (a) *Holocentrus* erythrophores, (b) angelfish melanophores, (c) crustacean chromatophores, (d) scorpion retinula cells, (e) crayfish retinula cells.

Whether transport is selective for pigment granules is a matter of dispute. *Holocentrus* erythrophores possess a number of large mitochondria which appear to maintain their position in the cell periphery regardless of the state of pigment granule distribution (Porter, 1973; Porter and McNiven, 1982). A morphometric analysis of angelfish melanophores suggests that roughly half of the small, rod-shaped mitochondria cotranslocate with melanosomes during aggregation (Schliwa and Euteneuer, 1978a). Egner (1971) presented evidence for an accumulation of mitochondria just distal to the mass of aggregated pigment granules if melanophores are kept in the

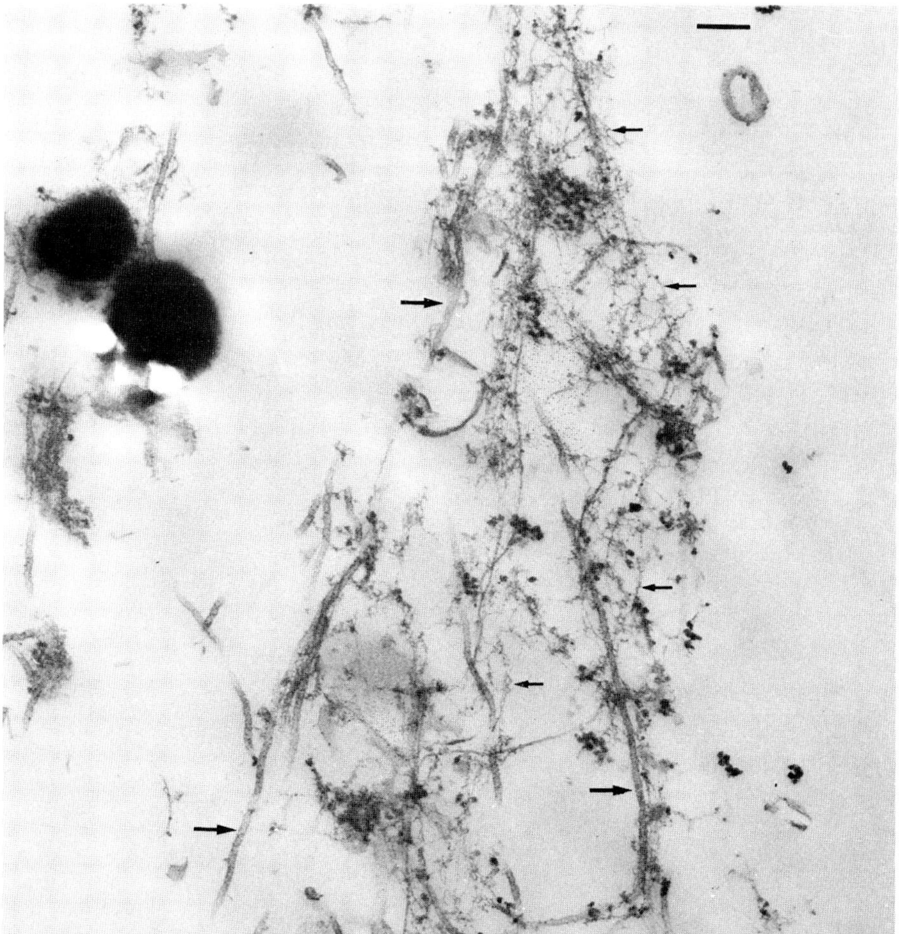

Figure 12. Thin section of an isolated angelfish melanophore extracted with Triton X-100 under microtubule-stabilizing conditions. In addition to microtubules (large arrows), an open network of 6–7 nm filaments is revealed in the cell periphery (small arrows). Bar = 0.2 μm.

aggregated state for 4–5 min. This aspect of intracellular transport, of identifying mechanisms of force transduction, warrants more detailed studies.

Relatively little is known about the distribution and possible function of actin or intermediate filaments in chromatophores. Thin filaments about 6 nm in diameter have been demonstrated in the cortical cytoplasm of some fish chromatophores (Fig. 12) and have been identified as actin by the heavy meromyosin (HMM) decoration technique (Obika *et al.*, 1978b; Lo *et al.*, 1980; Schliwa *et al.*, 1981) or by immunofluorescence microscopy (Schliwa *et al.*, 1981). Actin filaments do not appear to be an abundant cytoskeletal component of fish chromatophores; nothing is known about their organization in amphibian or invertebrate pigment cells. Nevertheless, an actomyosin system has frequently been invoked in granule transport on the basis of cytochalasin B experiments (see Section 3.1.1). Intermediate filaments are quite abundant in amphibian and some fish chromatophores (McGuire and Moellmann, 1972; Junqueira *et al.*, 1977) where they extend parallel to the long axes of the cell processes. They are rather sparse, however, in angelfish melanophores and altogether absent in *Holocentrus* erythrophores, and they have not been observed in invertebrate chromatophores. These observations have been taken to indicate that they are not universally involved in the mechanism of granule movements in chromatophores.

Recent studies of critical point-dried whole mount preparations of isolated pigment cells have added a new dimension to the analysis of intracellular transport phenomena. They provide evidence for the existence of an extensive three-dimensional network of irregular filamentous strands which have been given the name microtrabecula (Wolosewick and Porter, 1976); the entire system is called the microtrabecular network or lattice (MTL; Wolosewick and Porter, 1979). Pigment granules appear suspended in this network, the components of which are the sole, visible means of interaction between the pigment granules and other components of the cell such as microtubules and the plasma membrane (Byers and Porter, 1977). Its possible role in intracellular transport will be discussed in Section 3.1.5.

2.3. Particle Movement and Cytoplasmic Streaming in Some Protozoans

Intracellular movements of cytoplasmic organelles are prominent in many protozoa but were studied experimentally in only a few instances. I will focus on two systems, heliozoan axopodia and retriculopodial networks of foraminifera. For a discussion of intracellular transport phenomena in other unicellular organisms, see Tucker (1974, 1979), Dustin (1978), Hyams and Stebbings (1979), and Suchard and Goode (1982).

2.3.1. *Heliozoa*

Heliozoans possess slender, stiff extensions from the cell body, called axopodia, that function in locomotion and feeding (Fig. 13). Streaming of particles (mostly kinetocysts, but also mitochondria and other inclusions of unknown nature; Fig. 14) along these axopodia has been observed as early as

Figure 13. Part of the cell body and four axopods of the heliozoan, *Actinosphaerium eichhornii*, photographed in Nomarski Interference Contrast. In live specimen, particles (arrows) on the axopodia are seen to move along the long axis both towards and away from the cell body. Bar = 20 μm.

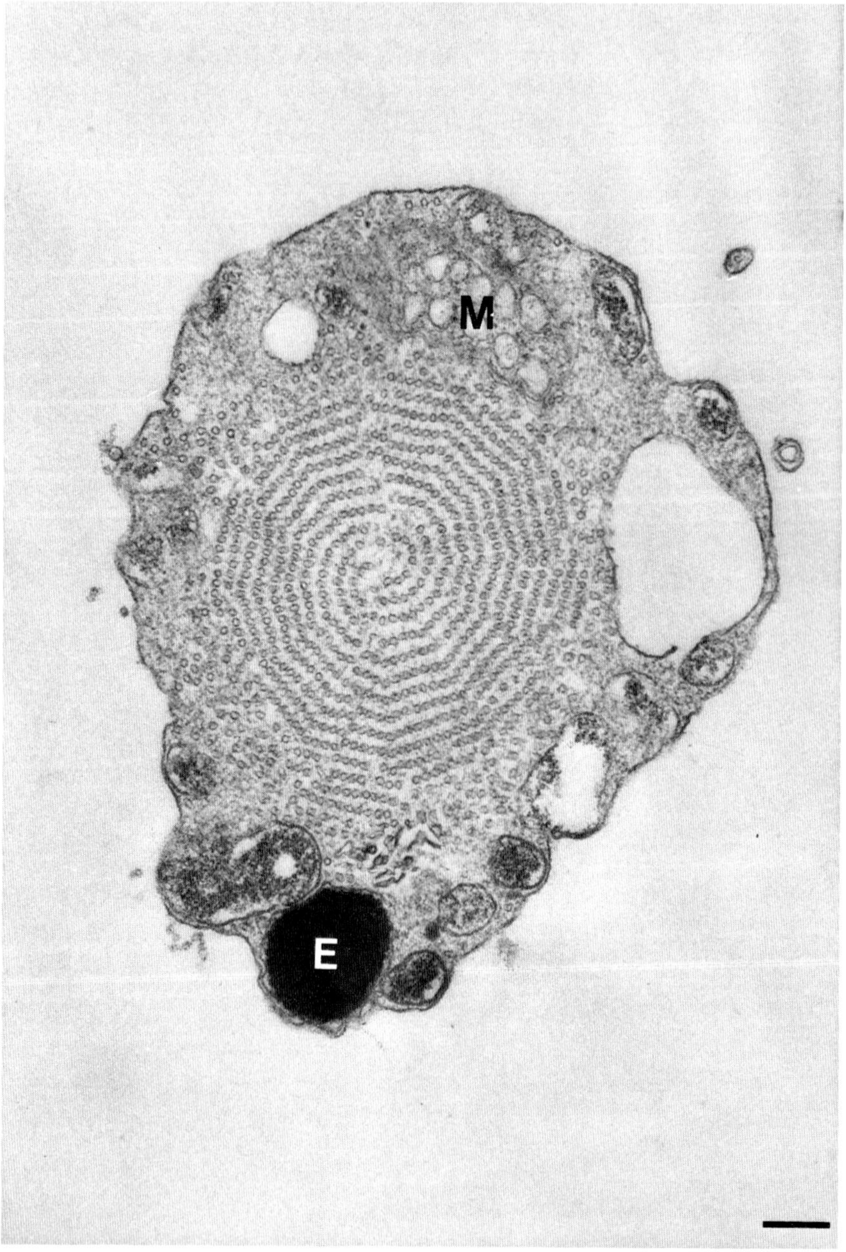

Figure 14. Transverse section of a heliozoan axopodium, showing double-spiral pattern of the microtubular axoneme (slightly distorted in this axopodium) and several membrane bounded organelles, including an extrusome (E) and a mitochondrion (M). These probably are the organelles seen as "particles" in the light microscope. Bar = 0.2 μm.

1858 by Claparède and Lachmann. Particle movement occurs at a rate of about 1 μm/sec (Fitzharris *et al.*, 1972) and is bidirectional and saltatory in nature.

Axopodia are supported by a central mictrotubule bundle, the axoneme. In organisms such as *Echinospherium* and *Actinophrys*, it consists of two interlocking coils of microtubules linked by cross-bridges (Kitching, 1964; Tilney and Porter, 1965), whereas other species may show different patterns of microtubule organization (Bardele, 1977). The role of the microtubule axoneme in the maintenance of axopodial integrity has been established through thorough experimental analyses (Tilney *et al.*, 1966; Tilney and Porter, 1967; Roth and Shigenaka, 1970; Schliwa, 1976). On the basis of their striking arrangement a role for axonemal microtubules in particle movement along axopodia was soon suspected (Kitching, 1964). This notion has been challenged by Edds (1975a,b) in a simple experiment that involves the insertion of a glass capillary through the cell body of a heliozoan in order to create an artificial axopodium. Particle movements along this glass axopodium continue in the presence of colchicine concentrations sufficient to induce degradation of all other axopodia. Even though some uncertainties about the validity of this experiment remain, it constitutes the single most convincing evidence against a direct involvement of microtubules in transport. Edds (1975b) further provided evidence for the presence of actin filaments in cytoplasmic extracts of these organisms and describes a thicker filament type which he thinks may be myosin. No experimental evidence, however, supports the notion of an involvement of an actomyosin system in force generation. Interestingly, a related type of particle motility occurs in association with the cell membrane on the outside of axopodial extensions. Bacteria (Troyer, 1975) or latex beads (Bloodgood, 1978) adhering to the external surface of axopodia are transported bidirectionally and at approximately the same rate as internal organelles, suggesting that the plasma membrane is part of the motile machinery responsible for the movement of particles both on the outside and on the inside of the membrane. This notion receives substantial support from the observation that the kinetocysts appear firmly anchored to the plasmalemma at specific attachment sites (Bardele, 1976; Davidson, 1976). Bidirectional saltatory particle movement has also been observed in association with the outer surface of *Chlamydomonas* flagella (Bloodgood, 1977).

2.3.2. *Foraminifera*

Foraminifera are rhizopods that extend a branching an anastomosing network of pseudopodia or slender filopodia (sometimes several millimeters long) from their cell body (Fig. 15). This "reticulopodial network" (RPN) constantly changes its overall organization by filopodial extension, retraction, fusion, or branching. Within the RPN there is constant bidirectional streaming of both cytoplasm and particles at velocities of up to 10 μm/sec (Rinaldi and Jahn, 1964). The identity of these particles is uncertain; ultrastructural

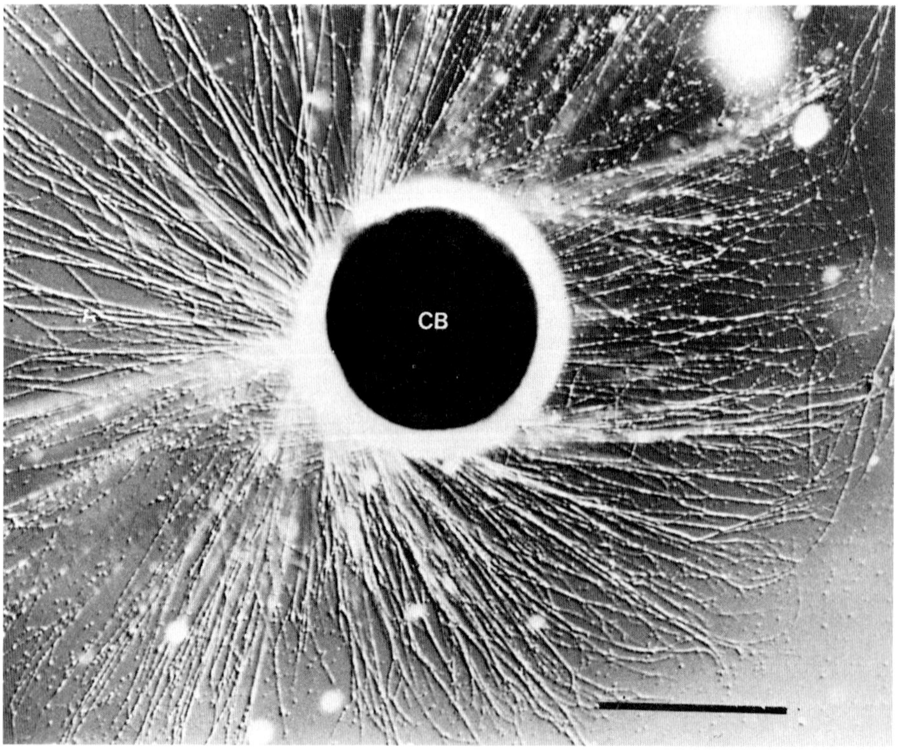

Figure 15. Survey micrograph of living *Allogromia* as seen with Nomarski Interference Contrast. An extensive interconnected reticulopodial network extends from the cell body (CB). Numerous particles are associated with the filopodia. These particles move at velocities often exceeding 10 μm/sec. Bar = 100 μm. Reprinted with permission from Travis and Allen(1981).

studies show a variety of vesicular inclusions, including digestive vacuoles, dense bodies, coated vesicles, and oblong vacuoles (McGee-Russel and Allen, 1971; Travis and Allen, 1981). The cytoplasm of the RPN shows abundant microtubules that extend parallel to the long axis of the filopodia (Fig. 16), either singly or in bundles linked by cross-bridges (McGee-Russel and Allen, 1971; McGee-Russel, 1974; Travis and Allen, 1981). Highly sensitive video-enhanced light microscopy of flattened areas of the RPN visualize what is believed to be the movement of small microtubule bundles and possibly even single microtubules (Allen *et al.*, 1981a,b). These studies suggest that microtubule bundles can move past each other, splay, or unzip laterally. Particles move only when associated with microtubules; if they "fall off" a microtubule they remain stationary unless "picked up" again by another microtubule bundle (Allen *et al.*, 1981b).

In addition to microtubules, the cytoplasm of the RPN contains abundant helical paracrystalline filaments (Hauser and Schwab, 1974; Travis and Allen, 1981) and thin filaments (5 nm) that extend parallel to microtubules (Fig. 17),

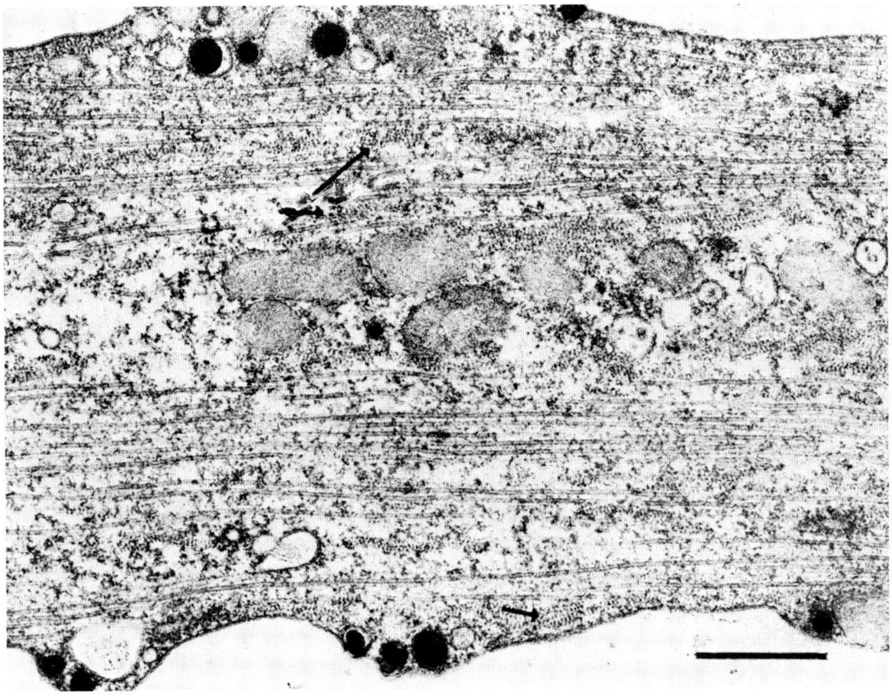

Figure 16. Thin section of a portion of an *Allogromia* filopod. Microtubules appear to be organized in distinct tracts or bundles, delineating channels where membrane-bounded vesicles of varying shape and electron density reside. Helical paracrystalline material (arrows) may be seen throughout this filopodium. Bar = 1 μm. Reprinted with permission from Travis and Allen (1981).

sometimes in close proximity to them (Travis and Allen, 1981). The helices are believed to represent an intermediate or alternate assembly form of tubulin; the identity of the 5 mm filaments which apparently do not bind heavy meromyosin is not known. Whether either of these cytoskeletal elements plays a role in cytoplasmic movements of the RPN remains enigmatic.

2.4. Cells in Culture

Intracellular movement of cytoplasmic inclusions of cultured cells has been the subject of relatively few thorough studies. The analysis of particle motility in cells in culture has advantages and disadvantages in comparison to the previously discussed systems. The advantages are that cultured cells are easier to maintain and handle, and that they can be grown in sufficient quantity to allow biochemical analysis. The disadvantages are that organelle movements are less conspicuous than in other systems specialized for intracellular transport, e.g., pigment cells, and that their cellular and cytoskeletal architecture is more complex, variable, and dynamic.

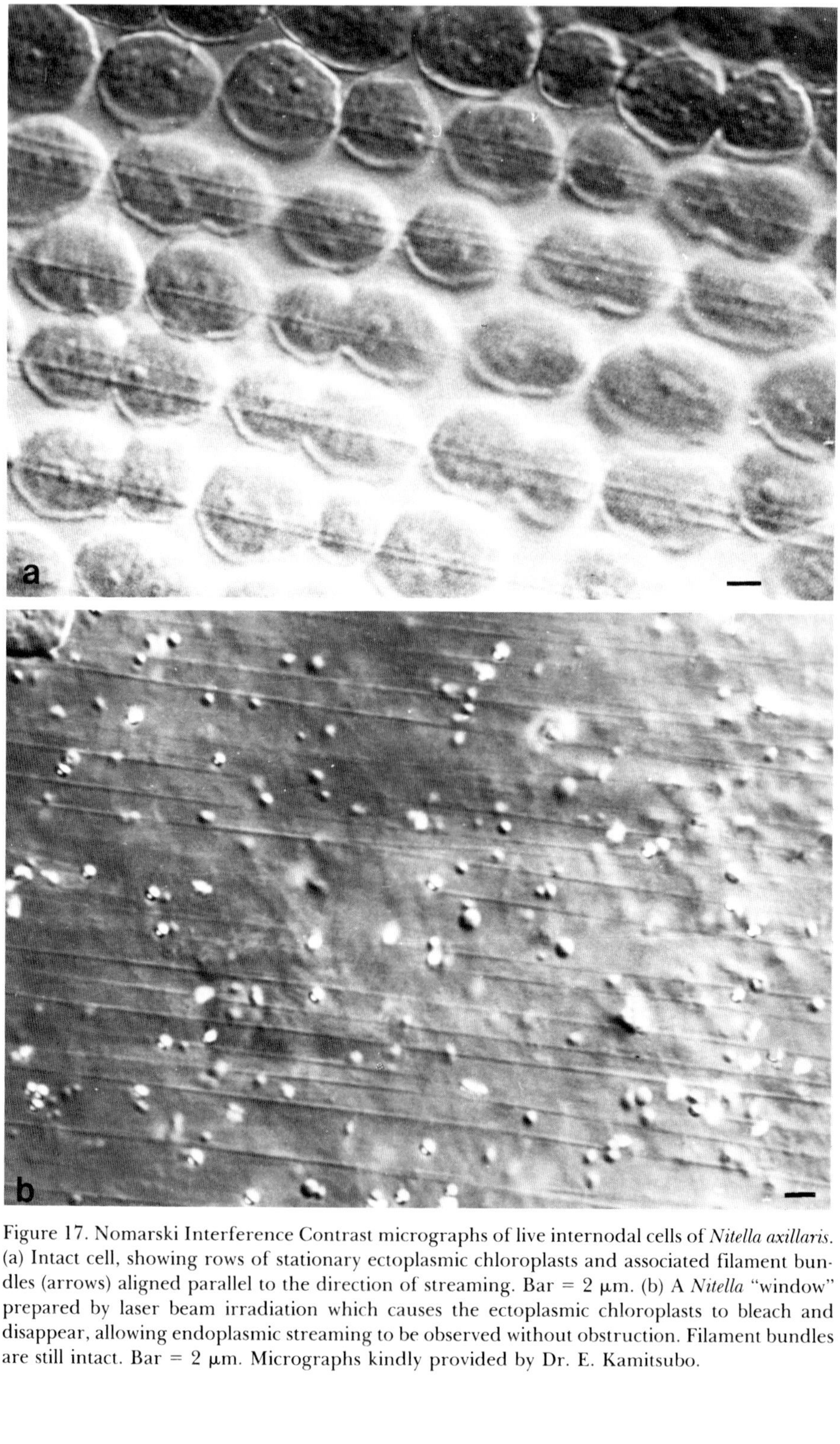

Figure 17. Nomarski Interference Contrast micrographs of live internodal cells of *Nitella axillaris*. (a) Intact cell, showing rows of stationary ectoplasmic chloroplasts and associated filament bundles (arrows) aligned parallel to the direction of streaming. Bar = 2 μm. (b) A *Nitella* "window" prepared by laser beam irradiation which causes the ectoplasmic chloroplasts to bleach and disappear, allowing endoplasmic streaming to be observed without obstruction. Filament bundles are still intact. Bar = 2 μm. Micrographs kindly provided by Dr. E. Kamitsubo.

As in the axon, two types of cellular components are observed to undergo rapid, saltatory movements; mitochondria and "particles." The latter include a variety of organelles, such as lysosomes, lipid droplets, pinocytotic vesicles, and ingested carbon particles or latex beads (phagosomes; Freed and Lebowitz, 1970; Wagner and Rosenberg, 1973; Wang and Goldman, 1978; Bell *et al.*, 1980). Particles as well as mitochondria are not moving constantly, as, for instance, pigment granules in chromatophores. Many of them may remain stationary for long periods of time. In well-spread cells the majority of movements is directed either toward or away from the cell center. Analyses by time-lapse cinematography demonstrate that the velocity of saltatory movements ranges from 0.5 to 5.0 μm/sec with an average speed of 1–2 μm/sec. Saltations usually do not exceed a few micrometers (Freed and Lebowitz, 1970; Wang and Goldman, 1978), but occasionally a particle may travel all the way from the centrosphere to the periphery and back in a continuous, smooth fashion reminiscent of streaming (Bell *et al.*, 1980). Prominent organelle movements toward and away from the spindle poles in mitotic cells can probably be regraded as a specialized variant of the general movements of cytoplasmic constituents observed in interphase cells (Rebhun, 1972). There is a strong correlation between the occurrence of saltations and the presence of microtubules. The distribution and directionality of saltations strongly conforms with the organization of the microtubule system (Freed and Lebowitz, 1970). Microtubule depolymerization leads to cessation of saltatory movements, but some organelles may still be carried through the cytoplasm passively by cytoplasmic streaming (Bhisey and Freed, 1971; Wang and Goldman, 1978). Cytochalasin B, though having a dramatic effect on the protrusive activity of the cell periphery and on cell shape, does not interfere with saltatory particle transport (Bell *et al.*, 1980). An involvement of the third major filament type, intermediate filaments, in organelle transport has been inferred on the basis of their codistribution with microtubules (Wang and Goldman, 1978; Wang *et al.*, 1979; Wang and Choppin, 1981).

2.5. Nutritive Tubes of Hemipteran Insects

A somewhat unusual transport system exists in the ovarioles of hemipteran insects. The developing oocytes are connected to a group of nutritive cells by tube-like channels, about 20 μm in diameter and up to several millimeters long packed with thousands of parallel microtubules (Macgregor and Stebbings, 1970). Oocytes, which themselves show very little synthetic activity, are supplied with various materials (mostly ribosomes) through these nutritive tubes. The system is unusual in several respects: (1) transport occurs at a slow rate, comparable to that of SC in axons (about 1 mm/day; Macgregor and Stebbings, 1970; Mays, 1972), (2) transport is unidirectional (from the nutritive cells to the occyte, and (3) the tubes do not contain actin, myosin, or dynein at levels which would be detectable by gel electrophoresis (Hyams and Stebbings, 1978). No distinct cross-bridges are observed between the microtubules and the transported material by electron microscopy.

2.6. Cytoplasmic Streaming in Plant Cells

Rotational cytoplasmic streaming is the most conspicuous form of intracellular movement in plant cells and probably the most striking example of intracellular transport of all eukaryotic cells. It therefore comes as no surprise that this form of intracellular transport has been described long before related motile activities were discovered in other cells types (Corti, 1774). Recent reviews of the literature were provided by Hepler and Palevitz (1974), Allen and Allen (1978), Seitz (1979), Kamiya (1981), and Buckley (1981).

Streaming in plant cells has been studied most successfully in the giant internodal cells of *Chara* and *Nitella* where it is restricted to a comparatively narrow zone between the stationary, chloroplast-containing ectoplasm and the large central vacuole. The flowing endoplasm follows a spiral path along the long axis of the cell, turns around at the end, and streams back at the opposite side of the cell. It carries with it nuclei, small particles about 0.5 μm in diameter called spherosomes, and a variety of other cytoplasmic structures which all move at approximately the same rate, namely 40–100 μm/sec. Occasionally, particles can move at even higher rates. Countertransport of particles in close proximity to one another, as it is characteristic for many types of saltatory movements in other cell types, is uncommon.

The now classical studies by Kamiya and Kuroda (1956, 1958) led to the view that the force for streaming was generated at the interface between the stationary cortex and the streaming endoplasm. This idea received further support from the observation of light microscopically detectable subcortical fibrils (Fig. 17a) at the presumptive site of generation of shearing forces (Kamitsubo, 1966, 1972b). These fibrils, when studied by electron microscopy, (Fig. 18) consist of bundles of microfilaments (Nagai and Rebhun, 1966; Pickett-Heaps, 1967) identified as actin by the HMM decoration method (Palevitz *et al.*, 1974; Palevitz and Hepler, 1975), supporting the long-suspected involvement of actomyosin in this form of motility. The polarity of the actin filaments is uniform (Palevitz *et al.*, 1974), with arrowheads pointing opposite to the direction of streaming (Kersey *et al.*, 1976). Using Kamitsubo's (1972a) window technique which produces chloroplast-free areas of cytoplasm (Fig. 17b) well suited for light microscopic analysis, Allen (1974) also observed the presence of filament bundles in the endoplasm of *Nitella* cells that appeared to undergo oscillatory or undulating movements. These fibrils were proposed to be anchored in the cortical cytoplasm or possibly even arise by branching from the subcortical fibrils. They are proposed to contribute substantially to the generation of the motive force for endoplasmic streaming (bulk flow of cytoplasm) by active undulations. Although microtubules are present in the cortical cytoplasm of *Nitella* (Green, 1969; Hepler and Palevitz, 1974), they are found next to the cell wall in the stationary ectoplasm, and thus are not appropriately located to participate in streaming. Rather, they seem to be involved in morphogenetic processes and cell wall formation, as in many other plant cells (reviewed in Hepler and Palevitz, 1974).

Phenomenologically different but clearly related forms of intracellular

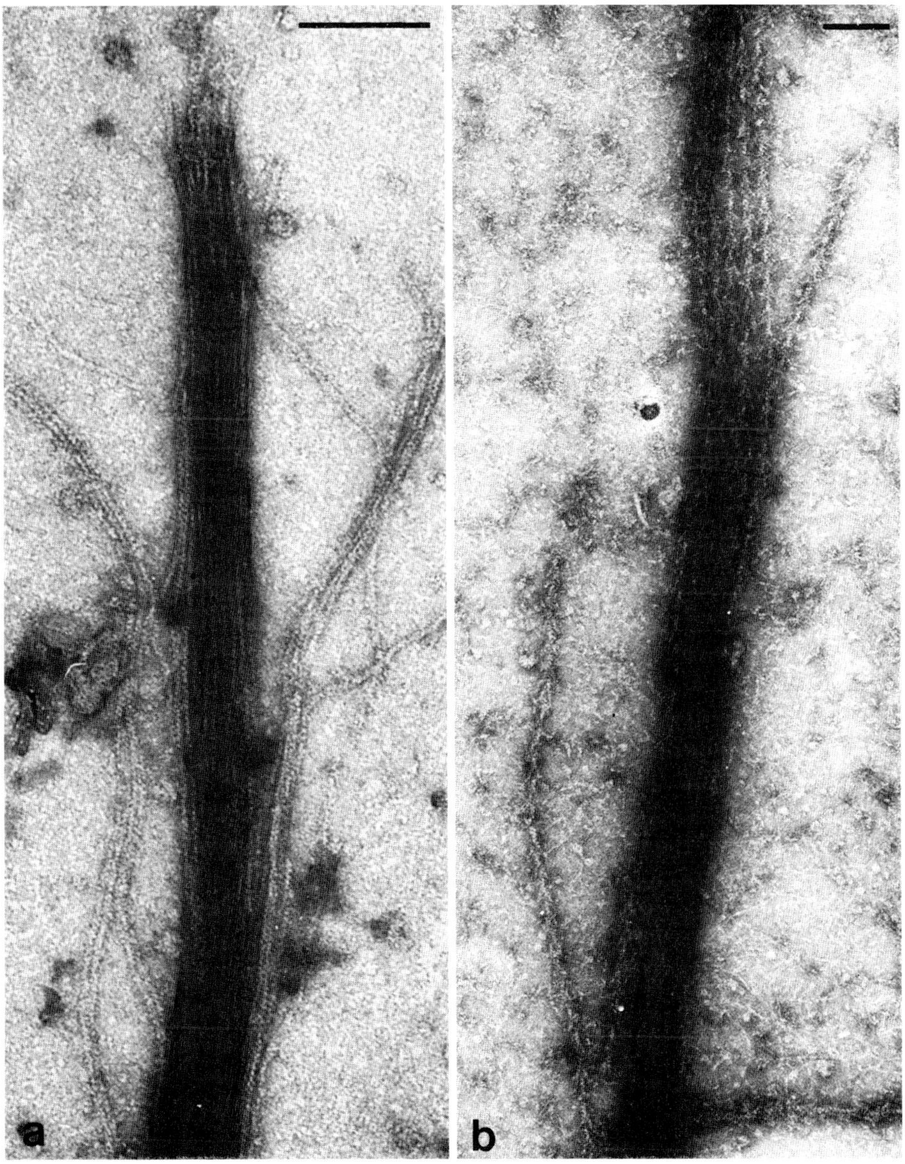

Figure 18. Filament bundles in a cytoplasmic suspension obtained from streaming *Nitella flexilis* observed after negative staining with uranyl acetate. (a) Untreated bundle. Bar = 0.1 μm. (b) Another filament bundle reacted with heavy meromyosin to reveal filament polarity. All filaments within this bundle appear to have the same polarity (arrowheads are pointing up). Bar = 0.1 μm. Reprinted with permission from Palevitz *et al.* (1974).

movement are expressed in other plant cells and have been studied in some detail. Movement of chloroplasts, nuclei, and other cytoplasmic constituents along the stalk of the giant marine alga *Acetabularia* proceeds in numerous tracks or channels (the "multistriated" type of streaming according to Kamiya, 1959) in a thin layer of cytoplasm apposed to the cell wall, while the center of the stalk is occupied by a large central vacuole limited by a tonoplast. Streaming in opposite directions may be observed in different channels at the same time (Nagai and Fukui, 1981). Since the distribution of chloroplasts in the cell apex and the rhizoid shows diurnal changes (Koop *et al.*, 1978), there is a rhythmic change in the proportion of chloroplast streams directed towards the apex or the base, respectively. The cytoplasmic regions between these channels normally remain quiescent, although particles may change tracks occasionally.

Puiseux-Dao (1979) and Koop and Kiermayer (1980a,b) distinguish between two different transport systems: (1) a system of channels along which chloroplasts move at a rate of 1–2 μm/sec, and (2) a second system which they named "headed streaming bands." This second system is characterized by a cytoplasmic droplet (head structure) leading a band of cytoplasm streaming at a velocity of 3–10 μm/sec (Fig. 19). The headed streaming bands move along the same filamentous channels as the chloroplasts, carrying with them small droplets, polyphosphate granules, and secondary nuclei. Since head droplets are moving at a higher rate than chloroplasts, they frequently overtake them on the same filament. All movements cease in the presence of 10 μm cytochalasin B, while movements of the headed streaming bands are differentially sensitive to microtubule inhibitors. Unfortunately, it has not yet been possible to preserve the ultrastructure of *Acetabularia* cytoplasm well enough to allow an analysis of the precise distribution of filaments and microtubules.

Microtubules are believed to play a role in cytoplasmic streaming in another marine alga, *Caulerpa*. In this alga, a network of endoplasmic strands traverses the cell body and also apposes the cell wall. Streaming of chloroplasts and other constituents is observed in these strands at an average rate of 4 μm/sec. Ultrastructural studies demonstrate bundles of microtubules in the cytoplasm (Dawes and Rhamstine, 1967; Sabnis and Jacobs, 1967) which appear to be oriented parallel to the axis of the endoplasmic streams (Dawes and Barilotti, 1969). Chloroplasts are frequently found in close proximity to these bundles (Fig. 20), their long axes aligned to them (McDonald, unpublished). No microfilaments were reported.

The pattern of cytoplasmic movement observed in *Caulerpa* is closely related to the phenomenon termed "circulation" that can be observed in many plant cells, for instance the hair cells of *Tradescantia* and *Gloxinia*, or the parenchymal cells of *Allium* (Allen and Allen, 1978). The streaming pattern of circulating particles is much more variable in space and time than in the cell types discussed previously. Streaming occurs bidirectionally in the cell cortex and along the many anastomosing strands that traverse the central vacuole. Strands may change their location, split, fuse, or cease transport altogether. The rate of particle movement is usually more variable than in other cell

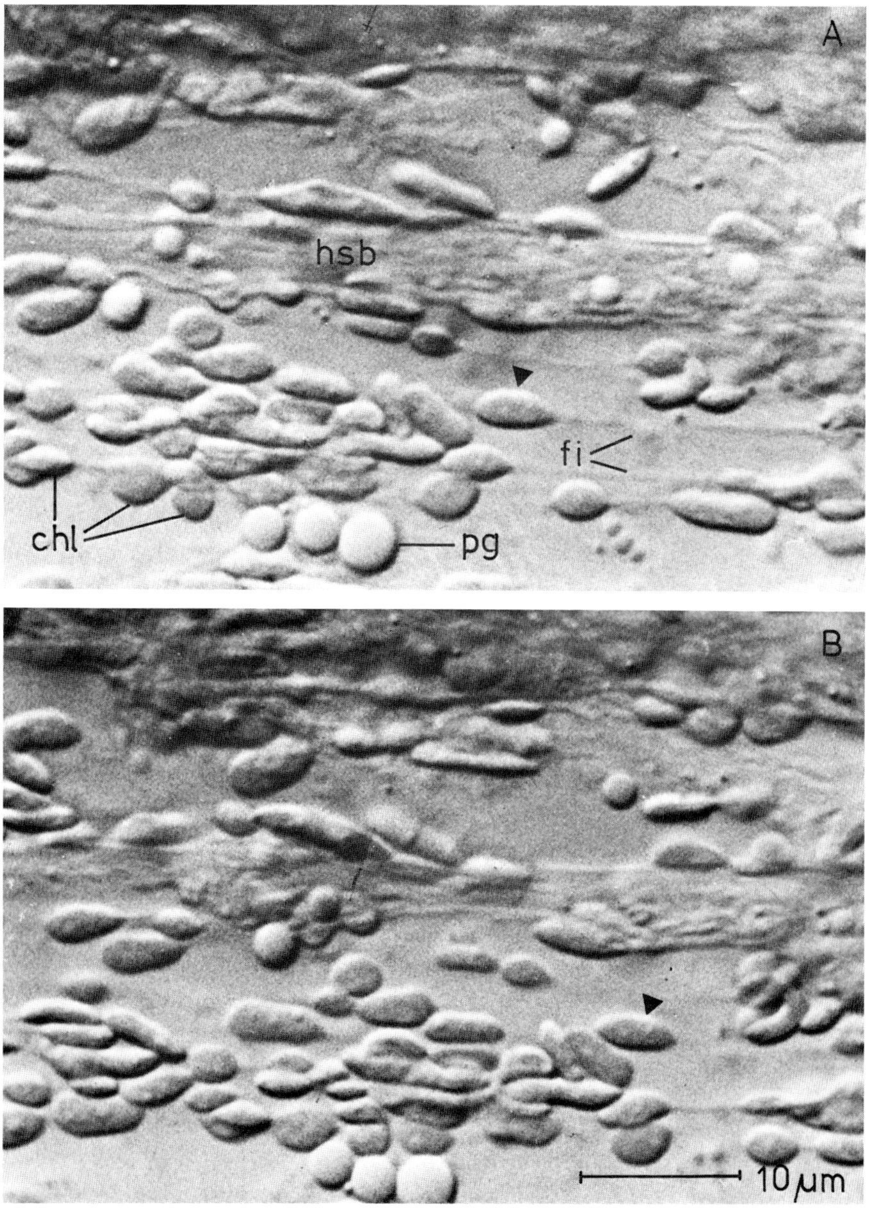

Figure 19. The two systems of intracellular transport in Acetabularia are represented by cytoplasmic filaments (fi) and headed streaming bands (hsb). (A,B) The two light micrographs of the same region are photographed at a 10-sec interval. A chloroplast moving along a cytoplasmic filament is indicated by an arrowhead. (chl) Chloroplasts; (pg) polyphosphate granule. Bar = 10 µm. Reprinted with permission from Koop and Kiermayer (1980b).

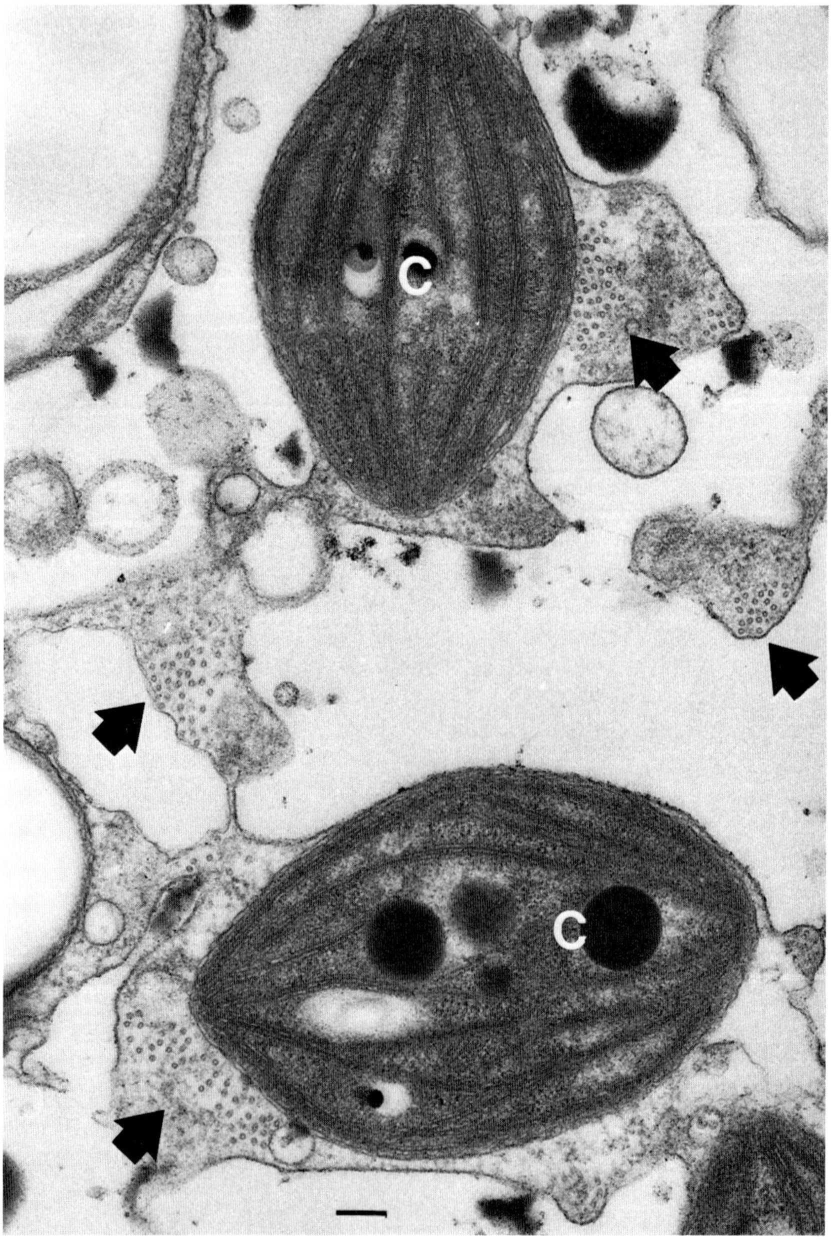

Figure 20. Electron micrograph of a thin section of the marine alga, *Caulerpa vanbossae*. The highly vacuolated cell body is traversed by cytoplasmic strands in which bundles of microtubules (arrows) reside. Chloroplasts (C) are frequently found in close association with these microtubule bundles. Bar = 0.2 μm. Micrograph kindly provided by Dr. Kent McDonald.

types, ranging from values far less than 1 μm/sec up to 30 μm/sec. Transvacuolar strands and streaming peripheral endoplasm contain bundles of microfilaments (Fig. 21) resembling actin (O'Brien and Thiman, 1966; Parthasarathy and Mühletaler, 1972). The location of these filament bundles, some of which extend the entire length of the cell, seems to be correlated with the direction of cytoplasmic streaming (O'Brien and McCully, 1970; Parthasarathy and Pesacreta, 1980).

Any cellular contractile machinery based on actin can only work if myosin is present as well. Unfortunately, the organization and distribution of myosin in relation to that of actin filaments and the site of force generation is not known for any of the plant cells in which intracellular transport has been observed. Myosin has been isolated from *Nitella* cells (Kato and Tonomura, 1977), but its organizational state (monomer vs. polymer) and location in the cell are unknown. Ultrastructural studies by Nagai and Hayama (1979) reveal the presence of endoplasmic organelles associated with the microfilament bundles of *Nitella*. These membrane-bounded vesicular structures show ordered arrays of globular bodies or bridges which connect them to the microfilament bundles. It is speculated that these bridges might represent myosin molecules whose interaction with actin filament bundles is responsible for generating the streaming force for transport. Williamson's (1979) search for presumptive myosin resulted in the identification of more regular rod-shaped filaments which are exclusively located in the endoplasm. The filaments are Ca^{2+} sensitive, and their disappearance induced by 10^{-4} M Ca^{2+} coincides with the arrest of cytoplasmic streaming. Using a rapid freezing–shallow etching technique, Allen (1980) identified delicate filaments 4–5 nm in diameter in the endoplasm that she believes represent cytoplasmic myosin.

3. Transport Mechanisms

The foregoing sections served to introduce some of the cell systems most frequently used to study forms of intracellular motility. Evidently, considerable diversity exists with respect to the size of transported elements and the range, velocity, and pattern of movement. Nevertheless, the cell systems discussed here fall into the two broad categories mentioned in the Introduction, namely, "saltations" (Sections 2.1–2.5), and "streaming" (Section 2.6). The following sections will review experiments designed to decipher the molecular mechanism(s) operating in these two forms of intracellular motility.

3.1. Category I: Saltations

3.1.1. *Microtubules: Scaffold or Motor?*

In addition to ample morphological evidence, e.g., abundance of microtubules, congruence of microtubule orientation and the direction of particle movements, links between microtubules and organelles, a vast body of experi-

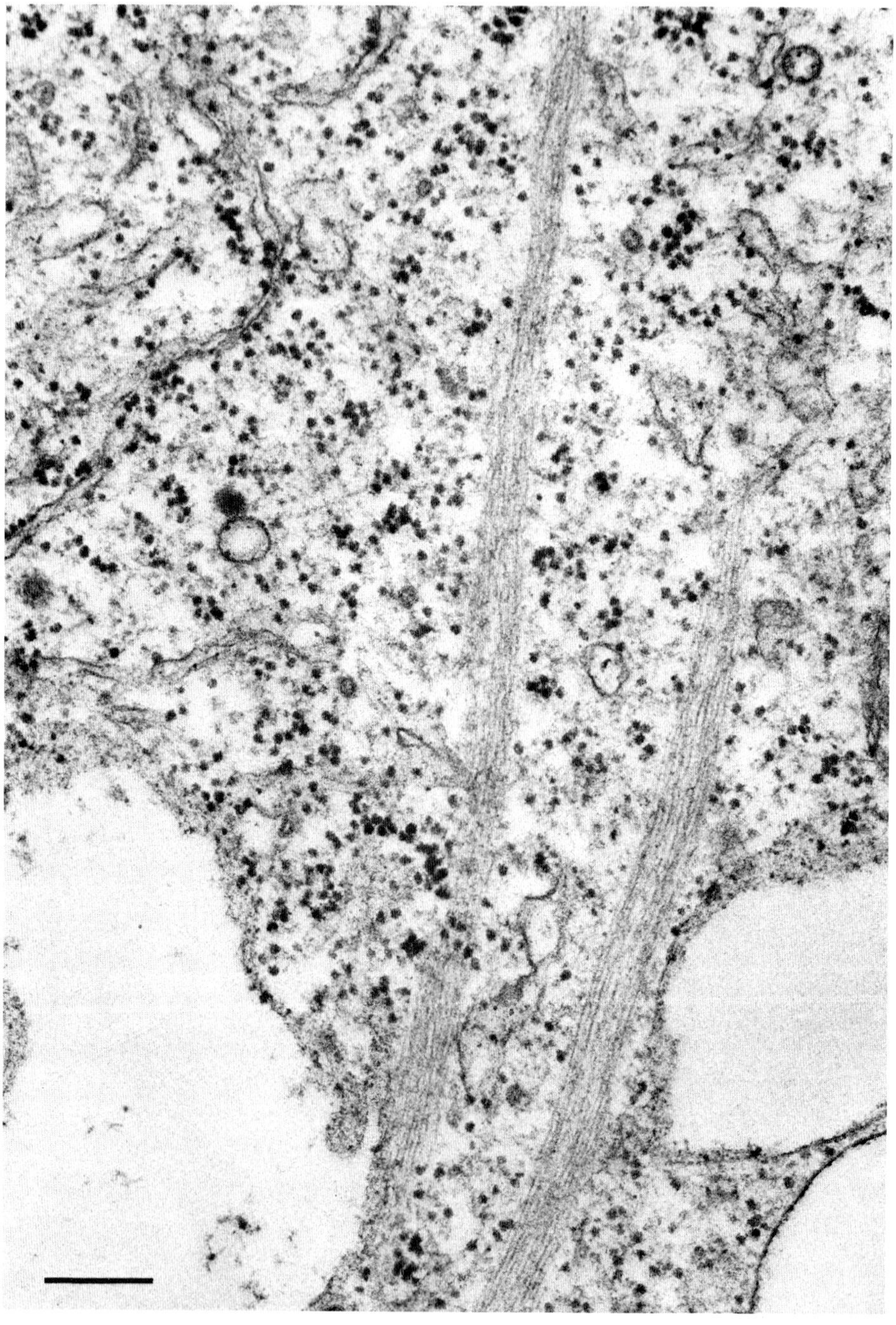

Figure 21. Two microfilament bundles traversing the cytoplasm in a phloem parenchyma cell of the kidney bean, *Phaseolus vulgaris*. Bar = 0.2 μm. Micrograph kindly provided by Dr. Barry Palevitz.

mental, mostly pharmacological, evidence for microtubule participation in organelle transport in a variety of cell systems has accumulated (for overviews, see Stephens and Edds, 1976; Hyams and Stebbings, 1979; Schliwa, 1982b). In teleost chromatophores, microtubule disruption by colchicine, vinblastine, or low temperature results in a disorganization of the parallel alignment of granules (Murphy and Tilney, 1974; Schliwa and Euteneuer, 1978b) and leads to a decrease in the rate of granule movements, often to the extent that all translocations appear to have ceased (Wikswo and Novales, 1969; Schliwa and Bereiter-Hahn, 1973b; Murphy and Tilney, 1974; Junqueira *et al.*, 1974). Microtubules of some cell types, especially those of teleosts, may show a remarkable stability against treatment with colchicine, even when used at millimolar concentrations for several hours (Schliwa and Bereiter-Hahn, 1973b; Schliwa and Euteneuer, 1978b; Burnside *et al.*, 1982). In amphibian melanophores, antimicrotubular agents such as colchicine, colcemid, vinblastine, and podophyllotoxin inhibit aggregation (Malawista, 1965, 1971a) but may have little effect on, or even enhance, dispersion (Malawista, 1965; Fisher and Lyerla, 1974). Unusually high concentrations of colchicine (25 mM) are necessary to produce significant inhibition of aggregation in crustacean chromatophores (Lambert and Crowe, 1973, 1976; Fingerman *et al.*, 1975; Frixione *et al.*, 1979). However, inhibitor studies on amphibian or invertebrate pigment cells were not complemented by parallel fine structural investigations so that no information on the deployment of microtubules in drug-treated cells is available.

Regarding the question of microtubule involvement in fast axoplasmic transport, a situation similar to that in chromatophore studies prevails. At about the same time the chromatophore system underwent a rigorous analysis, inhibitor studies were also performed on a variety of neuronal cell types, both vertebrate and invertebrate. Following Kreutzberg's (1969) observation of an inhibition of axonal transport in mammalian nerve cells by local administration of colchicine, many studies have shown that antimicrotubular agents also interfere with fast axonal transport (Karlsson and Sjöstrand, 1969; Banks *et al.*, 1971a,b; Banks and Till, 1975; Dahlström, 1971a; Edström and Mattson, 1972a; Fernandez *et al.*, 1970; Fink *et al.*, 1973; James *et al.*, 1970; Paulson and McClure, 1974, 1975; for recent reviews, see Samson, 1976; Hanson and Edström, 1978; Grafstein and Forman, 1980; Brady and Lasek, 1982a). Some studies were able to correlate inhibition of transport with a decrease or even complete loss of microtubules in treated axons at the ultrastructural level (Banks *et al.*, 1971a; Friede and Ho, 1977; Hammond and Smith, 1977). Application of microtubule inhibitors not only affects transport in the anterograde, but also movement of material in the retrograde direction (Abe *et al.*, 1974; Edström and Hanson, 1973b; McLean *et al.*, 1976; Hendry *et al.*, 1974; Stoeckel *et al.*, 1975). Differences in the sensitivity to antimicrotubular agents of the transport in one vs. the opposite direction were suspected (Bunt and Lund, 1974) but could not be confirmed in other studies.

A second approach that circumvents some of the problems involved in the use of inhibitors takes advantage of the anisotropy of the nerve axon which makes it possible to manipulate experimentally restricted and well-defined regions of a nerve fiber in a reversible manner while leaving the main portion of an axon undisturbed. Thus, a reversible cold block has been used successfully by Brimijoin and his colleagues (Brimijoin, 1975; Brimijoin and Helland, 1976; Brimijoin and Wiermaa, 1978) to analyze the structural characteristics of axonal transport. Brimijoins stop-flow technique causes local accumulation of material at the site of the cold block. When released, the material thus accumulated continues to move along the axon and allows accurate determination of transport velocities. Brimijoin *et al.* (1979) and Tsukita and Ishikawa (1980) who used a similar technique, have shown that a local cold block leads to the depolymerization of microtubules in this region. Anterogradely moving components accumulate proximally, while a different set of retrogradely moving vesicles accumulates distally to the microtubule-depleted portion of the axon (see Figs. 3 and 4).

Most of these studies thus seem indeed to suggest a strong correlation between microtubule integrity and the "saltatory" type of motility of cellular constituents. This has led to the still widely held belief that microtubules are intimately involved in this form of motility, possibly even by providing the motive force. However, it should not be overlooked that these studies still left the most important, fundamental question unanswered. Are microtubules directly driving intracellular movements through specific, microtubule-associated components that interact with the organelles? The uncertainties about how best to interpret these data were reinforced by a number of observations which do not seem to fit the suspected direct involvement of microtubules in transport:

1. In three different cell systems, the arrest of axonal transport by colchicine does not cause a decrease in the number of axonal microtubules (Fernandez *et al.*, 1970; Karlsson *et al.*, 1971; Flament-Durand and Dustin, 1972). To explain these observations, Paulson and McClure (1975) suggested a model in which tubulin molecules are able to leave the wall of intact microtubules, become complexed with colchicine and thus are rendered unable to re-enter the microtubule. Such a microtubule with a disturbed surface lattice would still appear intact in electron micrographs, but would be unable to function properly. As yet, however, no compelling evidence that tubulin subunits can leave (or reenter) a microtubule along its length has been obtained. Another possibility is that colchicine blocks transport by a mechanism unrelated to microtubule depolymerization and tubulin binding, a view that does not appear so far-fetched in the light of the numerous potential side effects of this compound (Dustin, 1978).

2. Colchiceine, a naturally occurring analog of colchicine, has been found to block axonal transport and yet exhibits negligible tubulin binding activity (Schönharting *et al.*, 1977). This observation as well has been interpreted to mean that colchicine and its analogues may act at a site unrelated to tubulin, at least in these *in vivo* situations.

3. At the other extreme, there are reports claiming the persistence of intracellular transport after partial or complete removal of microtubules.

Teleost pigment cells treated with colchicine at low temperature and rewarmed to room temperature are unable to reform their prominent microtubule system. This combined treatment is necessary since microtubules of these cells are surprisingly colchicine-resistant, but cold-labile (Schliwa and Bereiter-Hahn, 1973b; Schliwa *et al.*, 1978). Subsequent treatment with an appropriate aggregating stimulus induces slow and incomplete but nevertheless significant aggregation of pigment granules which can be reversed by a dispersing agent (Schliwa and Euteneuer, 1978b) (Fig. 22). Similar observations were made by Obika *et al.* (1978a) on *Fundulus* melanophores. Fast

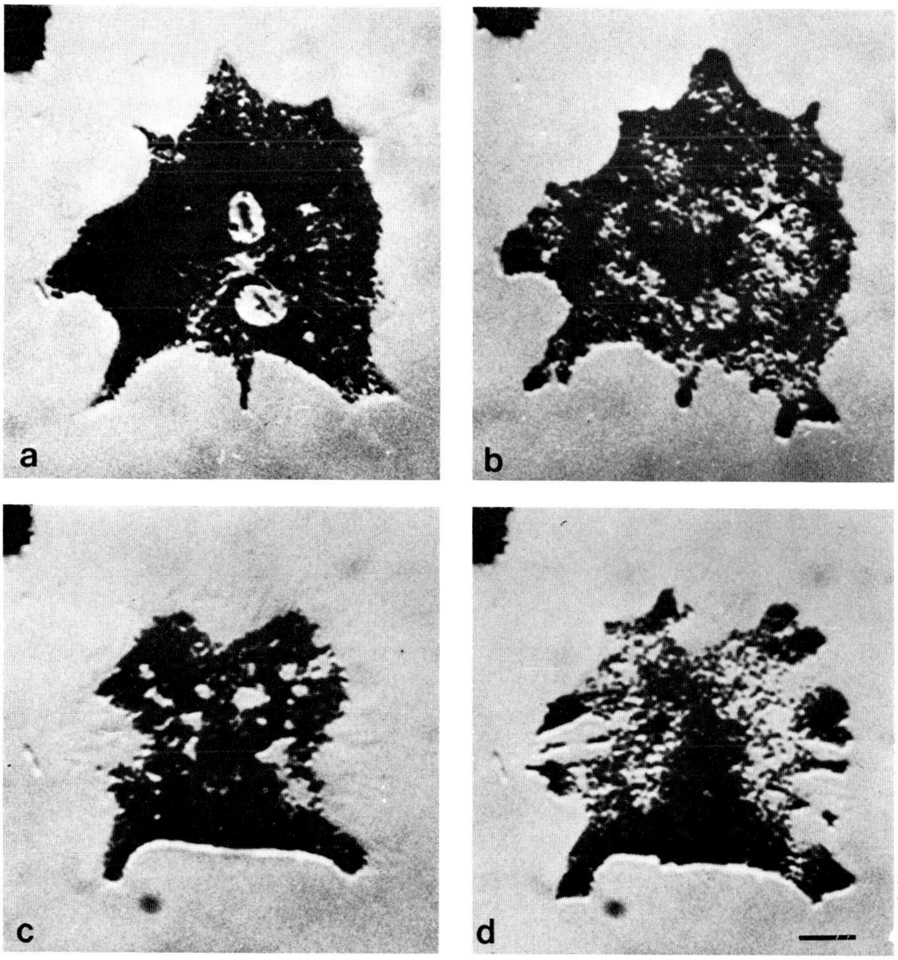

Figure 22. Isolated angelfish melanophore 6 hr after attachment to a coverslip. (a) Untreated cell. (b) After 30 min in 0°C in the presence of 5×10^{-5} M colchicine, note loss of granule alignment. (c) After 10 min recovery from cold treatment still in the presence of colchicine, an aggregating stimulus has caused the granules to form local clusters and migrate towards the cell center. (d) Ten min after a dispersing stimulus was applied, slow dispersion has occurred. Bar = 5 μm. Reprinted with permission from Schliwa and Euteneuer (1978b).

axonal transport in the rat sciatic nerve is reported to continue undisturbed in the presence of 75 mM external Ca^{2+} which causes complete disassembly of all microtubules (Brady *et al.*, 1980). Likewise, a substantial reduction in the number of microtubules is reported to be without effect on transport (Byers, 1974). These intriguing observations are interesting enough to warrant similar analyses in other systems.

4. In a related but probably more compelling experiment using the heliozoan *Actinophrys* as a model system, Edds (1975a) has studied particle movements along axopodia in an attempt to determine whether their persistence depends upon an intact microtubule axoneme. Since colchicine will cause axopodial retraction, thus dissolving the organelle in which to study movement, he decided to push a microneedle through the cell body of these organisms to create an artificial axopodium supported by a glass axoneme. Particle movements along the artificial axopodium continue even in the presence of concentrations of colchicine sufficient to induce retraction of native axopodia. Earlier, Tilney (1968) had already observed that particle movements continue in the cortex of the cell body after all axopodia had been dissolved in the presence of colchicine.

5. Although antimicrotubule agents interfere with particle saltations in many cultured cells (Wagner and Rosenberg, 1973; Freed and Lebowitz, 1970), there are reports that microtubule depletion may just alter the microscopical appearance of movement (Bhisey and Freed, 1971; Wang and Goldman, 1978). Some organelles seem to travel through the cytoplasm in a smooth, continuous fashion more reminiscent of cytoplasmic streaming.

Thus, the studies on microtubule-associated intracellular movements reviewed up to this point leave essentially two possibilities for consideration as to the extent of microtubule participation in motive force production:

1. Transport is essentially microtubule-based and absolutely dependent on their integrity. The motive force is delivered by microtubule-associated enzymes such as HMW MAPs or dynein-like molecules that interact with the transported organelle. Alternatively, such an enzyme could also be associated with the particle surface and interact with the microtubules. Following the disassembly of the microtubules by experimental manipulation, an actomyosin-dependent, but microtubule-independent, form of movement is expressed adventitously which, however, does not necessarily reflect an intimate involvement of this other force-producing system in the expression of movement under normal conditions.

2. A second component or system, e.g., actomyosin, is involved that actively interacts with both the microtubules (as elements providing directionality) and the organelles to be propelled.

3.1.2. *Microtubules as a Scaffold*

The possibility of a cooperative interaction between microtubules and an actomyosin system where the former constitute the guidelines while the latter represent the motor has received considerable attention. This concept has been incorporated into some prominent theories about intracellular movements (Ochs, 1972a). It should therefore be discussed in some detail.

To demonstrate the participation of an actomyosin system in saltatory movements, the first step is to demonstrate the presence of its most prevalent component, actin filaments, at sites where transport occurs. For many years, these attempts have suffered from the difficulty inherent in recognizing F-actin in micrographs of thin sections. This, as is now known, is due to the low visibility of F-actin embedded in a dense cytoplasmic matrix, as well as the instability of unprotected actin filaments, i.e., filaments not complexed with tropomyosin, towards fixation and dehydration (Maupin-Szamier and Pollard, 1978; Small, 1981). These problems were overcome in part by the use of gentle cell rupture-, lysis-, and glycerination-techniques which remove much of the obscuring cell constituents. However, in these experiments, better visiblity of actin filaments has been gained at the expense of possible unwarranted displacement or rearrangement of actin induced by the technique applied. The results should therefore be interpreted with caution. Nevertheless, this approach has been instrumental in demonstrating the presence of actin filaments in cell systems where this demonstration has met with difficulty before (see the chapters on chromatophores, neurons, and heliozoa in Section 2).

The predominantly peripheral location of actin filaments demonstrated by these techniques (Metuzals and Tasaki, 1978; Obika *et al.*, 1978a; Schliwa and Euteneuer, 1978a,b; Kuczmarski and Rosenbaum, 1979b) seems to argue against a direct role in organelle transport. However, none of these studies can rule out the existence of a highly labile, hard-to-fix actin filament system involved in the generation of the motive force. Therefore, attempts have been made using alternative approaches, including pharmacology, and more recently also microinjection and cell permeabilization, in order to approach the question of a functional involvement of an actin-based system.

The most widely used drugs to study actin-dependent motility are the mold metabolites cytochalasin B and its derivatives (Carter, 1967, 1972). These compounds were widely used in the late sixties and seventies even though their precise mechanism of action was not understood at that time. Until the early seventies, available evidence indicated that it affects the integrity of filament networks, presumably by "disrupting" actin filaments (Wessels *et al.*, 1971). Its mechanism of action in molecular terms has been understood only recently. Cytochalasins bind to the "fast-growing" end of an actin filament, i.e., that end at which subunits are added at a high rate. In doing so they block further addition of subunits at this end and thus inhibit (or at least slow down) polymerization (Lin *et al.*, 1980; Brown and Spudich, 1979; MacLean-Fletcher and Pollard, 1980). In addition, there is evidence that cytochalasins may fragment or break actin filaments (Hartwig and Stossel, 1979; Schliwa, 1982a). Thus, cytochalasins will interfere with cellular processes which require ordered actin assembly and/or intact three-dimensional actin networks.

Cytochalasin B has been applied to a number of pigment cell types, with diverging results. It inhibits both aggregation and dispersion in some invertebrate chromatophores (Lambert and Crowe, 1973; Robison and Charlton, 1973; Dambach and Weber, 1975; Lambert and Fingerman, 1978a), while those of amphibians may show an inhibition of dispersion only (Malawista, 1971b; McGuire and Moellmann, 1972; Novales and Novales, 1972; Magun, 1973; Fisher and Lyerla, 1974), and sometimes even enhancement of aggregation (Lyerla and Novales, 1972; Koyama and Takeuchi, 1980). One study reports that cytochalasin B blocks aggregation of iridophores at the same concentration that causes inhibition of melanophore dispersion (Novales and Novales, 1972). The physiological basis for these diverging cellular reactions is not understood. Adequate judgment of these studies is hampered by our ignorance of the organization of actin filaments in relation to granule transport in these cell types.

Fish chromatophores are either not significantly affected at all by cytochalasins (M. Beckerle, personal communication; M. Schliwa, unpublished observations), or may even show an acceleration of both aggregation and dispersion (Ohta, 1974). The retinal pigment epithelium of the sunfish appears to be an exception in that cytochalasin B will prevent pigment dispersion and cause aggregation (Burnside *et al.*, 1982), a reaction more reminiscent of the response of amphibian pigment cells. The apparent lack of an inhibitory action on integumental fish chromatophores is consistent with the observation that actin filaments are found predominantly in a subcortical location and not in association with moving pigment granules (Schliwa *et al.*, 1981).

Whether cytochalasin B blocks axonal transport is a matter of some dispute. The drug is apparently without effect on fast anterograde transport in frog sciatic nerve (Abe *et al.*, 1973; Anderson *et al.*, 1972), the optic nerve of the rat (Crooks and McClure, 1972), and the hypogastric nerve of the cat (Banks *et al.*, 1973). The same is true for retrograde transport in the frog nerve (Edström and Hanson, 1973b). On the other hand, there are reports describing an inhibitory action of cytochalasin B on fast transport in dorsal root ganglia and sciatic nerves of the rat (Crooks and McClure, 1972), crayfish nerve axons (Fernandez and Samson, 1973), and the Retzius cell of the leach (Isenberg *et al.*, 1980). The results of the latter three studies should, however, be viewed with some caution. Unusually high concentrations of cytochalasin B (200–400 μM) were necessary to observe an inhibitory effect on transport in the sciatic nerve (Crooks and McClure, 1972). In the other two studies, cytochalasin B was injected either into the ganglia of the crayfish nerve cord (Fernandez and Samson, 1973) or directly into the cell body of the Retzius cell. In these experiments, an effect on the initiation of transport in the cell body or the axon hillock, rather than inhibition of transport itself along the axon, could be the cause of the observed inhibition. In the light of these uncertainties, it is the belief of many investigators that an intact actin filament system is not absolutely required for fast transport.

Cytochalasin B, at levels that induce profound morphological effects, fails to inhibit the movement of ingested particles in cultured cells (Bell *et al.*, 1980). The drug has not been applied to protistan cells such as heliozoa or foraminifera, even though an actomyosin-like contractile system has been implicated in particle movement on the basis of morphological studies (Edds, 1975b; Bardele, 1976).

While cytoskeletal inhibitors such as colchicine or cytochalasin B can easily cross the cell membrane, a number of other compounds, including antibodies, are virtually impermeable and have to be introduced into cells by other means. Recently, the technique of microinjection has become widely used to deliver to the cell interior compounds or proteins that normally do not permate the cell membrane. The study of intracellular transport has already profited from this new technique because a number of additional actin-active compounds can now be applied.

Schliwa *et al.* (1979b) employed microinjection in a test of the involvement of actin in intracellular transport in chromatophores. Microinjection of DNase I, which avidly binds to G-actin and even causes depolymerization of F-actin (Hitchcock *et al.*, 1976), does not interfere with either aggregation or dispersion of pigment granules in cultured fish melanophores (Fig. 23). The injected compound does, however, induce a dramatic change in cellular morphology, indicating that the cell has been successfully impaled and that the injected molecules are active in the cytoplasm. Within 30–60 min the cells change from a rounded, discoid shape to a highly arborized, stellate morphology. Such a change could be explained by an interference of DNase I with the organization of the subcortical actin filament network present in this cell type (Schliwa *et al.*, 1981). The mushroom toxin phalloidin, which stabilizes actin filaments and induces actin polymerization (Wieland, 1977), thus exerting an effect opposite that of DNase I, also fails to interfere with the processes of pigment aggregation and dispersion. This compound does not induce cell arborization but will lead to the formation of filamentous aggregates that can be stained with antibodies against actin, similar to the aggregates observed in microinjected tissue culture cells (Wehland *et al.*, 1977). Essentially, similar observations were made by Beckerle and Porter (1981) in microinjection experiments employing DNase I and phalloidin on cultured *Holocentrus* erythrophores. In addition, these authors used a form of heavy meromyosin (HMM) that was chemically modified with *N*-ethylmaleimide (NEM-HMM). NEM-HMM will bind to actin filaments, but unlike its unmodified counterpart, cannot be dissociated from it by ATP (Meeusen and Cande, 1979). Thus it will inhibit actin–myosin interaction necessary for force generation. This probe, like DNase I and phalloidin, does not interfere with normal granule movements. The experiments, taken together, provide a strong argument against an involvement of an actin network and/or a contractile actomyosin system in granule transport of fish chromatophores.

The technique of microinjection has also been employed in tests of the role of actin in axonal transport. In addition to cytochalasin B, Isenberg *et al.* (1980) and Goldberg *et al.* (1980) injected a number of actin-active molecules directly into the perikarya of giant invertebrate neurons, the Retzius cell of

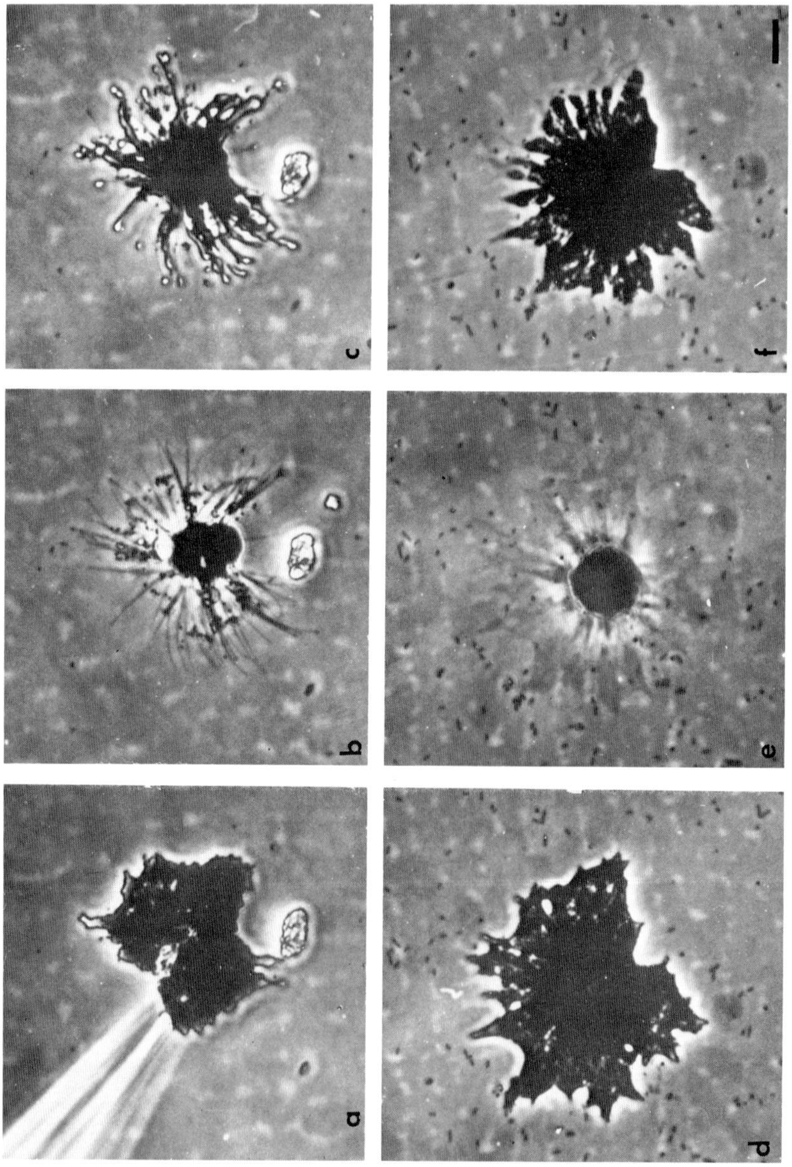

Figure 23. Light micrographs of angelfish melanophores injected with either DNase I (a–c) or bovine serum albumin (d–f). (a) and (d), cells before injection. Two hours postinjection, cells were challenged with 10^{-4} M adrenaline (b and e), followed by 10^{-4} M atropine to induce redispersion (c and f). Both DNase I and albumin-injected cells are able to aggregate and disperse their pigment granules. Note, however, the marked change in cell shape induced by DNase I injection. Bar = 5 μm.

the leech and the giant cerebral neuron of *Aplysia*, respectively. Transport was monitored by following the movement of radioactively labeled materials derived from microinjected labeled precursors along the axon. The compounds injected included DNase I, subfragment 1 (S1) of HMM, filamin (Wang *et al.*, 1975), antibodies against α-actinin, phalloidin, and an actin-depolymerizing protein present in rabbit sera. In both the leech (Isenberg *et al.*, 1980) and the *Aplysia* neuron (Goldberg *et al.*, 1980), DNase I, like cytochalasin B, decreased the fast transport of labeled materials substantially. Filamin and the actin depolymerizing protein also had an inhibitory effect. Phalloidin, HMM-S1, and antibodies against α-actinin were ineffective. Both studies came to the conclusion that actin might play an important role in fast axoplasmic transport. However, this interpretation has not been without criticism. Injection of the probe(s) into the cell soma might interfere with processes of organelle assembly and mobilization which would then manifest as a decrease in the amount of labeled material appearing further down the axon; the transport mechanism itself could be unaffected. In an attempt to resolve this ambiguity, Goldberg (1982) has essentially repeated some of his earlier injection experiments, with the important modification that injections were made directly into the axon. Fast transport of ^{3}H-fucose-labeled material is completely blocked following intra-axonal injection of DNase I, whereas NEM-HMM is without effect. In addition, dihydrocytochalasin B applied externally to the axon does not influence transport. The conclusion from these results was that actin filaments may play a structural role in maintaining an ordered arrangement of axoplasmic components, rather than a role in force generation. Presumptive depolymerization of actin filaments by DNase I is thought to cause severe disordering of the axoplasm; NEM-HMM and dihydrocytochalasin B would be without effect because they presumably do not cause actin disassembly. Even though these experiments seem to provide a strong argument against the involvement of a contractile system in fast axonal transport, they need to be repeated and further substantiated using other systems. That DNase I does in fact depolymerize actin filaments needs to be shown directly. Another point of uncertainty concerns the lack of effect of dihydrocytochalasin B. If actin filaments do indeed maintain the structural integrity of the axon, as suggested by the DNase experiment, and if their structural integrity is a prerequisite for a functional transport machinery, then cytochalasin B should not be without influence. Although it does not depolymerize F-actin, its profound effect on the structure of actin-based gels (Hartwig and Stossel, 1979; MacLean-Fletcher and Pollard, 1980) can be as devastating to network integrity as complete depolymerization of actin filaments.

3.1.3. Microtubules as the Motor

An approach that has provided a quantal jump in our understanding of the two paradigms of motile machineries, muscle and cilia, has begun to be applied to the study of more complex cellular systems only recently. As in muscle or cilia (Hanson and Huxley, 1955; Summers and Gibbons, 1971), *permeabilized cell models* may facilitate the identification of the components involved in intracellular transport and the mechanisms that regulate their activity. In fact, this approach has been used with considerable success in the study of a rather complex motile machinery, the mammalian mitotic spindle (see Cande, 1980, for an overview). The use of permeabilized cell models, like microinjection, allows the application of impermeable compounds and, in addition, provides a means to test nucleotide requirements, calcium dependency, and pH regulation, parameters difficult to determine (and even more difficult to vary experimentally) in intact cells. Cell models have to meet three basic requirements: (1) they should retain the molecule(s) or cell structure(s) presumed to be involved in the motile event, (2) they should retain the functional activity of these components, and (3) the plasma membrane should be fully permeable. The task of meeting these basic requirements has been facilitated in recent years by the availability of microtubule-preserving buffers and gentle detergents. Nevertheless, we are still at the beginning of what seems to be a promising period of research.

Gartz (1970) was the first to attempt reactivation of an intracellular transport system other than the mitotic spindle. He used glycerinated skin pieces of *Xenopus* tadpoles and showed that in the melanophores contained therein the area covered by pigment granules decreases by an average of 14% if ATP is added to these preparations. No attempt was made to document successful permeabilization and/or the presence of microfilaments. Although the effect of ATP shows that something became permeable, it can not be decided whether the decrease in pigment-covered area results from a reactivation of a microtubule-associated mechanochemical enzyme, or actomyosin, or both. Not even an effect totally unrelated to pigment cells, such as contraction of fibroblasts surrounding the pigment cells, can be excluded. Gartz (1970) interpreted his results as an indication for the involvement of an actomyosin system in pigment granule movements.

The past two years have seen a rising interest in the use of permeabilized cells to study intracellular transport. The cell types used in these studies include chromatophores, neurons, and cultured cells. In most cases, the approach is based upon detergent permeabilization in the presence of a stabilizing buffer. Clark and Rosenbaum (1982) show that permeabilized *Fundulus* melanophores are accessible to tubulin antibodies and ferritin molecules. Detergent-treated melanophores still require adrenalin stimulation to initiate pigment aggregation, but the response progressively diminishes with increasing time of incubation in the lysis medium. Surprisingly, no exogenous ATP is required to sustain granule movement. On the other hand, exogenous ATP has proved to be indispensable for saltatory movements in detergent-permeabilized cultured fibroblasts (Forman, 1982), erythrophores (Stearns and Ochs, 1982), and crustacean axons (Adams, 1982; Forman *et al.*, 1982). Although not rigorously tested, none of the reactivated model systems appear to require the presence of significant amounts of calcium ions since the extraction media used in these studies contained 10 mM EGTA.

Permeabilized cell models have been employed in recent efforts to explore the possibility of a direct role for microtubules in saltatory transport by addressing the question of an involvement of a dynein-like molecule. In these experiments, the ciliary dynein ATPase inhibitor, orthovanadate (which does not cross cell membranes), and the adenine derivative, erythro-9-[3-2(hydroxynonyl)] adenine (EHNA), which is permeable, were used as probes. Movement of particles in permeabilized cell models of melanophores (Clark and Rosenbaum, 1982), erythrophores (Stearns and Ochs, 1982), fibroblasts (Forman, 1982), and lobster giant axons (Forman *et al.*, 1982) is reported to subside in the presence of 10–100 μM vanadate. In melanophores and lobster axons, EHNA has the same effect. In addition to the studies on extracted cells, vanadate has been used in microinjection experiments. When injected into cultured erythrophores (Fig. 24), vanadate at an estimated effective concentration of 3.7 μM stops shuttle movements and inhibits aggregation (Beckerle and Porter, 1982). Injection into the axon of the *Aplysia* giant cerebral neuron (estimated intra-axonal concentration >80 μM) did not, however, interfere with fast transport. Injections of vanadate into the cell body of the Retzius cell of the leech (Isenberg *et al.*, 1980) were inhibitory only at high concentrations (1000 μM in the needle) but ineffective at lower concentrations (100 μM). In both erythrophores and the *Aplysia* neuron, EHNA, which easily crosses the cell membrane, inhibited transport when used at millimolar concentrations in the bathing medium.

How specific are the two inhibitors, orthovanadate and EHNA? There can be little doubt that both are specific and effective inhibitors of ciliary dynein ATPase activity. However, vanadate at concentrations used in some of the studies (around 100 μM) is also known to inhibit ATPases other than dynein (Simons, 1979), while EHNA may affect a variety of other enzymes (Pennigroth *et al.*, 1982). The conclusion that both these compounds affect transport through an inhibition of a dynein-like ATPase may therefore be premature; this possibility is recognized by most of the investigators using these compounds. There is also some conflicting evidence as to the effectiveness of microinjected vanadate which may inhibit transport in some systems (Beckerle and Porter, 1982) but not in others (Isenberg *et al.*, 1980; Goldberg, 1982). Even more puzzling is the observation that the ATPase inhibitors vanadate and EHNA inhibit a cell process in chromatophores, pigment aggregation, which reportedly does not require ATP and hence an ATPase (Junqueira *et al.*, 1974; Saidel, 1977; Luby and Porter, 1980). Finally, no other independent evidence exists for the presence of a cytoplasmic dynein-like molecule in these cell systems. Clearly, more work is required before definitive conclusions can be reached. It is equally obvious, however, that this approach bears great potential for the future.

3.1.4. The Importance of Microtubule Polarity

The best-studied motile machineries are striated muscle and cilia. In both systems, the intrinsic molecular polarity of the prevalent structural components, actin filaments and microtubule doublets, respectively, is of prime im-

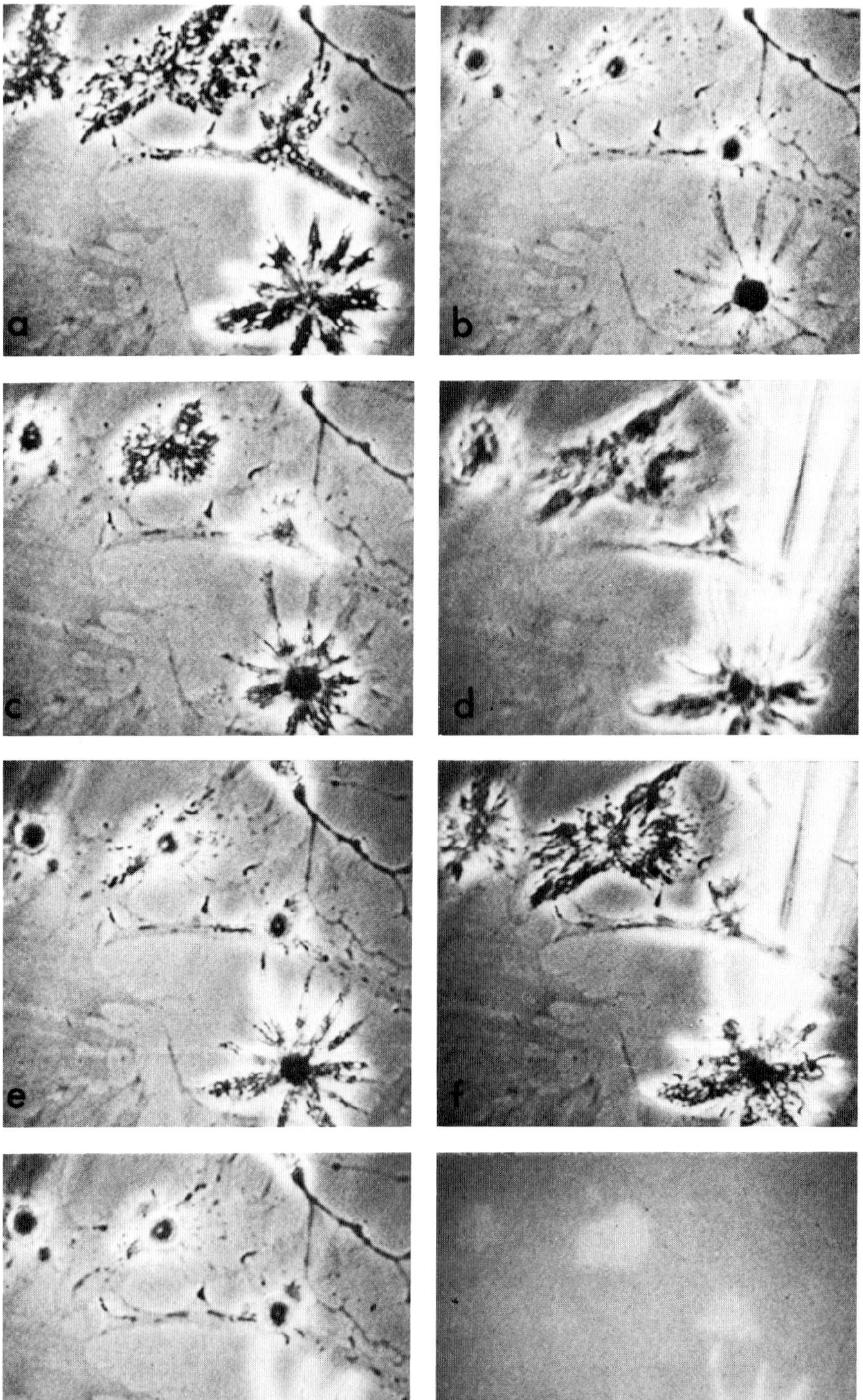
a
b
c
d
e
f
g
h

portance for the functioning of the machinery. In muscle, the generation of force depends on the sliding of thin filaments (actin) past thick filaments (myosin) mediated by the cyclic binding of myosin heads to actin. The polarity of actin filaments is always the same relative to the position of other components such as myosin filaments and the Z-line. This polarity is conveniently visualized by the binding of HMM, which results in an arrowhead or chevron-like pattern (Huxley, 1963). The method of HMM decoration has been applied successfully to nonmuscle cells (Ishikawa *et al.*, 1968) and has served as a convenient tool for both the identification of actin filaments and the demonstration of their structural polarity in a wide variety of cells.

For a long time, a similarly convenient marker for the polarity of microtubules was not at hand, despite a clear need for a direct assay applicable to a variety of cell types where microtubules are implicated in motile activities and where their intrinsic polarity might be of importance for certain features of the motile event. For instance, it is theoretically possible that movement of organelles in one direction is associated with one set of microtubules, while a different set of microtubules with opposite polarity is responsible for movement in the opposite direction. Thus, microtubule polarity would be as important for intracellular transport as actin polarity is for muscle contraction. Indeed, some theories of intracellular movements assigned a prominent role to microtubule polarity (Bikle *et al.*, 1966; Schmitt, 1968; McIntosh *et al.*, 1969; Ochs, 1972a; Gross, 1975). In fact, if two sets of microtubules having opposite polarities could be demonstrated in systems such as chromatophores, neurons, or the spindle, it would provide an indirect argument for a transport mechanism directly dependent on microtubules.

There are indirect methods to determine microtubule polarity, but they are tedious and can only be used in a limited number of systems (McIntosh *et al.*, 1980). Recently, however, a direct, convenient method has been developed that has already been used successfully to determine microtubule polarity in a number of different cell types. The technique is based on the addition of enantiomorphic protofilament sheets added to pre-existing microtubules; when viewed in cross section, they reveal microtubule polarity (Heidemann and McIntosh, 1980). The method has been used to determine the polarity of spindle microtubules (Euteneuer and McIntosh, 1980, 1981b, 1982) and other cell systems where organelle movements are very prominent: angelfish melanophores (Euteneuer and McIntosh, 1981a), nerve axons (Burton and

Figure 24. Phase-contrast micrographs of four *Holocentrus* erythrophores in culture. The cells are coupled, that is, they disperse (a), aggregate (b), and redisperse (c) their pigment granules in synchrony. (d) The cells are in the dispersed state and one cell is microinjected with vanadate (250 μM) in injection buffer containing fluorescein-labeled bovine serum albumin. The injected cell is rapidly inhibited in its intracellular transport function whereas the three other cells continue to aggregate and disperse their pigment (e,f). Touching the microinjected cell with the micropipette (f) results in aggregation of the three coupled, uninjected cells, but the vanadate-injected cell does not respond to the stimulus with pigment movement (g). A fluorescence image (h) of the four cells illustrates that the cell on the lower right has been successfully injected. Slight autofluorescence of the pigment granules can be seen in the other erythrophores. Bar = 5 μm. Reprinted with permission from Beckerle and Porter (1982).

Paige, 1981; Filliatreau and DiGiamberardino, 1981; Heidemann *et al.*, 1981; Fig. 25), and heliozoan axopodia (Euteneuer and McIntosh, 1981a) (Fig. 26). The somewhat surprising outcome of these studies is that microtubules in all these systems are of uniform polarity with respect to the cell center and/or the microtubule-organizing center, their plus ends (or fast-growing ends) being distal. Thus, prominent bidirectional transport takes place in the presence of one set of microtubules with uniform intrinsic polarity. The observations rule out microtubule polarity as a determinant for the directionality of organelle movements. They do not rule out, however, an involvement of microtubules in general, even though the extent of that involvement has been substantially narrowed down. If transport is indeed microtubule-based, directionality is provided by some component(s) associated with microtubules or interacting with them; the polarity of subunits within the microtubule wall is irrelevant. This significant finding should not be overlooked in considerations of the nature of the mechanochemical transducer proposed to interact with microtubules and the transported particle.

3.1.5. *An Alternative Way to Think about Microtubule-Associated Transport*

In 1968, Green published a paper in which she discussed possible mechanisms of granule movements in *Fundulus* melanophores. On the basis of her careful light microscopic observations of the behavior of granules during their movement toward and away from the cell center, she came to the conclusion that the pigment granules behave as if they are contained within a structural continuum or elastic matrix which, through its contraction during aggregation and re-expansion during dispersion, mediates granule movements. At any instant, the distribution of granules is the function of a dynamic equilibrium between aggregative and dispersive forces developed by and within this matrix component. At that time, her interpretation seemed incompatible with the popular belief, derived mainly from biochemical work, that cells are, in general, containers for a poorly structured cytosol in which membrane-bounded organelles float. In retrospect, however, her view appears almost prophetic. Years later, Porter and his associates (Byers and Porter, 1977; Luby and Porter, 1980; Beckerle and Porter, 1981; Porter and McNiven, 1982; Schliwa, 1979) investigated the structural organization of the chromatophore cytoplasm by whole-cell high-voltage electron microscopy and described the existence of a substructure of fine strands comprising an irregular three-dimensional lattice or meshwork (the microtrabecular lattice) that connects pigment granules to one another, to microtubules, and to the plasmalemma and internal membranes. The pigment granules appear as if they are contained in, or are an integral part of, the system of microtrabecular strands (Fig. 27). The morphological disposition of this network fulfills the structural requirements of Green's "elastic continuum." The view that Green's elastic continuum and Porter's microtrabecular lattice are identical is further substantiated by the observation that the microtrabecular system undergoes characteristic and seemingly predictable changes in morphology in conjunction with aggregation and dispersion of the pigment granules. They shorten and thicken when

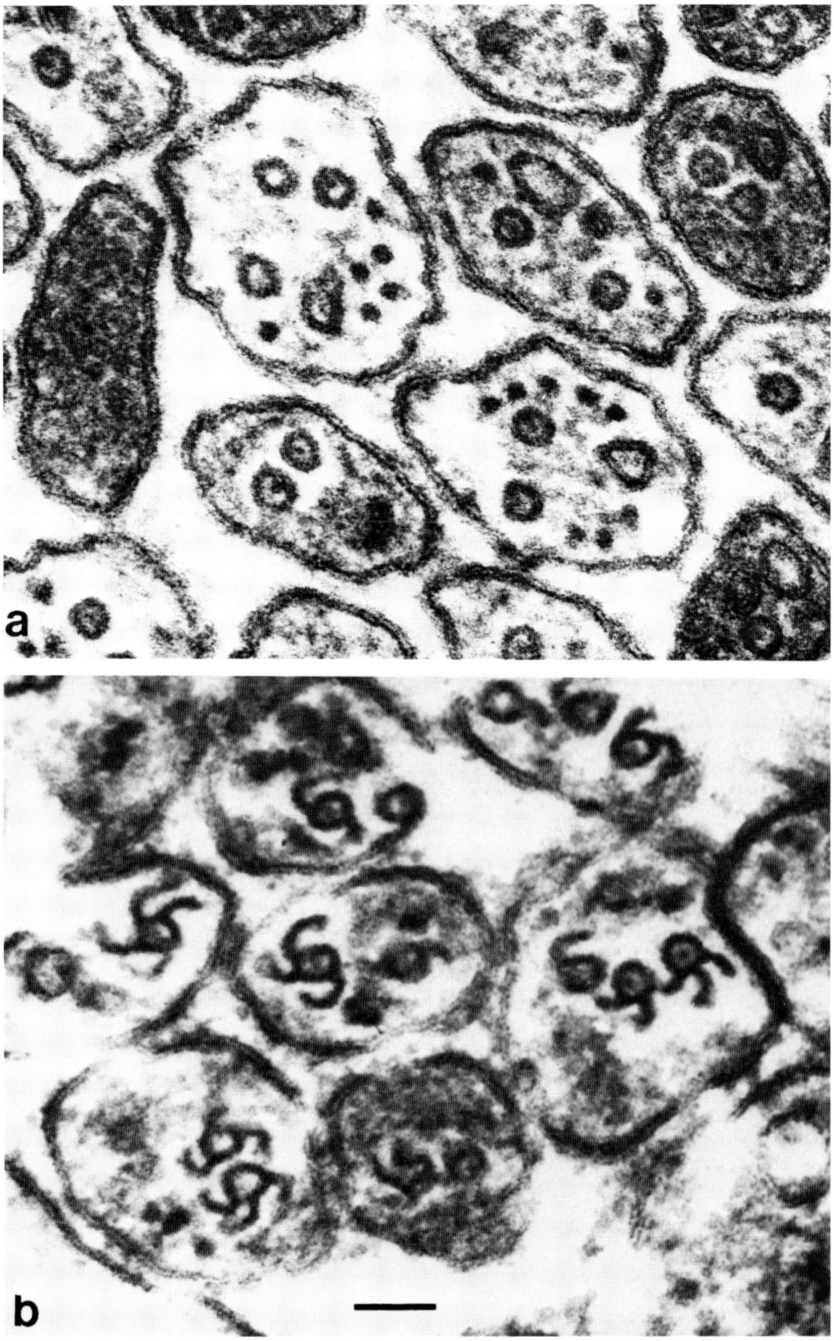

Figure 25. Cross sections of frog olfactory nerves. (a) Untreated nerve. Each of the small, unmyelinated axons usually contains only two or three microtubules and a variable number of neurofilaments. (b) A nerve incubated under conditions that allow the formation of polarity-revealing "hooks" which consist of curved protofilament sheets that attach to pre-existing microtubules. The viewer is looking toward the cell bodies located in the olfactory epithelium and all "hooks" seen curve clockwise, indicating that all microtubules in the olfactory axons appear to be of uniform polarity. Bar = 0.5 μm. Micrograph kindly provided by Dr. Paul Burton.

Figure 26. An axopodium of the heliozoan *Actinosphaerium* decorated with polarity-revealing hooks. The double-spiral pattern of the microtubules is slightly distorted. We are looking from the tip toward the base of the axopodium, hooks are predominantly curved clockwise. Bar = 0.5 μm. Micrograph kindly provided by Dr. Ursula Euteneuer.

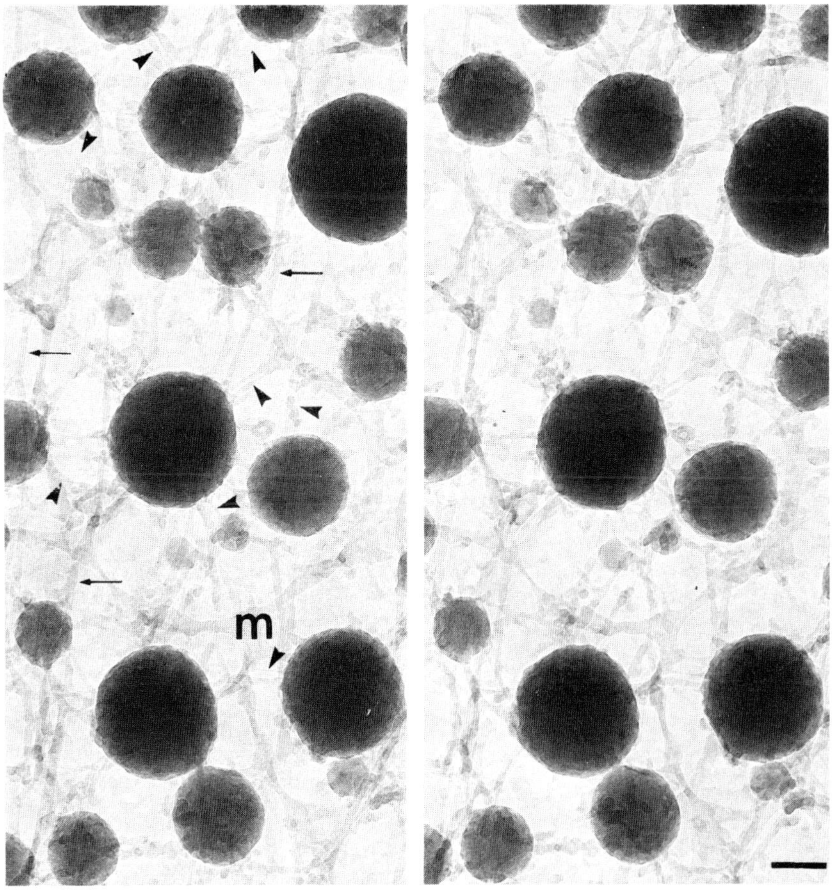

Figure 27. Stereo pair of part of an isolated *Holocentrus* erythrophore prepared by critical point-drying and photographed in a 1000 KV high-voltage electron microscope. A three-dimensional network of filamentous strands (microtrabeculae; arrowheads) interconnect granules, membranes, and microtubules (arrows). Part of an intracellular tubular membrane system may be seen at m. Bar = 0.1 μm. Reprinted with permission from Luby-Phelps and Schliwa (1982).

granules are transported towards the cell center and appear to re-form or elongate during dispersion. The entire cytoplasmic compartment (the "cytoplast") would appear to behave as a unit structure in which every pigment granule has a defined position in relation to its neighboring pigment granules. Indeed, evidence supporting this view has been presented recently. Identifiable pigment granules in erythrophores return to the same location relative to the whole pigment complex during successive phases of spontaneous aggregation and dispersion (Porter and McNiven, 1982). They retain their position even though the cell undergoes quite dramatic changes in overall morphology, from discoid to spheroid and back to discoid again (Gartz, 1970; Schliwa and Euteneuer, 1978b; Porter and McNiven, 1982). The bulk of the microtrabecular lattice moves together with the pigment, thereby changing its structural configuration. In doing so, it is believed to provide the motive force (Byers and Porter, 1977; Luby and Porter, 1980; Porter and McNiven, 1982).

A morphologist may be satisfied with the identification of a structural component that seems to meet the requirements for a translocative machinery whose morphological transformations in space and time correlate with the motile phenomenon under study. The cellular biochemist, however, is not. He would like to know the composition of the network and, if possible, pinpoint the molecules that actually provide the motive force. The properties of the elastic continuum described by Porter and associates seem to match best with those of a reversibly contractile actin-based gel (Taylor and Condeelis, 1979). Unfortunately, the evidence for an extensive actin-based gel in chromatophores is inconclusive, if not meager (Schliwa *et al.*, 1981). We need to consider, however, that the properties of this putative actin gel may be such that we are simply unable to preserve and fix it adequately at this point. A second possibility, that a dynein-like molecule is part of the granule-moving machinery, has been discussed in Section 3.1.3. This area of research is in its very early stages, and the question of dynein involvement and its relation to microtrabecular organization requires more detailed consideration than it has received to this point. So, at present, as Porter and McNiven (1982) put it, the microtrabecula is "the ghost in the machine," but, quite like the structure of the microtrabeculae themselves, the ghost is difficult to describe in quantitative terms.

It is unclear to what extent the findings based on whole-mount preparations of erythrophores can be extended to other intracellular transport systems. Certain features of the erythrophore model are quite unique and not easily applicable to other cell types. For instance, particles in nerve axons, heliozoa, or foraminifera probably never return to the same location in the cell as part of a cyclical restructuring of the cytoplast, and it even remains to be tested whether granules in other chromatophores do so. Nevertheless, a microtrabecular system has been invoked in other forms of intracellular motility, notably axonal transport (Ellisman and Porter, 1980). Here, a morphologically homogeneous microtrabecular system has been described that interconnects microtubules, neurofilaments, membrane-bounded organelles, and the plasmalemma. Particles are proposed to be transported by sequential making and breaking of trabecular links. The view of a homogeneous matrix component has been challenged recently on the basis of rapid-freezing–shallow-etching studies of axoplasm (Schnapp and Reese, 1982). Transport is thought to occur in channels delineated by microtubules embedded in a loose granular matrix. The structure of these microtubule domains clearly differs from that of nearby cytoplasmic domains where neurofilaments reside (Fig. 28). Even though this study probably provides us with a more accurate view of the structural organization of the axoplasm than previously achieved by any other technique, it still does not allow to distinguish between the different possibilities currently considered as the molecular machinery mediating transport, and how this molecular machinery relates to the structural organization of the axoplasm.

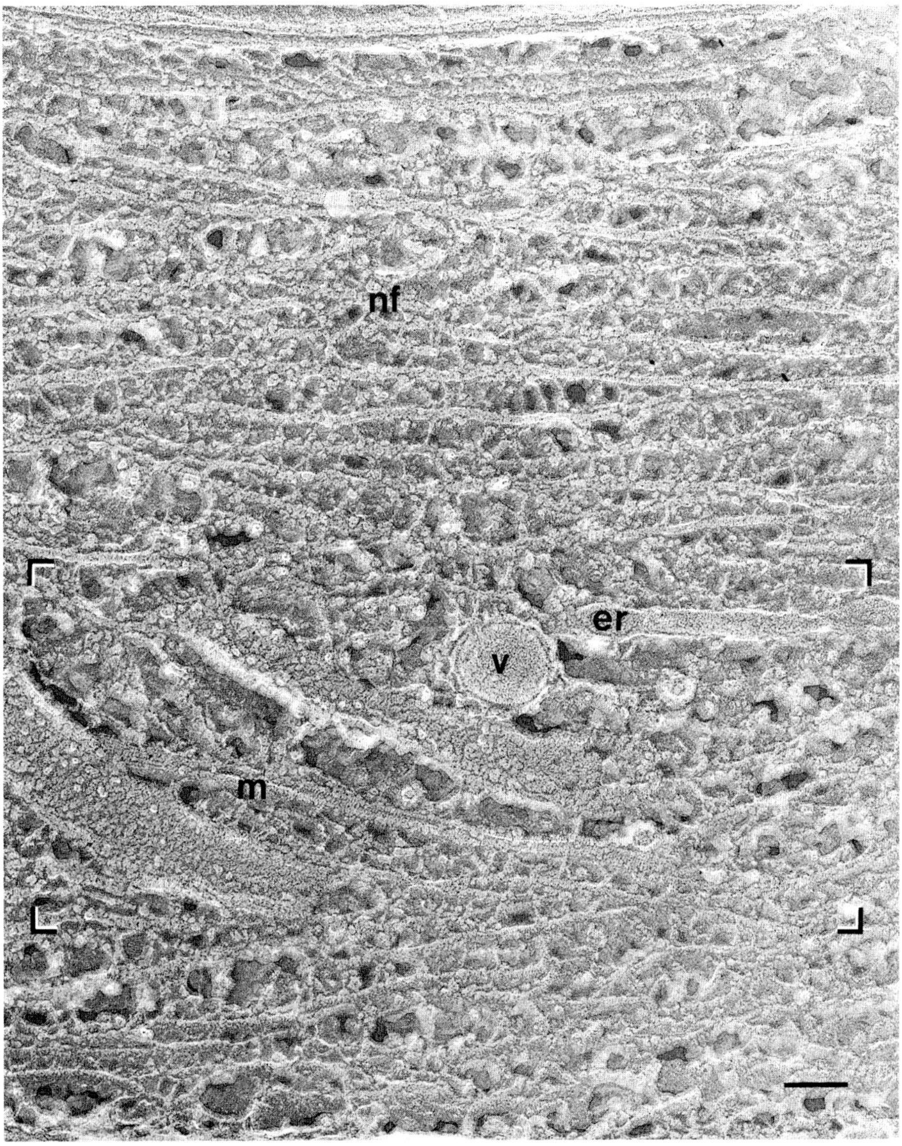

Figure 28. Part of an axon in the turtle optic nerve prepared by direct freezing, freeze-fracturing, minimal etching produced by sublimation of about 10 nm of ice, and rotary shadowing. Parallel longitudinally oriented neurofilaments (nf) are interconnected by cross-bridging filamentous and granular material. A cisterna of endoplasmic reticulum (er) and a vesicular organelle (v) with a smooth E-fracture face lie in a longitudinal domain of granular axoplasm (delineated by brackets) which is distinct from the neurofilament domain. A microtubule (m) located in this domain has been fractured open, exposing its inner surface. Bar = 0.1 μm. Reprinted with permission from Schnapp and Reese (1982).

3.1.6. Regulation of Microtubule-Associated Transport

A different approach toward understanding the mechanisms of microtubule-dependent transport is to study possible regulatory mechanisms. It appears that the only cell system where some insights into regulatory mechanisms have been gained is the chromatophore. It is well established that pigment dispersion is accompanied by a rise in the level of cyclic AMP (Bitensky and Burstein, 1965; Abe *et al.*, 1969), and that externally applied cyclic AMP, its dibutyryl analog, or phosphodiesterase inhibitors (or combinations thereof) will induce granule dispersion in aggregated chromatophores (Novales and Davies, 1967; Novales and Fujii, 1969; Byers and Porter, 1977). That this response is brought about by an increase in the level of intracellular cyclic AMP, and not just an action of these different agents on the cell membrane, has been shown by iontophoretic injection of cyclic AMP which will induce immediate dispersion (Geschwind *et al.*, 1977). Thus, cyclic AMP may be the intracellular second messenger mediating the outward movement of pigment granules. It is not known, however, how this control is achieved. One possibility is that cyclic AMP promotes the assembly of a labile population of microtubules, this assembly process somehow being coupled to dispersion. Such a function would be consistent with the observation of an increased number of microtubules in the dispersed state (Porter, 1973; Schliwa and Euteneuer, 1978a). A second, possibly related, function is that it influences the state of phosphorylation of some component(s) involved in granule movements. Consistent with this proposal is the recent finding of changes in the state of phosphorylation of several polypeptides resolved in two-dimensional polyacrylamide gels (Lynch *et al.*, 1982). One way by which cyclic AMP-dependent phosphorylation could regulate cytoskeletal components is suggested by studies of *in vitro*-reassembled brain microtubules. Here, a cyclic AMP-dependent protein kinase copurifies with microtubules through several cycles of assembly and disassembly, indicating specific association with microtubules. This protein kinase phosphorylates MAP_2 (Vallee, 1980), one of the major high-molecular-weight proteins that form lateral projections extending from the microtubule surface (Murphy and Borisy, 1975). MAP_2 in turn seems to mediate the interaction of microtubules with actin filaments (Griffith and Pollard, 1978) in a phosphorylation-dependent manner; if phosphorylated, no such interaction is observed (Selden and Pollard, 1982). Whether these *in vitro* studies bear any relationship to the *in vivo* processes in chromatophores is completely unknown, but thinking along these lines may lead to the development of meaningful new experiments.

Less is known about regulatory mechanisms involved in pigment aggregation. Recently, Luby-Phelps and Porter (1982) set out to investigate the possible role of calcium transients in the initiation of centripetal granule transport. The rationale for this study was that the aggregation response to adrenaline is mediated by α-adrenergic receptors (Fujii and Miyashita, 1975; Matsumoto *et al.*, 1978), the stimulation of which elicits an increase in the concentration of cytoplasmic free calcium in a number of different cell systems (Putney, 1978). Luby-Phelps and Porter (1982) showed that an experimentally induced rise in intracellular calcium to a calculated concentration of 5 μM will lead to pigment aggregation in the absence of other stimuli. Experimentally elevated cytoplasmic calcium will also induce aggregation in erythrophores of a crustacean (Lambert and Fingerman, 1978b), but the level of calcium required was not calculated.

Attempts to demonstrate similar regulatory roles for cyclic AMP and/or calcium in other cell systems with prominent intracellular organelle movements have proven elusive. Cyclic AMP is without effect on axonal transport (Edström, 1977), and studies on the influence of elevated or reduced levels of calcium on permeabilized axons (Adams, 1982) or extruded axoplasm (Brady and Lasek, 1982b) failed to demonstrate a calcium sensitivity up to the point where calcium induces destruction of the cytoskeleton. On the other hand, fast axonal transport in desheathed but otherwise intact peroneal nerves is dependent on adequate (millimolar) levels of extracellular calcium (Ochs *et al.*, 1977; Stearns, 1982). Information for other systems is not available at this point. It is possible, though entirely speculative, that neurons, cultured cells, etc. do not possess a regulatory system at all, and that transport machinery is "turned on" all the time. Chromatophores clearly require a regulator or switch that allows the organism to assess the state of pigment distribution at will and without delay.

3.1.7. Involvement of Intermediate Filaments in Microtubule-Associated Transport

Intermediate filaments were occasionally proposed to play a role as mediators of organelle movements in some cell types, including neurons (Willard and Simon, 1981), cultured cells (Wang and Goldman, 1978), and chromatophores (McGuire and Moellmann, 1972; Junqueira *et al.*, 1977). This proposal has been made mainly on the basis of their abundance in cell regions where transport occurs, and the colinearity of filament extension and the direction of organelle movements. However, it is probably fair to state that there is no direct evidence for their role as guiding elements, let alone their participation in the generation of the motive force. By the same token, there is also no strong evidence against their involvement in some cells, at least. The inaccessibility of this, the third major filament type of eukaryotic cells, to *in vivo* experimentation is due to the lack of a specific inhibitor or depolymerizing agent analogous to the cytochalasins or microtubule inhibitors. However, vimentin filament collapse induced by microinjection of specific antibodies is reported to be without effect on directed saltatory motility (Gawlitta *et al.*, 1981). In addition, recently a compound has become known that induces a reorganization of axonal neurofilaments without interference with fast axonal transport (Papasozomenos *et al.*, 1981). β,β′-imidopropionitrile (IDPN) leads to a segregation of all neurofilaments towards the periphery of the axon, while microtubules, smooth endoplasmic reticulum, and mitochondria occupy the central core. Rapidly transported materials identified by electron-microscopic autoradiography are clearly located in the central core region in IDPN-treated neurons, but uniformly distributed throughout the axon in untreated cells (Papasozomenos *et al.*, 1982). This finding is consistent with models of fast transport viewing the microtubule as an important component in force production. While IPDN experiments still do not provide a strong argument against the possible involvement of intermediate filaments in some transport systems, they seem to support the conjecture that, if anything, their role is purely structural or supportive. Beyond doubt, however, they are not universally required in intracellular transport since many cell systems with well-developed forms of organelle movement do not express intermediate filaments at all (heliozoa, foraminifera, crustacean neurons, *Holocentrus* erythrophores).

3.2. *Category II: Cytoplasmic Streaming*

Of the cell types introduced in Section 2.6., the internodal cells of *Chara* and *Nitella* are by far the best-studied model systems for cytoplasmic streaming. Both morphological and experimental evidence is now overwhelmingly in favor of a transport mechanism based on force generation at or near the bundles of actin filaments located at the ectoplasm–endoplasm interface. Whether the mechanism of force generation involves an interaction between these actin filaments and myosin is not rigorously proven experimentally, but remains an almost inescapable conclusion that has been reached by many investigators. The residual uncertainty about this point resides largely with our ignorance about the nature and location of myosin molecules in these cells, as well as other cell types exhibiting cytoplasmic streaming. In fact, only one report seems to exist demonstrating the biochemical extraction of myosin from *Nitella* (Kato and Tonomura, 1977). Two ultrastructural studies suggest the presence of filamentous components reminiscent of myosin molecules (Williamson, 1979; Allen, 1980). Consequently, until this ambiguity is settled, the nature of the force-generating mechanism of streaming has to be considered essentially unresolved. Nevertheless, several lines of very strong indirect evidence now exist which are consistent with the hypothesis that streaming is generated by actin–myosin interactions.

3.2.1. *Indispensability of Actin*

The presence of actin filaments next to the stationary cortical chloroplast files is now firmly established. The coincidence of their location with the site where streaming appears most vigorous (Kamiya and Kuroda, 1956) and their molecular unidirectionality (Kersey *et al.*, 1976) are consistent with the hypothesis that the motive force for streaming is generated at or near them. Cytochalasin B reversibly inhibits streaming (Bradley, 1973; Nagai and Kamiya, 1977; Williamson, 1972; Nothnagel *et al.*, 1981), and this inhibitory action coincides with the disappearance of actin-containing bundles as seen by NBD-phallacidin staining (Nothnagel *et al.*, 1981). Fluorescently labeled HMM having a partially disabled ATPase activity, a property that causes it to remain bound to actin filaments even in the presence of ATP, likewise inhibits streaming (Nothnagel *et al.*, 1982). Phalloidin, which stabilizes actin filaments, does not inhibit streaming, but does modify the inhibitory effect of cytochalasin B (Nothnagel *et al.*, 1981). Thus, intact actin filament bundles at the ectoplasm–endoplasm interface appear to be absolutely required for force generation.

3.2.2. Localization of Putative Myosin Molecules

Several lines of indirect evidence suggest a location of putative myosin molecules in the motile endoplasm, either as freely movable components of the cytoplasmic matrix, or associated with streaming particles. If characean cells are subjected to mild centrifugation, the motile endoplasm collects at one pole of the cell and thus is effectively separated from the stationary ectoplasm, including actin filament bundles. If this centrifugation step is performed in a double-chambered cuvette, then the centrifugal and centripetal portions of the cell can be treated selectively with inhibitors. In a second centrifugation step the two components can be brought into direct contact again. Chen and Kamiya (1975) used this technique to differentially treat the two halves of the cell with *N*-ethylmaleimide(which inhibits the actin-activated ATPase activity of myosin) and cytochalasin B. Treatment of endoplasm with NEM causes an irreversible loss of streaming, while NEM applied to the cell cortex is without effect. In similar experiments, cytochalasin B was inhibitory only if applied to the cortical half of the cell, while the reciprocal experiment, treatment of the endoplasm, had no effect. Even though there are some problems with the interpretation of this experiment (see discussion by Allen and Allen, 1978), it can be taken as indirect evidence for an endoplasmic location of a myosin-like ATPase.

A different set of structural and experimental evidence suggests an association of a myosin-like component with moving particles. Filamentous structures different from actin filaments have been seen in association with vesicular components (Nagai and Hayama, 1979; Allen, 1980), some of which exhibit ATP-dependent attachment to, and movement along, microfilament bundles (Kamitsubo, 1972b; Williamson, 1975). An unusual phenomenon possibly related to the question of the localization of force-generating enzymes has been observed in isolated endoplasmic droplets. Chloroplasts in these droplets may rotate rapidly and independently at a frequency of up to one revolution per second (Jarosch, 1956a, 1976; Kuroda, 1964). Optical inspection reveals a stream of buffer opposite that of the rotation in the immediate vicinity of the chloroplast. If chloroplasts are held with a microneedle, liquid in their vicinity is seen to stream vigorously. These observations are consistent with the idea of a shear force generated at the chloroplast surface. Such a force could be brought about by myosin molecules associated with the chloroplast surface that interact with actin filaments in the droplet. It is equally plausible, however, to envision an interaction of actin filaments attached to the chloroplast surface with soluble myosin.

It is an open issue at present whether the interaction of either soluble or membrane-associated myosin with actin filament bundles can create a drag force large enough to account for the massive bulk flow of endoplasm. The question of the site of force generation in relation to the localization of putative myosin molecules has recently been addressed in a theoretical consideration of hydrodynamic models. Nothnagel and Webb (1982) analyzed three models of viscous coupling between myosin and endoplasm. They came to the conclusion that individual myosin molecules moving along the actin bundles cannot exert enough viscous pull to explain the observed streaming velocities. Likewise, myosin attached to moving endosomes, though improving viscous coupling to the endoplasm, is not sufficient. The only model that easily explains the observed streaming profile is one in which myosin is attached to an extensive network or gel extending into and throughout the endoplasm. The entire network is proposed to be pulled forward as the attached myosin slides along the actin bundles. This model should stimulate further studies on the structural organization of the endoplasm. If proven correct, it would also incorporate Allen's (1974) observation of a prominent endoplasmic filament system.

3.2.3. Perfused and Permeabilized Cell Models

Cell models where the cell interior has been made accessible to otherwise impermeant compounds have been instrumental in increasing the understanding of streaming in plant cells. Contrary to microtubule-based transport systems where such studies have just begun (see Section 3.1.3), cell models permeable to a variety of molecules or proteins have been used for almost 10 years in characean cells.

The unique geometrical disposition of the giant internodal cells of *Nitella* or *Chara* has led to the development of a vacuolar-perfusion technique (Tazawa, 1964, 1968) that allows study of various aspects of rotational streaming in these cells. In these experiments, the two ends of the cell are cut off and the vacuolar contents are replaced by a solution capable of sustaining cytoplasmic flow for many days. This experimental situation has been carried one step further by removing the tonoplast (vacuole membrane) with a perfusate containing EGTA (Williamson, 1975; Tazawa *et al.*, 1976). Williamson (1975, 1980) developed a stop–go reactivated model in which he demonstrated the dependence of streaming on the level of ATP (see also Shimmen, 1978; Shimmen and Tazawa, 1982). The presence of Mg^{2+} ions is required, and Ca^{2+} ions are inhibitory at greater than 10^{-6} M (Williamson, 1975). On the basis of their vacuole perfusion experiments, Donaldson (1972) and Williamson (1975) concluded that the force for organelle movement is generated in the immediate vicinity of the cortical actin fibrils. As yet, no experimental system has been devised to unequivocally document force generation by endoplasmic fibrils, as suggested by the model of Allen (1974).

A modification of the vacuolar perfusion technique has been used to determine the motive force for streaming. The flow of the perfusate exerts a shear force upon the tonoplast which tends to decelerate streaming on one side of the cell. By determining the flow rate necessary to bring streaming to a standstill, Tazawa (1968) calculated the force produced at the ectoplasm–endoplasm interface to be 1.4–2 dyn/cm^2. This calculation is based on the assumption that at zero streaming, the shear force produced by the flowing perfusate equals that produced by the cellular motor. His estimations are in good agreement with other calculations based on different experimental designs (Kamiya and Kuroda, 1958, 1973).

3.2.4. Isolated Cytoplasmic Droplets and Cytoplasm in Vitro

Cytoplasm squeezed out of a dissected characean cell into isotonic medium will re-form a limiting membrane and, under proper conditions, will survive and be active for many hours. The motile phenomena observed in these isolated droplets will depend on the presence or absence of ectoplasmic components. Pure endoplasm will show only poor signs of motile activity, whereas droplets containing fibrils (probably derived from subcortical filament bundles) show a surprising wealth of movements. The fibrils often take on the form of closed circles or polygons and show rotational or translational motility accompanied by a counterstream of cytoplasm in their immediate vicinity (Jarosch, 1956b; Kamitsubo, 1966). Small particles are seen to move along these fibrils in a saltatory fashion. This line of investigation has been taken one step further first by Kuroda and Kamiya (1975) and later by Higashi-Fujime (1980) who studied the movement of isolated fibrils and rows of chloroplasts in demembranated cytoplasmic droplets in an artificial medium. In the presence of ATP, chains of chloroplasts linked to filament bundles show a variety of translational movements at respectable rates (around 10 μm/sec) if the concentration of calcium ions is kept below 10^{-7} M (Higashi-Fujime, 1980). Movements of the fibrils, which consist of bundles of actin filaments of uniform polarity, continue even after all the endoplasm and chloroplasts are washed out. Especially in the light of the latter observation, it is difficult to envision an interaction between fibrils and soluble myosin molecules as the source of motive force for these *in vitro* movements. While such a model would appear reasonable to explain the movement of polygonal fibers in membrane-bounded droplets, it does not appear applicable to the movement of fibrils after the putative pool of soluble myosin has been washed away. Higashi-Fujime (1980) discusses a number of possible models, none of which completely explains all the observed phenomena. The most plausible explanation at this point is that fibers are propelled by interaction with a protein or proteins attached to the slide. In fact, moving fibrils were observed only very close to the glass surface (within about 1-μm distance).

Even though no straightforward explanation for these *in vitro* movements is available, it is immediately apparent that this line of experimentation bears great potential for the future. It suggests that eventually we may be able to reconstruct a motile machinery *in vitro* from its constituent elements.

3.2.5. Control Mechanisms

Little is known about the factors that control and regulate cytoplasmic streaming. Perfusion studies and experiments with isolated cytoplasmic droplets or demembranated cytoplasm all agree in one important point, however, that movement is inhibited by elevated levels of calcium ions (Tazawa *et al.*, 1976; Hayama and Tazawa, 1980; Williamson, 1975; Hayama *et al.*, 1979; Higashi-Fujime, 1980). The site of calcium action and the mechanism of inhibition are not known.

3.3. On the Relationship between Saltations and Streaming

A number of criteria can be, and indeed have been, used to distinguish the phenomena of cytoplasmic streaming and saltatory transport. The most valuable criterion in terms of phenomenology probably relates to the fact that in streaming, the entire cytoplasm moves, carrying with it whatever particulate materials are embedded in it regardless of size and shape. Saltatory transport is selective. It encompasses the movement of individual particles or organelles, not entire portions of the cytoplasm. In fact, the most characteristic feature of saltatory movement is that even closely apposed particles may behave quite differently. With regard to structural associations with cytoskeletal elements, saltations are almost exclusively observed in the presence of (and even in close spatial relationship to) microtubules, whereas streaming is dependent on microfilaments only. Or is it?

How well defined is the border between these two phenomena? A closer look reveals some interrelationships, the study of which may be beneficial for an understanding of either phenomenon. Several investigators noticed the occurrence of "saltatory" movements of individual particles in plant cells normally characterized by vigorous streaming. This form of particle transport always occurs in close spatial relationship to the fibrils. In chloroplast-free windows of *Nitella*, Kamitsubo (1972a) observed particles in the interfibrillary zone which, when in contact with a fibril, were rapidly transported for some distance. In excised epidermal cells, cytoplasmic streaming ceases for a short time due to some traumatic effect. Motility is regained in a sequence that starts with the saltation of individual particles via long, directed movements and finally bulk cytoplasmic flow (Jarosh, 1956b). One possible way to explain this phenomenon is that particles interact directly with microfilaments, or bundles thereof, that during excision become disorganized or detached from presumptive anchorage points. Similar observations were made by Allen (1974) and others in *Nitella*. Re-initiation of streaming after cell stimulation begins as a series of single rapid saltations of particles (spherosomes, mitochondria) along subcortical and endoplasmic fibrils, followed by gradual resumption of movement in surrounding cytoplasmic domains. Similarly, saltatory movements of small particles is the prevailing form of motility observed in isolated cytoplasmic droplets of endoplasm (Jarosch, 1976; Kuroda, 1964). Thus, saltations are observed in systems for which no participation or requirement of microtubular components has been demonstrated. It seems saltations of individual particles can also occur in association with actin filament bundles.

Conversely, there appear to be examples of microtubule-associated cytoplasmic streaming in some cell types. The smooth, continuous flow of pigment granules towards the cell center during aggregation in chromatophores resembles more a streaming process than a saltatory process. In fact, if aggregation were not terminated after granules have reached the cell center, this process would probably be indistinguishable from, e.g., streaming in *Nitella*. On the basis of their observation of the movement of submicroscopic particles in squid giant axons by video-enhanced contrast microscopy, Allen *et al.* (1982) came to the conclusion that "the fundamental processes in the mechanisms of organelle movement in axonal transport are not saltatory but continuous." This conclusion is based on the observation of an almost uniform, continuous movement of myriads of small (30–50 nm) vesicles superimposed upon the more intermittent movements of larger organelles. The most promising system, however, for a study of the relationship between streaming and microtubules appears to be the marine alga *Caulerpa*. This is the best-documented cell system so far where microtubules have been shown to be present in regions of cytoplasmic bulk flow. Moreover, streaming in this organism is inhibited by colchicine and nocodazole, but not by cytochalasin B (Pestri and Cande, unpublished observations).

One conclusion that can be derived from these observations is that microtubule- and microfilament-associated transport have more in common than previously assumed. Both organelles seem to be capable of generating rapid, saltatory movement of particles in close contact with their surfaces, and both can generate smooth, continuous bulk flow of cytoplasm. Which of these phenomena prevails depends on a number of conditions. Quite clearly, each system works more efficiently in one mode than in the other. Saltations are far more prevalent in microtubule-based cellular machineries, while streaming is more often found associated with prominent actin bundles. To understand the relationship between streaming and saltations, more knowledge on the nature and organization of the mechanochemical link between microtubules or microfilaments, respectively, and the transported material is required. We lack this knowledge at present.

3.4. Reconstituted Model Systems

In the past few years, attempts were made to understand the biochemical basis of the interaction between intracellularly transported organelles and the presumptive cytoskeletal machinery by studying their association *in vitro*. Burridge and Phillips (1975) demonstrated an association of purified actin and myosin with secretory (chromaffin) granules and granule membranes in a sedimentation assay and further demonstrated structural association of actin filaments with granule membrane ghosts by electron microscopy. These interactions seem to be sensitive to levels of free calcium above 0.1 μM (Fowler and Pollard, 1982). In related studies, Ostlund *et al.* (1977) demonstrated binding of pituitary secretory granules to muscle actin filaments *in vitro*. Sherline *et al.* (1977), also using pituitary granules, showed specific binding to microtubules as well, a process that seems to depend on MAPs. Suprenant and Dentler (1982) later showed that binding of pancreatic secretory granules to microtubules requires the presence of MAPs and that this association is abolished by ATP and increased by cyclic AMP.

While these studies demonstrate specific binding of intracellular vesicles to actin filaments or microtubules or both, the possible role of this phenomenon in the intracellular translocation of these organelles along the fibers or tubules is difficult to estimate. For instance, the levels of cytoplasmic ATP are high enough to prevent any association of vesicles with microtubules, if the conditions in a living cell are anywhere near those used by Suprenant and Dentler (1982) in their *in vitro* studies. It is tempting to speculate, however, that the binding observed *in vitro* represents one step in the series of steps that leads to the translocation of a vesicle along a microtubule. The verification of this hypothesis will depend on the demonstration of an enzymatic activity (similar to, e.g., flagellar dynein) that is capable of cyclic interaction with the two components. Attempts to demonstrate the presence of such an activity in brain MAP preparations have failed so far (Murphy and Wallis, 1981). It is entirely possible, however, that such an activity is associated with the granule membranes and not with the microtubules. We may simply not have looked in the right place yet.

A slightly different and also very promising approach has been taken by Oplatka, Tirosh, and associates, and by Yano and colleagues. Based on model systems of cytoplasmic contractibility devised by Allen and colleagues more than 20 years ago (see summary by Taylor and Condeelis, 1979), these investigators reported on the occurrence of cytoplasmic streaming-like motility *in vitro* in solutions containing actin filaments, heavy meromyosin, and ATP. Oplatka and Tirosh (1973) and Oplatka *et al.* (1974) observed active streaming of such a mixture in microcapillaries. Although their earlier work has been criticized because the streaming they described was said to be indistinguishable from passive flow due to the capillary phenomenon, they have recently demonstrated that streaming can occur against gravity (Tirosh and Oplatka, 1982). Yano (1978) has taken this approach one step further by attempting to create a situation more closely resembling that in, e.g., a characean cell. Streaming was observed in a circular slit between two concentric cylinders with unidirectionally aligned actin filaments attached to the walls of this cylinder if HMM and ATP were present in the solution bordered by the cylinder walls. The solution in the cylindrical slot streamed at a constant rate of about 10 μm/sec for up to 30 min (see also Yano *et al.*, 1978). A modification of this technique resulted in the construction of a rotary motor driven by the ATP-dependent interaction of HMM in solution with actin filaments attached to the "rotor" (Yano *et al.*, 1982). The experiments clearly show that soluble HMM and stationary actin filaments can produce a force against the medium strong enough to drive a rotor or to sustain streaming of a solution of HMM. Although the applicability of this model to the situation, e.g., in a streaming *Nitella* cell, remains to be demonstrated (see discussion by Nothnagel and Webb, 1982). The general design of this experimental approach suggests promising trends for the future.

4. Summary and Outlook

A brief, by no means comprehensive, overview has been presented of some of the cell systems most frequently used to study intracellular organelle motility. What has been learned in the past one or two decades about the cytoplasmic structures involved and the mechanisms operating? In 1963, David W. Bishop stated in his introductory remarks to a session on saltatory motion at the first Conference devoted entirely to nonmuscle cell motility (Allen and Kamiya, 1964), ". . . the material presented at this Symposium suggests that heroic efforts are being made to bring unity to a chaotic situation which, in reality, may better be viewed as a packet of processes." Since then, the efforts have not become less heroic, and some order is beginning to emerge from chaos. The situation appears to be different for the cytoplasmic streaming field than for the saltatory motility area. There can be little double now that streaming is based on an actomyosin-like mechanism. Important details of the organization of the force-producing machinery remain to be solved, however. Where is the mechanochemical transducer, myosin, localized (dispersed in the cytoplasm, or attached to vesicles and endomembranes, or both), and how is it organized (monomer or filament)? Is force produced entirely at the ectoplasm–endoplasm interface, as suggested by some, or is it generated in the endoplasm as well, as proposed by others? And how is the machinery regulated? Though these are still very basic questions to be resolved, this area of research is in a better position than that attempting to unravel the mysteries of microtubule-associated transport. Admittedly, a great deal has been learned about the way different cell types organize their cytoplasm for saltatory motility, the behavior of particles under different experimental conditions, and the factors which induce, modify, or extinguish this activity. Yet, the most important question still remains unsettled: which molecules constitute the motor? The history of the study of microtubule-associated movements has seen the pendulum swing from theories strongly invoking microtubules to those favoring actomyosin and recently back again to microtubule-based models. Theories involving a contractile system cooperating with microtubules, though popular in the early seventies (see summary by Rebhun, 1972), are again on the wane. Recent micro-injection experiments on neurons and chromatophores seem to provide rather strong evidence against a direct role for actin. Nonetheless, other experiments show that microtubules are dispensable for certain forms of motility and thus indirectly point to the existence of a microtubule-independent component. Edds' (1975a) experiment on the artificial axopodium along which transport proceeds undisturbed seems to be the most informative in this respect, and equivalent experiments should be attempted in other systems. Is it conceivable that neither microtubules nor actin are involved in some forms of intracellular motility? This possibility has been discussed repeatedly from a purely theoretical standpoint. The motile machineries that have been considered as the most likely candidates for a novel cytoplasmic mechanochemical transducer are the spasmonemes of vorticellids (Amos, 1975) or the myonemes of *Stentor* (Huang and Pitelka, 1973), both of which exhibit calcium-dependent contractile activity. It needs to be emphasized, however, that the idea of an involvement of a novel contractile machinery in metazoan cells is not backed by any experimental evidence yet. This thought seems to surface only because of the lack of strong evidence for or against conventional actomyosin or microtubular mechanisms. It deserves to be put to a serious test.

On the other hand, the possibility of an involvement of a dynein-like ATPase has found increasing support in the recent microinjection and cell model studies. So far, however, the evidence for its involvement essentially relies on the use of the two inhibitors, vanadate and EHNA, the specificities of which for nonciliary dyneins remain to be established more convincingly. At the same time, the search for cytoplasmic dynein ATPases by biochemical techniques needs to be reinforced; promising first steps have already been undertaken (Pratt, 1980). The cell models require increased attention. Presently, they are at best only able to sustain motility for some time after permeabilization. Ideally, one should be able to stop and restart motility at will by appropriate experimental manipulations. It is hoped that the future will see some major improvements in this respect.

ACKNOWLEDGMENTS. I am grateful to Drs. Beth Burnside, Ursula Euteneuer, and Kent McDonald for helpful suggestions and comments on the manuscript. I also thank Drs. Mary Beckerle, Paul Burton, Ursula Euteneuer, David Forman, Nobutake Hirokawa, Harunori Ishikawa, Eiji Kamitsubo, Ulrich Koop, Katherine Luby-Phelps, Kent McDonald, Mark McNiven, Barry Palevitz, and Jeffrey Travis for kindly providing illustrative material.

References

Abe, K., Robinson, G. A., Liddle, G. W., Butcher, R. W., Nicholson, W. E., and Baird, C. E., 1969, Role of cyclic AMP in mediating the effects of MSH, norepinephrine, and melatonin on frog skin color, *Endocrinology* **85:**674–682.

Abe, T., Haga, T., and Kurokawa, M., 1973, Rapid transport of phosphatidylcholine occurring simultaneously with protein transport in the frog sciatic nerve, *Biochem. J.* **136:**731–740.

Abe, T., Haga, T., and Kurokawa, M., 1974, Retrograde axoplasmic transport: Its continuation as anterograde transport, *FEBS Lett.* **47:**272–275.

Adams, R. J., 1982, Organelle movement in axons depends on ATP, *Nature* **297:**327–329.

Allen, N. S., 1974, Endoplasmic filaments generate the motive force for rotational streaming in *Nitella*, *J. Cell Biol.* **63:**270–287.

Allen, N. S., 1980, Cytoplasmic streaming and transport in the characean alga *Nitella*, *Can. J. Bot.* **58:** 786–796.

Allen, N. S., and Allen, R. D., 1978, Cytoplasmic streaming in green plants, *Annu. Rev. Biophys. Bioeng.* **7:**497–526.

Allen, R. D., and Kamiya, N., 1964, *Primitive Motile Systems in Cell Biology*, Academic Press, New York.

Allen, R. D., Travis, J. L., Allen, N. S., and Yilmaz, H. 1981a, Video-enhanced contrast polarization (AVEC-POL) microscopy: A new method applied to the detection of birefringence in the motile reticulopodial network of *Allogromia laticollaris*, *Cell Motil.* **1**:275–289.

Allen, R. D., Allen, N. S., and Travis, J. L., 1981b, Video-enhanced contrast, differential interference contrast (AVEC-DIC) microscopy: A new method capable of analyzing microtubule-related motility in the reticulopodial network of *Allogromia laticollaris*, *Cell Motil.* **1**:291–302.

Allen, R. D., Metuzals, J., Tasaki, I., Brady, S. T., and Gilbert, S. P., 1982, Fast axonal transport in squid giant axon, *Science* **218**:1127–1129.

Amos, W. B., 1975, Contraction and calcium binding in the vorticellid ciliates, in: *Molecules and Cell Movement* (S. Inoue and R. E. Stephens, eds.), pp. 411–436, Raven Press, New York.

Anderson, K.-E., Edström, A., and Mattson, H., 1972, Effects of cytochalasin B on uptake of glucosamine, leucine, and sulfate into nerve cells: Incorporation into glycoproteins and rapid axonal transport, *Brain Res.* **43**:343–353.

Bagnara, J. T., and Hadley, M. E., 1973, *Chromatophores and Color Change*, Prentice Hall, Englewood Cliffs.

Ballowitz, E., 1914, Über die Pigmentströmung in den Farbstoffzellen und die Kanälchenstruktur des Protoplasmas, *Pflügers Arch. ges. Physiol.* **157**:165–210.

Banks, P., and Till, R., 1975, A correlation between the effects of anti-mitotic drugs on microtubule assembly in vitro and the inhibition of axonal transport in noradrenergic neurons, *J. Physiol.* **252**:283–294.

Banks, P., Mayor, D., Mitchell, M., and Tomlinson, D., 1971a, Studies on the translocation of noradrenalin-containing vesicles in post-ganglionic sympathetic neurons in vitro. Inhibition of movement by colchicine and vinblastine and evidence for the involvement of axonal microtubules, *J. Physiol.* **216**:625–639.

Banks, P., Mayor, D., and Tomlinson, D. R., 1971b, Further evidence for the involvement of microtubules in the intra-axonal movement of noradrenaline storage granules, J. Physiol. **219**:755–761.

Banks, P., Mayor, D., and Mraz, P., 1973, Cytochalasin B and the intra-axonal movement of noradrenaline storage vesicles, *Brain Res.* **49**:417–421.

Bardele, C. F., 1976, Particle movement in heliozoan axopods associated with lateral displacement of highly ordered membrane domains, *Z. Naturforsch.* **31C**:190–194.

Bardele, C. F., 1977, Comparative study of axopodial microtubule patterns and possible mechanisms of pattern control in the centrohelidan heliozoa *Acanthocystis*, *Raphidiophrys*, and *Heterophrys*, *J. Cell Sci.* **25**:205–232.

Barondes, S. H., 1967, Axoplasmic transport, *Neurosci. Res. Prog. Bull.* **5**:307–419.

Beckerle, M. C., and Porter, K. R., 1981, Intracellular motility in erythrophores examined by microinjection, *J. Cell Biol.* **91**:302a.

Beckerle, M. C., and Porter, K. R., 1982, Inhibitors of dynein activity block intracellular transport in erythrophores, *Nature* **295**:701–703.

Bell, E., Merrill, C., and Verna, J. M., 1980, Movements of intracellular particles in relation to the protrusions of cell processes, *J. Cell Biol.* **87**:328a.

Berl, S., Puszku, S., and Nicklas, W. J., 1973, Actomyosin-like protein in brain, *Science* **179**:441–446.

Berlinrood, M., McGee-Russel, S. M., and Allen, R. D., 1972, Pattern of particle movements in nerve fibers in vitro. An analysis by photokymography and microscopy, *J. Cell Sci.* **11**:875–886.

Bhisey, A. N., and Freed, J. J., 1971, Altered movement of endosomes in colchicine-treated cultured macrophages, *Exp. Cell Res.* **64**:430–438.

Bikle, D., Tilney, L. G., and Porter, K. R., 1966, Microtubules and pigment migration in the melanophores of *Fundulus heteroclitus*, *Protoplasma* 61:322–345.

Bitensky, M. W., and Burstein, S. R., 1965, Effects of cyclic adenosine monophosphate and melanocyte stimulating hormone on frog skin in vitro, *Nature* **208**:1282–1284.

Black, M. M., and Lasek, R. J., 1979, Axonal transport of actin: Slow component b is the principal source of actin for the axon, *Brain Res.* **171**:401–413.

Black, M. M., and Lasek, R. J., 1980, Slow components of axonal transport: Two cytoskeletal networks, *J. Cell Biol.* **86**:616–623.

Bloodgood, R. A., 1977, Motility occurring in association with the surface of the *Chlamydomonas* flagellum, *J. Cell Biol.* **75:**983–989.

Bloodgood, R. A., 1978, Unidirectional motility occurring in association with the axopodial membrane of *Echinosphaerium nucleophilum, Cell Biol. Int. Rep.* **2:**171–176.

Bradley, M. O., 1973, Microfilaments and cytoplasmic streaming: Inhibition of streaming with cytochalasin, *J. Cell Sci.* **12:**327–343.

Brady, S. T., and Lasek, R. J., 1981, Nerve-specific enolase and creatine phosphokinase in axonal transport: Soluble proteins and the axoplasmic matrix, *Cell* **23:**515–523.

Brady, S. T., and Lasek, R. J., 1982a, Axonal transport: A cell biological method for studying proteins that associate with the cytoskeleton, *Methods Cell Biol.* **25:**365–398.

Brady, S. T., and Lasek, R. J., 1982b, Fast axonal transport in isolated axoplasm: Roles of neurofilaments and microtubules, *J. Cell Biol.* **95:**330a.

Brady, S. T., Crothers, S. D., Nosal, C., and McClure, W. O., 1980, Fast axoplasmic transport in the presence of high Ca^{2+}:Evidence that microtubules are not required, *Proc. Natl. Acad. Sci. USA* **77:**5909–5913.

Brady, S. T., Tydell, M., Heriot, K., and Lasek, R. J., 1981, Axonal transport of calmodulin: A physiologic approach to long-term associations between proteins, *J. Cell Biol.* **89:**607–614.

Brady, S. T., Lasek, R. J., and Allen, R. D., 1982, Fast axonal transport in extruded axoplasm from squid giant axon, *Science* **218:**1129–1130.

Bray, D., and Bunge, M. B., 1981, Serial analysis of microtubules in cultured rat sensory axons, *J. Neurocytol.* **10:**589–605.

Bray, D., and Gilbert, D., 1981, Cytoskeletal elements in neurons, *Annu. Rev. Neurosci.* **4:**505–523.

Bretscher, A., and Weber, K., 1978, Tropomyosin from brain contains two polypeptide chains of slightly different molecular weights, *FEBS Lett.* **85:**145–148.

Breuer, A. C., Christian, C. M., Henkart, M., and Nelson, P. G., 1975, Computer analyses of organelle translocation in primary neuronal cultures and continuous cell lines, *J. Cell Biol.* **65:**562–576.

Brimijoin, S., 1975, Stop-flow: A new technique for measuring axonal transport, and its application to the transport of dopamine-β-hydroxylase, *J. Neurobiol.* **6:**379–394.

Brimijoin, S., and Helland, L., 1976, Rapid retrograde transport of dopamine-β-hydroxylase as examined by the stop-flow technique, *Brain Res.* **102:**217–228.

Brimijoin, S., and Wiermaa, M. J., 1978, Rapid orthograde and retrograde transport of acetylcholinesterase as characterized by the stop-flow technique, *J. Physiol.* **285:**129–142.

Brimijoin, S., Olsen, J., and Rosenson, R., 1979, Comparison of the temperature-dependence of rapid axonal transport and microtubules in nerves of the rabbit and bullfrog, *J. Physiol.* **287:**303–314.

Broadwell, R. D., and Brightman, M. W., 1979, Cytochemistry of undamaged neurons transporting exogenous protein in vivo, *J. Comp. Neurol.* **185:**31–74.

Brown, S. S., and Spudich, J. A., 1979, Cytochalasin inhibits the rate of elongation of actin filament fragments, *J. Cell Biol.* **83:**657–662.

Brücke, E., 1852, Untersuchungen über den Farbenwechsel des afrikanischen Chamäleons, *Denkschr. Akad. Wiss. Wien, math. nat. Kl.* **4:**179.

Buckley, I. K., 1981, Fine-structural and related aspects of non-muscle cell motility, in: *Cell and Muscle Motility* (R. M. Dowben and J. W. Shay, eds.), pp. 135–203, Plenum Publishing, New York.

Bunt, A. H., and Haschke, R. H., 1978, Features of foreign proteins affecting their retrograde transport in axons of the visual system, *J. Neurocytol.* **7:**665–678.

Bunt, A. H., and Lund, R. D., 1974, Vinblastine-induced blockage of orthograde and retrograde axonal transport of protein in retinal ganglion cells, *Exp. Neurol.* **45:**288–297.

Burnside, B., Adler, R., and O'Connor, P., 1982, Retinomotor pigment migration in the teleost retinal pigment epithelium: I. Roles for actin and microtubules in pigment granule transport and cone movement, *J. Invest. Ophthalmol.*, in press.

Burridge, K., and Phillips, J. H., 1975, Association of actin and myosin with secretory granule membranes, *Nature* **254:**526–528.

Burton, P. R., and Fernandez, H. L., 1973, Delineation by lanthanum staining of filamentous elements associated with the surfaces of axonal microtubules, *J. Cell Sci.* **12:** 12:567–583.

Burton, P. R., and Hinkley, R. E., 1974, Further electron microscopic characterization of axoplasmic microtubules of the ventral nerve cord of the crayfish, *J. Submicroscop. Cytol.* **6:**311–326.

Burton, P. R., and Kirkland, W. L., 1972, Actin detected in mouse neuroblastoma cells by binding of HMM, *Nature New Biol.* **239:**244–246.

Burton, P. R., and Paige, J. L., 1981, Polarity of axoplasmic microtubules in the olfactory nerve of the frog, *Proc. Natl. Acad. Sci. USA* **78:**3269–3273.

Byers, H. R., and Porter, K. R., 1977, Transformations in the structure of the cytoplasmic ground substance in erythrophores during pigment aggregation and dispersion. I. A study using whole-cell preparations in stereo high voltage electron microscopy, *J. Cell Biol.* **75:**541–558.

Byers, M. R., 1974, Structural correlates of rapid axonal transport: Evidence that microtubules may not be directly involved, *Brain Res.* **75:**97–113.

Cande, W. Z., 1980, Physiology of chromosome movement in lysed cell models, in: *International Cell Biology* (H. G. Schweiger, ed.), pp. 382–391, Springer, Berlin.

Carlsson, L., Nyström, L. E., Sundkist, I., Markey, F., and Lindberg, U., 1977, Actin polymerizability is influenced by profilin, a low molecular weight protein in non-muscle cells, *J. Mol. Biol.* **115:**465–483.

Carter, S. B., 1967, Effects of cytochalasins on mammalian cells, *Nature* **213:**261–264.

Carter, S. B., 1972, The cytochalasins as research tools in cytology, *Endeavour* **31:**77–82.

Chalfie, M., and Thomson, J. N., 1979, Organization of neuronal microtubules in the nematode *Caenorhabditis elegans*, *J. Cell Biol.* **82:**278–289.

Chang, C. M., and Goldman, R. D., 1973, The localization of actin-like fibers in cultured neuroblastoma cells as revealed by HMM binding, *J. Cell Biol.* **57:**867–874.

Chen, J. C. W., and Kamiya, N., 1975, Localization of myosin in the internodal cell of *Nitella* as suggested by differential treatment with N-ethylmaleimide, *Cell Struct. Funct.* **1:**1–9.

Claparéde, J., and Lachmann, W., 1858–1859, *Études sur les Infusoires et les Rhizopodes*, Geneve.

Clark, T. G., and Rosenbaum, J. L., 1982, Pigment particle translocation in detergent-permeabilized melanophores of *Fundulus heteroclitus*, *Proc. Natl. Acad. Sci. USA* **79:**4655–4659.

Cooper, P. D., and Smith, R. S., 1974, The movement of optically detectable organelles in myelinated axons of *Xenopus laevis*, *J. Physiol.* **242:**77–97.

Corti, B., 1774, Osservatione Microscopisce Sulla Tremella e sulla Circulazione del Fluido in una Planto Acquaguola, Lucca, Italy.

Crooks, R. F., and McClure, W. O., 1972, The effect of cytochalasin B on fast axoplasmic transport, *Brain Res.* **45:**643–646.

Dahlström, A., 1971a, Effects of vinblastine and colchicine on monoamine containing neurons of the rat, with special regard to the axoplasmic transport of amine granules, *Acta Nueorpathol. Suppl.* **5:**226–237.

Dahlström, A., 1971b, Axoplasmic transport (with particular respect to adrenergic neurons), *Phil. Trans. R. Soc. London Ser. B* **261:**325–358.

Dahlström, A., and Häggendahl, J., 1966, Studies on the transport and life-span of amine storage granules in a peripheral adrenergic neuron system, *Acta Physiol. Scand.* **67:**278–288.

Dahlström, A., and Häggendahl, J., 1967, Studies on the transport and life-span of amine storage granules in the adrenergic neuron system of the rabbit sciatic nerve, *Acta Physiol. Scand.* **69:**153–157.

Dambach, M., and Weber, W., 1975, Inhibition of pigment movement by cytochalasin B in the chromatophores of the sea urchin, *Centrostephanus longispinus*, *Comp. Biochem. Physiol.* **50C:**49–52.

Davidson, L. A., 1976, Ultrastructure of the membrane attachment sites of the extrusomes of *Ciliophrys marina*, *Cell Tissue Res.* **170:**353–365.

Dawes, C. J., and Barilotti, D. C., 1969, Cytoplasmic organization and rhythmic streaming in growing blades of *Caulerpa prolifera*, *Am. J. Bot.* **56:**8–15.

Dawes, C. J., and Rhamstine, E., 1967, An ultrastructural study of the giant green algal coenocyte, *Caulerpa prolifera, J. Phycol.* **3:**117–126.
Dentler, W. L., Granett, S., and Rosenbaum, J. L., 1975, Ultrastructural localization of the high molecular weight proteins (MAPs) associated with in vitro assembled brain microtubules, *J. Cell Biol.* **65:**237–241.
Donaldson, I. G., 1972, Cyclic longitudinal fibrillar motion as a basis for steady rotational protoplasmic streaming, *J. Theoret. Biol.* **37:**75–91.
Dumas, M., Schwab, M. E., and Thoenen, H., 1979, Retrograde axonal transport of specific macromolecules as a tool for characterizing nerve terminal membranes, *J. Neurobiol.* **10:**179–197.
Dustin, P., 1978, *Microtubules,* Springer Verlag, Berlin.
Edds, K. T., 1975a, Motility in *Echinosphaerium* I. An analysis of particle motions in the axopodia and a direct test of the involvement of the axoneme, *J. Cell Biol.* **66:**145–155.
Edds, K. T., 1975b, Motility in *Echinosphaerium* II. Cytoplasmic contractility and its molecular basis, *J. Cell Biol.* **66:**156–164.
Edström, A., 1977, Rapid axonal transport in vitro. Effects of derivatives of cyclic AMP and other agents acting on the cyclic AMP system, *J. Neurobiol.* **8:**371–380.
Edström, A., and Hanson, M., 1973a, Temperature effects on fast axonal transport of proteins in vitro in frog sciatic nerves, *Brain Res.* **58:**345–354.
Edström, A., and Hanson, M., 1973b, Retrograde axonal transport of proteins in vitro in frog sciatic nerves, *Brain Res.* **61:**311–320.
Edström, A., and Mattsson, H., 1972a, Fast axonal transport in vitro in the sciatic system of the frog, *J. Neurochem.* **19:**205–221.
Edström, A., and Mattsson, H., 1972b, Rapid axonal transport in vitro in the sciatic system of the frog of fucose-, glucosamine- and sulphate-containing material, *J. Neurochem.* **19:**1717–1729.
Egner, O., 1971, Zur Physiologie der Melanosomenverlagerung in den Melanophoren von *Pterophyllum scalare CUV, u. VAL., Cytobiologie* **4:**262–292.
Elam, J. S., and Agranoff, B. W., 1971, Transport of proteins and sulfated mucopolysaccharides in the goldfish visual system, *J. Neurobiol.* **2:**379–390.
Elam, J. S., Goldberg, J. M., Radin, N. S., and Agranoff, B. W., 1970, Rapid axonal transport of sulfated mucopolysaccharide proteins, *Science* **170:**458–460.
Ellisman, M. A., and Porter, K. R., 1980, Microtrabecular structure of the axoplasmic matrix: Visualization of cross-linking structures and their distribution, *J. Cell Biol.* **87:**464–479.
Euteneuer, U., and McIntosh, J. R., 1980, Polarity of midbody and phragmoplast microtubules, *J. Cell Biol.* **87:**509–515.
Euteneuer, U., and McIntosh, J. R., 1981a, Polarity of some motility-related microtubules, *Proc. Natl. Acad. Sci. USA* **78:**372–376.
Euteneuer, U., and McIntosh, J. R., 1981b, Structural polarity of kinetochore microtubules in PtK_1 cells, *J. Cell Biol.* **89:**338–345.
Euteneuer, U., Jackson, W. T., and McIntosh, J. R., 1982, Polarity of spindle microtubules in *Haemanthus* endosperm, *J. Cell Biol.* **94:**644–653.
Fernandez, H. L., and Samson, F. E., 1973, Axoplasmic transport: Differential inhibition by cytochalasin B, *J. Neurobiol.* **4:**201–206.
Fernandez, H. L., Huneeus, F. C,., and Davison, P. F., 1970, Studies on the mechanism of axoplasmic transport in the crayfish cord, *J. Neurobiol.* **1:**395–409.
Fernandez, H. L., Burton, P. R., and Samson, F. E., 1971, Axoplasmic transport in the crayfish nerve cord. The role of fibrillar constituents of neurons, *J. Cell Biol.* **51:**176–192.
Filliatreau, G., and DiGiamberardino, L., 1981, Microtubule polarity in myelinated axons as studied after decoration with tubulin, *Biol. Cell* **42:**69–72.
Fingerman, M., 1965, Chromatophores, *Phys. Rev.* **45:**296–339.
Fingerman, M., Fingerman, S. W., and Lambert, D. T., 1975, Colchicine, cytochalasin B, and pigment movements in ovarian and integumentary erythrophores of the prawn, *Palaemonetes vulgaris, Biol. Bull. (Woods Hole)* **149:**165–177.
Fink, B. R., Byers, M. R., and Middaugh, M. E., 1973, Dynamics of colchicine effects on rapid axonal transport and axonal morphology, *Brain Res.* **56:**299–312.

Fink, D. J., and Gainer, H., 1980a, Axonal trasnport of proteins: A new view using in vivo covalent labeling, *J. Cell Biol.* **85:**175–186.

Fink, D. J., and Gainer, H., 1980b, Retrograde axonal transport of endogenous proteins in sciatic nerve demonstrated by covalent labeling in vivo, *Science* **208:**303–305.

Fisher, M., and Lyerla, T. A., 1974, The effect of cytochalasin B on pigment dispersion and aggregation in perfused Xenopus laevis tailfin melanophores, *J. Cell Physiol.* **83:**117–130.

Fitzharris, T. P., Bloodgood, R. A., and McIntosh, J. R., 1972, Particle movement in the axopodia of *Echinosphaerium:* Evidence concerning the role of the axoneme, *J. Mechanochem. Cell Motil.* **1:**117–124.

Flamant-Durand, J., and Dustin, P., 1972, Studies on the transport of secretory granules in the magnocellular hypothalamic neurons. I. Action of colchicine on axonal flow and neurotubules in the paraventricular nuclei, *Z. Zellforsch. Mikrosk. Anat.* **130:**440–454.

Forman, D. S., 1982, Vanadate inhibits saltatory organelle movement in a permeabilized cell model, *Exp. Cell Res.* **141:**139–147.

Forman, D. S., McEwen, B. S., and Grafstein, B., 1971, Rapid transport of radioactivity in goldfish optic nerve following injections of labeled glucosamine, *Brain Res.* **28:**119–130.

Forman, D. S., Grafstein, B., and McEwen, B. S., 1972, Rapid axonal transport of [^{3}H]fucosyl glycoproteins in the goldfish optic system, *Brain Res.* **48:**327–342.

Forman, D. S., Padjen, A. L., and Siggins, G. R., 1977a, Axonal transport of organelles visualized by light microscopy: Cinemicrographic and computer analysis, *Brain Res.* **136:**197–213.

Forman, D. S., Padjen, A. L., and Siggins, G. R., 1977b, Effect of temperature on the rapid retrograde transport of microscopically visible intra-axonal organelles, *Brain Res.* **136:**215–226.

Forman, D. S., Brown, K. J., Adelman, M. R., and Livengood, D. R., 1982, Nucleotide specificity of reactivation of saltatory movement in permeabilized giant axons, *J. Cell Biol.* **95:**323a.

Fowler, V. M., and Pollard, H. B., 1982, Chromaffin granule membrane-F-actin interactions are calcium sensitive, *Nature* **295:**336–338.

Freed, J. J., and Lebowitz, M. M., 1970, The association of a class of saltatory movements with microtubules in cultured cells, *J. Cell Biol.* **45:**334–354.

Friede, R. L., and Ho, K.-C., 1977, The relation of axonal transport of mitochondria with microtubules and other axoplasmic organelles, *J. Physiol.* **265:**507–519.

Frixione, E., Arechiga, H., and Tsutsumi, V., 1979, Photomechanical migrations of pigment granules along the retinula cells of the crayfish, *J. Neurobiol.* **10:**573–590.

Fujii, R., 1969, Chromatophores and pigments, in: *Fish Physiology* (W. S. Hoar and D. J. Randall, eds.), pp. 307–353, Academic Press, New York.

Fujii, R., and Miyashita, Y., 1975, Receptor mechanisms in fish chromatophores, *Comp. Biochem. Physiol.* **51C:**171–178.

Garner, J., and Lasek, R. J., 1981, Clathrin is axonally transported as part of slow component b: The axoplasmic matrix, *J. Cell Biol.* **88:**172–178.

Gartz, R., 1970, Adaptationsmorphologie der Melanophoren von Krallenfroschlarven, *Cytobiologie* **2:**220–234.

Gawlitta, W., Osborn, M., and Weber, K., 1981, Coiling of intermediate filaments induced by microinjection of a vimentin-specific antibody does not interfere with locomotion and mitosis, *Eur. J. Cell Biol.* **26:**83–90.

Geschwind, I. I., Horowitz, J. M., Mikuckis, G. M., and Dewey, R. D., 1977, Iontophoretic release of cyclic AMP and dispersion of melanosomes within a single melanophore, *J. Cell Biol.* **74:**928–943.

Glenney, J. R., Glenney, P., Osborn, M., and Weber, K., 1982, An F-actin and calmodulin-binding protein from isolated intestinal brush borders has a morphology related to spectrin, *Cell* **28:**843–854.

Goldberg, D. J., 1982, Microinjection into an identified axon to study the mechanism of fast axonal transport, *Proc. Natl. Acad. Sci. USA* **79:**4818–4822.

Goldberg, D. J., Harris, D. A., Lubit, B. W., and Schwartz, J. H., 1980, Analysis of the mechanism of fast axonal transport by intracellular injection of potentially inhibitory macromolecules: Evidence for a possible role for actin filaments, *Proc. Natl. Acad. Sci. USA* **77:**7448–7452.

Grafstein, B., 1967, Transport of protein by goldfish optic nerve fibers, *Science* **157:**196–198.

Grafstein, B., and Forman, D. S., 1980, Intracellular transport in neurons, *Physiol. Rev.* **60:**1167–1283.

Grafstein, B., Miller, J. A., Ledeen, R. W., Haley, J., and Specht, S. C., 1975, Axonal transport of phospholipid in goldfish optic system, *Exp. Neurol.* **46:**261–281.

Gras, H., and Weber, W., 1977, Light-induced alterations in cell shape and pigment displacement in chromatophores of the sea urchin *Centrostephanus longispinus*, *Cell Tissue Res.* **182:**165–176.

Green, L., 1968, Mechanism of movements of granules in melanocytes of *Fundulus heteroclitus*, *Proc. Natl. Acad. Sci. USA* **59:**1179–1186.

Green, P. B., 1969, Cell morphogenesis, *Annu. Rev. Plant Physiol.* **20:**365–394.

Griffith, L. M., and Pollard, T. D., 1978, Evidence for actin filament microtubule interaction mediated by microtubule-associated proteins, *J. Cell Biol.* **78:**958–965.

Gross, G. W., 1975, The microstream concept of axoplasmic and dendritic transport, *Adv. Neurol.* **12:**283–296.

Hammond, G. R., and Smith, R. S., 1977, Inhibition of the rapid movement of optically detectable axonal particles by colchicine and vinblastine, *Brain Res.* **128:**227–242.

Hanson, J., and Huxley, H. E., 1955, The structural basis of contraction in striated muscle, *Symp. Soc. Exp. Biol.* **9:**228–264.

Hanson, M., and Edström, A., 1978, Mitosis inhibitors and axonal transport, *Int. Rev. Cytol. Suppl.* **7:**373–402.

Hartwig, J. H., and Stossel, T. P., 1979, Cytochalasin B and the structure of actin gels, *J. Mol. Biol.* **134:**539–553.

Hauser, M., and Schwab, D., 1974, Mikrotubuli und helikale Mikrofilamente im Cytoplasma der Foraminifere *Allogromia laticollaris ARNOLD*, *Cytobiologie* **9:**263–279.

Hayama, T., and Tazawa M., 1980, Ca^{2+} reversibly inhibits active rotation of chloroplasts in isolated cytoplasmic droplets of *Chara*, *Protoplasma* **102:**1–9.

Hayama, T., Shimmen, T., and Tazawa, M., 1979, Participation of Ca^{2+} in cessation of cytoplasmic streaming induced by membrane excitation in Characeae internodal cells, *Protoplasma* **99:**305–321.

Heidemann, S. R., and McIntosh, J. R., 1980, Visualization of the structural polarity of microtubules, *Nature* **286:**517–519.

Heidemann, S. R., Landers, J. M., and Hamborg, M. A., 1981, Polarity orientation of axonal microtubules, *J. Cell Biol.* **91:**661–665.

Heidenhain, M., 1907, *Plasma und Zelle*, G. Fischer Verlag, Jena.

Hendrickson, A. E., 1972, Electron microscopic distribution of axoplasmic transport, *J. Comp. Neurol.* **144:**381–398.

Hendry, I. A., Stockel, K., Thoenen, H., and Iversen, L. L., 1974, The retrograde axonal transport of nerve growth factor, *Brain Res.* **68:**103–121.

Hepler, P. K., and Palevitz, B. A., 1974, Microtubules and microfilaments, *Annu. Rev. Plant Physiol.* **25:**309–362.

Heslop, J. P., 1975, Axonal flow and fast transport in nerves, *Adv. Comp. Physiol. Biochem.* **6:**75–163.

Higashi-Fujime, S., 1980, Active movement in vitro of bundles of microfilaments isolated from *Nitella* cell, *J. Cell Biol.* **87:**569–578.

Hirokawa, N., 1982, Cross-linker system between neurofilaments, microtubules, and membraneous organelles in frog axons revealed by the quick-freeze, deep-etching method, *J. Cell Biol.* **94:**129–142.

Hitchcock, S. E., Carlsson, L., and Lindberg, U., 1976, DNase I-induced depolymerization of actin filaments, in: *Cell Motility* (R. Goldman, T. Pollard, and J. Rosenbaum, eds.), pp. 545–559, Cold Spring Harbor Laboratory, New York.

Hoffmann, P. N., and Lasek, R. J., 1975, The slow component of axonal transport. Identification of major structural polypeptides of the axon and their generality among mammalian neurons, *J. Cell Biol.* **66:**351–366.

Huang, B., and Pitelka, D., 1973, The contractile process in the ciliate, *Stentor coerulens*. I. The role of microtubules and filaments, *J. Cell Biol.* **57:**704–722.

Huxley, H. E., 1963, Electron microscope studies on the structure of natural and synthetic protein filaments from striated muscle, *J. Mol. Biol.* **7**:281–308.

Hyams, J. S., and Stebbings, H., 1978, The mechanism of microtubule associated cytoplasmic transport. Isolation and preliminary characterization of microtubule transport system, *Cell Tissue Res.* **196**:103–116.

Hyams, J. S., and Stebbings, H., 1979, Microtubule associated cytoplasmic transport, in: *Microtubules* (K. Roberts and J. S. Hyams, eds.), pp. 487–530, Academic Press, London.

Inoue, S., 1981, Video image processing greatly enhances contrast, quality and speed in polarization-based microscopy, *J. Cell Biol.* **89**:346–356.

Isenberg, G., Schubert, P., and Kreutzberg, G. W., 1980, Experimental approach to test the role of actin in axonal transport, *Brain Res.* **194**:588–593.

Ishikawa, H., Bischoff, R., and Holtzer, H., 1968, Formation of arrowhead complexes with heavy meromyosin in a variety of cell types, *J. Cell Biol.* **43**:312–335.

James, K. A. C., Bray, J. J., Morgan, I. G., and Austin, L., 1970, The effect of colchicine on the transport of axonal protein in the chicken, *Biochem. J.* **117**:767–771.

Jarlfors, U., and Smith, D. S., 1969, Association between synaptic versicles and neurotubules, *Nature* **224**:710–711.

Jarosch, R., 1956a, Plasmaströmung und Chloroplastenrotation bei Characeen, *Phyton* **6**:87–107.

Jarosch, R., 1956b, Die Impulsrichtungsänderungen bei der Induktion der Protoplasmaströmung, *Protoplasma* **47**:478–486.

Jarosch, R., 1976, Dynamisches Verhalten der Aktinfibrillen von Nitella auf Grund schneller Filamentrotation, *Biochem. Physiol. Pflanz.* **170**:111–131.

Jeffrey, P. L., and Austin, L., 1973, Axoplasmic transport, *Prog. Neurobiol.* **2**:207–255.

Jokusch, H., Jokusch, B., and Burger, M. M., 1979, Nerve fibers in culture and their interactions with non-neural cells visualized by immunofluorescence, *J. Cell Biol.* **80**:629–641.

Junqueira, L. C. U., Raker, E., and Porter, K. R., 1974, Studies on pigment migration in the melanophores of the teleost, *Fundulus heteroclitus*, *Arch. Histol. Jpn.* **36**:339–366.

Junqueira, L. C. U., Reinach, F., and Salles, L. M. M., 1977, The presence of spontaneous and induced filaments in the melanophores of three species of teleosts, *Arch. Histol. Jpn.* **40**:435–443.

Kamitsubo, E., 1966, Motile protoplasmic fibrils in cells of characeae. II. Linear fibrillar structure and its bearing on protoplasmic streaming, *Proc. Jpn. Acad.* **42**:640–643.

Kamitsubo, E., 1972a, A "window technique" for detailed observation of characean cytoplasmic streaming, *Exp. Cell Res.* **74**:613–616.

Kamitsubo, E., 1972b, Motile protoplasmic fibrils in cells of the Characeae, *Protoplasma* **74**:53–70.

Kamiya, N., 1959, Protoplasmic streaming, *Protoplasmatolgia* **8**:3a.

Kamiya, N., 1981, Physical and chemical basis of cytoplasmic streaming, *Annu. Rev. Plant Physiol.* **32**:205–236.

Kamiya, N., and Kuroda, K., 1956, Velocity distribution of the protoplasmic streaming in *Nitella* cells, *Bot. Mag.* **69**:544–554.

Kamiya, N., and Kuroda, K., 1958, Measurement of the motive force of the protoplasmic rotation in *Nitella*, *Protoplasma* **50**:144–148.

Kamiya, N., and Kuroda, K., 1973, Dynamics of cytoplasmic streaming in a plant cell, *Biorheology* **10**:179–187.

Karlsson, J. O., and Sjöstrand, J., 1969, The effect of colchicine on axonal transport of protein in optic nerve and tract of the rabbit, *Brain Res.* **13**:617–619.

Karlsson, J. O., and Sjöstrand, J., 1971a, Synthesis, migration, and turnover of protein in retinal ganglion cells, *J. Neurochem.* **18**:749–767.

Karlsson, J. O., and Sjöstrand, J., 1971b, Rapid intracellular transport of fucose-containing glycoproteins in retinal ganglion cells, *J. Neurochem.* **18**:2209–2216.

Karlsson, J. O., Hansson, H. A., and Sjöstrand, J., 1971, Effect of colchicine on axonal transport and morphology of retinal ganglion cells, *Z. Zellforsch. Mikrosk. Anat.* **115**:265–283.

Kato, T., and Tonomura, Y., 1977, Identification of myosin in *Nitella flexilis*, *J. Biochem.* **82**:777–782.

Kersey, Y. M., Hepler, P. K., Palevitz, B. A., and Wessels, N. K., 1976, Polarity of actin filaments in Characean algae, *Proc. Natl. Acad. Sci. USA* **73**:165–167.

Kim, H., Binder, L. I., and Rosenbaum, J. L., 1979, The periodic association of MAP_2 with brain microtubules in vitro, *J. Cell Biol.* **80:**266–276.
Kirschner, M. W., 1978, Microtubule assembly and nucleation, *Int. Rev. Cytol.* **54:**1–71.
Kitching, J. A., 1964, The axopods of the sun animacule, *Actinophrys sol*, in: *Primitive Motile Systems in Cell Biology* (R. D. Allen and N. Kamiya, eds.), pp. 445–455, Academic Press, New York.
Koop, H. U., and Kiermayer, O., 1980a, Protoplasmic streaming in the giant unicellular green alga, *Acetabularia mediterranea.* I. Formation of intracellular transport systems in the course of cell differentiation, *Protoplasma* **102:**147–166.
Koop, H. U., and Kiermayer, O., 1980b, Protoplasmic streaming in the giant unicellular green alga, *Acetabularia mediterranea.* II. Differential sensitivity of movement systems to substances acting on microfilaments and microtubules, *Protoplasma* 102:295–306.
Koop, H. U., Schmid, R., Heunert, H. H., and Miethaler, B., 1978, Chloroplast migrations: A new circadian rhythm in *Acetabularia, Protoplasma* **97:**301–310.
Kopenec, A., 1949, Farbwechsel der Larve von *Corethra plumicornis, Z. vergl. Physiol.* **31:**490–505.
Koyama, Y., and Takeuchi, T., 1980, Differential effect of cytochalasin B on the aggregation of melanosomes in cultured mouse melanoma cells, *Anat. Rec.* **196:**449–459.
Kreutzberg, G. W., 1969, Neuronal dynamics and axonal flow. IV. Blockage of intra-axonal enzyme transport by colchicine, *Proc. Natl. Acad. Sci. USA* **62:**722–728.
Kristensson, K., 1978, Retrograde transport of macromolecules in axons, *Annu. Rev. Pharmacol. Toxicol.* **18:**97–110.
Kuczmarski, E., and Rosenbaum, J. L., 1979a, Chick brain actin and myosin. Isolation and characterization, *J. Cell Biol.* **80:**341–355.
Kuczmarski, E., and Rosenbaum, J. L., 1979b, Studies on the organization and localization of actin and myosin in neurons, *J. Cell Biol.* **80:**356–371.
Kuroda, K., 1964, Behavior of naked cytoplasmic drops isolated from plant cells, in: *Primitive Motile Systems in Cell Biology* (R. D. Allen and N. Kamiya, eds.), pp. 31–41, Academic Press, New York.
Kuroda, K., and Kamiya, N., 1975, Active movement of *Nitella* chloroplasts in vitro, *Proc. Jpn. Acad.* **51:**774–777.
Lambert, D. T., and Crowe, J. H., 1973, Colchicine and cytochalasin B: Effects on pigment granule translocation in melanophores of *Uca pugilator, Comp. Biochem. Physiol.* **45A:**11–16.
Lambert, D. T., and Crowe, J. H., 1976, Colchicine, cytochalasin B, cyclic AMP, and pigment granule translocation in melanophores of *Uca pugilator* and *Hemigrapsus oregonensis, Comp. Biochem. Physiol.* **53C:**115–122.
Lambert, D. T., and Fingerman, M., 1978a, Colchicine and cytochalasin B: A further characterization of their actions on crustacean chromatophores using the ionophore A 23187 and thiol reagents, *Biol. Bull.* **155:**563–575.
Lambert, D. T., and Fingerman, M., 1978b, Evidence implicating calcium as the second messenger for red-pigment concentrating hormone in the prawn *Palaemonetes pugio, Physiol. Zool.* **52:**497–508.
Lasek, R. J., 1968, Axoplasmic transport in cat dorsal root ganglion cells: As studied with [^{3}H]-L-leucine, *Brain Res.* **7:** 360–377.
Lasek, R. J., and Hoffmann, P. N., 1976, The neuronal cytoskeleton, axonal transport, and axonal growth, in: *Cell Motility* (R. Goldman, T. Pollard, and J. Rosenbaum, eds.), pp. 1021–1050, Cold Spring Harbor Conf. Cell Prolif., Cold Spring Harbor.
LaVail, J. H., and LaVail, M. M., 1974, The retrograde intraaxonal transport of horseradish peroxidase in the chick visual system: A light and electron microscopic study, *J. Comp. Neurol.* **157:**303–358.
LaVail, J. H., Rapisardi, S., and Sugino, I. K., 1980, Evidence against the smooth endoplasmic reticulum as a continuous channel for the retrograde axonal transport of horseradish peroxidase, *Brain Res.* **191:**3–20.
LeBeux, Y. J., and Willemot, J., 1975, An ultrastructural study of the microfilaments in rat brain by means of heavy meromyosin labeling. I. The perikaryon, dendrites, and the axon. *Cell Tissue Res.* **160:**1–36.
Leestma, J. E., and Freeman, S. S., 1977, Computer assisted analysis of particulate axoplasmic flow in organized CNS tissue cultures, *J. Neurobiol.* **8:**453–467.

Levine, J., and Willard, M., 1981, Fodrin: Axonally transported polypeptides associated with the internal periphery of many cells, *J. Cell Biol.* **90**:631–643.
Lin, D. C., Tobin, K. D., Gramet, M., and Lin, S., 1980, Cytochalasins inhibit nuclei-induced actin polymerization by blocking filament elongation, *J. Cell Biol.* **84**:455–460.
Lo, S. J., Tchen, T. T., and Taylor, J. D., 1980, Hormone-induced filopodium formation and movement of pigment into newly formed filopodia, *Cell Tissue Res.* **210**:371–382.
Lorenz, T., and Willard, M., 1978, Subcellular fractionation of intra-axonally transported polypeptides in the rabbit visual system, *Proc. Natl. Acad. Sci. USA* **75**:505–509.
Lubinska, L., 1964, Axoplasmic streaming in regenerating and in normal nerve fibers, in: *Progress in Brain Research. Mechanisms of Neural Regeneration.* (M. Singer and J. P. Schade, eds.), pp. 1–66, Elsevier, Amsterdam.
Lubinska, L., 1975, On axoplasmic flow, *Int. Rev. Neurobiol.* **17**:241–296.
Luby, K. J., and Porter, K. R., 1980, The control of pigment migration in isolated erythrophores of *Holocentrus ascensionis.* I. Energy requirements, *Cell* **21**:13–23.
Luby-Phelps, K., and Porter, K. R., 1982, The control of pigment migration in isolated erythrophores of *Holocentrus ascensionis.* II. The role of calcium, *Cell* **29**:441–450.
Luby-Phelps, K., and Schliwa, M., 1982, Pigment migration in chromatophores: A model system for intracellular particle transport, in: *Axoplasmic Transport* (D. G. Weiss, ed.), pp. 15–26, Springer-Verlag, Berlin, Heidelberg.
Lyerla, T. A., and Novales, R. R., 1972, The effect of cyclic AMP and cytochalasin B on tissue cultured melanophores of *Xenopus laevis, J. Cell Physiol.* **80**:243–251.
Lynch, T. J., Taylor, J. D., and Tchen, T. T., 1982, Phosphorylation of organelle proteins during pigment translocation, *J. Cell Biol.* **95**:331a.
Macgregor, H. C., and Stebbings, H., 1970, A massive system of microtubules associated with cytoplasmic movement in telotrophic ovarioles, *J. Cell Sci.* **6**:431–449.
MacLean-Fletcher, S., and Pollard, T. D., 1980, Mechanism of action of cytochalasin B on actin, *Cell* **20**:329–341.
Magun, B., 1973, Two actions of cyclic AMP on melanosome movement in frog skin, Dissection by cytochalasin B, *J. Cell Biol.* **57**:854–858.
Malawista, S. E., 1965, On the action of colchicine, *J. Exp. Med.* **122**:361–384.
Malawista, S. E., 1971a, The melanocyte model: Colchicine-like effects of other antimitotic agents, *J. Cell Biol.* **49**:848–855.
Malawista, S. E., 1971b, Cytochalasin B reversibly inhibits melanin granule movement in melanocytes, *Nature* 234:354–355.
Matsumoto, J., Watanabe, Y., Obika, M., and Hadley, M. E., 1978, Mechanisms controlling pigment movement within swordtail erythrophores in primary cell culture, *Comp. Biochem. Physiol.* **61A**:509–517.
Matus, A., Bernhardt, R., and Hugh-Jones, T., 1981, HMWP proteins are preferentially associated with dendritic microtubules in brain, *Proc. Natl. Acad. Sci. USA* **78**:3010–3014.
Maupin-Szamier, P., and Pollard, T. D., 1978, Actin filament destruction by osmium tetroxide, *J. Cell Biol.* **77**:837–852.
Mays, U., 1972, Stofftransport im Ovar von *Pyrrhocoris apterus L., Z. Zellforsch.* **123**:395–410.
McGee-Russel, S. M., 1974, Dynamic activities and labile microtubules in cytoplasmic transport in the marine foraminiferan, *Allogromia, Symp. Soc. Exp. Biol.* **28**:157–189.
McGee-Russel, S. M., and Allen, R. D., 1971, Reversible stabilization of labile microtubules in the reticulopodial network of *Allogromia, Adv. Cell Mol. Biol.* **1**:153–184.
McGuire, J., and Moellmann, G., 1972, Cytochalasin B: Effects on microfilaments and movement of melanin granules within melanocytes, *Science* **175**:642–644.
McIntosh, J. R., Hepler, P. K., and Van Wie, D. G., 1969, Model for mitosis, *Nature* **224**:659–663.
McIntosh, J. R., Euteneuer, U., and Neighbors, B., 1980, Intrinsic polarity as a factor in microtubule function, in: *Microtubules and Microtubule Inhibitors 1980* (M. DeBrabander and J. DeMey, eds.), pp. 357–371, Elsevier, Amsterdam.
McLean, W. G., Frizell, M., and Sjöstrand, J., 1976, Labelled proteins in rabbit vagus nerve between the fast and slow phases of axonal transport, *J. Neurochem.* **26**:77–82.
McNiven, M., and Porter, K. R., 1981, The microtubule-organizing center in erythrophores: Its three-dimensional structure and behavior during pigment motion, *J. Cell Biol.* **91**:334a.

Meeusen, R. L., and Cande, W. Z., 1979, N-ethylmaleimide modified heavy meromyosin. A probe for actomyosin interactions, *J. Cell Biol.* **82:**57–65.

Metuzals, J., 1969, Configuration of a filamentous network in the axoplasm of the squid (*Loligo pealli L.*) giant nerve fiber, *J. Cell Biol.* **43:**480–505.

Metuzals, J., and Tasaki, I., 1978, Subaxolemmal filamentous network in the giant nerve fiber of the squid and its possible role in excitability, *J. Cell Biol.* **78:**597–622.

Miani, N., 1960, Proximo-distal movement along the axon of protein synthesized in the perikaryon of regenerating neurons, *Nature* **189:**541.

Miani, N., 1963, Analysis of the somato-axonal movement of phospholipids in the vagus and hypoglossal nerves, *J. Neurochem.* **10:**859–874.

Murphy, D. B., and Borisy, G. G., 1975, Association of high molecular weight proteins with microtubules and their role in microtubule assembly in vitro, *Proc. Natl. Acad. Sci. USA* **72:**2696–2700.

Murphy, D. B., and Tilney, L. G., 1974, The role of microtubules in the movement of pigment granules in teleost melanophores, *J. Cell Biol.* **61:**757–779.

Murphy, D. B., and Wallis, K. T., 1981, The ATPase activity of neuronal microtubules is associated with membrane vesicles, *J. Cell Biol.* **91:**47a.

Nadelhaft, I., 1974, Microtubule densities and total numbers in selected axons of the crayfish abdominal nerve cord, *J. Neurocytol.* **3:**73–86.

Nagai, R., and Fukui, S., 1981, Differential treatment of Acetabularia with cytochalasin B and N-ethylmaleimide with special reference to their effects on cytoplasmic streaming, *Protoplasma* **109:**79–89.

Nagai, R., and Hayama, T., 1979, Ultrastructural aspects of cytoplasmic streaming in *Characean* cells, in: *Cell Motility: Molecules and Organization* (S. Hatano, H. Ishikawa, and H. Sato, eds.), pp. 321–337, University of Tokyo Press, Tokyo.

Nagai, R., and Kamiya, N., 1977, Differential treatment of *Chara* cells with cytochalasin B with special reference to its effect on cytoplasmic streaming, *Exp. Cell Res.* **108:**231–237.

Nagai, R., and Rebhun, L. I., 1966, Cytoplasmic microfilaments in streaming Nitella cells, *J. Ultrastruct. Res.* **14:**571–585.

Nauta, H. J. W., Kaiserman-Abramof, I. R., and Lasek, R. J., 1975, Electron microscopic observations of horseradish peroxidase transported from the caudoputamen to the substantia nigra in the rat: Possible involvement of the agranular reticulum, *Brain Res.* **85:**373–384.

Nothnagel, E. A., and Webb, W. W., 1982, Hydrodynamic models of viscous coupling between motile myosin and endoplasm in *Characean* algae, *J. Cell Biol.* **94:**444–454.

Nothnagel, E. A., Barak, L. S., Sanger, J. W., and Webb, W. W., 1981, Fluorescence studies on modes of cytochalasin B and phallotoxin action on cytoplasmic streaming in *Chara, J. Cell Biol.* **88:**364–372.

Nothnagel, E. A., Sanger, J. W., and Webb, W. W., 1982, Effects of exogenous proteins on cytoplasmic streaming in perfused *Chara* cells, *J. Cell Biol.* **93:**735–742.

Novales, R. R., and Davies, W. J., 1967, Melanin-dispersing effect of adenosine-3′5′-monophosphate on amphibian melanophores, *Endocrinology* **81:**283–290.

Novales, R. R., and Davies, W. J., 1969, Cellular aspects of the control of physiological color changes in amphibians, *Am. Zool.* **9:**479–488.

Novales, R. R., and Fujii, R., 1969, A melanin-dispersing effect of cyclic adenosine monophosphate on *Fundulus* melanophores, *J. Cell. Physiol.* **75:**133–135.

Novales, R. R., and Novales, B. J., 1972, Effect of cytochalasin B on the response of the chromatophores in isolated frog skin to MSH, theophylline, and dibutyryl-cyclic AMP, *Gen. Comp. Endocrinol.* **19:**363–366.

Obika, M., Lo, S. J., Tchen, T. T., and Taylor, J. D., 1978a, Ultrastructural demonstration of hormone-induced movement of carotenoid droplets and endoplasmic reticulum in Xanthophores of the goldfish, *Carassius auratus, Cell Tissue Res.* **190:**409–416.

Obika, M., Menter, D. G., Tchen, T. T., and Taylor, J. D., 1978b, Actin microfilaments in melanophores of *Fundulus heteroclitus.* Their possible involvement in melanosome migration, *Cell Tissue Res.* **193:**387–397.

O'Brien, T. P., and McCully, M. E., 1970, Cytoplasmic fibers associated with streaming and saltatory particle movement in *Herculaneum mantegazzianum*, *Planta* **94:**91–94.

O'Brien, T. P., and Thiman, K. V., 1966, Intracellular fibers in oat coleoptile cells and their possible significance in cytoplasmic streaming, *Proc. Natl. Acad. Sci. USA* **56:**888–894.

Ochs, S., 1972a, Fast transport of materials in mammalian nerve fibers, *Science* **176:**252–260.

Ochs, S., 1972b, Rate of fast axoplasmic transport in mammalian nerve fibers, *J. Physiol.* **227:**627–645.

Ochs, S., Johnson, J., and Ng, M.-H., 1967, Protein incorporation and axoplasmic flow in motoneuron fibres following intra-cord injection of labelled leucine, *J. Neurochem.* **14:**317–331.

Ochs, S., Worth, R. M., and Chan, S.-J., 1977, Calcium requirement for axoplasmic transport in mammalian nerve, *Nature* **270:**748–750.

Ohta, T., 1974, Movement of pigment granules within melanophores of an isolated fish scale. Effects of cytochalasin B on melanophores, *Biol. Bull. (Woods Hole)* **146:**258–266.

Oplatka, A., and Tirosh, R., 1973, Active streaming in actomyosin solutions, *Biochim. Biophys. Acta* **305:**684–688.

Oplatka, A., Gadasi, H., Tirosh, R., Lamed, Y., Muhlrad, A., and Liron, N., 1974, Demonstration of mechanochemical coupling in systems containing actin, ATP and non-aggregating active myosin derivates, *J. Mechanochem. Cell Motil.* **2:**295–306.

Ostlund, R. E., Leung, J. T., and Kipnis, D. M., 1977, Muscle actin filaments bind pituitary secretory granules in vitro, *J. Cell Biol.* **73:**78–87.

Palevitz, B. A., and Hepler, P. K., 1975, Identification of actin in situ at the ectoplasm–endoplasm interface of *Nitella.* Microfilament-chloroplast association, *J. Cell Biol.* **65:**29–38.

Palevitz, B. A., Ash, J. F., and Hepler, P. K., 1974, Actin in the green algae *Nitella, Proc. Natl. Acad. Sci. USA* **71:**363–366.

Papasozomenos, S. C., Autilio-Gambetti, L., and Gambetti, P., 1981, Reorganization of axoplasmic organelles following β,β′-iminodipropionitrile administration, *J. Cell Biol.* **91:**866–871.

Papasozomenos, S. C., Yoon, M., Crane, R., Autilio-Gambetti, L., and Gambetti, P., 1982, Redistribution of proteins of fast axonal transport following administration of β,β′-iminodipropionitrile: A quantitative autoradiographic study, *J. Cell Biol.* **95:**672–675.

Parthasarathy, M. V., and Mühlethaler, K., 1972, Cytoplasmic microfilaments in plant cells, *J. Ultrastruct. Res.* **38:**46–62.

Parthasarathy, M. V., and Pesacreta, T. C., 1980, Microfilaments in plant vascular cells, *Can. J. Bot.* **58:**807–815.

Paulson, J. C., and McClure, W. O., 1974, Microtubules and axoplasmic transport, *Brain Res.* **73:**333–337.

Paulson, J. C., and McClure, W. O., 1975, Microtubules and axoplasmic transport. Inhibition of transport by podophyllotoxin: An interaction with microtubule protein, *J. Cell Biol.* **67:**461–467.

Pennigroth, S. M., Cheung, A., Bouchard, P., Gagnon, C., and Bordin, C. W., 1982, Dynein ATPase is inhibited selectively in vitro by erythro-9-[3-2(hydroxynonyl)]adenine, *Biochem. Biophys. Res. Commun.* **104:**234–240.

Pickett-Heaps, J. D., 1967, Ultrastructure and differentiation in *Chara sp.* I. Vegetative cells, *Austr. J. Biol. Sci.* **20:**539–551.

Pomerat, C. M., Hendelman, W. J., Raiborn, C. W., and Massey, J. F., 1967, Dynamic activities of nervous tissues in vitro, in: *The Neuron* (H. Hydén, ed.), pp. 119–178, Elsevier, Amsterdam.

Porter, K. R., 1973, Microtubules in intracellular locomotion, *Ciba Found. Symp.* **14:**149–166.

Porter, K. R., and McNiven, M. A., 1982, The cytoplast: A unit structure in chromatophores, *Cell* **29:**23–32.

Pratt, M. M., 1980, The identification of a dynein ATPase in unfertilized sea urchin eggs, *Dev. Biol.* **74:**364–378.

Puiseux-Dao, S., 1979, Movements cytoplsmiques et morphogenese chez l'Acetabularia mediterranea, *Biol. Cellulaire* **34:**83–90.

Puszkin, S., Berl, S., Puszkin, E., and Clarke, D. D., 1968, Actomyosin-like protein isolated from mammalian brain, *Science* **161:**170–171.

Puszkin, S., Nicklas, W. J., and Berl, S., 1972, Actomyosin-like protein in brain: Subcellular distribution, *J. Neurochem.* **19:**1319–1333.

Putney, J. W., 1978, Stimulus-permeability coupling: Role of calcium in the receptor regulation of membrane permeability, *Pharmacol. Rev.* **30:**209–245.

Raine, C. S., Ghetti, B., and Shelanski, M. L., 1971, On the association between microtubules and mitochondria within axons, *Brain Res.* **34:**389–393.

Rambourg, A. and Droz, B., 1980, Smooth endoplasmic reticulum and axonal transport, *J. Neurochem.* **35:**16–25.

Rebhun, L. I., 1972, Polarized intracellular particle transport: Saltatory movements and cytoplasmic streaming, *Int. Rev. Cytol.* **32:**93–137.

Rinaldi, R. A., and Jahn, T. L., 1964, Shadowgraphs of protoplasmic movement in *Allogromia laticollaris* and a correlation of this movement to striated muscle contraction, *Protoplasma* **58:**369–390.

Robison, W. G., and Charlton, J. S., 1973, Microtubules, microfilaments, and pigment granule movement in the chromatophores of *Palaemonetes vulgaris*, *J. Exp. Zool.* **186:**297–304.

Ross, J., Olmsted, J. B., and Rosenbaum, J. L., 1975, The ultrastructure of mouse neuroblastoma cells in tissue culture, *Tissue Cell* **7:**107–136.

Roth, L. E., and Shigenaka, Y., 1970, Microtubules in the heliozoan axopodium II. Rapid degradation by cupric and nickelous ions, *J. Ultrastruct. Res.* **31:**356–374.

Sabnis, D. D., and Jacobs, W. P., 1967, Cytoplasmic streaming and microtubules in the coenocytic marine alga *Caulerpa prolifera*, *J. Cell Sci.* **2:**465–472.

Saidel, W. M., 1977, Metabolic energy requirements during teleost melanophore adaptations, *Experientia* **33:**1573–1574.

Samson, F. E., 1976, Pharmacology of drugs that affect intracellular movement, *Annu. Rev. Pharmacol. Toxicol.* **16:**143–159.

Scheele, R. B., and Borisy, G. G., 1979, In vitro assembly of microtubules, in: *Microtubules* (K. Roberts and J. S. Hyams, eds.), pp. 175–254, Academic Press, London.

Schliwa, M., 1976, The role of divalent cations in the regulation of microtubule assembly. In vivo studies on microtubules of the heliozoan axopodium using the ionophore A 23187, *J. Cell Biol.* **70:**527–540.

Schliwa, M., 1978, Microtubular apparatus of melanophores. Three-dimensional organization, *J. Cell Biol.* **76:**605–614.

Schliwa, M., 1979, Stereo high voltage electron microscopy of melanophores. Matrix transformations and the effects of cold and colchicine, *Exp. Cell Res.* **118:**323–340.

Schliwa, M., 1981, Microtubule-dependent intracellular transport in melanophores, in: *International Cell Biology 1980–1981* (H. G. Schweiger, ed.), pp. 275–285, Springer Verlag, Berlin-Heidelberg.

Schliwa, M., 1982a, Action of cytochalasin D on cytoskeletal networks, *J. Cell Biol.* **92:**79–91.

Schliwa, M., 1982b, Chromatophores: Their use in understanding microtubule-dependent intracellular transport, *Methods Cell Biol.* **25:**285–312.

Schliwa, M., and Bereiter-Hahn, J., 1973a, Pigment movements in fish melanophores: Morphological and physiological studies. II Cell shape and microtubules, *Z. Zellforsch.* **147:**107–125.

Schliwa, M., and Bereiter-Hahn, J., 1974b, Pigment movements in fish melanophores: Morphological and physiological studies. III. The effects of colchicine and vinblastine, *Z. Zellforsch.* **147:**127–148.

Schliwa, M., and Euteneuer, U., 1978a, Quantitative analysis of the microtubule system in isolated fish melanophores, *J. Supramol. Struct.* **8:**177–190.

Schliwa, M., and Euteneuer, U., 1978b, A microtubule-independent component may be involved in granule transport in pigment cells, *Nature* **273:**556–558.

Schliwa, M., Osborn, M., and Weber, K., 1978, Microtubule system of isolated fish melanophores as revealed by immunofluorescence microscopy, *J. Cell Biol.* **76:**229–236.

Schliwa, M., Euteneuer, U., Herzog, W., and Weber, K., 1979a, Evidence for rapid structural and functional changes of the melanophore microtubule-organizing center upon pigment movements, *J. Cell Biol.* **83**:623–632.

Schliwa, M., Wehland, J., and Weber, K., 1979b, Localization and organization of actin in a fish melanophore, and a functional test of its involvement in intracellular transport, *J. Cell Biol.* **83**:315a.

Schliwa, M., Weber, K., and Porter, K. R., 1981, Localization and organization of actin in melanophores, *J. Cell Biol.* **89**:267–275.

Schmitt, F. O., 1968, Fibrous proteins—neuronal organelles, *Proc. Natl. Acad. Sci. USA* **66**:1092–1101.

Schnapp, B. J., and Reese, T. S., 1982, Cytoplasmic structure in rapid-frozen axons, *J. Cell Biol.* **94**:667–679.

Schönharting, H., Breer, H., Rahmann, H., Siebert, G., and Rösner, H., 1977, Colchiceine, a novel inhibitor of fast axonal transport without tubulin binding properties, *Eur. J. Cell Biol.* **16**:106–117.

Schonbach, J., Schonbach, C., and Cuénod, M., 1971, Rapid phase of axoplasmic flow and synaptic proteins: An electron microscopical autoradiographic study, *J. Comp. Neurol.* **141**:485–498.

Schwab, M. E., 1977, Ultrastructural localization of nerve growth factor-horseradish peroxidase (NGF-HRP) coupling product after retrograde axonal transport in adrenergic neurons, *Brain Res.* **130**:190–196.

Schwab, M. E., and Thoenen, H., 1978, Selective binding, uptake, and retrograde transport of tetanus toxin by nerve terminals in the rat iris, *J. Cell Biol.* **77**:1–13.

Schwab, M. E., Suda, K., and Thoenen, H., 1979, Selective retrograde transsynaptic transfer of a protein, tetanus toxin, subsequent to its retrograde axonal transport, *J. Cell Biol.* **82**:798–810.

Schwartz, J. H., 1979, Axonal transport: Components, mechanisms and specificity, *Annu. Rev. Neurosci.* **2**:467–504.

Seitz, K., 1979, Cytoplasmic streaming and cyclosis of chloroplasts, in: *Physiology of Movements* (W. Haupt and G. Feinlieb, eds.), pp. 150–169, Springer Verlag, Berlin.

Selden, S. C., and Pollard, T. D., 1982, Phosphorylation of microtubule-associated proteins regulates their interaction with actin filaments, *J. Cell Biol.* **95**:348a.

Sherline, P., Lee, Y.-C., and Jacobs, L. S., 1977, Binding of microtubules to pituitary secretory granules and secretory granule membranes, *J. Cell Biol.* **72**:380–389.

Shimmen, T., 1978, Dependency of cytoplasmic streaming on intracellular ATP and Mg^{2+} concentrations, *Cell Struct. Funct.* **3**:113–121.

Shimmen, T., and Tazawa, M., 1982, Cytoplasmic streaming in the cell model of *Nitella*, *Protoplasma* **112**:101–106.

Simons, T. J. B., 1979, Vanadate—new tool for biologists, *Nature* **281**:337–338.

Small, J. V., 1981, Organization of actin in the leading edge of cultured cells: Influence of osmium tetroxide and dehydration on the ultrastructure of actin meshworks, *J. Cell Biol.* **91**:695–705.

Smith, D. S., 1971, On the significance of cross-bridges between microtubules and synaptic vesicles, *Phil. Trans. R. Soc. London Ser. B* **261**:365–405.

Smith, D. S., Järlfors, U., and Cayer, M. L., 1977, Structural cross-bridges between microtubules and mitochondria in central axons of an insect (*Periplaneta americana*), *J. Cell Sci.* **27**:235–272.

Smith, R. S., 1972, Detection of organelles in myelinated nerve fibers by dark field microscopy, *Can. J. Physiol. Pharmacol.* **50**:467–469.

Smith, R. S., 1980, The short term accumulation of axonally transported organelles in the region of localized lesions of single melinated axons, *J. Neurocytol.* **9**:39–65.

Smith, R. S., and Koles, Z. J., 1976, Mean velocity of optically detected intra-axonal particles measured by a cross-correlation method, *Can. J. Physiol. Pharmacol.* **54**:859–869.

Sotelo, C., and Riche, D., 1974, The smooth endoplasmic reticulum and the retrograde and fast orthograde transport of horseradish peroxidase in the nigro-striatonigral loop, *Anat. Embryol.* **146**:209–218.

Stearns, M. E., 1982, High voltage electron microscopy studies of axoplasmic transport in neurons: a possible regulatory role for divalent cations, *J. Cell Biol.* **92**:765–776.

Stearns, M. E., and Ochs, R., 1982, A functional in vitro model for studies of intracellular motility in permeabilized erythrophores, *J. Cell Biol.* **94:**727–739.

Stephens, R. E., and Edds, K. T., 1976, Microtubules: Structure, chemistry and function, *Physiol. Rev.* **56:**709–777.

Stoeckel, K., Schwab, M., and Thoenen, H., 1975, Specificity of retrograde transport of nerve growth factor (NGF) in sensory neurons: A biochemical and morphological study, *Brain Res.* **89:**1–14.

Stone, G. C., Wilson, D. L., and Hall, M. E., 1978, Two dimensional gel electrophoresis of proteins in rapid axoplasmic transport, *Brain Res.* **144:**287–302.

Suchard, S. J., and Goode, D., 1982, Microtubule-dependent transport of secretory granules during stalk secretion in a peritrich ciliate, *Cell Motil.* **2:**47–71.

Summers, K. E., and Gibbons, I. R., 1971, Adenosine triphosphate-induced sliding of tubules in trypsin-treated flagella of sea urchin sperm, *Proc. Natl. Acad. Sci. USA* **68:**3092–3096.

Suprenant, K. A., and Dentler, W. L., 1982, Association between endocrine pancreatic secretory granules and in vitro-assembled microtubules is dependent upon microtubule-associated proteins, *J. Cell Biol.* **93:**164–174.

Tani, E., and Ametani, T., 1970, Substructure of microtubules in brain nerve cells as revealed by ruthenium red, *J. Cell Biol.* **46:**159–165.

Taylor, D. L., and Condeelis, J. S., 1979, Cytoplasmic structure and contractility in amoeboid cells, *Int. Rev. Cytol.* **56:**57–144.

Tazawa, M., 1964, Studies on Nitella having artificial cell sap. I. Replacement of the cell sap with artificial solutions, *Plant Cell Physiol.* **5:**33–43.

Tazawa, M., 1968, Motive force of the cytoplasmic streaming in *Nitella, Protoplasma* **65:**207–222.

Tazawa, M., Kikuyama, M., and Shimmen, T., 1976, Electric characteristics and cytoplasmic streaming of *Characeae* cells lacking tonoplast, *Cell Struct. Funct.* **1:**165–176.

Thoenen, H., and Kreutzberg, G. W., 1981, The role of fast transport in the nervous system, *Neurosci. Res. Prog. Bull.* **20:**(1).

Tilney, L. G., 1968, Studies on the microtubules in heliozoa. IV. The effect of colchicine on the formation and maintenance of the axopodia and the redevelopment of pattern in *Actinosphaerium nucleofilum* (*Barrett*), *J. Cell Sci.* **3:**549–562.

Tilney, L. G., and Porter, K. R., 1965, Studies on the microtubules in heliozoa. I. Fine structure of *Actinosphaerium* with particular reference to axial rod structure, *Protoplasma* **60:**317–344.

Tilney, L. G., and Porter, K. R., 1967, Studies on the microtubules in heliozoa. II. The effect of low temperature on these structures in the formation and maintenance of axopodia, *J. Cell Biol.* **34:**327–343.

Tilney, L. G., Hiramoto, Y., and Marsland, D., 1966, Studies on the microtubules in heliozoa. III. A pressure analysis of the role of these structures in the formation and maintenance of the axopodia of *Actinosphaerium nucleofilum, J. Cell Biol.* **29:**77–95.

Tirosh, R., and Oplatka, A., 1982, Active streaming against gravity in glass microcapillaries of solutions containing acto-heavy meromyosin and native tropomyosin, *J. Biochem.* **91:**1435–1440.

Travis, J. L., and Allen, R. D., 1981, Studies on the motility of the Foraminifera. I. Ultrastructure of the reticulopodial network of *Allogromia laticollaris* (*Arnold*), *J. Cell Biol.* **90:**211–221.

Troyer, D. S., 1975, Possible involvement of the plasma membrane in saltatory particle movement in heliozoan axopods, *Nature* **254:**696–698.

Tsukita, S., and Ishikawa, H., 1980, The movement of membraneous organelles in axons. Electron microscopic identification of anterogradely and retrogradely transported organelles, *J. Cell Biol.* **84:**513–530.

Tsukita, S., and Ishikawa, H., 1981, The cytoskeleton in myelinated axons: Serial section study, *Biomed. Res.* **2:**424–437.

Tucker, J. B., 1974, Microtubule arms and cytoplasmic streaming and microtubule bending and stretching of intertubule links in the feeding tentacle of the suctorian ciliate *Tokophrya, J. Cell Biol.* **62:**424–437.

Tucker, J. B., 1979, Spatial organization of microtubules. in: *Microtubules* (K. Roberts and J. S. Hyams, eds.), pp. 315–357, Academic Press, London.

Tytell, M., Black, M. M., Garner, J. A., and Lasek, R. J., 1981, Axonal transport: Each major rate component reflects the movement of distinct macromolecular complexes, *Science* **214:**179–181.
Vallee, R. B., 1980, Structure and phosphorylation of microtubule-associated protein 2 (MAP 2), *Proc. Natl. Acad. Sci. USA* **77:**3206–3210.
Wagner, R. C., and Rosenberg, M. D., 1973, Endocytosis in Chang liver cells: The role of microtubules in vacuole orientation and movement, *Cytobiology* **7:**20–27.
Wallach, D., Davies, P. J. A., and Pastan, I., 1978, Purification of mammalian filamin, *J. Biol. Chem.* **254:**10250–10255.
Wang, E., and Choppin, P. W., 1981, Effect of vanadate on intracellular distribution and function of 10 nm filaments, *Proc. Natl. Acad. Sci. USA* **78:**2363–2367.
Wang, E., and Goldman, R. D., 1978, Functions of cytoplasmic fibers in intracellular movements in BHK-21 cells, *J. Cell Biol.* **79:**708–726.
Wang, E., Cross, R. K., and Choppin, P. W., 1979, Involvement of microtubules and 10 nm filaments in the movement and positioning of nuclei in syncytia, *J. Cell Biol.* **83:**320–337.
Wang, K., Ash, F., and Singer, S. J., 1975, Filamin: A new high molecular weight protein found in smooth muscle and nonmuscle cells, *Proc. Natl. Acad. Sci. USA* **72:**4483–4487.
Weber, W., and Dambach, M., 1972, Amöboid bewegliche Pigmentzellen im Epithel des Seeigels *Centrostephanus longispinus*, *Z. Zellforsch.* **133:**87–102.
Wehland, J., Osborn, M., and Weber, K., 1977, Phalloidin-induced actin polymerization in the cytoplasm of cultured cells interferes with cell locomotion and growth, *Proc. Natl. Acad. Sci. USA* **74:**5613–5617.
Weiss, D., 1982, *Axoplasmic Transport*, Springer Verlag, Berlin.
Weiss, P. A., and Hiscoe, H. B., 1948, Experiments on the mechanism of nerve growth, *J. Exp. Zool.* **107:**315–395.
Weiss, P. A., and Mayr, R., 1971a, Neuronal organelles in neuroplasmic ("axonal") flow. I. Mitochondria, *Acta Neuropathol. Suppl.* **5:**187–197.
Weiss, P. A., and Mayr, R., 1971b, Neuronal organelles in neuroplasmic ("axonal") flow. II. Neurotubules, *Acta Neuropathol. Suppl.* **5:**198–206.
Wessels, N., Spooner, B., Ash, J., Bradley, H., Luduena, M., Taylor, E., Wrenn, J., and Yamada, K., 1971, Microfilaments in cellular and developmental processes, *Science* **171:**135–143.
Wieland, T., 1977, Modification of actins by phallotoxins, *Naturwissenschaften* **64:**303–309.
Wikswo, M. A., and Novales, R. R., 1969, The effect of colchicine on migration of pigment granules in the melanophores of *Fundulus heteroclitus*, *Biol. Bull.* **137:**228–237.
Willard, M., 1977, The identification of two intra-axonally transported polypeptides resembling myosin in some respects in the rabbit visual system, *J. Cell Biol.* **75:**1–11.
Willard, M., and Simon, C., 1981, Antibody decoration of neurofilaments, *J. Cell Biol.* **89:**198–205.
Willard, M., Cowan, W. M., and Vagelos, P. R., 1974, The polypeptide composition of intra-axonally transported proteins: Evidence for four transport velocities, *Proc. Natl. Acad. Sci. USA* **71:**2183–2187.
Williamson, R. E., 1972, A light microscope study of the action of cytochalasin B on the cells and isolated cytoplasm of the characeae, *J. Cell Sci.* **10:**811–819.
Williamson, R. E., 1975, Cytoplasmic streaming in *Chara:* A cell model activated by ATP and inhibited by cytochalasin B, *J. Cell Sci.* **17:**655–668.
Williamson, R. E., 1979, Filaments associated with the endoplasmic reticulum in the streaming cytoplasm of *Chara corallina*, *Eur. J. Cell Biol.* **20:**177–183.
Williamson, R. E., 1980, Actin in motile and other processes in plant cells, *Can. J. Bot.* **58:**766–772.
Wilson, D. L., and Stone, G. C., 1979, Axoplasmic transport of proteins, *Annu. Rev. Biophys. Bioeng.* **8:**27–45.
Wolosewick, J. J., and Porter, K. R., 1976, Stereo high voltage electron microscopy of whole cells of the human diploid line WI-36, *Am. J. Anat.* **147:**303–323.
Wolosewick, J. J., and Porter, K. R., 1979, The microtrabecular lattice of the cytoplasmic ground substance. Artifact or reality? *J. Cell Biol.* **82:**114–139.

Wuerker, R. B., and Kirkpatrick, J. B., 1972, Neuronal microtubules, neurofilaments, and microfilaments, *Int. Rev. Cytol.* **33**:45–75.
Yano, M., 1978, Observations of steady streamings in a solution of Mg-ATP and acto-heavy meromyosin from rabbit skeletal muscle, *J. Biochem.* **83**:1203–1204.
Yano, M., Yamada, T., and Shimizu, H., 1978, Studies on the chemo-mechanical conversion in artifically produced streamings, *J. Biochem.* **84**:277–284.
Yano, M., Yamamoto, Y., and Shimizu, H., 1982, An actomyosin motor, *Nature* **299**:557–559.
Zatz, M., and Barondes, S. H., 1971, Rapid transport of fucosyl glycoproteins to nerve endings in mouse brain, *J. Neurochem.* **18**:1125–1133.
Zenker, W., and Hohberg, E., 1973, A α-nerve fibre: Number of neurotubules in the stem fibre and in the terminal branches, *J. Neurocytol.* **2**:143–148.

An addendum for this chapter may be found on page 403.

2

Organization and Function of Stress Fibers in Cells in Vitro and in Situ

A REVIEW

H. Randolph Byers, Glenn E. White, and Keigi Fujiwara

1. Introduction

The purpose of this review is to present the tremendous body of research on stress fibers, which has grown exceedingly rapidly in the last 7 or 8 years, due to immunofluorescent techniques, with both a technical and functional perspective. The first section is primarily a chronology of technical innovations which have enabled better observation and characterization of stress fibers. This section also reviews the numerous contractile-associated proteins, actin-binding proteins, regulator proteins, and other proteins shown to localize to stress fibers in a characteristic distribution. The second section discusses the wide variety of roles for stress fibers that have been set forth, including cell spreading, cell adhesion, cell locomotion, contraction, isometric contraction, cell surface compartmentalization, differentiation, cell transformation, tumorigenicity, and morphogenesis. In order to help interpret the significance of the many purported roles of stress fibers, it is important to ask whether stress fibers *in vitro* are pure artifacts and whether stress fibers exist in cells *in situ*. This review demonstrates that the stress fiber is fundamentally a light microscopic term, and in order to avoid confusion with other microfilament bundle-containing structures seen in the electron microscope, such as circumferential microfilament bundles, contractile rings, microvilli, microspikes, and

---

H. Randolph Byers, Glenn E. White, and ***Keigi Fujiwara*** • Department of Anatomy, Harvard Medical School, Boston, Massachusetts 02115.

rootlet structures, it is necessary to establish criteria for the identification of stress fibers *in situ*. Finally, using these criteria, this paper presents two recently introduced models which exhibit stress fibers in cells in tissues: the fibroblast, called a scleroblast in the fish scale, and the endothelial cells of avian and mammalian vasculature.

2. Toward a Definition of Stress Fibers

2.1. Light Microscopy

Cytoplasmic fibers termed "tension striae" or "stress fibers," were first described in a variety of nonmuscle cultured cells using bright field microscopy (Lewis and Lewis, 1924). The origin of the fibers were believed to be caused by the development of tension within the cytoplasm, hence their name. With the use of phase-contrast microscopy and time-lapse cinematography, investigators noted that the parallel, linear patterns of stress fibers changed slowly over a number of hours. Stress fibers were usually found oriented parallel with the long axis of spread bipolar cells, and occurred in "sets" or groups of fibers running parallel to one another or converging slightly towards the poles of well-spread multipolar cells (Rose and Cattoni, 1963; Buckley and Porter, 1967). Stress fibers were found primarily in spreading or well-spread cells following several days of culturing, and occasionally developed complex geometric patterns or "mosaic patterns" described as predominantly triangular (Rose and Cattoni, 1963). It is clear that some of these patterns correspond to the "geodesic" patterns observed in cultured cells years later following specific immunofluorescent staining for contractile proteins (Lazarides and Burridge, 1975; Lazarides, 1975a; see Section 2.3). It was noted that fluorescently labeled stress fibers often branched into smaller cytoplasmic filaments that were not detected by phase-contrast microscopy.

Advances in light microscopic techniques have provided additional information in the distribution, role, and composition of stress fibers in a variety of cells. The reduced depth of focus of Nomarski differential interference optics has been used to better situate stress fibers in the cytoplasm. The stress fibers were found in greater number in the basal or substrate region of the ectoplasm or cell cortex, thus suggesting a role in cell adhesion (Goldman, 1971; Goldman and Knipe, 1973; Goldman *et al.*, 1975; see Section 3.4). Further support of such a role came with the advent of interference-reflection microscopy which correlated cell substrate contacts or adhesion sites with the distribution of stress fibers (Izzard and Lochner, 1976; Heath and Dunn, 1978; see Section 3.4).

Finally, stress fibers in living cells exhibit birefringence (Hughes and Swann, 1948; Goldman *et al.*, 1975; Goldman, 1971), which supports the fixed, electron microscopic observations that they are composed of oriented filamentous elements (see Section 2.2).

2.2. Electron Microscopy

Early whole cell preparations revealed large cytoplasmic fibrils similar in size and orientation to stress fibers (Porter *et al.*, 1945; Bang and Gey, 1948). Fibroblasts, fixed briefly in OsO_4 vapors and shadowed with chromium revealed stress fibers that demonstrated a substructure of finer filaments (Porter, 1953). Thin-sectioned fibroblasts frequently displayed cytoplasmic filaments about 6 nm in diameter which were organized in parallel array just beneath the plasma membrane (Goldberg and Green, 1964).

In a definitive study correlating stress fibers in living cells with structures observed in sectioned fibroblast preparations, stress fibers were shown to be composed of bundles of microfilaments about 7.5 nm in diameter (Buckley and Porter, 1967). However, thick or myosin-like filaments were not observed in or near stress fibers in cells in culture (Buckley and Porter, 1967). The identification of the microfilaments in nonmuscle cells as actin came with the advent of heavy meromyosin (HMM) labeling first used by Huxley (1963) and later adopted by other investigators to study nonmuscle cells and tissues (Ishikawa *et al.*, 1969; Spooner *et al.*, 1973). These studies revealed that bundles or "sheaths" or actin filaments resided primarily in the cortical regions of cultured nonmuscle cells. Stress fibers, as identified by phase-contrast or Nomarski optics, when glycerinated, labeled with HMM, and flat embedded for electron microscopy (EM), often revealed thickened and "fuzzy" filaments (Goldman, 1975).

Although successful determination of actin polarity in glycerinated microvilli was possible in thin-sectioned embedded material (Mooseker and Tilney, 1975), labeling the actin filaments in stress fibers with HMM or myosin subfragment (S-1) often failed to give polarity of the actin filaments. Fortunately, the advent of tannic acid in fixative after S-1 decoration of microfilament bundles greatly enhanced the arrowhead pattern in thin-sectioned material. Bundles were thus shown to contain filaments with opposite polarity (Begg *et al.*, 1978), a significant finding if one invokes an actomyosin-mediated mechanism for the generation of force (see Section 3.3).

Tannic acid and HMM labeling in ethanol extracted cells have revealed decorated actin filaments pointing away from junctions between cells and away from attachment plaques (Sanger and Sanger, 1980). Such ultrastructural work, when combined with observations obtained from immunofluorescence work (see Section 2.3) have prompted some investigators to propose stress fiber models based on the muscle sarcomere (Gordon, 1978; Sanger and Sanger, 1980; see Section 3.3).

As mentioned earlier, searches for myosin-like filaments in stress fibers were largely unsuccessful (Ishikawa, 1974, 1979; Buckley and Raju, 1976) until recently (see below), and it required the development of immunofluorescent techniques to determine the distribution of myosin along the stress fiber (see Section 2.3). However, periodic electron-dense material can be visualized along the microfilament bundles of nonmuscle cells (Ishikawa, 1974; Gordon, 1978; Porter *et al.*, 1979; Rathke *et al.*, 1979), particularly after the addition of tannic acid to the fixative (Goldman *et al.*, 1979; Sanger and Sanger, 1980; Maupin and Pollard, 1983). Myosin-like thick filaments have recently been identified in thin sections of stress fibers using tannic acid glutaraldehyde and low exposure to OsO_4 (Maupin and Pollard, 1983). In addition, periodicity of the registered densities along stress fibers of tannic acid preparations correlate reasonably well with immunofluorescent labeling for myosin and α-actinin (see Section 2.3). The electron densities are localized to microfilament bundles which are closely associated to the inner surface of the plasma membrane, suggesting attachment sites for microfilaments (Ishikawa, 1974, 1979; Maupin and Pollard, 1983; see Section 3.4). Early speculation suggested that the regions of increased density along stress fibers may be analogous to skeletal muscle Z-lines (Wessells *et al.*, 1973), and some immunofluorescent studies have supported this view (see Section 2.3).

The use of fixed whole cell preparations in the electron microscope has allowed visualization of the fine structure of stress-fiber systems *in toto* (Buckley and Porter, 1967, 1975). Use of stereo high voltage electron microscopy of critical-point dried whole cells, has revealed continuity of the microfilaments of stress fibers with a matrix of randomly oriented microfilaments (Buckley, 1975). This three-dimensional matrix of microfilaments and associated fine filaments of varying diameter have been termed the microtrabeculae (Wolosewick and Porter, 1976, 1979; Porter *et al.*, 1979). Not only does the stress fiber often fan out into microfilaments at terminal regions, but multiple lateral connections via the microtrabeculae with other components of the cytoskeleton can be seen. These other components may be more randomly arranged actin filaments, microtubules, or 10-nm filaments, which may run parallel or cross over the microfilament bundles. Likewise, microtrabeculae are often seen interconnecting mitochondria and endoplasmic reticulum in association with stress fibers.

Recent high resolution images of stress fibers showing subunits of actin filaments has been accomplished using platinum replicas of freeze-dried cytoskeletons (Heuser and Kirschner, 1980) and negative staining of whole cells (Small and Celis, 1978; Small and Langanger, 1981). Both of these techniques require extraction of the cells with Triton X-100 before further processing, which may remove minor proteins associated with the actin filaments. The problems with OsO_4 disruption of actin filaments is well known (Maupin-Szamier and Pollard, 1978; Small and Langanger, 1981), and addition of tannic acid with low exposure to OsO_4 has improved preservation of actin and myosin-like filaments in stress fibers in sectioned specimens (Maupin and Pollard, 1983). Methods for the preservation of stress fibers in whole cells or tissues before extraction of the cytoplasm and preservation of ultrastructure and antigenicity are badly needed.

2.3. Immunofluorescent Localization of Cytoplasmic Contractile Proteins and Other Proteins along Stress Fibers

Several cytoplasmic contractile proteins showing characteristic distributions have been visualized along stress fibers of cultured cells using immunofluorescent techniques. These proteins include actin, myosin, α-actinin, and tropomyosin. Calmodulin and myosin light-chain kinase (MLCK) proteins which could be involved in Ca^{2+}-dependent regulation of actomyosin contraction in nonmuscle cells have also been localized along stress fibers. Two other proteins, filamen and vinculin, are also localized to specific domains along stress fibers. Many other proteins, including extracellular matrix, cell surface, and cytoplasmic proteins, as shown by similar techniques, appear to be influenced by the distribution of stress fibers (see Section 3).

Stress-fiber patterns of cultured cells are beautifully stained after incubation with anti-actin and indirect immunofluorescent labels (Lazarides and Weber, 1974). The stress fibers are brightly and continuously labeled along their entire length when compared to phase contrast images. Direct anti-actin immunofluorescent labeling and fluorescein labeled S-1 give similar results (Herman and Pollard, 1979). Myosin, on the other hand, whether by indirect (Weber and Gröschel-Stewart, 1974) or direct immunofluorescent labeling (Fujiwara and Pollard, 1976), reveals either continuous, periodic, or intermittant staining patterns. Likewise, tropomyosin (Lazarides, 1975b) and α-actinin (Lazarides and Burridge, 1975; Lazarides, 1976a; Fujiwara *et al.*, 1978) are distributed in a periodic or intermittant pattern along stress fibers. Using simultaneous staining with antisera to α-actinin and tropomyosin (Lazarides and Burridge, 1975) and antisera to myosin and α-actinin (Gordon, 1978), the stress fibers no longer displayed periodicities. This suggested that these proteins are alternately distributed along the stress fiber similar to the organization of sarcomeres of myofibrils (Lazarides and Burridge, 1975; Gordon, 1978). However, there is evidence that the disposition of contractile proteins are not always so regularly arranged (Zigmond *et al.*, 1979; Fujiwara and Pollard, 1980). Many stress fibers still show intermittant staining following antimyosin and anti α-actinin or antitropomyosin and anti α-actinin staining. Indeed, within the same cell, some stress fibers often show only tropomyosin staining and others show only myosin staining following double direct immunofluorescent staining (Fujiwara and Pollard, 1980). Likewise, similar noncorrelating patterns can be found following α-actinin and myosin localization (Fig. 1). These variations in stress-fiber staining patterns likely reflect a dynamic process of addition or loss of contractile associated proteins during the formation and disappearance of stress fibers in living cultured cells (see Section 3.1). Nevertheless, there is accumulating evidence that some stress fibers have all the contractile proteins organized in a fashion to permit their contraction in extracted or laser manipulated cell model systems (Isenberg *et al.*, 1976; Kreis and Birchmeier, 1980; see Section 3.3).

Insights into the regulation of the cytoplasmic contractile proteins along stress fibers have followed immunofluorescent localization of calmodulin (Dedman *et al.*, 1978) and MLCK (De Lanerolle *et al.*, 1981; Guerriero *et al.*, 1981) to stress fibers. The staining pattern of MLCK appears periodic and is in close association with the myosin-staining pattern (De Lanerolle *et al.*, 1981). Thus, Ca^{2+}-dependent regulation of actomyosin interaction in stress fibers appears possible (see Section 3.3).

Other muscle-associated proteins, the M-line proteins, including a

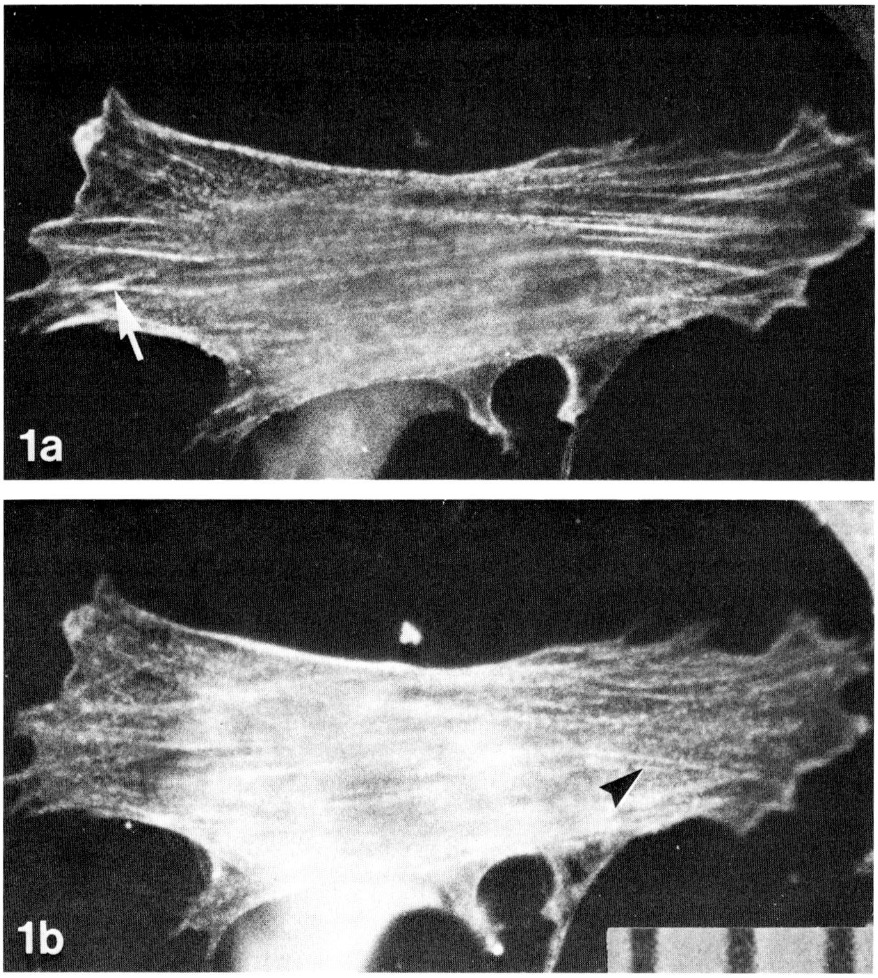

Figure 1. Chick embryo fibroblast labeled with fluorescein conjugated anti-α-actinin and rhodamine conjugated antimyosin (a). Distribution of α-actinin. Arrow indicates stress-fiber staining with no corresponding image in myosin-staining image (b). Likewise, some myosin-staining stress fibers (arrowhead) do not show corresponding α-actinin staining. Scale: 1 division = 10 μm.

160–165-kd protein, is also periodically distributed along stress fibers in non-muscle cells (Schollmeyer *et al.*, 1976). The precise function of this M-line protein remains to be elucidated. Antibodies to another M-line associated protein, creatine phosphokinase (CPK), have been reported to localize along stress fibers in fibroblasts (Fuseler *et al.*, 1981), whereas other antibodies to CPK do not appear to bind stress fibers but localize to intermediate filaments (Eckert *et al.*, 1980). The significance of these findings needs to be further clarified.

Despite the ability of isolated stress fibers to contract *in vitro,* the role of contraction of the stress fiber in the motility of cultured cells is controversial (see Burridge, 1981, for review; see Section 3.3). In fact, they are most prominent in well-spread cells that are not actively migrating (Buckley and Porter, 1967; Herman, *et al.*, 1981; Lewis *et al.*, 1982). The localization of the majority of the microfilament bundles in the basal cortex of well-spread cells (Buckley and Porter, 1967; Buckley, 1975; Goldman and Follet, 1969; Wolosewick and Porter, 1976; Osborn *et al.*, 1978) has implicated a role in cellular adhesion. It is clear that stress fibers frequently terminate or coincide with focal cell surface contacts with the substrate as visualized by reflection interference optics (Heath and Dunn, 1978; Wehland *et al.*, 1979, see Section 3.4).

Isolation and immunofluorescent localization of a 130-kd protein, named vinculin, has revealed that this protein is associated with substrate adhesion plaques (focal contacts), cell–cell contacts, and the termini of stress fibers in nonmuscle cells (Geiger, 1979). Vinculin, isolated from smooth muscle, appears in close association with α-actinin near the cell membrane. Immunoelectron-microscopic evidence suggests that vinculin is closer to the membrane than α-actinin, indicating a potential role in anchorage of the actin bundles to the cell membrane (Geiger *et al.*, 1980; see Section 3.4).

Another stress-fiber associated protein, also originally isolated from chicken gizzard, has been named filamen (Wang *et al.*, 1975). This 250-kd protein has the capacity to cross-link actin filaments into bundles *in vitro* (Wang and Singer, 1977). Two other proteins with identical molecular weight and similar characteristics, actin-binding protein (ABP) from macrophages (Hartwig and Stossel, 1981) and high molecular weight protein from cultured kidney cells (Schloss and Goldman, 1979), can also laterally cross-link actin filaments into actin bundles. Filamen is uniformly distributed along stress fibers as revealed by immunofluorescence techniques (Wang *et al.*, 1975). In comparison to myosin and actin, however, filamin is more diffusely distributed throughout the cytoplasm (Heggeness *et al.*, 1977). The distribution of the other two 250-kd actin-binding proteins along stress fibers as well as a number of other actin-binding proteins (see review in Schliwa, 1981) needs to be elucidated. The definitive role of the actin binding-proteins in sol-gel and/or actin filament to actin bundle transitions needs further study.

Many other proteins, intracellular, extracellular, and membrane proteins and other macromolecules have been shown in association with stress fibers by immunofluorescent techniques. These will be discussed in another section in relation to the potential function of the stress fiber (Section 3).

2.4. Other Techniques for Actin Localization in Stress Fibers

Soon after the localization of actin in stress fibers using anti-actin and indirect immunofluorescent methods (Lazarides and Weber, 1974), other methods for identification of actin in nonmuscle cells were developed. These alternative methods are basically divided into three categories, those which rely on the well-characterized binding of HMM or S-1 to actin filaments, those which rely on other molecules which display specific and high affinity binding to actin, and *in vivo* localization methods (see Section 2.5).

Soon after permeabilized cell models became available, fluorescently tagged HMM or S-1 revealed stress-fiber patterns analogous to the fluorescent antibody methods (Sanger, 1975; Goldman *et al.*, 1976; Schloss *et al.*, 1977; Herman and Pollard, 1979). An alteration on this theme involved covalent linkage of biotin with HMM and then use of fluorescein-avidin, thus utilizing biotin's high affinity binding with avidin, which is reported to improve stress-fiber visualization in cultured cells over that of indirect anti-actin immunofluorescence (Heggeness and Ash, 1977).

The second general method developed to localize actin did not involve use of antibodies or HMM, but made use of other proteins which bind actin. For example, stress-fiber patterns are readily visualized, following DNase I binding to actin, followed by rabbit anti-DNase I and fluorescein labeled goat anti-rabbit serum or by direct fluorescent microscopy with rhodamine-conjugated DNase I (Wang and Goldberg, 1978). Actin in stress fibers can also be labeled by the use of phalloidin, a poisonous derivative from the mushroom *Amanita phalloides,* which binds to F-actin with high affinity (see Wieland, 1977; Wieland and Faulstich, 1978, for reviews). Phalloidin is bicyclic hepatopeptide, and after conjugation with fluorescein (FL-phalloidin), it is still of sufficiently small molecular weight to penetrate formaldehyde-fixed, nondetergent-treated cells (Wulf *et al.*, 1979; Verderame *et al.*, 1980). Such staining reveals the typical stress-fiber patterns in well-spread cells as well as occasional polygonal arrays and concentric fibers and cell ruffles. The low molecular weight of FL-phalloidin complex has opened possibilities for studying the distribution of actin *in vivo* (see next section).

2.5. *Localization of Specific Stress-Fiber Proteins in Vivo*

Much of our knowledge of the behavior of stress fibers in cultured cells has come from simple observation or time-lapse phase-contrast cinematographic studies of stress fibers in living cells (see Section 2.1). On the other hand, fine structural studies and all earlier specific localization procedures for contractile associated proteins which implicate function are carried out on fixed, dead cells. In recent years, two methods have been developed whereby one may localize contractile-associated proteins in the living cell (see Taylor and Wang, 1980; Kreis and Birchmeier, 1982, for reviews). The first has employed microinjection of fluorescently labeled antibodies and the other has relied on diffusion, pinocytosis, of a small molecular weight molecule, Nitrobenzoxadiazole-phallacidin (Barak *et al.*, 1981; see Section 2.5.2).

2.5.1. *Microinjection Studies*

Microinjection of fluorescently labeled actin (Kreis *et al.*, 1979), α-actinin (Feramisco, 1979; Kreis and Birchmeier, 1980), tropomyosin (Wehland and Weber, 1980), and vinculin (Burridge and Feramisco, 1980) into living cultured cells with glass capillaries reveals that earlier immunofluorescent studies were valid in localizing these proteins in a characteristic distribution in the stress fibers. Although periodic staining is detectable for tropomyosin and α-actinin, and their distribution is often alternating along the stress fiber (Feramisco and Blose, 1980), no evidence of active contraction of the stress fibers has been reported. However, microinjection of fluorescein-labeled α-actinin, followed by permeabilization procedures, and treatment with Mg^{2+} and ATP reveals that stress fibers are able to contract (Kreis and Birchmeier, 1982; see Section 3.3). Insight on the mobility of actin in the cytoplasm has become available by combination of time-lapse video recording and fluorescence photobleaching recovery (Kreis *et al.*, 1982). Mobility studies of other proteins will certainly add to the understanding of the role of stress fibers in cell motility and adhesion.

2.5.2. *NBD-Phallicidin*

Another method that visualizes stress fibers in living cells has made use of the actin binding low molecular weight phallotoxin, phallacidin, which can be conjugated with the fluorochrome nitrobenzoxadiazole (Barak *et al.*, 1980, 1981). Living cells that are mildly permeabilized with lysolecithin and treated with nitrobenzoxadiazole-phallacidin (NBD-Ph) display typical stress fiber patterns, polygonal networks, ruffles, and other actin-containing structures (Barak *et al.*, 1980). Indeed, *in vivo* staining of stress fibers can be obtained without permeabilization procedures with higher, but nontoxic concentrations of NBD-Ph (Barak *et al.*, 1981). Unfortunately, the fluorescent signal is weak, increasing NBD-Ph levels causes toxicity questions of viability and cell shape changes, and the accumulating vesicular fluorescent signal soon overwhelms the actin fiber signal. Also, the mechanism by which NBD-Ph is taken into the cell is unclear, although evidence suggests it is by fluid-phase pinocytosis (Barak *et al.*, 1981).

Further studies on the distribution of actin and other stress-fiber-associated proteins in relation to cell cycles and cell behavior are needed to fully understand how stress fibers form and how they function (see Section 3.1).

2.6. *Immunoelectron Microscopy*

Studies on the distribution of contractile proteins in nonmuscle cells in tissue culture have largely been hampered by loss of antigenicity during fixation for ultrastructural preservation and nonspecific staining due to presumably residual-free aldehydes. Another problem is obtaining a label with sufficient electron density to contrast against the usual staining of tissues for EM. Finally, the label or reaction product should be within reasonable distance to the original antigen. Certainly, none of these pitfalls have been entirely overcome to date, however, stress fibers (as well as microtubules) have been useful in initial studies as they are well identified ultrastructurally and display specific distributions of contractile proteins.

Whole cell preparations of cultured cells reveal prominent staining of actin bundles following indirect immunoferritin labeling for actin (Webster *et al.*, 1978). However, due to the small size of ferritin, in order to obtain enough electron density for low magnification identification of stress fibers, the immunoperoxidase method is more useful (Henderson and Weber, 1979). Although the resolution and visualization of the three-dimensional arrangement of the stress fibers is improved using stereo-immunoelectron microscopy, little new information is obtained studying anti-actin distribution. Studies on the distribution of myosin, α-actinin, and tropomyosin will undoubtedly give more information with the improved resolution of the techniques outlined above.

The distribution of myosin in stress fibers by electron microscopy has recently been investigated using purified antimyosin coupled with ferritin (Herman and Pollard, 1981). It was found that ferritin antimyosin concentrated in the dense regions of stress fibers which had a periodicity of about 0.5 μm, correlating well with light microscopic observations. It was also reported that the ferritin antibody was seen associated but not exclusively with 10 to 15 nm filaments in Hela stress fibers (Herman and Pollard, 1981).

Insight into the organization of proteins involved in the cell-substratum or cell–cell contact site at stress-fiber termini has also been made available by double immunoelectron microscopic labeling experiments on vertical ultrathin frozen sections (Chen and Singer, 1982). Differences in the distribution of α-actinin, vinculin, and fibronectin (LETS) at these sites showed a consistent distribution suggested by their specific functional roles in cell adhesion (see Section 3.4). However, preservation of ultrastructure in frozen sections is relatively poor, and methods for preservation of cytoskeletal details as well as antigenicity remains a major problem.

2.7. Problems in Definition: What Is a Stress Fiber?

The term "tension striae" or stress fiber was originally ascribed to a single component of a system of cytoplasmic fibers that were readily identified in tissue culture cells by phase-contrast microscopy (Lewis and Lewis, 1924; Buckley and Porter, 1967). Other imaginative terms used were cytoplasmic "bars" in so-called "mosaic" cells (Rose and Cattoni, 1963). As electron microscopists prepared whole cell mounts and embedded and thin-sectioned stress fibers, the term microfilament bundles was utilized, reflecting the greater resolution of substructure (Buckley and Porter, 1967; Goldman and Knipe, 1973). Following the HMM binding studies, microfilament bundles were more accurately termed bundles of actin filaments or simply "actin bundles" (Ishikawa *et al.*, 1969). However, with the spectacular growth of information obtained from immunofluorescent microscopy on nonmuscle cells, the term stress fibers began to have new significance and were frequently termed actin "cables." As mentioned earlier, these fibers often show intermittant staining for α-actinin, tropomyosin, and myosin which sometimes appears to be alternatively arranged along the stress fiber, similar to the myofibril (Gordon, 1978; Sanger and Sanger, 1980; see Sections 2.3 and 3.3). Although they occasionally have a sarcomere-like organization, they do not appear to actively contract *in vivo* (see Section 2.5.1), hence, it would be too broad a generalization to call stress fibers myofibrils in nonmuscle cells.

The term stress fiber has become useful to denote a linear structure seen in the light microscope that has good correlation with microfilament bundles by electron microscopy, and often has other cytoplasmic contractile proteins besides actin by immunofluorescence methods. Nevertheless, studies on stress fibers in cultured cells has progressed to the point where clarification of terminology may be necessary. There are many different actin-containing structures in cultured cells as demonstrated by immunofluorescence microscopy. Many of these structures have parallel arrays of actin filaments when examined by EM. These include microvilli, microspikes or filopodia, stress fibers, polygonal nets, and contractile and trailing fibroblast tails (see Section 4). Other structures containing actin filament bundles have been reported that do not fall into the usual stress fiber description of sets of linear, parallel, or slightly convergent cytoplasmic fibrils. Arcs or circles that are concentric with the cell margin have been observed in spreading living cells by interference reflection and have been shown to contain curved microfilament bundles (Kaiho and Sato, 1978; Masayoshi and Sato, 1978; Heath, 1981; Soranno and Bell, 1982). Arcs stain with anti-actin and are postulated, on the basis of birefringence studies, to be compressional waves of the microfilament network (Soranno and Bell, 1982). Clearly, the localization of other contractile proteins in arcs needs to be investigated further before they are excluded or included in the stress fiber category.

Another problem with terminology is that actin bundles not visualized in the light microscope but having stress fiber-like characteristics have been reported. This includes the periodic antimyosin staining of the microfilament sheath (Zigmond *et al.*, 1979) in the upper cortex of fibroblasts, also reported as a "sheet-like arrangement" of actin filaments (Fallon and Nachmias, 1980). These structures are not seen by phase contrast presumably due to the uniform thickness of the sheet. However, the arrangement of the contractile proteins appears essentially identical to stress fibers. But as they are in the upper cortex of the cell, they presumably serve more of a structural role than one of adhesion (see Section 3.4). Nevertheless, it would be interesting to determine the distribution of vinculin relative to the sheath. Because of the sensitivity of immunofluorescent staining, stress fibers detected by direct antimyosin labeling may not show a corresponding phase dense stress fiber. The question then arises whether cytoplasmic fibrils cannot be termed stress fibers unless seen by phase contrast, or whether one's point of reference should now be shifted to immunofluorescent techniques. We favor the latter viewpoint as immunofluorescence microscopy has permitted a more sensitive and precise characterization of the stress fiber with respect to its protein composition. Thus, a stress fiber by definition must be a bundle of actin filaments in cells that occurs in sets of parallel or slightly convergent fibers which generally display to varying degrees intermittent myosin, α-actinin, or tropomyosin-staining images. We favor intermittent as opposed to periodic as many stress fibers within a set pattern may show continuous myosin staining or interuptions in the staining without a specific periodicity. Others display highly regular periodicities. Finally, not every stress fiber within a particular cell contains myosin, tropomyosin, or α-actinin, but the presence, absence, or proportion of these contractile-associated proteins will likely vary with the dynamic process of formation or disruption of stress fibers and reflect the specific function of the stress fiber (see Section 3).

Thus, a stress fiber is not simply a "microfilament bundle" or "actin cable" but may be a structure that is composed of a framework of parallel actin filaments held together with actin-binding proteins and contains other contractile-associated proteins. But what is its role in nonmuscle cells? As mentioned earlier, although they have the machinery to contract, they have not been observed to contract *in vivo* (see review in Burridge, 1981). The following section will discuss the various proposed roles for stress fibers.

3. Role of Stress Fibers

This section will not attempt to review all the literature on the subject but will concentrate on citing more recent evidence of theories supporting formation of stress fibers and their role in cell locomotion, cell spreading, cell adhesion, and generation of isometric contraction. The role of stress fibers in cell transformation and tumorigenicity in relation to fibronectin will be discussed. Finally, the stress fiber's potential contribution to cell surface compartmentalization and tissue morphogenesis will also be reviewed.

3.1. Formation, Maintenance, and Disruption of Stress Fibers

Although the mechanism of stress-fiber formation is largely unknown, there is likely a dynamic interaction among actin monomers, actin filaments, actin associated proteins, and other components. The kinetics of such an interaction is certain to be complex. First, Mg^{2+} and ATP is involved in the initial formation of actin filaments, and second, proteins such as tropomyosin and α-actinin will influence the stability of the actin filaments. Third, myosin oligmers or monomers, actin-binding proteins such as filamen (Geiger, 1979), and other actin-binding proteins (see review in Schliwa, 1981), will influence the formation and stability of lateral actin associations or "bundling." Fourth, local α-actinin and vinculin concentrations may influence nucleation and/or attachment of actin filaments near the cell membrane at focal substrate contacts. Finally, intrinsic or extrinsic forces distorting actin gels will have the effect of rapidly changing local actin filament concentrations (see below). It would be a mistake to classify the stress fiber as a static structure as it is only relatively stable when compared to the dynamics of most nonmuscle cellular motility phenomena (see review in Buckley, 1981).

A large variety of factors influencing the expression of stress fibers have been studied in cultured cells including cell adhesion, the effect of nucleotides, metabolic inhibitors, hormones, drugs, and other metabolically active agents. The role of cell adhesion, fibronectin, and transformation will be discussed in other sections (see Section 3.4).

The formation of stress fibers in spreading cells appears to utilize actin already present in the cytoplasm as inhibitors of protein synthesis do not prevent the formation of stress fibers (Goldman and Knipe, 1973). Maintenance of stress fibers also appears to be an energy-requiring process. Following certain metabolic inhibitors, there is loss of stress fibers without cell rounding (Bershadsky *et al.*, 1980). The mechanism of loss of stress fibers on induced cell rounding or during the normal cell cycle is unknown. There is evidence which suggests that the loss of stress fibers causes cell rounding and the loss of cell adhesions results in loss of stress fibers. Likewise, similar reasoning can be applied to stress fiber formation. Do stress fibers cause cell spreading and adhesion or does cell adhesion induce cell spreading and stress-fiber formation? The answers will likely not rest with one or the other being correct, as there is evidence to support both hypotheses. First, with respect to stress-fiber disruption, cytochalasin B treatment induces cell rounding (Norberg *et al.*, 1975; Weber *et al.*, 1976), and trypsin or plasmin induces cell rounding with concomitant, reversible loss of stress fibers (Pollack and Rifkin, 1975a,b). Likewise, with respect to the formation of stress fibers, 3′5′dibutyrl-cyclic AMP is reported to promote formation of stress fibers with an associated flattened morphology (Willingham and Pastan, 1975; Bloom and Lockwood, 1980), whereas increasing substrate adhesivity with addition of fibronectin also promotes stress-fiber formation (Willingham *et al.*, 1977).

Thus, it appears that both substrate adhesivity and intrinsic metabolic factors influence the formation of stress fibers. The sequence of events and the relative contribution of stress-fiber formation and cell-substrate attachment to cell shape changes needs further clarification (see Section 3.4).

Other agents reported to decrease stress-fiber distribution include epidermal growth factor (Schlessinger and Geiger, 1981), dimethyl sulfoxide, Ca^{2+} ionophore (Osborn and Weber, 1980), and a carbamate herbicide (Oliver *et al.*, 1978). Treatment reported to enhance stress-fiber formation include fibronectin (Willingham *et al.*, 1977), sodium butyrate and dexamethasone (Der *et al.*, 1981), and hormones. Hormones have a variety of influences on stress fibers. To name only a few, follicle-stimulating hormone (FSH) stimulates granulosa cell differentiation and loss of stress fibers (Albertini and Herman, 1983, this volume), whereas thyrotropin (TSH) stimulates formation in cultured thyroid cells (Westermark and Porter, 1982). Prostaglandin E_2 or parathyroid hormone (PTH) have been reported to cause a dramatic loss in stress-fiber patterns in cultured bone cells (Beertsen *et al.*, 1982). Much study will be required to determine the nature and site of action of these many components that influence stress fibers *in vitro*.

It is important perhaps to distinguish between at least two different kinds of stress fibers. The first type, much more numerous, appear in the basal cortex region of the cell (Goldman and Knipe, 1973) and are presumably involved in cell adhesion. The second type, fewer in number in the upper cell cortex (Buckley and Porter, 1967; Zigmond *et al.*, 1979; Osborn *et al.*, 1978), are often orthogonally oriented with respect to the first type. The role of the upper cortical stress fibers is less clear, but more of a structural role is suggested.

Formation of the basal stress-fiber system appears on initial cell spreading, often as a radial pattern (Lazarides, 1976b; Soranno and Bell, 1982). Subsequent generation of stress fibers may be caused by the production of tension in a randomly organized net of cytoplasmic microfilaments. Retraction fibers (Harris, 1973; Revel *et al.*, 1974) and "tails" (Chen, 1981), are also believed to be distortions of the cytoplasmic actin gel, and they develop birefringence as do stress fibers. Micromanipulation experiments (Fischer, 1946; Chen, 1979) suggest that stress fibers and tails are under considerable tension. Release or rupture of tails results in rapid loss of birefringence (Chen, 1981), whereas severance of stress fibers by laser microbeam experiments (Strahs and Berns, 1979) causes little or no retraction. Thus, with the formation of stress fibers comes greater stability than simple distortion of an elastic actin gel. The likely answer is that stress fibers are often anchored along multiple points either with the cytoskeletal elements or with the cell membrane, and severance of a stress fiber between such points would not produce a dramatic recoil or contraction.

In the formation of stress fibers, it is likely that actin-binding proteins (see review in Schliwa, 1981) such as filamin (Wang and Singer, 1977) cause lateral linkage of actin filaments. Such cross-linkage will likely occur where actin filaments are in close proximity, such as along lines of tension in the cytoplasm where the distorted actin "mesh" would thus stabilize into a more linear array. Such a cross-linkage of a stretched random net would give rise to the antiparallel orientation of stress fibers (Begg *et al.*, 1978; Sanger and Sanger, 1980). Incorporation of myosin, α-actinin, and tropomyosin would then further stabilized the structure or permit the intrinsic generation of tension to counteract or resist the original distorting force. The mechanism by which these other proteins become so organized is as speculative as theories on the formation of myofibrils.

It is easier to understand the mechanism of the formation of tension between migrating cytoplasm and focal adhesions (see Section 3.4) on the substrate than the formation of tension in the upper cortex, where the fibers are often orthogonally oriented relative to the substrate fibers.

One of the more interesting cytoskeletal phenomena in cultured cells is the development of polygonal networks whose phase-dark linear elements have the same dimension as stress fibers. These actin-containing polygonal or "geodesic" dome-like structures have been proposed as intermediate in the formation of stress fibers (Lazarides and Burridge, 1975; Lazarides, 1976a; Lazarides and Revel, 1979; Gordon and Bushnell, 1979). Immunofluorescent-staining data reveal that α-actinin is concentrated at the vertices or intersections of spokes of the polygonal networks (Lazarides and Burridge, 1975; Lazarides, 1976a) whereas tropomyosin is excluded from them. Myosin is also excluded from the vertices (Osborn *et al.*, 1978). Often, stress fibers are seen radiating out from the more lateral vertices into the cell margins demonstrating that the vertices may serve as a nucleation center for stress fibers. Ultrastructural studies reveal several sets of actin bundles converging on the vertices which have no recognizable substructure (Perry *et al.*, 1981; Heuser and Kirschner, 1980; Rathke *et al.*, 1979).

The polygonal network is often situated above the nucleus primarily in the upper cortex region as confirmed by stereo-immunofluorescent microscopy (Osborn *et al.*, 1978) and electron microscopy (Gordon and Bushnell, 1979; Perry *et al.*, 1981) suggesting more of a structural role than cell adhesion. Nevertheless, as mentioned above, stress fibers often extend from vertices into the cell margins where the entire geodesic dome-like structure may be anchored. Such transition zones between the polygonal network to linear stress fibers has led investigators to suggest that the networks are intermediate in the organization of contractile proteins into stress fibers during cell spreading (Lazarides and Burridge, 1975; Lazarides, 1976a; Gordon and Bushnell, 1979). However, although the polygonal networks have been seen in a wide variety of nonmuscle cells, their frequency in each cell line varies greatly, and no extensive correlation with frequency of stress fiber formation has been undertaken. Also, cultured cells often exhibit extensive stress-fiber patterns within 24–48 hr, whereas, a greater number of polygonal networks are often seen following several days of culturing (Ireland and Voon, 1981). Time-lapse recording studies have provided evidence that polygonal networks do not appear in migrating cells, and do not appear to be essential intermediates in stress-fiber formation (Ireland and Voon, 1981). Perhaps the most important aspect of the polygonal patterns is to recognize that they are dynamic structures (Ireland and Voon, 1981) and, thus, they may be intermediate structures in the formation of stress fibers during respreading of cells (Gordon and Bushnell, 1979), but they may not be essential for cell spreading. Clearly, more *in vivo* investigations are needed to determine the relationship among polygonal networks, stress fiber formation, and cell spreading.

3.2. Cell Locomotion

The relationship of stress fibers to cells that are spreading, actively migrating, or undergoing slow, random locomotion has been recently reviewed (Buckley, 1981). In this review it was emphasized that a distinction could be made in the number and size of stress fibers in actively migrating cells moving in one direction and cells exhibiting slower, "to and fro" maneuvers, including "zig-zag" or oscillating behavior. In migrating cells, few if any stress fibers are detected by phase contrast microscopy except in "tails" or posterior ends which appear to have tenacious adhesion sites at the substratum (Buckley and Porter, 1967; Chen, 1979, 1981). In contrast, the oscillating type of cell locomotion is frequently seen in cells that are more spread out on the substrate with extensive stress-fiber patterns. Early time-lapse studies revealed that the rate of locomotion of well-spread cells is slow (0.1–0.75 μm/min), similar to the rate of changes observed in stress-fiber patterns (Buckley and Porter, 1967).

Utilization of the interference reflection microscope enables visualization of focal and close contacts of the cell with the substratum. The focal contacts are frequently linear and have been shown to be immediately subjacent to the end of stress fibers (Izzard and Lochner, 1976; Heath and Dunn, 1978; Badley *et al.*, 1978). These linear focal contacts are presumably involved in adhesion (see Section 3.4) as they are the closest contacts the cell makes with the substratum and they are seen to resemble stress-fiber patterns in well-spread stationary cells. The anterior portion of locomoting cells form focal contacts which do not advance with the leading point of cells (Abercrombie *et al.*, 1976; Izzard and Lochner, 1976, 1980). Since the focal contacts and stress fibers do not glide over the substrate, shearing must take place between the substrate contacts and the cell body. Cytoplasmic streaming in amoeboid movement has long been postulated to be the result of relatively stationary ectoplasm organized into linear arrays of actin filaments, and a more dynamic endoplasm, composing in part of a loose network of actin filaments. Free fibrils, composed of actin bundles, in isolated *Nitella* cytoplasm, are capable of translocating in the direction of fiber orientation if a medium containing Mg^{2+}-ATP is perfused into the model system (Higashi-Fujime, 1980). Not surprisingly, with our knowledge of muscle motility, the shear forces are generated parallel to the direction of the actin bundles. It is possible that such shear forces develop over stress fibers, indeed, the direction of locomotion of leading edges of spreading cells are frequently parallel to the direction of stress fibers (Soranno and Bell, 1982). It would make sense that well-spread cells, stationary or in oscillating locomotatory patterns, often exhibit crossed or complex stress-fiber patterns. This permits opposition of the streaming forces with the cytoplasm with the resultant generation of tension along the stress fiber. In order to prevent rupture of a "tail" (Chen, 1979, 1981) or stress fibers composed of antiparallel actin filaments (Begg *et al.*, 1978), recruitment of α-actinin, myosin, and other proteins would take place. Hence, a "resistive" or intrinsic tension could then develop, leading to the fully-developed stress fiber with intermittant myosin, α-actinin, and tropomyosin distribution. Thus, a dynamic, "tug of war" over control of the cytoplasm would occur in well-spread cells and, if the forces are well balanced, would lead to a relatively stationary cell. Slight perturbations in this balance would lead to oscillations (Wohlfarth-Bottermann and Fleischer, 1976), and slow locomotion of the cell over the substrate as the stress fibers slowly changed (Harris, 1973). Thus, it appears that stress fibers are not involved as contractile elements in active, rapid migration, but actually are used to resist cell locomotion by their involvement in anchoring the cytoplasm to the substrate by development of isometric contraction (see Section 3.3).

Certainly, many observations are consistent with the viewpoint that stress fibers may resist active displacement of the cytoplasm over the substrate. First, the most highly motile cells contain cortical actin filaments in parallel arrays, such as macrophages (De Petris *et al.*, 1962; Reaven and Axline, 1973), physarum (Wohlfarth-Bottermann, 1964), and amoeba (Nachmias, 1964; Bowers and Korn, 1968; Wohlman and Allen, 1968; Taylor and Condeelis, 1979; Taylor, 1976), but do not contain stress fiber-staining structures. Similarly, fibroblasts which are rapidly migrating *in vitro* exhibit few or no stress fibers as demonstrated by phase contrast, interference reflection, or by immunofluorescence microscopy (Badley *et al.*, 1980; Couchman and Rees, 1979a,b; Lewis *et al.*, 1982; Soranno and Bell, 1982). Careful time-lapse recording studies correlating stress fibers with cell behavior, morphology, and immunofluorescent-staining patterns have revealed that cellular motility is in fact associated with lack of stress fibers and diffuse staining of anti-actin and antimyosin (Herman *et al.*, 1981; Lewis *et al.*, 1982). It is found that cells which exhibit a mixture of diffuse staining and stress fiber patterns are stationary or exhibit slow locomotion following slow shape changes.

During initial cell spreading, formation of a radial stress-fiber pattern often occurs (Soranno and Bell, 1982). As the cells are initially spreading, it appears that a 360° "tug of war" of the cytoplasm is taking place with a radial stress-fiber pattern developing. Once the cell is relatively flattened, one side of the cell develops a fan-shaped assembly of stress fibers and the cell begins to slowly locomote in the direction of the sum of the vectors of the "fan," suggesting the balance over the control of the cytoplasm has shifted towards one side of the cell. Often, stress fibers perpendicular to these initial fibers may develop, and depending on the orientation of the stress fibers, the cell may begin to follow a zig-zag track (Albrect-Buehler, 1977b), possibly representing a history of the balance of stress-fiber sets or systems over the control of anchoring the cytoplasm.

In summary, stress fibers appear to be absent in cells that are rapidly migrating, are found in spread cells that are undergoing oscillations or slow locomotion, and are frequent in well-spread cells. Thus, they may be considered to be elements which resist cell detachment and rapid locomotion, hence their role in cell adhesion appears more important.

3.3. The Generation of Isometric Contraction in Stress Fibers

Although there are no available direct evidence on the generation of isometric contraction in stress fibers, there is a growing body of indirect observations which are leading to such a conclusion. The potential for stress-fiber contraction has recently been reviewed (Burridge, 1981), with the conclusion that stress fibers have not been observed to actively contract *in vivo*. However, the similarities of the organization of contractile proteins in stress fibers to the myofibrils has already been pointed out (see Section 2.3).

The potential of isolated stress fibers to contract *in vitro* has been demonstrated by Isenberg *et al.* (1976) who used microlaser dissection techniques to separate stress fibers from the surrounding cytoplasm and then stimulated them to contract using Mg^{2+} and ATP. The potential for a sliding filament mechanism in stress-fiber contraction is supported by fluorescein α-actinin

(FL α-actinin) microinjection studies (Kreis and Birchmeier, 1980). Following cell permeabilization and perfusion with Mg^{2+} or ATP, the nonfluorescent bands along the stress fiber can be stimulated to shorten. It must be emphasized that shortening has not yet been observed in living, microinjected cells. It appears that although stress fibers have the necessary proteins to enable them to actively contract, their role in living culture cells appears to be more of a resisting isometric contraction during the development of tension following adherence of the cell to the substratum (Willingham *et al.*, 1977; Burridge, 1981).

3.4. Cell Adhesion

Stress fibers are often located in the basal cortex in regions of cell substrate contacts as revealed by electron microscopic sections (Buckley and Porter, 1967; Spooner *et al.*, 1971; Goldman, 1971; Goldman and Knipe, 1973; Goldman *et al.*, 1979; Perdue, 1973). Prior to cell attachment, bundles of microfilaments are not found in suspended cells and the process of attachment leads to formation of microspikes and then appearance of microfilament bundles (Bragina *et al.*, 1976). Intermittent cell contacts with the substrate often appeared electron dense and were termed plaques (Abercrombie *et al.*, 1971), or "attachment sites" (Revel and Wolken, 1973) and frequently had associated microfilaments that appeared to emanate from the specializations (Abercrombie *et al.*, 1971; Brunk *et al.*, 1976; Goldman *et al.*, 1976).

The interference reflection microscopy technique on living cells produces images representing the distance between the cell and the substratum in leading lamellae, and in regions near the edge or cell margins of well-spread, moving or stationary cells (Abercrombie and Dunn, 1975; Izzard and Lochner, 1976; Couchman and Rees, 1979a,b; Curtis, 1964). It was noted that the focal contacts were coincident with and had the same dimensions as the ends of cytoplasmic fibers. Stereo high voltage electron micrographs confirmed that the ends of microfilament bundles were coincident with the punctate or linear focal contacts seen by interference reflection microscopy (Heath and Dunn, 1978). The pattern of linear focal contacts in living cells corresponds closely with α-actinin distribution and they are subjacent to the actin staining in stress fibers (Wehland *et al.*, 1979). The development of new focal contacts by spreading cells reveals that a succession of the new focal contacts are formed ahead of previous ones which do not advance (Izzard and Lochner, 1980). Finally, comparison of differential interference images with the development of focal contacts suggests that linear fibrous elements appear before focal contacts arise, followed by development of stress fibers (Izzard and Lochner, 1980).

In the last several years, characterization of adhesion plaques or focal contacts by identification of its components by biochemical and immunocytochemical means has provided increasing information on the role of stress fibers in cell adhesion.

Recent studies have investigated the relationships of fibronectin (see review in Hynes, 1981), α-actinin (Lazarides and Burridge, 1975; Schollmeyer *et al.*, 1976), and a 130-kd protein called vinculin (Geiger, 1979; Geiger *et al.*, 1980; Burridge and Feramisco, 1980) to adhesion sites and stress fibers.

Fibronectin in serum or secreted fibronectin by normal fibroblasts appears to mediate cell adhesion in cultured cells (Grinnel and Feld, 1979; Klebe, 1974; Pearlstein, 1976). The addition of the glycoprotein fibronectin (LETS) to transformed cells in culture increases their adhesion to the substrate as well as increases the number of stress fibers (Ali *et al.*, 1977; Willingham *et al.*, 1977). Double label immunofluorescent evidence suggests that the initial fibronectin distribution on the cell surface of cultured cells is oriented with the distribution of stress fibers (Mautner and Hynes, 1977; Hynes and Destree, 1978; Heggenness *et al.*, 1978). The extracellular fibronectin is reported to form a colinear transmembrane association with submembranous microfilaments, termed the fibronexus (Singer, 1979a), which is associated with electron dense plaques or focal contacts.

Recently, a 130-kd protein isolated from chicken gizzard and named vinculin has been shown to localize at the termini or microfilament bundles at focal contact sites (Geiger, 1979; Geiger *et al.*, 1980). Microinjection of fluorescently labeled vinculin into living cells reveals its distribution on dorsal termini of stress fibers in association with fibronectin and as elongated focal patches on the ventral stress-fiber adhesion sites (Burridge and Feramisco, 1980). Using a double-label immunofluorescence technique, an association among extracellular fibronectin, fibers of vinculin and linear focal contacts has been reported. On the other hand, investigators have failed to find coincidence of fibronectin and focal contacts (Avnur and Geiger, 1981a,b; Badley *et al.*, 1980; Birchmeier *et al.*, 1980; Chen and Singer, 1982; Fox *et al.*, 1980). However, association of fibronectin and vinculin with focal contacts appears to be dependent on low serum concentrations (0.3% FBS), and lack of fibronectin focal contact correlation is postulated to be due to rapid flux of focal contacts induced by high serum levels (Singer, 1982). Still more recent evidence using double immunoelectron microscopic labeling of fibronectin, vinculin, or α-actinin concluded that fibronectin was excluded from focal adhesion sites, but was often localized immediately adjacent to focal adhesion sites (close contacts) and in extracellular matrix sites, which were subjacent to intracellular labeling for α-actinin and vinculin or α-actinin alone (Chen and Singer, 1982). Thus, the relationship among actin filaments, α-actinin, vinculin, or extracellular matrix molecules is increasingly complex, and the different compositions of transmembrane assembly of these elements will likely show functional differences in cell adhesion properties.

3.5. Cell Transformation

In the last decade there has been a tremendous amount of research on stress-fiber distribution in relation to cell transformation and tumorigenicity. Initial electron-microscopic and phase-contrast microscopic studies suggested that virally-transformed fibroblasts exhibited a reduction in number and thickness of stress fibers and microfilament-membrane associations (Ash *et al.*, 1976; McNutt *et al.*, 1973; Goldman and Knipe, 1973; Goldman *et al.*, 1974, 1976; Gruenskin *et al.*, 1975; Robbins, 1975; Wang and Goldberg, 1978; Wickus *et al.*, 1975); myosin was reported to decrease by 50% (Ostlund *et al.*, 1974). Introduction of immunofluorescent staining techniques has permitted visualization of the entire stress-fiber system of many cells, thus permitting greater sampling. Loss of stress-fiber staining in cells after transformation by DNA (SV40; Osborn and Weber 1975; Pollack *et al.*, 1975) or RNA (Rous sarcoma virus; Ash *et al.*, 1976; Edelman and Yahara, 1975) viruses exhibit temperature-sensitive re-expression of stress fibers. The decrease or loss of stress fibers in transformed cells was concomitant with a decrease in fibronectin distribution (see Section 3.4; Mautner and Hynes, 1977; Hynes and Destree, 1978). It was noted that in general, virally transformed cells are more rounded in cell shape, presumably due to decreased cell adhesiveness. As discussed in previous sections, it is difficult to determine whether decreased cell adhesiveness is secondary to reduced stress fiber formation or that fewer stress fibers lead to diminished adhesiveness. Some evidence suggests that decreased adhesiveness gives rise to reduced stress fibers as adding fibronectin to transformed cells enhances adhesion and development of stress fibers (Ali *et al.*, 1977; Willingham *et al.*, 1977).

A wide variety of transformed cells have been examined for correlative stress-fiber and fibronectin studies. Temperature-sensitive virally transformed cells, when grown at nonpermissive temperatures, reveal stress fibers by anti-actin, anti α-actinin, antimyosin, and antitropomyosin staining. Switching to permissive temperatures results in decrease in stress-fiber pattern and loss of fibronectin (Boschek *et al.*, 1981). Likewise, hybrids between malignant and nonmalignant cells exhibit decreased stress fibers (Watt *et al.*, 1978) and decreased fibronectin (Marshall *et al.*, 1978). Similar observations have been reported in chemical induced transformation (Rifkin *et al.*, 1979; Leavitt *et al.*, 1982), spontaneously transformed cells (Tucker *et al.*, 1978), and microinjection of cytoplasm of RSV (Rous sarcoma virus) transformed cells into normal cells (McClain *et al.*, 1978).

Cells from pre-existing tumors have been obtained and their stress-fiber pattern and fibronectin pattern may be compared and then correlated with tumorigenicity. The rat mammary tumor cell, rama 25, develops cell surface fibronectin when it is stimulated to differentiate *in vitro,* while stress-fiber bundles increase and tumorigenicity decreases (Wharburton *et al.*, 1981). Teratocarcinoma cells in culture do not exhibit extensive stress-fiber patterns, yet when they are stimulated to differentiate under certain culture conditions, their tumorigenicity decreases as they begin to exhibit anchorage-dependent growth and formation of stress fibers (Paulin *et al.*, 1978).

Fibroblast cells, obtained by skin biopsy and grown in culture have been reported to exhibit fewer stress fibers in patients with high risk for colon cancer than in fibroblasts from related normal subjects (Kopelovich *et al.*, 1977, 1980). Studies correlating actin patterns with tumorigenicity in nude mice (Shih *et al.*, 1975; Freedman and Shin, 1974) have led investigators to propose that actin organization be used as an *in vitro* assay for tumorigenicity (Pollack *et al.,* 1982).

The mechanisms by which cell transformation induces loss or reduction of stress fibers and fibronectin is unknown, although important clues are developing rapidly. In virally transformed cells, the viral src gene product ($pp60^{src}$) is a 60-kd phosphoprotein with tyrosine kinase activity which may phosphorylate a cytoskeletal protein at adhesion sites (David-Pfeuty and Singer, 1980; Rohrschneider, 1980; Shriver and Rohrschneider, 1981), and evidence is suggestive that the cytoskeletal protein may be vinculin (see review in Hynes, 1982). Terminal amino acid alterations in $pp60^{src}$ in a recovered avian sarcoma virus (ASV) lead to decreased membrane association and decreased *in vivo* tumorigenicity. Other investigators working with chemically transformed fibroblasts containing a mutant β-actin gene demonstrated decreased stress fibers and fibronectin. A new mutant, (X) β-actin, was isolated which exhibited further decrease in stress fibers and fibronectin with a concomitant incremental increase in tumorigenicity (Leavitt *et al.,* 1982). Whatever the mechanisms of tumorigenicity, these studies reveal that multiple sites may be involved in the development of malignant cells.

Caution must be exercised in the interpretation of changes in cytoskeletal protein distribution in transformed cells. Initial reports of altered tubulin distribution in transformed cells was likely due to superimposition of microtubules in less well-spread cells. Only a few years later did studies reveal no significant differences (Brinkley *et al.,* 1980; De Mey *et al.,* 1978). Although the case for decreased stress fibers and fibronectin in transformed vs. normal cells is substantial, there is equally substantial and growing information which suggests the predictive value of such correlations may be decreasing as exceptions to the principle are being found. First, there are published accounts of transformed cells with known tumorigenicity which have normal stress fibers (Karsenti, 1978). In addition, cell hybridization studies have developed tumorigenic hybrids with high amounts of fibronectin expressed and several transformed, nontumorigenic hybrids (Der and Stanbridge, 1978). Other tumorigenic hybrids exhibit reduced stress-fiber patterns but can be stimulated to develop them with agents such as sodium butyrate (Der *et al.,* 1981). There are also reports that transformed cells with decreased fibronectin and stress fibers can be stimulated with retinoic acid to undergo anchorage-dependent growth without the return of cytoplasmic fibrils or fibronectin (Mukherjee *et al.,* 1982). Transformed and normal epithelial cell lines of salivary gland, bladder (Wigley and Summerhayes, 1979), and mammary gland (Yang *et al.,* 1980; Brinkley *et al.,* 1980) exhibited no correlation among stress fibers, fibronectin, and tumorigenicity. Similar inconsistencies were found in rat liver cultures (Bannikov *et al.,* 1982) and studies of anchorage independence and tumorigenicity in established cell lines (Celis *et al.,* 1978). Finally, in a study of eight hybrid clones, no relationship between stress-fiber pattern and tumorigenicity was found (Celis *et al.,* 1979).

Lack of correlation of cell transformation and stress-fiber pattern may be present in only certain cell types, such as epithelial cells. Nevertheless, there is still a considerable controversy to resolve before stress fibers may be reliably used for assessment of tumorigenicity.

3.6. Organization of Cell Surface Domains

That stress fibers are associated with specific domains of the cell membrane for cell adhesion is well established and discussed in previous sections. In addition to their codistribution with fibronectin or their termination at focal contacts with vinculin, stress fibers have been reported to show a linear correlation with other membrane-associated proteins including various receptors, known cell surface antigens and proteins associated with pinocytosis such as clathrin.

The cell surface lectin concanavalin A, has been reported to induce alignment of concanavalin A surface receptors with stress fibers (Ash and Singer, 1976). Likewise, three membrane proteins, β_2 macroglobulin (part of the HLA complex), an aminopeptidase, and Na^+, K^+ ATPase are normally diffusely distributed except for linear "gaps" in their distribution which appear to correlate with stress fibers. When these integral membrane proteins are cross-linked by specific bivalent antibodies, fluorescent patch formation is aligned with stress fibers (Ash *et al.*, 1977). Immunoelectron microscopic labeling of antibody-induced clustering and endocytosis of HLA antigens reveals a transmembrane association of patches with microfilament bundles (Huet *et al.*, 1980). Thus, the domain of antigen-antibody internalization appears to correlate with the orientation of stress fibers in well-spread cells. Indeed, freeze-etch studies confirm that aggregates of cortical pinocytotic vesicles are linearly ordered along longitudinal ridges and furrows where submembranous microfilament bundles are seen by thin sections. Transformation with reported reduction in the microfilament bundles, results in loss of longitudinal ridges and randomly scattered pinocytotic vesicle aggregates (Singer, 1979b).

Other evidence for internalization sites being related to stress fibers has come from immunocytochemical localization of coated pits and vesicles in fibroblasts which reveal linear patterns of fluorescent dots on well-spread cells. These dots correlate with immunoperoxidase labeling of phase-contrast microscopy of stress-fiber patterns (Anderson *et al.*, 1978). In addition, immunofluorescent visualization of clathrin suggests that the linear arrays of fluorescent dots appears to be in the limited cytoplasmic spaces between stress fibers (Kartenbeck *et al.*, 1981).

Other receptors have been reported to show colinear arrangements with stress-fiber pattern. Beaded, linear acetylcholine receptor clusters are coincident with vinculin and focal contact sites (Bloch and Geiger, 1980). Receptors for ricin (a carbohydrate) have also been localized along part of the length of the associated stress fiber and treatment with cytochalasin or trypsin results in loss of the "ricin lines" (Badley *et al.*, 1980). Reversible alterations in insulin binding capacity has been reported following cytochalasin B-induced loss of stress fibers (increase in insulin binding) (Raizida *et al.*, 1981). Although the relationships of the distribution of insulin receptors and stress fibers has not been determined, an interesting hypothesis may implicate alteration in endocytosis and hence receptor turnover.

Studies correlating membrane receptors with stress-fiber distribution should ideally be done with living cells. As an example, immunofluorescent staining for cholera toxin receptor ganglioside GM distribution is uniform in living cells, whereas after fixation and Triton X-100 extraction of cells, fluorescent dots aligned with stress fibers. Thus, caution must be taken when interpreting the interaction of stress fibers and membrane protein receptors following Triton X-100 extraction (Strenli *et al.*, 1981).

Despite the care needed while interpreting the codistribution of stress fibers and cell surface proteins, it appears that stress fibers may demarcate specific domains on the cell surface by at least two general mechanisms. One, by specific transmembrane linkage, which may be implicated in the distribution of fibronectin, adhesion sites and certain receptors. Second, by steric considerations which may exclude certain cell surface receptors or sites of membrane internalization from areas of the membrane associated stress fibers.

3.7. Morphogenesis

The previous sections have provided evidence that stress fibers are capable of organizing the cell surface into specific domains, including cell adhesion sites and coordinating certain extracellular matrix proteins. This capability, coupled with evidence on their potential cytoskeletal and structural role in regulating cell activity and cell shape, provides a basis for speculation on the role of stress fibers in the morphogenesis of cells and tissues. Although few investigators would quarrel with the role microfilament bundles play in development of microvilli (Overton and Shoup, 1964), contractile rings (Schroeder, 1973), and circumferential apical bands, it may be questioned whether stress fibers participate in development or even whether they exist in cells in tissues (see Section 4). Thus, the notion that stress fibers contribute to development is conjectural and based on *in vitro* observations.

Perhaps the most dramatic potential for the stress fiber in morphogenesis is the relationship of the stress-fiber pattern to cell shape. The development of anisometry on cell spreading is clearly related to the intracellular pattern of stress fibers. Bipolar cells have longitudinal fibers whereas mutlipolar cells have converging stress fibers at each pole. Membrane ruffling is rapid and transitory whereas stress fibers are relatively stable structures which appear to maintain the cell shape (Buckley and Porter, 1967). That stress fibers appear to influence the path of slow locomotion of daughter cells, and that the stress-fiber patterns are mirror images of one another (Albrect-Buehler, 1977a,b) has interesting implications for development.

One of the more important developmental events in histogenesis is the elaboration of the extracellular matrix. Stress fibers have been implicated in the orientation of matrices of fibronectin *in vitro* (Mautner and Hynes, 1977;

Hynes and destree, 1978; see review in Hynes, 1982). Thus, the coordination of fibronectin, collagen, and glycosaminoglycans (Hedman *et al.*, 1982; see review in Ruoslahti and Engvall, 1980) may be influenced by orientation of stress fibers in fibroblasts.

The most perplexing problem, however, is whether stress fibers appear in normal fibroblasts *in situ*. Indeed, fibroblasts exhibiting stress fiber patterns *in vitro*, when placed in hydrated collagen gels, do not develop stress-fiber staining patterns and electron micrographs of those fibroblasts have not revealed stress fibers (Tomasek *et al.*, 1982). The following section will review the evidence for stress fibers in cells *in situ*.

4. Do Stress Fibers Exist in Cells in Situ?

The implications of the role of the stress fiber in adhesitivity, tumorigenicity, and morphogenesis are provocative. Given the extraordinary amount of research on stress fibers in the last 10 years, it is essential to reconfirm *in vitro* observations with *in situ* models before conclusions on the significance of the stress-fiber studies can be made.

4.1. Microfilament Bundles in Cells in Situ

Longitudinally oriented fine filaments were observed in nonmuscle cells *in situ* in early electron-microscopic studies on the core filaments in the microvilli of intestinal absorptive cells (Palay and Karlin, 1959; McNabb and Sandborn, 1964). The elegant studies by Mooseker and Tilney (1975) on isolated brush borders determined by HMM binding that these microfilament bundles were composed of actin. A few years later it was found that tropomyosin, α-actinin, myosin, and the actin-binding protein, filamen, were localized only in the terminal web region of isolated intestinal absorptive cells and not along the microfilament bundles of the microvilli (Bretscher and Weber, 1978). In contrast to the composition of the microvillar core, all these proteins are associated with stress fibers in tissue culture cells. It is thus clear that microvillar actin bundles cannot be considered as a stress fiber *in situ*.

Besides the microvillar-related structures such as stereocilium (Flock and Cheung, 1977; DeRosier *et al.*, 1980; Tilney *et al.*, 1980) and filopodia, there are many published examples where microfilament bundles are found in nonmuscle cells *in situ*. To name only a few, they have been shown in Sertoli cells (Franke *et al.*, 1978), in iridophores of reptiles (Rohrlich and Porter, 1972), in cone cells of the teleost retina (Burnside, 1978), in cuticle-forming cells of insects (Overton, 1966), and in the plant, *Nitella* (Nagai and Rebhun, 1966). However, none of these microfilament bundles have been shown to have the same composition and arrangement of contractile-associated proteins as stress fibers in cultured cells.

Microfilament bundles are associated with the Zonula adherens in intestinal epithelial cells by electron-microscopy (Hull and Staehelin, 1979) and appear to be circumferentially localized by immunofluorescent staining (Bretscher and Weber, 1978). Retinal pigmented epithelium *in situ* also have circumferential microfilament bundles (CMBs) at the zonula adherens region (Owaribe and Masuda, 1982). Glycerinated epithelial sheets and isolated CMBs undergo Mg^{2+}-ATP dependent contraction and electrophoretic gels show actin and myosin as major components of the CMBs (Owaribe *et al.*, 1981; Owaribe and Masuda, 1982). Epithelial cells *in vitro* demonstrate stress-fiber patterns, but when grown to cofluency, appear to form CMBs and occluding junctions. Anti-actin staining of these epithelial sheets reveals a continuous outline of the polygonal, confluent cells (Meza *et al.*, 1980). Many immunofluorescent studies on the localization of contractile proteins in tissues reveal peripheral staining (Fagraeus *et al.*, 1973; Biberfeld *et al.*, 1974; Trencheu *et al.*, 1974; Toh *et al.*, 1979) and corneal epithelial cells also reveal circumferential actin staining (Gordon *et al.*, 1982). In other studies, isolated cell–cell adhesion junctions reveal continuous vinculin staining, continuous actin and punctate α-actinin staining (Shriver and Rohrschneider, 1981). Although CMBs have not been observed contracting *in vivo*, they contract when isolated *in vitro* following Mg^{2+} and ATP perfusion. These studies suggest that CMBs are close relatives to the stress fiber and at least CMBs are known to exist in cells *in situ*. Further studies showing the distribution of myosin, α-actinin, and tropomyosin are needed to determine how similar CMBs are to stress fibers. Nevertheless, differences are evident between CMBs and stress fibers as the stress fiber is involved more with cell–substrate adhesion than cell–cell adhesion. In addition, stress fibers can be found in the upper cortical regions, apparently not related to adhesion. Finally, the overall patterns are different as the stress fiber is usually linear and occurs in sets of parallel or convergent fibers whereas CMBs demarcate the periphery of cells.

4.2. Immunofluorescent Localization of Cytoplasmic Contractile Proteins in Tissues

The distribution of several contractility-related proteins has been visualized in a variety of tissues using cryostat sections or en bloc preparations and immunofluorescence microscopy (see review in Gröschel-Stewart, 1980). Actin has been shown to be localized in liver and thymus tissues (Fagraeus *et al.*, 1973; Toh *et al.*, 1979), thyroid (Biberfeld *et al.*, 1974), intestinal absorptive cells (Drenckhahn and Gröschel-Stewart, 1980), specific regions in central nervous system tissues (Toh *et al.*, 1976; Benzonana *et al.*, 1979), ocular tissue (Kibbelaar *et al.*, 1980; Gordon *et al.*, 1981), and testis (Franke *et al.*, 1978; Gröschel-Stewart and Unsicker, 1977). Myosin has been localized in ocular tissue (Drenckhahn and Gröschel-Stewart, 1977), brain tissues (Gröschel-Stewart *et al.*, 1977), glandular tissue (Drenckhahn *et al.*, 1977), vascular smooth muscle, endothelium of blood vessels, fibroblasts, myoepithelial cells of breast tissue (Becker and Nachman, 1973; Burkle *et al.*, 1979), intestinal epithelium (Drenckhahn *et al.*, 1980), ovaries (Amsterdam *et al.*, 1977), and testis (Gröschel-Stewart and Unsicker, 1977). α-Actinin is localized to the junctional complex of intestinal epithelial cells (Bretscher and Weber, 1978; Craig and Pardo, 1979). None of these cryostat sections of tissues or other preparations treated with fluorescent antibodies has resolved a filamentous distribution of actin, myosin, or other contractile-associated proteins resembling stress fibers.

In order to reliably identify stress fibers in cells *in situ,* the following criteria may be used to avoid confusion with circumferential microfilament bundles and other specialized microfilament bundle-containing structures. First, the presence of several linear fibers which stain for actin and have the dimensions of stress fibers in cells *in vitro.* Second, these fibers are assembled in convergent or parallel sets. Finally, some of them should demonstrate intermittent myosin and α-actinin staining, similar to patterns of stress fibers in cells *in vitro.* Using these criteria, the following sections will present observations on the identification of stress fibers in fibroblasts and endothelial cells *in situ.*

Until recently, examples of the classic parallel or convergent stress-fiber patterns commonly observed by immunocytochemical methods in cultured cells had not been observed in cells *in situ.* Our laboratory among others has recently identified stress fibers in fibroblasts in specific regions of the fish scale (Byers and Fujiwara, 1982) and in vascular endothelial cells (Rogers and Kalnins, 1981; Wong *et al.,* 1983; White *et al.,* 1983; see Sections 4.3 and 4.4).

4.3. Stress Fibers in Fibroblasts in Situ

In order to determine whether structures such as stress fibers actually exist in cells *in situ,* the goldfish scale can be used as a model for epifluorescent microscopy. On the surface of the collagenous fibrillary plate of the fish scale fibroblast cells, called scleroblasts, reside as a thin monolayer (Junqueira *et al.,* 1970; Onozato and Watabe, 1979; Zylberberg and Nicolas, 1982). Since the scleroblasts on the underside of the scale are not covered by epidermal cells, they can be directly visualized after indirect immunofluorescent staining for actin, α-actinin, and myosin (Byers and Fujiwara, 1982).

For orientation purposes, an illustration of the fish scale and a cube cut from the scale are shown in Fig. 2. Tiers of orthogonally arranged collagen fibers make up the plies of the fibrillary plate and a flat monolayer of scleroblasts reside on the under and upper side of this plate. These scleroblasts are responsible for the production of collagen fibrils of the fibrillary plate (Junquiera *et al.,* 1970; Onozato and Watabe, 1979).

Phase-contrast microscopy of the fish scales does not reveal stress-fiber patterns or scleroblast cell boundaries, whereas Nomarski differential interference contrast images reveal the uppermost collagen fiber of the fibrillary plate and cell nuclei and cell boundaries of the scleroblasts (Fig. 3). Cell boundaries can also be determined by staining for microtubule distribution (Byers *et al.,* 1980; Fig. 4).

Immunofluorescent localization for actin reveals linear fibers residing in sets of two, three, or four that are seen just outside and penetrating into the bright, epidermal staining region on the edge of the scale. The fibers have a

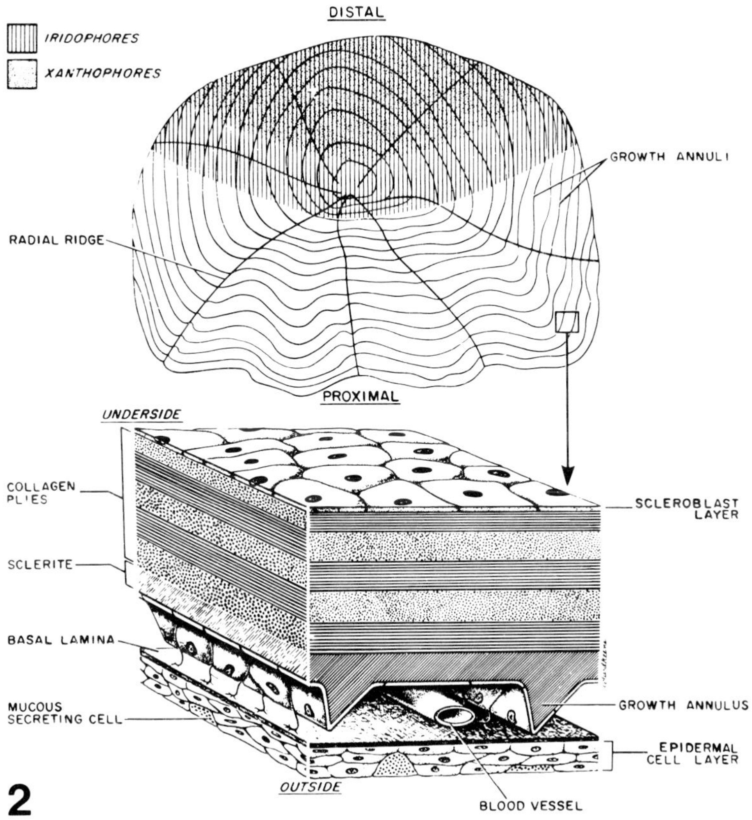

2

Figure 2. Illustration of the structure of the goldfish scale. Upper drawing of the whole scale reveals growth annuli, radial ridges, and the distribution of zanthrophores and iridophores. The lower drawing is an imaginary cube cut from the proximal region of the scale. The fibrillary plate is composed of orthogonally oriented plies of collagen and an outer layer called sclerite (partially mineralized collagen). Fibroblasts, called scleroblasts, reside as a monolayer on both sides of the fibrillary plate, but the underside of the scale is better visualized by epifluorescence microscopy as it is relatively flat. Printed with permission from the *Journal of Cell Biology*.

similar pattern and dimensions as stress fibers in cells *in vitro*. The fibers reside in sets of two, three, or four and are often convergent or nearly parallel (Fig. 5, arrowheads). These fibers are oriented approximately perpendicular to the edge of the scale and in contrast to microtubules, do not consistently align with the collagen fibers (Byers *et al.*, 1980). The majority of scleroblasts, residing in the flat, central regions of the scale do not exhibit stress-fiber patterns after immunofluorescent localization for actin (Fig. 6b). What is observed are single actin fibers that occasionally intersect into triads with forming angles of roughly 120° (Fig. 6b, arrows), and the anti α-actinin staining is intermittent along the length of these fibers. The orientation of these fibers does not coincide with the orientation of collagen fibers (Fig. 5a, arrows). By scanning multiple areas of the scale it becomes evident that the pattern

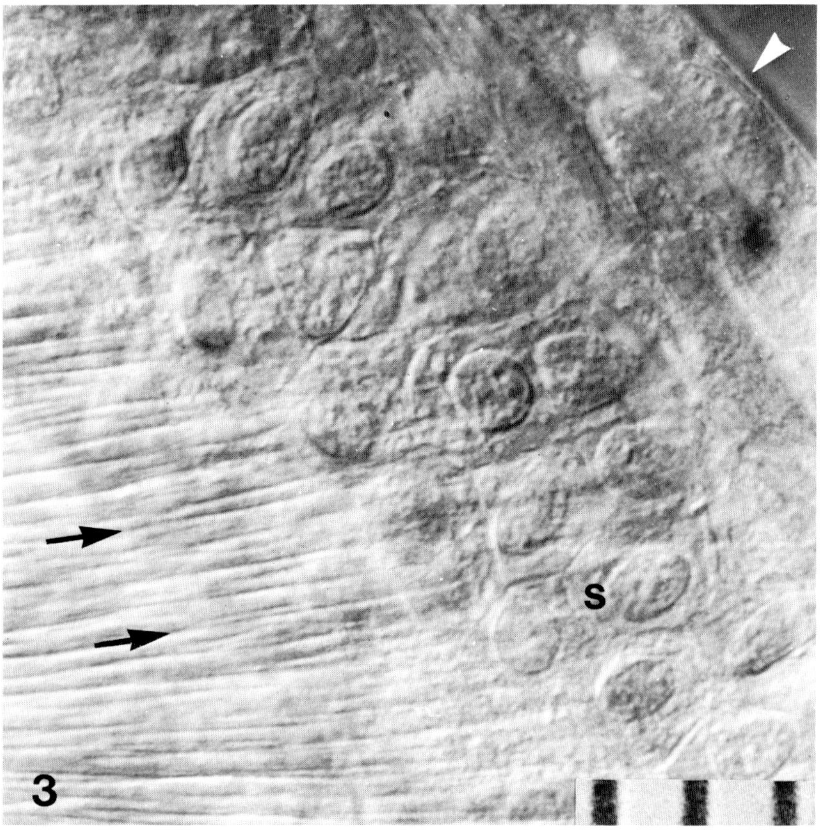

Figure 3. Nomarski differential interference contrast image of scleroblasts (s) near the edge of the scale (arrowhead). Nuclei and cell boundaries can be detected, but no stress-fiber pattern is evident. The uppermost ply of parallel collagen fibers is in intimate contact with the scleroblast layer (arrows). Scale: 1 division = 10 μm.

formed by these fibers is roughly hexagonal, with the same dimensions and patterns of the cell–cell contacts seen by Nomarski differential interference optics (Fig. 3). Presumably, these actin fibers represent microfilament bundles parallel to cell–cell contacts (see last section). Thus, it appears that the microfilament bundle may not be entirely circumferential, but peripheral segments at cell–cell contacts may also exist.

Anti α-actinin staining in scleroblasts near the edge of the scale is similar to the anti-actin staining pattern in the scleroblasts (Fig. 7). These fluorescent fibers, oriented nearly perpendicular to the edge of the scale, exhibit intermittant staining along their length.

Likewise, antimyosin staining in scleroblasts *in situ* reveals fluorescent fibers that are found near the edge of the scale and show no consistent correlation to the orthogonally arranged collagen fibers. Although most of the fibers show continuous fluorescence along their length, some fibers demonstrate intermittent staining with a periodicity of approximately 1–2 μm. The scleroblasts over the radiating ridges of the scale exhibit antimyosin staining fibers that are oriented perpendicular to this structure and cross over it (Fig. 8b).

The identification of stress fibers in fibroblasts *in situ* near the edges of the fishscale or over radiating ridges may reflect a greater need for cells to adhere to these regions. Greater shear forces may be expected to occur at these sides where a more solid medium (the fibrillary plate) moves within a softer medium (loose connective tissue). Quantitation of such shear will likely prove difficult, nevertheless, *in vitro* models studying the relation of shear to endothelial cells have noted an increase in stress fibers in cells exposed to extrinsic shear (White *et al.*, 1982). The next section reviews the identification of stress fibers in endothelial cells *in situ*.

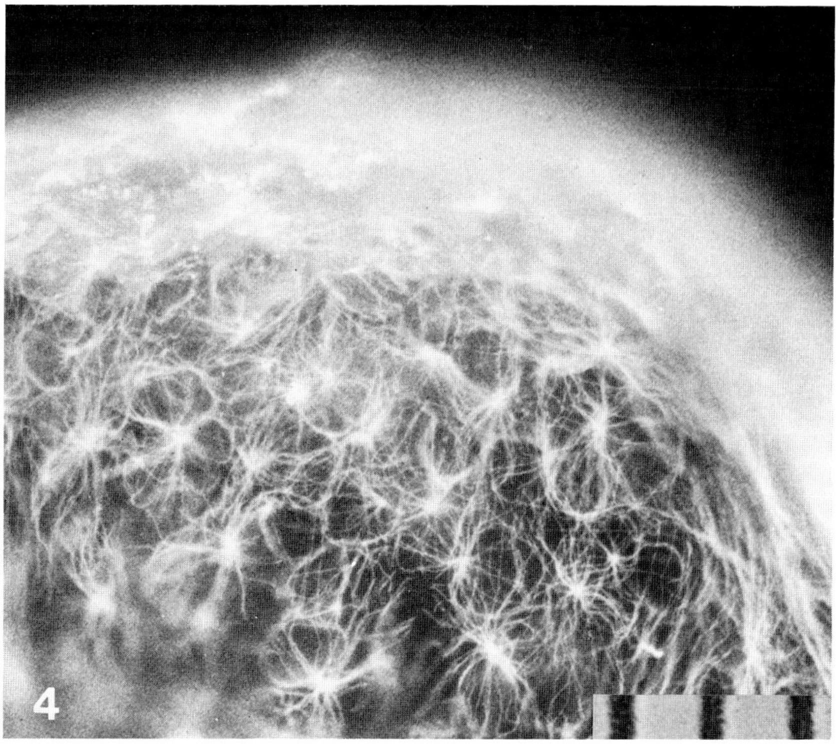

Figure 4. Epifluorescence image of scleroblasts on the edge of the goldfish scale following immunofluorescent staining for microtubules. The microtubules emanate from a central microtubule organizing center to the cell periphery. The general cell shape and cell boundaries can thus be determined. Scale: 1 division = 10 μm.

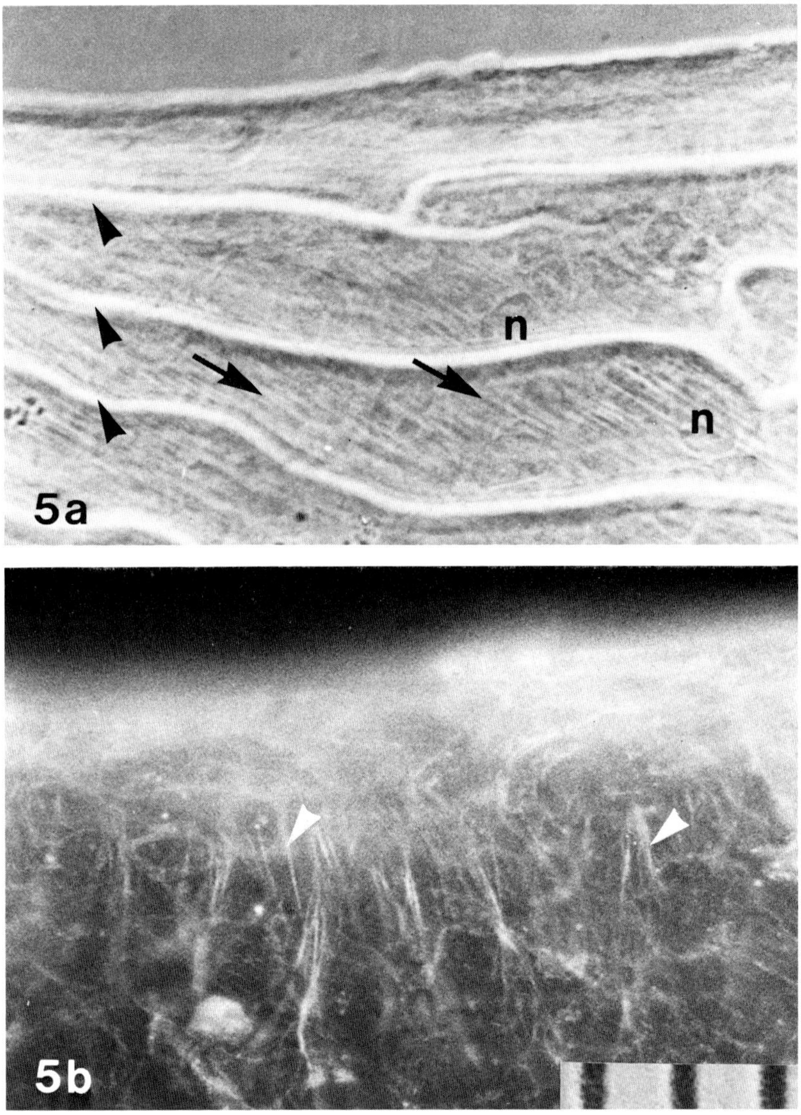

Figure 5. Phase contrast (a) and anti-actin stress fiber-staining pattern (b) in scleroblasts *in situ.* (a) Phase contrast images of goldfish scales reveals parallel phase-dark striations (arrows) representing collagen fibers and broad phase-bright growth annuli (arrowheads). Scleroblast nuclei are occasionally visualized (n). (b) Following paraformaldehyde fixation and Triton X-100 permeabilization procedures, scales can be stained sequentially with affinity purified anti-actin and fluorescein labeled goat anti-rabbit antibodies (Fl-GAR). A diffuse background fluorescence due to epidermal cells at a lower focal plane illuminates the entire scale and reduces the contrast of the fluorescent signal of the scleroblasts. Arrowheads indicate parallel and convergent actin-staining fibers. The epidermal cell layer which is slightly exposed on the underside of the edge of the scale, gives rise to bright staining. Scale: 1 division = 10 μm.

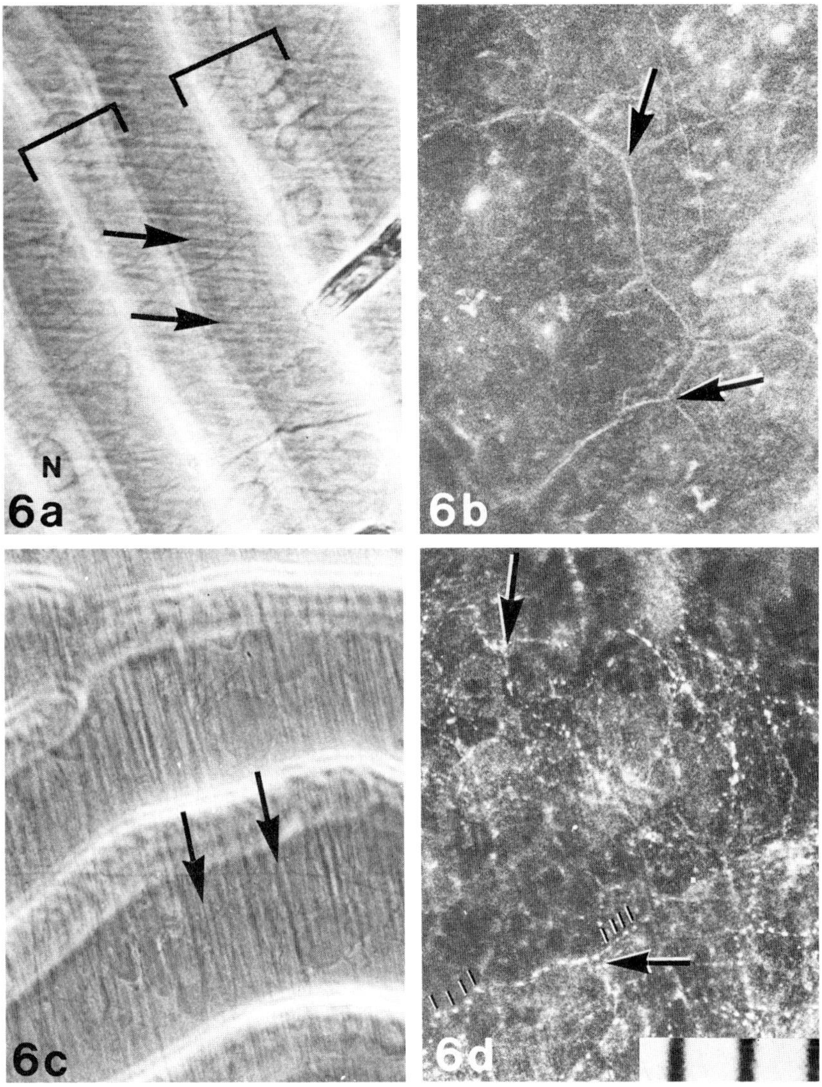

Figure 6. Distribution of actin and α-actinin in central regions of the scale. (a,b) Phase-contrast images of scleroblasts in the central regions of the fish scale reveal nuclei (n), out of focus broad growth annuli (brackets), and the parallel phase-dark striations (arrows) that represent the uppermost layer of collagen fibers of the orthogonally organized collagen of the fibrillary plate of the scale. (c) Anti-actin staining pattern reveals fibers that intersect at angles of roughly 120° (arrows). (d) Anti α-actinin staining pattern is similar to the actin pattern (arrows) except the fiber staining is periodic (bars). The dimensions of these fibers are similar to the outlines of scleroblasts, determined by Nomarski differential interference contrast. Scale: 1 division = 10 μm.

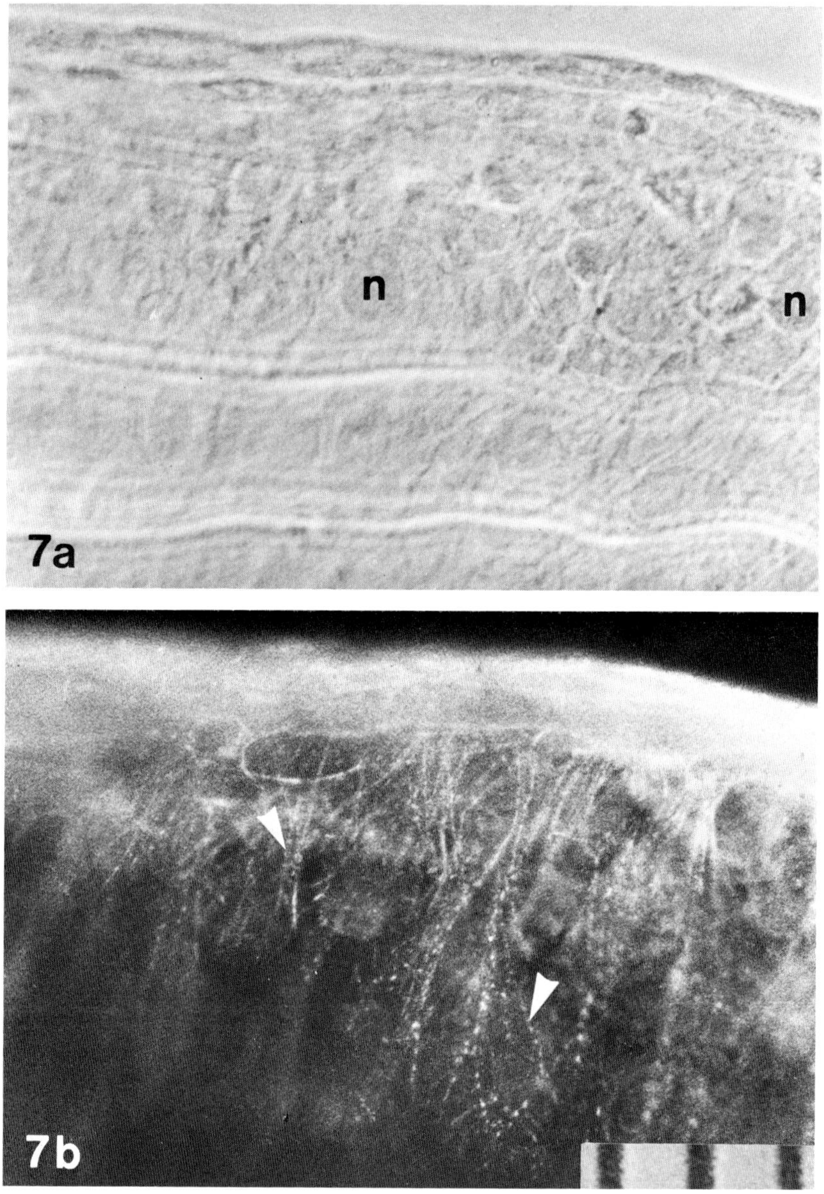

Figure 7. Phase-contrast image and stress-fiber staining patterns following α-actinin localization. (a) Striations and cell nuclei (n) are visible by phase-contrast microscopy. (b) Similar to the actin-staining patterns, α-actinin is localized in parallel as well as convergent fibers (arrowheads) and the periodicity of the fibers is approximately 1–2 μm. The point of convergence of the fibers is directed either toward or away from the edge of the scale. The orientation of these fibers is not consistently defined to the direction of the growth annuli nor to the phase-dark striations due to collagen fibers, but are usually perpendicular to the edge of the scale. Scale: 1 division = 10 μm.

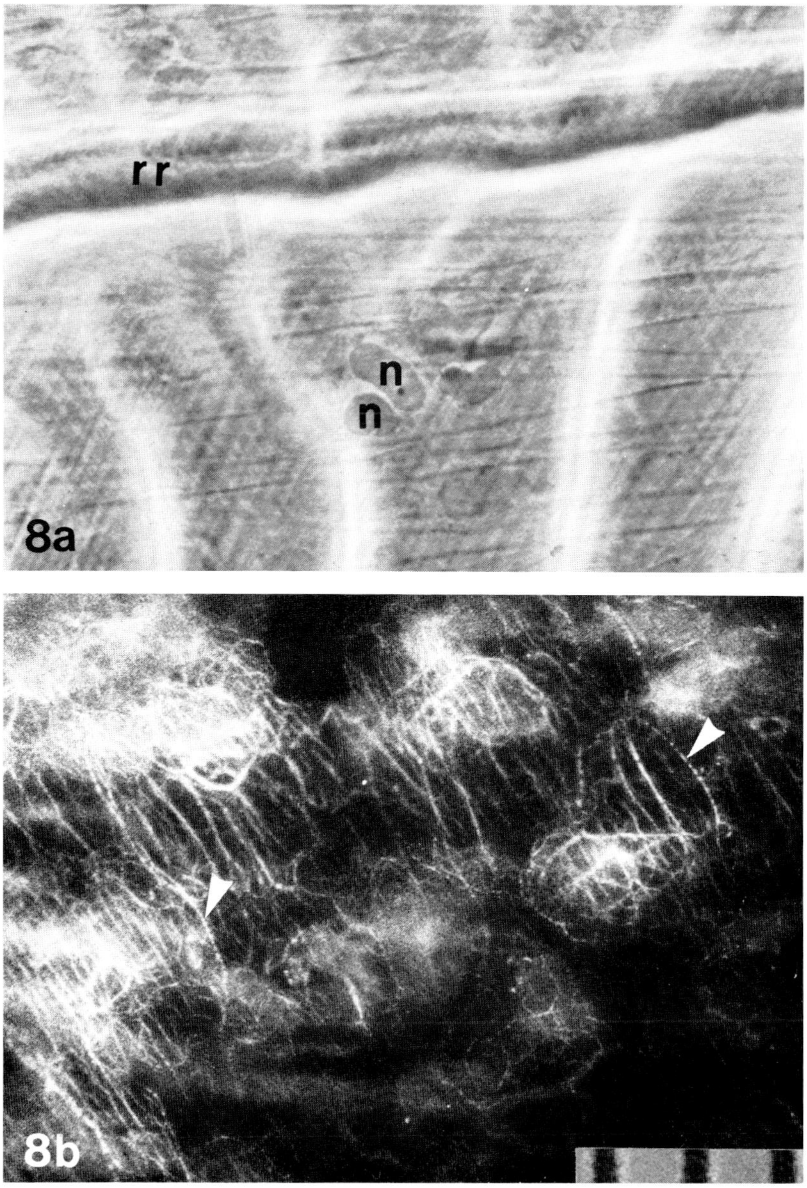

Figure 8. (a) Phase-contrast image of radial ridge (rr) region of fish scale. Cell nuclei (n). (b) Stress-fiber pattern following antimyosin staining. Nearly all stress fibers are oriented perpendicular to the radial ridge and intermittant staining is occasionally observed (arrowheads). Scale: 1 division = 10 μm.

4.4. Stress Fibers in Endothelial Cells in Situ

Vascular endothelial cells in blood vessels of normotensive animals demonstrate microfilament bundles (DeBruyn and Cho, 1974; see review in Hammersen, 1980), and periodic densities along the microfilament bundles, similar to the appearance of stress fibers in cultured cells, have also been reported (Bensch *et al.*, 1964; Röhlich and Olah, 1967). Other investigators report that cross-striated microfilament bundles are only found in endothelium of cerebral arteries (Giacomelli *et al.*, 1970) and aorta of hypertensive rats. *En face* preparations of endothelial cells of the aortic arch of hypertensive rats reveals prominent anti-actin staining at the periphery of most of the cells, whereas there is little staining of the cell peripheries of cells of normotensive rats (Gabbiani *et al.*, 1975).

These peripheral staining patterns have recently been included in discussions of stress fibers (Burridge, 1981), however, the aortic endothelial immunofluorescent pattern and phase images are more consistent with the recently characterized circumferential microfilament bundles (CMBs; see section 4.1). Indeed, ultrastructural studies support the notion that prominent endothelial microfilament bundles are longitudinally oriented at the level of the zonulae occludentes at the luminal border of the cell (Giacomelli *et al.*, 1970). Thus, the endothelial staining patterns reveal more a continuous cell–cell adhesion pattern (CMBs) than cell–substrate adhesion patterns (stress fibers). Recent observations on transepithelial permeability changes and CMB contraction studies (Meza *et al.*, 1980; Owaribe and Masuda, 1982, respectively) may have more potential than stress fibers for explaining the pathophysiologic mechanisms of increased vascular permeability or response to increased hydrostatic pressures.

Until recently, immunofluorescent studies have made use of en face preparations of endothelial sheets which have been stripped from the aorta (Gabbiani *et al.*, 1975; Wong *et al.*, 1983). Although these preparations reveal prominent circumferential actin staining of endothelial cells as in many other epithelial tissues, parallel cytoplasmic actin fibers resembling stress fibers can be detected (Wong *et al.*, 1983).

This laboratory has examined endothelial cells of rats and mice which were fixed by perfusion *in vivo* and stained and examined as whole mount preparations for epifluorescence microscopy (White *et al.*, 1982, 1983). Such methods are more likely to preserve cell morphology than earlier methods of stripping of the endothelium from the aorta before staining (Gabbiani *et al.*, 1975; Wong *et al.*, 1983).

Endothelial cells of the normal thoracic aorta of the mouse cannot be seen by phase-contrast microscopy in whole mounts. Indirect immunofluorescent staining for actin reveals prominent cortical or peripheral staining which clearly delineates the cell shape which is elongated along the direction of blood flow. In addition, fluorescent, linear fibers are readily detected in the cytoplasm of endothelial cells. These parallel fibers are oriented along the long axis of the cell and the direction of blood flow (Fig. 9). Most of the fibers

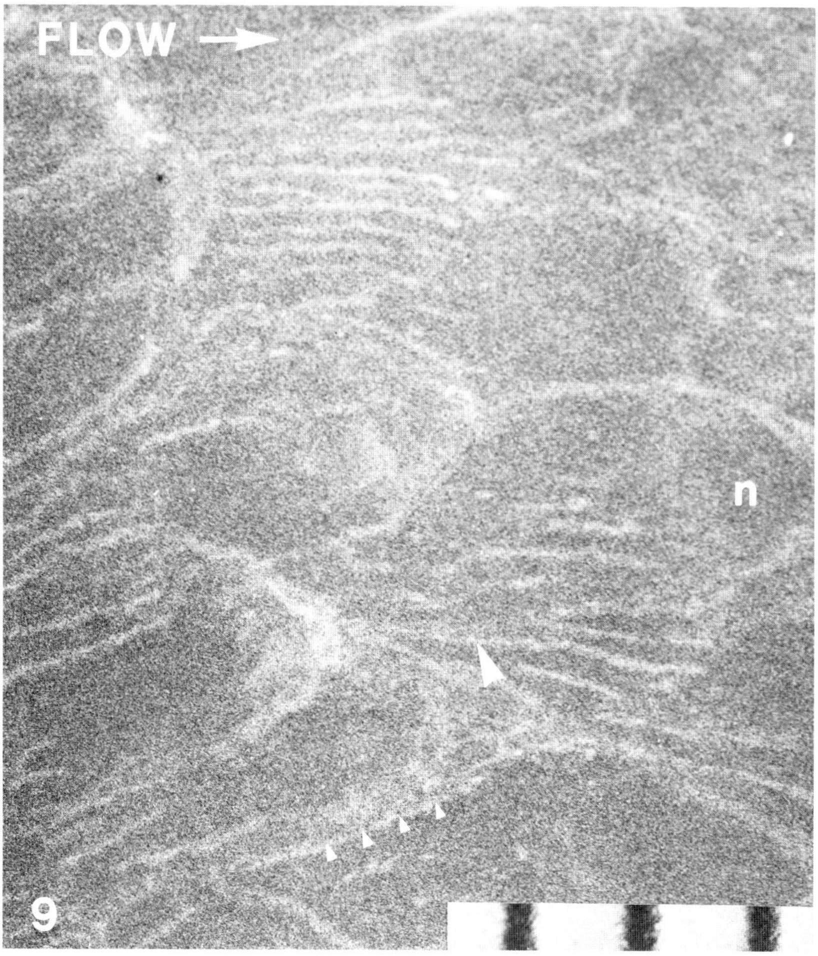

Figure 9. Endothelial cells *in situ* in the mouse thoracic aorta, stained for anti-actin distribution. Prominent staining of the periphery of the cells is apparent (small arrowheads), which outlines shape of the cells. The long axis of the cells is oriented along the direction of blood flow. Several parallel staining cytoplasmic fibers are also oriented along the direction of flow, and tend to be localized in the "upstream" pole of the cell. Many cells show only peripheral staining (small arrowheads). Cell nucleus (n). Scale: 1 division = 10 μm.

are located in the upstream half of the cell where the fibers often appear to converge slightly towards the pole of the cell.

Antimyosin staining of endothelial cells reveal a similar pattern to the actin staining except the fibers occasionally exhibit an intermittent or "banded" staining pattern, similar to stress fibers *in vitro* (Fig. 10).

Double immunofluorescent labeling of endothelial cells *in situ* shows colocalization for these proteins along the stress fiber (White *et al.*, 1983). While the majority of endothelial cells do not contain stress fibers, in some regions of the aorta as many as 40% of the cells have stress fibers.

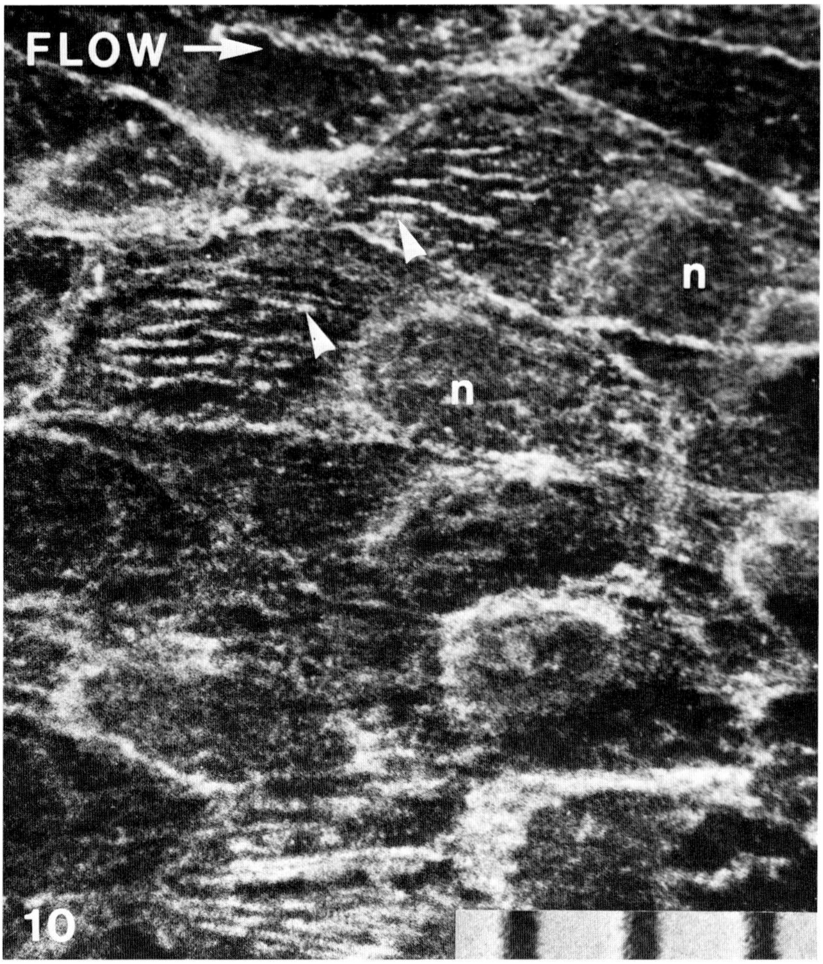

Figure 10. Endothelial cells *in situ* in the mouse thoracic aorta, following staining for antimyosin. The long axis of the cells, as determined by strong peripheral cytoplasmic staining, is oriented along the direction of blood flow. Stress fibers, also oriented along the long axis of the cell, often show periodic staining along their length (arrowheads) in the proximal half of the cell with respect to blood flow. Many cells do not show stress-fiber staining. Cell nuclei (n). Scale: 1 division = 10 μm.

Aortic endothelium of normal, nonhypertensive rats exhibits similar antimyosin staining patterns to the mouse. The descending thoracic aorta of the rat also reveals fine, wispy cytoplasmic fibers that are oriented with the direction of blood flow and occasionally show intermittent staining of fibrils following antimyosin staining (Fig. 11). The percentage of endothelial cells exhibiting stress fibers is considerably lower than in the mouse; only about 10% of the cells have stress fibers (White *et al.*, 1983). The significance of this difference between species is unknown. Aortic endothelial cells *in situ* of spontaneously hypertensive rats appear to have more numerous and prominent stress fibers than in normal rats, although further quantitation is needed (see Fig. 12). Inferior vena cava endothelial cells *in situ* also demonstrate stress fibers (Fig. 13). Thus, the expression of stress fibers does not appear to be related to absolute hydrostatic pressures, but may reflect other parameters which are more difficult to measure such as shear. A discussion on differences in distribution of stress fibers among normal rat and hypertensive aortic and inferior vena cava endothelial cells is detailed elsewhere (White *et al.*, 1983). Nevertheless, it is clear that endothelial cells *in situ* contain stress fibers which may enhance cell to substrate or basal lamina adhesion. Endothelial stress fibers may thus be related to cell adhesion and/or structural support in the face of shear forces (Wong *et al.*, 1983; White *et al.*, 1983).

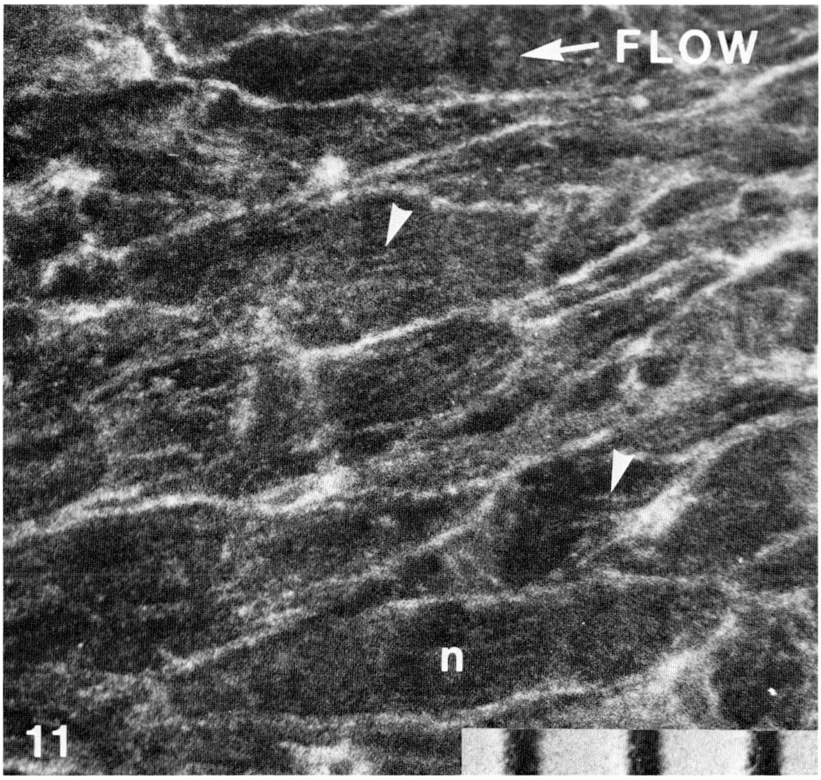

Figure 11. Endothelial cells stained for antimyosin distribution in the intact normal rat thoracic aorta. Fine, wispy, parallel stress fibers can be detected in addition to prominent peripheral staining (arrowheads). The cells and these fine fibers are oriented in the direction of the blood flow. Nuclei (n), are contrasted against a light, diffuse staining of the cytoplasm. Scale: 1 division $= 10\ \mu m$.

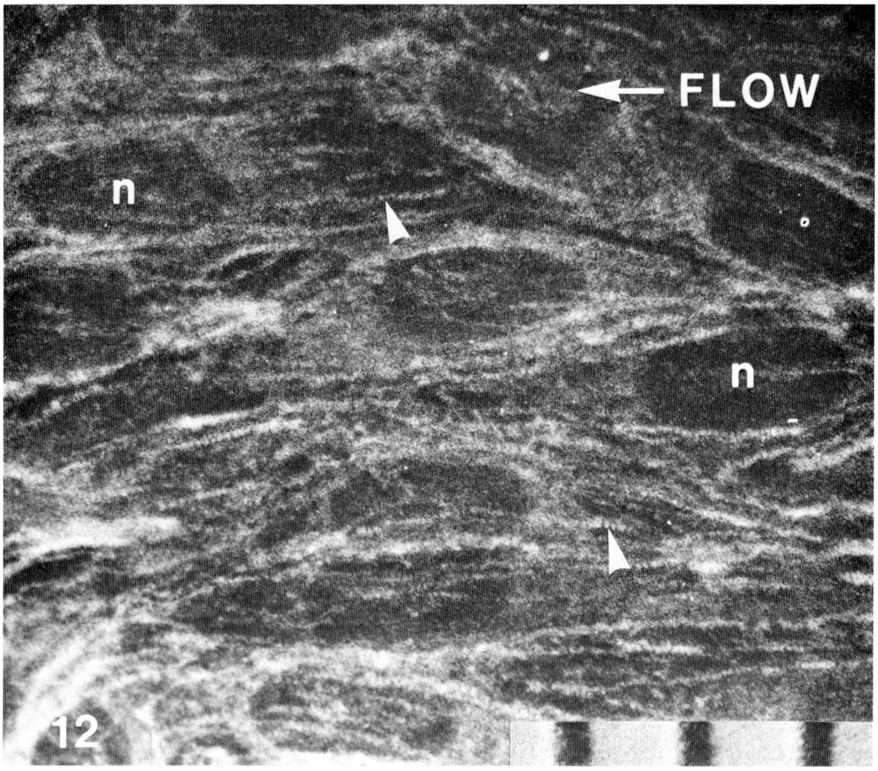

Figure 12. Immunofluorescent staining for antimyosin in endothelial cells residing on the aorta from a spontaneously hypertensive rat. Note the co-orientation of blood flow, long axis of the cells, and stress fibers, which show periodic staining and reside in the proximal pole of the cell (arrowheads). Cell nuclei (n) are contrasted against a diffuse background staining. Scale: 1 division = 10 μm.

5. Summary and Conclusions

We have reviewed the vast literature on stress fibers in cells *in vitro* and *in situ*. This review is by no means totally comprehensive, but it discusses most of the major and more recent papers on the subject. In this section, we discuss the significance of stress fibers in cells *in vitro*. Then, criteria for the definition of stress fibers is put forth in order to clarify the objectives of identifying stress fibers in cells *in situ*. The final section discusses the significance of finding stress fibers in cells *in situ*.

5.1. Significance of Stress Fibers in Cells in Vitro

Dissociation of tissues into cell culture systems produces by definition, an artifact. Partially digested or violently treated cells are exposed to artificial

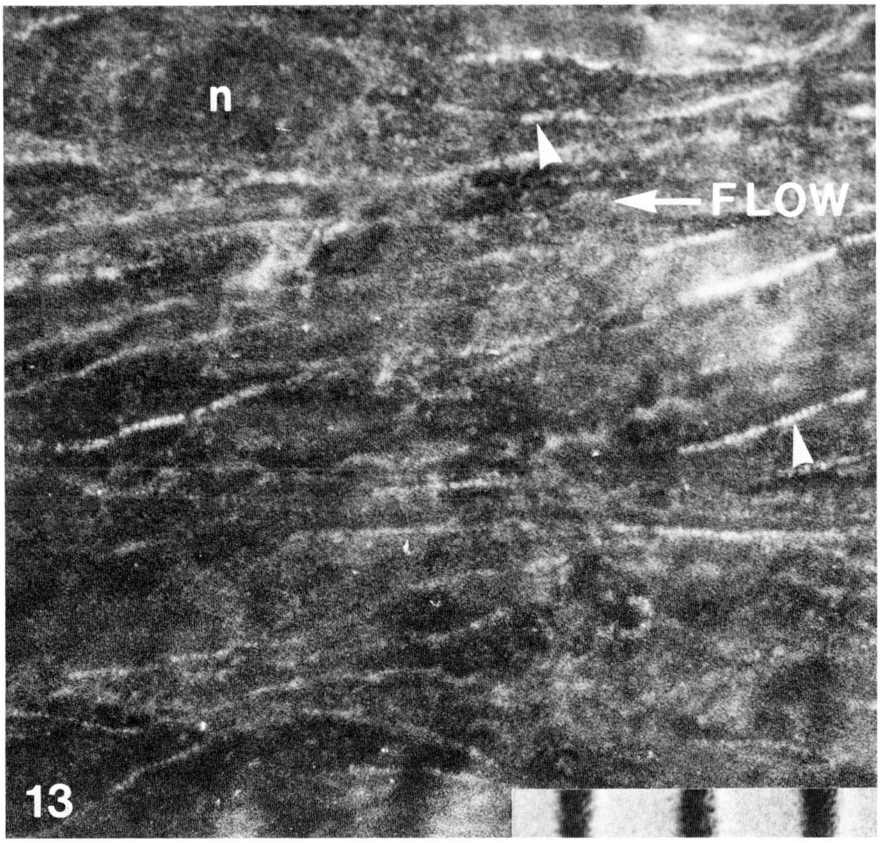

Figure 13. Endothelial cells *in situ* on the inferior vena cava of the spontaneously hypertensive rat. Antimyosin staining pattern reveals co-alignment of the direction of blood flow and stress fibers, which often show intermittant staining (arrowheads). Cell nuclei (n) are occasionally observed contrasted against the diffuse staining of the cytoplasm. Scale: 1 division = 10 μm.

media and substrates. Cell attachment, spreading, and formation of adhesion sites on artificial substrates may well be considered to produce artifactual structures such as stress fibers. It is likely that this mode of thinking caused relatively little interest in the cell biology of stress fibers over several decades. However, investigators interested in the fine structure of cytoplasm (Buckley and Porter, 1967) heralded the subsequent proliferation of data following immunofluorescent characterization of stress fibers (see Section 2.3).

Since stress fibers are composed of longitudinally oriented actin filaments, their potential role in cell motility provoked much investigation (see Section 3.2). It appears that stress fibers are not essential for rapid cell migration, as actively migrating cells contain few if any stress fibers (Herman *et al.*, 1981; Couchman and Rees, 1979a). They appear after cell spreading has begun, form an initial radial pattern (Lazarides, 1976b; Soranno and Bell, 1982), and then develop extensively during formation of focal contacts for cell adhesion. Finally, they are most prominent in well-spread cells that are relatively stationary. They are nevertheless dynamic structures in living cells, appearing and disappearing over time (Buckley and Porter, 1967). Reflecting this dynamic nature, stress fibers contain varying proportions of contractile-associated proteins (see Sections 2.3 and 3.3) and are occasionally organized similar to sarcomeres (see Section 3.3). Nevertheless, they appear not to actively contract, but likely generate isometric contraction between adhesion points on the ventral side of the cell (see review in Burridge, 1981). In contrast, the function of the dorsal stress fibers is unknown, but speculation centers on more of a structural or cytoskeletal role. The dorsal stress fibers may be used for cell surface compartmentalization of membrane proteins and organization of internalization sites (see Section 3.6).

Perhaps the most provocative speculation on the role of stress fibers stems from the relationship of fibronectin and stress fibers. Fibronectin is the extracellular protein which relates stress fibers with cell adhesion, transformation, and morphogenesis. Much has been recently learned about the relationship of stress fibers, α-actinin, vinculin, and fibronectin at focal and close contact or adhesion sites. Controversy still abounds on the question of whether fibronectin is involved with focal adhesion sites, or is peripheral to them (see Section 3.4). The prospects for cortical stress fibers controlling fibronectin orientation and subsequently collagen and other extracellular matrix components has great implications for tissue morphogenesis (see Section 3.7). Likewise, hormonal influences and receptors or cell membrane protein compartmentalization has implications for expression of differentiation. Finally, the role of fibronectin and stress fibers in cell transformation and tumorigenicity is in heated debate. Many studies find correlation of cell transformation, decrease in stress fibers and fibronectin, and associated increase in tumorigenicity, whereas other investigators find no correlation or multiple exceptions to these generalizations (see Section 3.5).

Indeed, because of the tremendous amount of data and the far ranging importance of many of the debates on interpretation, it has become essential to ask how "relevant" are stress fibers. Are they simply artifacts of tissue culture that bear no relation to cells *in situ*? To help answer this question, it is important to ask whether stress fibers exist in cells *in situ*. This brings up difficulties in definition of stress fibers, particularly in view of the lack of concern for strict adherence to any type of terminology. Thus, it is essential to set up criteria for the identification of stress fibers *in situ*.

5.2. Criteria for Identification of Stress Fibers in Cells in Situ

In the course of this review, it has become clear that the term stress fiber has undergone an evolution as its ultrastructural and biochemical composition has become known. First and foremost, stress fibers are a term derived from the observations on cultured cells using the light microscope, whether seen by bright field, phase contrast, polarizing Nomarski differential interference, interference reflection, or fluorescent optics.

The term is applied in electron microscopy, only when it is certain that the microfilament bundles being identified relate to the light microscope image. In the electron microscope, microfilament bundles or "actin cables" are not necessarily synonymous with stress fibers. As reviewed in section 4.1, there are many specialized structures or regions in cells that contain microfilament bundles that could be mistaken for a stress fiber, such as the microfilament arrays of the subcortical ectoplasm in amoeboid cells, circumferential microfilament bundles of epithelial cells, the contractile ring, and rootlets of microspikes or microvilli. Likewise, immunohistochemical studies in tissues which rely on only one contractile protein for localization may also lead to confusion of a stress fiber *in situ* with one of the aforementioned specialized structures. As reviewed in section 2.3, immunofluorescent studies have demonstrated that stress fibers contain varying amounts of contractile-associated proteins other than actin that differ from other microfilament bundle-containing structures. In order to identify with certainty stress fibers in cells *in situ*, the following criteria for their identification has been developed which agrees with what most authors have used during their descriptions of stress fibers over the past decade. First, stress fibers are linear cytoplasmic fibrils composed of bundles of actin. Second, they are often in sets or arrays of parallel or slightly convergent fibrils which are often oriented along the major axis of the cell or towards multiple poles. Third, they demonstrate varying degrees of α-actinin, myosin, or tropomyosin distribution which is often intermittent along the fiber. Specific immunoelectron microscopic localization may demonstrate intermittent contractile-associated proteins distributed along a microfilament bundle, however, appropriate sections must be taken to avoid confusion with other microfilament bundle-containing structures which also may contain intermittent distributions of contractile associated proteins. Use of such criteria may avoid confusion in the future as more model tissues for visualizing stress fibers in cells *in situ* are studied.

5.3. Significance of Stress Fibers in Cells in Situ

Using the criteria for defining stress fibers in cells *in situ*, stress fibers can be identified in fibroblasts and endothelial cells *in situ*.

The fish scale model (see Section 4.3) has established that stress fibers can be observed by immunofluorescence in fibroblast cells called scleroblasts. The intracellular distribution of these linear fibers is analogous to the sets of convergent and parallel fibers observed in cultured cells. In addition, the localization of myosin and α-actinin along fibers exhibits the intermittent staining often seen in cultured cells. The fibroblasts or scleroblasts exist as a "pseudo-epithelium" (without basement membrane) on the orthogonally arranged collagen fibers of the fibrillary plate and are responsible for collagen synthesis (Junqueira *et al.*, 1970; Onozato and Watabe, 1979). The stress fibers are not consistently oriented with respect to the uppermost collagen fibers as are microtubules (Byers *et al.*, 1980). It would be interesting to determine if the actin fibers are oriented with fibronectin *in situ*, as it is *in vitro*

(Mautner and Hynes, 1977; Hynes and Destree, 1978). The potential of these structures in the morphogenesis of such a highly organized tissue awaits further investigation.

It is interesting to note that while microtubule patterns are found in all scleroblasts, only the scleroblasts near the edge of the scale show stress fibers. The stress fibers are oriented perpendicular to the edges and the elevated radial ridges of the scale. The majority of scleroblasts in the central regions appear to contain peripheral or circumferential actin staining with periodic α-actinin staining. These are likely the circumferential microfilament bundles and represent sites of cell–cell adherence. With respect to cell–substrate adhesion it is clear that stress fibers are not essential for well-flattened discoid scleroblasts on the fibrillary plate. On the other hand, the fact that cells having stress fibers at the edges and ridges of scales may reflect a greater need for adhesivity at sites of greater shear, which would be expected to occur at semisolid (fibrillary plate)-viscoelastic (cells) interfaces. Indeed, there is evidence that increased shear forces create stress fibers oriented along the direction of shear in endothelial cells (White *et al.,* 1982).

Endothelial cells in the aorta of mice and rats clearly demonstrate cytoplasmic fibers in the proximal end of the cell relative to the direction of blood flow. These parallel slightly convergent fibers show intermittent staining following antimyosin labeling (White *et al.,* 1983). The position of these stress fibers on the proximal end of the cell, and their orientation, both suggest they may serve a role in adhesion and prevention of the endothelial cell from shearing off the vessel wall. There is a much more prominent circumferential staining pattern in the endothelial cells, presumably involved in cell–cell contacts, and pathological contraction of this ring may alter permeability across the endothelial cells (see Section 4.4). The majority of endothelial cells do not show stress fibers, but reside in regions where greater shear is suspected (White, unpublished). Likewise, experimental manipulation increases shear forces on endothelial cells *in vitro* (White *et al.,* 1982). Other endothelium exposed to high shear *in vivo* such as those on aortic valve or endocardium also show stress fibers (Wong *et al.,* 1983). However, although normal rats show stress fibers in aortic endothelial cells, there appears to be an increase in number of stress fibers in spontaneously hypertensive rats. The correlation to shear forces needs further work, however, as endothelial cells of rat inferior vena cava also show stress fibers, although fewer in number.

Other tissues exhibiting stress fibers *in situ* will likely be demonstrated. It appears that stress fibers often appear in association with pathologic phenomena. Certainly, "wounds" made in epithelial sheets *in vitro* are followed by appearance of stress fibers in cells adjacent to the "wound" (Gotlieb *et al.,* 1979). A similar process has now been identified in corneal epithelium *in situ.* Corneal epithelium normally only exhibits the circumferential microfilament bundles actin-staining pattern, but when a wound is inflicted, stress fibers appear in the epithelial cells at the wound edges (Gordon *et al.,* 1982). Finally, stress fibers are found in endothelium exposed to hypertension (White *et al.,* 1983) and there is evidence that stress fibers may exist in granulation tissue

(Gabbiani *et al.*, 1972).) Other pathologic as well as nonpathologic *in situ* systems that will help interpret the voluminous data on the stress fiber, fibronectin, and tumorigenicity relationships will be greatly valued.

ACKNOWLEDGMENTS. We would like to thank Liz Dreesen and Kay Cosgrove for help in the preparation of the manuscript and Peter Ley for help with the illustrations. This paper was funded by National Institutes of Health training grant GM 07753 to H.R.B. and National Science Foundation Grant PCM-8119171 to K.F.

References

Abercrombie, M., and Dunn, G. A., 1975, Adhesions of fibroblasts to substratum during contact inhibition observed by interference reflection microscopy, *Exp. Cell Res.* **92**:57–62.

Abercrombie, M., Heaysman, J. E. M., and Pogrum, S. M., 1971, The locomotion of fibroblasts in culture. IV. Electron microscopy of the leading lamella, *Exp. Cell Res.* **67**:359–367.

Abercrombie, M., Dunn, G. A., and Heath, J. P., 1976, Locomotion and contraction in non-muscle cells, in: *Contractile Systems in Non-Muscle Tissues* (S. V. Perry, A. Margreth, and R. S. Adelstein, eds.), pp. 3–11, Elsevier, Amsterdam.

Albertini, D. F., and Herman, B., 1983, Cell shape and membrane receptor dynamics: Modulation by cytoskeleton, in: *Cell and Muscle Motility*, Vol. 5 (R. M. Dowbin and J. W. Shay, eds.), Plenum Press, New York.

Albrecht-Buehler, G., 1977a, Daughter 3T3 cells: Are they mirror images of each other? *J. Cell Biol.* **72**:595–603.

Albrecht-Buehler, G., 1977b, Phagokinetic tracks of 3T3 cells: Parallels between the orientation of track segments and of cellular structures which contain actin or tubulin, *Cell* **12**:333–339.

Ali, I. U., Mautner, V., Lanza, R., and Hynes, R. O., 1977, Restoration of normal morphology, adhesion and cytoskeleton in transformed cells by addition of a transformation sensitive surface protein, *Cell* **11**:115–126.

Amsterdam, A., Lindner, H. R., and Gröschel-Stewart, U., 1977, Localization of actin and myosin in the rat oocyte and follicular wall by immunofluorescence, *Anat. Rec.* **187**:322–328.

Anderson, R. G. W., Vasile, E., Mello, R. J., Brown, M. S., and Golstein, J. L., 1978, Immunocytochemical visualization of coated pits and vesicles in human fibroblasts: Relation to low density lipoprotein receptor distribution, *Cell* **15**:919–933.

Ash, J. F., and Singer, S. J., 1976, Concanavalin-A-induced transmembrane linkage of concanavalin A surface receptors to intracellular myosin-containing filaments, *Proc. Natl. Acad. Sci. USA* **73**:4575–4579.

Ash, J. F., Vogt, P. K., and Singer, S. J., 1976, Reversion from transformed to normal phenotype by inhibition of protein synthesis in rat kidney cells infected with a temperature sensitive mutant of Rous sarcoma virus, *Proc. Natl. Acad. Sci. U.S.A.* **73**:3603–3607.

Ash, J. F., Louvard, D., and Singer, S. J., 1977, Antibody-induced linkages of plasma membrane proteins to intracellular actomyosin-containing filaments in cultured fibroblasts, *Proc. Natl. Acad. Sci. USA* **74**:5584–5588.

Avnur, Z., and Geiger, B., 1981a, Substrate-attached membranes of cultured cells: Isolation and characterization of ventral cell membranes and the associated cytoskeleton, *J. Mol. Biol.* **153**:361–379.

Avnur, Z., and Geiger, B., 1981b, The removal of extracellular fibronectin from areas of cell-substrate contact, *Cell* **25**:121–132.

Badley, R. A., Couchman, J. R., and Rees, D. A., 1980, Comparison of the cytoskeleton in migratory and stationary chick fibroblasts, *J. Muscle Res. Cell Motil.* **1**:5–14.

Badley, R. A., Lloyd, C. W., Woods, A., Carruthers, L., Allock, C., and Rees, D. A., 1978,

Mechanisms of cellular adhesion. III. Preparation and preliminary characterization of adhesions, *Exp. Cell Res.* **117:**231–244.

Bang, F. B., and Gey, G. O., 1948, A fibrillar structure in rat fibroblasts as seen by electron microscopy, *Proc. Soc. Exp. Biol. Med.* **69:**86–89.

Bannikov, G. A., Guelstein, V. I., Montesano, R., Tint, I. S., Tomatis, L., Troyanovsky, S. M., and Vasiliev, J. M., 1982, Cell shape and organization of cytoskeleton and surface fibronectin in non-tumorigenic and tumorigenic rat liver cultures, *J. Cell Sci.* **54:**47–67.

Barak, L. S., Yocum, R. R., Nothnagel, E. A., and Webb, W. W., 1980, Fluorescence staining of the actin cytoskeleton in living cells with 7-nitrobenz-2-oxa-1,3 diazole-phallacidin, *Proc. Natl. Acad. Sci. USA* **77:**980–984.

Barak, L. S., Yocum, R. R., and Webb, W. W., 1981, In vivo staining of cytoskeletal actin by autointernalization of nontoxic concentrations of nitrobenzoxadiazole-phallacidin, *J. Cell Biol.* **89:**368–372.

Becker, C. G., and Nachman, R. L., 1973, Contractile proteins of endothelial cells, platelets and smooth muscle, *Am. J. Pathol.* **71:**1–22.

Beertsen, W., Heersche, N. M., and Aubin, J. E., 1982, Free and polymerized tubulin in cultured bone cells and Chinese hamster ovary cells: The influence of cold and hormones, *J. Cell Biol.* **95:**387–393.

Begg, D. A., Rodewald, R., and Rebhun, L. I., 1978, The visualization of actin filament polarity in thin sections: Evidence for the uniform polarity of membrane-associated filaments, *J. Cell. Biol.* **79:**846–852.

Bensch, K. A., Gordon, G. B., and Miller, L., 1964, Fibrillar structures resembling leiomyofibrils in endothelial cells of mammalian pulmonary blood vessels, *Zeit. Zellforsch. Mik. Anat.* **63:**759–766.

Benzonana, G., Dreifuss, J. J., and Gabbiani, G., 1979, Actin is unevenly distributed in the pituitary gland, *Cell Tissue Res.* **200:**123–133.

Bershadsky, A. D., Gelfand, V. I., Svitkina, T. M., and Tint, I. S., 1980, Destruction of microfilament bundles in mouse embryo fibroblasts treated with inhibitors of energy metabolism, *Exp. Cell Res.* **127:**423–431.

Biberfeld, G., Fagraeus, A., and Lenke, R., 1974, Reaction of human smooth muscle antibody with thyroid cells, *Clin. Exp. Immunol.* **18:**371–377.

Birchmeier, C., Creis, T. E., Eppenberger, H. M., Winterhalter, K. H., and Birchmeier, W., 1980, Corrugated attachment membrane in WI-38 fibroblasts: Alternating fibronectin fibers and actin-containing focal contacts, *Proc. Natl. Acad. Sci. USA* **77:**4108–4112.

Bloch, R. J., and Geiger, B., 1980, The localization of acetylcholine receptor clustors in areas of cell-substrate contact in cultures of rat myotubes, *Cell* **21:**25–35.

Bloom, G. S., and Lockwood, A. H., 1980, Redistribution of myosin during morphological reversion of Chinese hamster ovary cells induced by db-cAMP, *Exp. Cell Res.* **129:**31–45.

Boschek, C. B., Jockusch, B. M., Friis, R. R., Back, R., Grundmann, E., and Bauer, H., 1981, Early changes in the distribution and organization of microfilament proteins during cell transformation, *Cell* **24:**175–184.

Bowers, B., and Korn, E. D., 1968, The fine structure of *acanthamoeba castallanii.* I. The trophozoite, *J. Cell Biol.* **39:**95–111.

Bradley, M. O., 1973, Microfilaments and cytoplasmic streaming: Inhibition of streaming by cytochalasin, *J. Cell Sci.* **12:**327–343.

Bragina, E. E., Vasiliev, J. M., and Gelfand, I. M., 1976, Formation of bundles of microfilaments during spreading of fibroblasts on the substrate, *Exp. Cell Res.* **97:**241–248.

Bretscher, A., and Weber, K., 1978, Localization of actin and microfilament-associated proteins in microvilli and terminal web of intestinal brush-border by immunofluorescence microscopy, *J. Cell Biol.* **9:**839–845.

Brinkley, B. R., Bell, P. T., Wible, L. J., Mace, M. L., Turner, D. S., and Cailleau, R. M., 1980, Variation in cell form and cytoskeleton in human breast carcinoma cells *in vitro, Cancer Res.* **40:**3118–3129.

Brunk, U., Schellens, J., and Westermark, B., 1976, Influence of epidermal growth factor (EGF) on ruffling activity, pinocytosis, and proliferation of cultivated human glia cells, *Exp. Cell Res.* **103:**295–302.

Buckley, I. K., 1974, Subcellular motility: A correlated light and electron microscopic study using cultured cells, *Tissue Cell* **6:**1–20.

Buckley, I. K., 1975, Three dimensional fine structure of cultured cells: Possible implications for subcellular motility, *Tissue Cell* **7:**51–72.

Buckley, I. K., 1981, Fine-structural and related aspects of nonmuscle-cell motility, in: *Cell and Muscle Motility,* Vol. 1 (R. M. Dowben and J. W. Shay, eds.), pp. 135–203, Plenum Press, New York.

Buckley, I. K., and Porter, K. R., 1967, Cytoplasmic fibrils in living cultured cells: A light and electron microscope study, *Protoplasma* **64:**349–380.

Buckley, I. K., and Porter, K. R., 1975, Electron microscopy of critical point dried whole cultured cells, *J. Microsc. (Oxford)* **104:**107–120.

Buckley, I. K., and Raju, T. R., 1976, Form and distribution of actin and myosin in non-muscle cells: A study using cultured chick embroy fibroblasts, *J. Microsc. (Oxford)* **107:**129–149.

Burkl, B., Mahlmeister, C., Gröschel-Stewart, U., Chamley-Campbell, J., and Campbell, G., 1979, Production of specific antibodies to contractile proteins and their use in immunofluorescence microscopy. III. Their use against human smooth muscle myosin, *Histochemistry* **60:**135–143.

Burnside, B., 1978, Thin (actin) and thick (myosinlike) filaments in cone contraction in the telost retina, *J. Cell Biol.* **78:**227–246.

Burridge, K., 1981, Are stress fibers contractile? *Nature* **294:**691–692.

Burridge, K., and Feramisco, J. R., 1980, Microinjection and localization of a 130K protein in living fibroblasts: A relationship to actin and fibronectin, *Cell* **19:**587–595.

Byers, H. R., and Fujiwara, K., 1982, Stress fibers in cells *in situ:* Immunofluorescent visualization with anti-actin, anti-myosin, and anti-alpha-actinin, *J. Cell Biol.* **93:**804–811.

Byers, H. R., Fujiwara, K., and Porter, K. R., 1980, Visualization of microtubules of cells *in situ* by indirect immunofluorescence, *Proc. Natl. Acad. Sci. USA* **77:**6657–6661.

Celis, J. E., Small, J. V., Andersen, P., and Celis, A., 1978, Microfilament bundles in cultured cells: Correlation with anchorage independence and tumorigenicity in Nude mice, *Exp. Cell Res.* **114:**335–348.

Celis, J. E., Small, J. V., Kaltoft, K., and Celis, A., 1979, Microfilament bundles in transformed mouse CLID X transformed CHO cell hybrids: Correlation with tumorigenicity in Nude mice, *Exp. Cell Res.* **120:**79–86.

Chen, W.-T., 1979, Induction of spreading during fibroblast movement, *J. Cell Biol.* **81:**684–691.

Chen, W.-T., 1981, Mechanism of retraction of the trailing edge during fibroblast movement, *J. Cell Biol.* **90:**187–200.

Chen, W.-T., and Singer, S. J., 1982, Immunoelectron microscopic studies of the sites of cell-substratum and cell-cell contacts in cultured fibroblasts, *J. Cell Biol.* **95:**205–222.

Couchman, J. R., and Rees, D. A., 1979a, Actomyosin organization for adhesion, spreading, growth and movement in chick fibroblasts, *Cell Biol. Int. Rep.* **3:**431–439.

Couchman, J. R., and Rees, D. A., 1979b, The behavior of fibroblasts migrating from chick heart explants: Changes in adhesion, locomotion and growth, and in the distribution of actomyosin and fibronectin, *J. Cell Sci.* **39:**149–165.

Craig, S. W., and Pardo, J. V., 1979, Alpha-actinin localization in thejunctional complex of intestinal epithelial cells, *J. Cell Biol.* **80:**203–210.

Creutz, C. E., 1977, Isolation, characterization and localization of bovine adrenal medullary myosin, *Cell Tissue Res.* **178:**17–38.

Curtis, A. S. G., 1964, The mechanism of adhesion of cells to glass: A study by interference reflection microscopy, *J. Cell Biol.* **20:**199–215.

David-Pfeuty, T., and Singer, S. J., 1980, Altered distribution of the cytoskeletal proteins vinculin and alpha-actinin in cultured fibroblasts transformed by Rous sarcoma virus, *Proc. Natl. Acad. Sci. USA* **77:**6687–6691.

DeBruyn, P. P. H., and Cho, Y., 1974, Contractile structures in endothelial cells of splenic sinusoids, *J. Ultrastruct. Res.* **49:**24–33.

Dedman, J. R., Welsh, M. J., and Means, A. R., 1978, Ca^{++}-dependent regulator. Production and characterization of a monospecific antibody, *J. Biol. Chem.* **253:**7515–7521.

De Lanerolle, P., Adelstein, R. S., Feramisco, J. R., and Burridge, K., 1981, Characterization of antibodies to smooth-muscle myosin kinase and their use in localizing myosin kinase in non-muscle cells, *Proc. Natl. Acad. Sci. USA* **78:**4738–4742.

De Mey, J., Joniau, M., De Brabander, M., Moens, W., and Geuens, G., 1978, Evidence for unaltered structure and *in vivo* assembly of microtubules in transformed cells, *Proc. Natl. Acad Sci. USA* **75:**1339–1343.

De Petris, S., Karlsbad, G., and Pernis, B., 1962, Filamentous structures in cytoplasm of normal mononuclear phagocytes, *J. Ultrastruct. Res.* **7:**39–55.

Der, C. J., and Stanbridge, E. J., 1978, Lack of correlation between the decreased expression of cell surface LETS protein and tumorigenicity in human cell hybrids, *Cell* **15:**1241–1251.

Der, C. J., Ash, J. F., and Stanbridge, E. J., 1981, Cytoskeletal and transmembrane interactions in the expression of tumorigenicity in human cell hybrids, *J. Cell Sci.* **52:**151–166.

DeRosier, D. J., Tilney, L. G., and Egelman, E., 1980, Actin in the inner ear: The remarkable structure of the stereocilium, *Nature* **287:**291–296.

Drenckhahn, D., and Gröschel-Stewart, U., 1977, Localization of myosin and actin in ocular nonmuscle cells. Immunofluorescence-microscopic, biochemical, and electron-microscopic studies, *Cell Tissue Res.* **181:**493–503.

Drenckhahn, D., and Gröschel-Stewart, U., 1980, Localization of myosin, actin, and tropomyosin in rat intestinal epithelium: Immunohistochemical studies at the light and electron microscopic levels, *J. Cell Biol.* **86:**475–482.

Drenckhahn, D., Gröschel-Stewart, U., and Unsicker, K., 1977, Immunofluorescence-microscopic demonstration of myosin and actin in salivary glands and exocrine pancreaus of the rat, *Cell Tissue Res.* **183:**273–279.

Drenckhahn, D., Steffens, R., and Gröschel-Stewart, U., 1980, Immunocytochemical localization of myosin in the brush border region of intestinal epithelium, *Cell Tissue Res.* **205:**163–166.

Eckert, B. S., Koons, S. J., Schantz, A. W., and Zobel, C. R., 1980, Association of creatine phosphokinase with the cytoskeleton of cultured mammalian cells, *J. Cell Biol.* **86:**1–5.

Edelman, G. M., and Yahara, I., 1975, Temperature-sensitive changes in surface modulating assemblies of fibroblasts transformed by mutants of Rous sarcoma virus, *Proc. Natl. Acad. Sci. USA* **73:**2047–2051.

Fagraeus, A., The, H., and Biberfeld, G., 1973, Reaction of human smooth muscle antibody with thymus medullary cells, *Nature (London) New Biol.* **246:**113–115.

Fallon, J. R., and Nachmias, V. T., 1980, Localization of cytoplasmic and skeletal myosins in developing muscle cells by double-label immunofluorescence, *J. Cell Biol.* **87:**237–247.

Feramisco, J. R., 1979, Microinjection of fluorescently labeled alpha-actinin into living fibroblasts, *Proc. Natl. Acad. Sci. USA* **76:**3967–3971.

Feramisco, J. R., and Blose, S. H., 1980, Distribution of fluorescently labeled alpha-actinin in living and fixed fibroblasts, *J. Cell Biol.* **86:**608–615.

Fischer, A., 1946, *Biology of Tissue Cells*, Gyldendalske Boghandel, Copenhagen.

Fleischer, M., and Wohlfarth-Bottermann, K. E., 1975, Correlation between tension force generation, fibrillogenesis and ultrastructure of cytoplasmic actomyosin during isometric and isotonic contractions of protoplasmic strands, *Cytobiologie* **10:**330–365.

Flock, A., and Cheung, H. C., 1977, Actin filaments in sensory hairs of inner ear receptor cells, *J. Cell Biol.* **75:**339–343.

Fox, C. H., Cottler-Fox, M. H., and Yamada, K. M., 1980, The distribution of fibronectin in attachment sites of chick fibroblasts, *Exp. Cell Res.* **130:**477–481.

Franke, W. W., Grund, C., Fink, A., and Weber, K., 1978, Location of actin in microfilament bundles associated with junctional specializations between sertoli cells and spermatids, *Biol. Cell.* **31:**7–14.

Freedman, V. H., and Shin, S., 1974, Cellular tumorigenicity in nude mice: Correlation with cell growth in semi-solid medium, *Cell* **3:**355–359.

Fujiwara, K., and Pollard, T. D., 1976, Fluorescent antibody localization of myosin in the cytoplasm, cleavage furrow, and mitotic spindle of human cells, *J. Cell Biol.* **71:**848–875.

Fujiwara, K., and Pollard, T. D., 1980, Relative disposition of myosin, alpha-actinin and tropomyosin in stress fibers, *J. Cell Biol.* **87**:222a.

Fujiwara, K., Porter, M. E., and Pollard, T. D., 1978, Alpha-actinin localization in the cleavage furrow during cytokinesis, *J. Cell Biol.* **79**:268–275.

Fuseler, J. W., Shay, J. W., and Feit, H., 1981, The role of intermediate (10-nm) filaments in the development and integration of the myofibrillar contractile apparatus in the embryonic mammalian heart, in: *Cell and Muscle Motility,* Vol. 1 (R. M. Dowben and J. W. Shay, eds.), pp. 205–259, Plenum Press, New York.

Gabbiani, G., Hirschel, B. J., Ryan, G. B., Statkov, P. R., and Majno, G., 1972, Granulation tissue as a contractile organ: A study of structure and function, *J. Exp. Med.* **135**:719–734.

Gabbiani, G., M. C. Badonnel, and G. Rona, 1975, Cytoplasmic contractile apparatus in aortic endothelial cells of hypertensive rats, *Lab. Invest.* **32**:227–234.

Geiger, B., 1979, A 130K protein from chicken gizzard: Its localization at the termini of microfilament bundles in cultured chicken cells, *Cell* **187**:193–205.

Geiger, B., Tokuyasu, K. T., Dutton, A. H., and Singer, S. J., 1980, Vinculin, an intracellular protein localized at specialized sites where microfilament bundles terminate at cell membranes, *Proc. Natl. Acad. Sci. USA* **77**:4127–4131.

Giacomelli, F., Weiner, J., and Spiro, D., 1970, Cross-striated arrays of filaments in endothelium, *J. Cell Biol.* **45**:188–192.

Goldberg, B., and Green, H., 1964, An analysis of collagen secretion by established mouse fibroblast lines, *J. Cell Biol.* **22**:227–258.

Goldman, R. D., 1971, The role of three cytoplasmic fibers in BHK-21 cell motility, I. Microtubules and the effect of colchicine, *J. Cell Biol.* **51**:752–762.

Goldman, R. D., 1975, The use of heavy meromyosin binding as an ultrastructural cytochemical method for localizing and determining the possible functions of actin-like microfilaments in non-muscle cells, *J. Histochem.* **23**:529–542.

Goldman, R. D., and Follett, E. A. C., 1969, The structure of the major cell processes of isolated BHK21 fibroblasts, *Exp. Cell Res.* **57**:263–276.

Goldman, R. D., and Knipe, D. M., 1973, Functions of cytoplasmic fibers in nonmuscle cell motility, *Cold Spring Harbor Symp. Quant. Biol.* **37**:523–534.

Goldman, R. D., Chang, C., and Williams, J., 1974, Properties and behavior of hamster embryo cells transformed by human adenovirus type 5, *Cold Spring Harbor Symp. Quant. Biol.* **39**:601–614.

Goldman, R. D., Lazarides, E., Pollack, R., and Weber, K., 1975, The distribution of actin in non-muscle cells: The use of actin antibody in the localization of actin within the microfilament bundles of mouse 3T3 cells, *Exp. Cell Res.* **90**:333–344.

Goldman, R. D., Yerna, M.-J., and Schloss, J. A., 1976, Localization and organization of microfilaments and related proteins in normal and virus-transformed cells, *J. Supramol. Struct.* **5**:155–183.

Goldman, R. D., Chojnacki, B., and Yerna, M.-J., 1979, Ulstrastructure of microfilament bundles in baby hamster kidney (BHK-21) cells: The use of tannic acid, *J. Cell Biol.* **80**:759–766.

Gordon, S. R., Essner, E., and Rothstein, A., 1982, *In Situ* demonstration of actin in normal and injured ocular-tissue using 7-nitrobenz-2-oxa-1,3 diazole phallacidin, *Cell Motil.* **2**:343–354.

Gordon, S. R., Essner, E., and Rothstein, H., 1981, The *in situ* localization of actin in occular tissues with 7-nitrobenz-2-oxa-1,3 diaxole phallacidin, *IRCS Med. Sci.* **9**:956–967.

Gordon, W. E., 1978, Immunofluroescent and ultrastructural studies of "sarcomeric" units in stress fibers of cultured non-muscle cells, *Exp. Cell Res.* **117**:253–250.

Gordon, W. E., and Bushnell, A., 1979, Immunofluorescent and ultrastructural studies of polygonal microfilament networks in respreading non-muscle cells, *Exp. Cell Res.* **120**:335–348.

Gotlieb, A. I., Heggeness, M. H., Ash, J. F., and Singer, S. J., 1979, Mechanochemical proteins, cell motility and cell-cell contacts: The localization of mechanochemical proteins inside cultured cells at the edge of an *in vitro* "wound," *J. Cell Physiol.* **100**:563–578.

Grinnel, F., and Feld, M. K., 1979, Initial adhesion of hyman fibroblasts in serum-free medium: Possible role of secreted fibronectin, *Cell* **17**:117–129.

Gröschel-Stewart, U., 1980, Immunochemistry of cytoplasmic contractile proteins, *Int. Rev. Cytol.* **65**:193–254.

Gröschel-Stewart, U., and Unsicker, K., 1977, Direct visualization of contractile proteins in peritubular cells of the guinea-pig testis using antibodies against highly purified actin and myosin, *Histochemistry* **51**:315–319.

Gröschel-Stewart, U., Unsicker, K., and Leonhardt, H., 1977, Immunohistochemical demonstration of contractile proteins in astrocytes, marginal glial and ependymal cells in rat diencephalon, *Cell Tiss. Res.* **180**:133–137.

Gruenstein, E., Rich, A., and Weihing, R., 1975, Actin associated with membranes from 3T3 mouse fibroblast and HeLa cells, *J. Cell Biol.* **64**:223–234.

Guerriero, V., Rowley, D. R., and Means, A. R., 1981, Production and characterization of an antibody to myosin light chain kinase and intracellular localization of the enzyme, *Cell* **27**:449–458.

Hammersen, F., 1980, Endothelial contractility: Does it exist, in: Advances in Microcirculation, Vol. 9, *Vascular Endothelium and Basement Membranes* (B. M. Alturo, ed.), pp. 99–134, Karger, Basel.

Harris, A. K., 1973, Behavior of cultured cells on substrata of variable adhesiveness, *Exp. Cell Res.* **77**:285–297.

Hartwig, J.-H., and Stossel, T. P., 1981, Structure of macrophage actin-binding protein molecules in solution and interacting with actin filaments, *J. Mol. Biol.* **145**:563–581.

Heath, J. P., 1981, Arcs, curved microfilament bundles beneath the dorsal surface of the leading lamellae of moving chick embryo fibroblasts, *Cell Biol. Int. Rep.* **5**:975–980.

Health, J. P., and Dunn, G. A., 1978, Cell to substratum contacts of chick fibroblasts and their relation to the microfilament system: A correlated interference-reflexion and high-voltage electron-microscope study, *J. Cell Sci.* **29**:197–212.

Hedman, K., Johansson, S., Vartio, T., Kjellen, L., Vaheri, A., and Hook, M., 1982, Structure of the pericellular matrix: Association of heparan and chondroitin sulfates with fibronectin-procollagen fibers, *Cell* **28**:663–671.

Heggeness, M. H., and Ash, J. F., 1977, Use of the avidin-biotin complexes for the localization of actin and myosin with fluorescence microscopy, *J. Cell Biol.* **73**:783–788.

Heggeness, M. H., Wang, K., and Singer, S. J., 1977, Intracellular distributions of mechanochemical proteins in cultured fibroblasts, *Proc. Natl. Acad. Sci. USA* **74**:3883–3887.

Heggeness, M. H., Ash, J. F., and Singer, S. J., 1978, Transmembrane linkage of fibronectin to intracellular actin-containing filaments in cultured human fibroblasts, *Ann. N.Y. Acad. Sci.* **312**:414–417.

Henderson, D., and Weber, K., 1979, Three-dimensional organization of microfilaments and microtubules in the cytoskeleton, *Exp. Cell Res.* **124**:301–316.

Herman, I. M., and Pollard, T. D., 1979, Comparison of purified anti-actin and fluorescent-heavy meromyosin staining patterns in dividing cells, *J. Cell Biol.* **80**:509–520.

Herman, I. M., and Pollard, T. D., 1981, Electron microscopic localization of cytoplasmic myosin with purified ferritin-antimyosin, *J. Cell Biol.* **88**:346–351.

Herman, I. M., Crisona, N. J., and Pollard, T. D., 1981, Relation between cell activity and distribution of cytoplasmic actin and myosin, *J. Cell Biol.* **90**:84–91.

Heuser, J. E., and Kirschner, M. W., 1980, Filament organization revealed in platinum replicas of freeze-dried cytoskeletons, *J. Cell Biol.* **86**:212–234.

Higashi-Fujime, S., 1980, Active movement *in vitro* of bundles of microfilaments isolated from Nitella cell, *J. Cell. Biol.* **87**:569–578.

Huet, C., Ash, J. F., and Singer, S. J., 1980, The antibody-induced clustering and endocytosis of HLA antigens on cultured human fibroblasts, *Cell* **21**:429–438.

Hughes, A. F., and Swann, M. M., 1948, Anaphase movements in the living cell, *J. Exp. Biol.* **25**:45–70.

Hull, B. E., and Staehelin, L. A., 1979, The terminal web, a reevaluation of its structure and function, *J. Cell Biol.* **81**:67–82.

Huxley, H. E., 1963, Electron microscope studies on the structure of natural and synthetic filaments from striated muscle, *J. Mol. Biol.* **7**:281–308.

Hynes, R. O., 1981, Relationships between fibronectin and the cytoskeleton, in: *Cell Surface Reviews,* Vol. 7 (G. Poste and G. L. Nicolson, eds.), pp. 99–137, Elsevier/North Holland, Amsterdam.

Hynes, R. O., 1982, Phosphorylation of vinculin in pp60^{src}: What might it all mean? *Cell* **28:**437–438.

Hynes, R. O., and Destree, A. T., 1978, Relationships between fibronectin (LETS protein) and actin, *Cell* **15:**875–886.

Ireland, G. W., and Voon, F. C. T., 1981, Polygonal networks in living chick embryonic cells, *J. Cell Sci.* **52:**55–69.

Isenberg, G., Rathke, P. C., Hülsmann, N., Franke, W. W., and Wohlfarth-Bottermann, K. E., 1976, Cytoplasmic actomyosin fibrils in tissue culture cells: Direct proof of contractability by visualization of ATP-induced contraction in fibrils isolated by laser microbeam dissection, *Cell Tissue Res.* **166:**427–443.

Ishikawa, H., 1974, Arrowhead complexes in a variety of cell types, in: *Exploratory Concepts in Muscular Dystrophy,* Vol. II (A. T. Milhorat, ed.), pp. 37–50, Excerpta Medica, Amsterdam.

Ishikawa, H., 1979, Identification and distribution of intracellular filaments, in: *Cell Motility: Molecules and Organization* (S. Hatano, H. Ishikawa, and H. Sato, eds.), pp. 417–444, University of Tokyo Press, Tokyo.

Ishikawa, H., Bischoff, R., and Holtzer, H., 1969, Formation of arrowhead complexes with heavy meromyosin in a variety of cell types, *J. Cell Biol.* **43:**312–328.

Izzard, C. S., and Lochner, L. R., 1976, Cell-to-substrate contacts in living fibroblasts: An interference reflexion study with an evaluation of the technique, *J. Cell Sci.* **21:**129–159.

Izzard, C. S., and Lochner, L. R., 1980, Formation of cell-to-substrate contacts during fibroblast motility: An interference-reflexion study, *J. Cell Sci.* **42:**81–116.

Junqueira, L. C. U., Toledo, A. M. S., and Porter, K. R., 1970, Observations on the structure of the skin of the teleost, *Funulus heteroclitus, Arch. Histol. Hpn. (Niigata, Jpn.)* **32:**1–15.

Kaiho, M. A., and Sato, A., 1978, Circular distribution of microfilaments in cells spreading in vitro, *Exp. Cell Res.* **113:**222–227.

Karsenti, E., Guilbert, B., Bornens, M., Avrameas, S., Whalen, R., and Pantaloni, D., 1978, Detection of tubulin and actin in various cell lines by an immunoperoxidase technique, *J. Histochem. Cytochem.* **26:**934–947.

Kartenbeck, J., Schmid, K., Muller, H., and Franke, W. W., 1981, Immunological identification and localization of clathrin and coated vesicles in cultured cells and in tissues, *Exp. Cell Res.* **133:**191–211.

Kibbelaar, M. A., Ramaekers, F. C. S., Ringens, P. J., Selten-Versteegen, A. M. E., Poels, L. G., Jap, P. H. K., van Rossum, A. L., Feltkamp, T. E. W., and Bloemendal, H., 1980, Is actin in eye lens a possible factor in visual accomodation? *Nature* **285:**506–508.

Klebe, R. J., 1974, Isolation of a collagen dependent cell attachment factor, *Nature (London)* **250:**248–251.

Kopelovich, L., Conlon, S., and Pollack, R., 1977, Defective organization of actin in cultured skin fibroblasts from patients with inherited adenocarcinoma, *Proc. Natl. Acad. Sci. USA* **74:**3019–3022.

Kopelovich, L., Lipkin, M., Blattner, W. A., Fraumeni, J. F., Jr., Lynch, H. T., and Pollack, R. E., 1980, Organization of actin-containing cables in cultured skin fibroblasts from individuals at high risk of colon cancer, *Int. J. Cancer* **26:**302–308.

Kreis, T. E., and Birchmeier, W., 1980, Stress fiber sarcomeres of fibroblasts are contractile, *Cell* **22:**555–561.

Kreis, T. E., and Birchmeier, W., 1982, Microinjection of fluorescently labeled proteins into living cells with emphasis on cytoskeletal proteins, *Int. Rev. Cytol.* **75:**209–227.

Kreis, T. E., Winterhalter, K. H., and Birchmeier, H., 1979, In vivo distribution and turnover of fluorescently labeled actin microinjected into human fibroblasts, *Proc. Natl. Acad. Sci. USA* **76:**3814–3818.

Kreis, T. E., Geiger, B., and Schlessinger, J., 1982, Mobility of microinjected rhodamine actin within living chicken gizzard cells determined by fluorescence photobleaching recovery, *Cell* **29:**835–845.

Lazarides, E., 1975a, Immunofluorescence studies on the structure of actin filaments in tissue culture cells, *J. Histochem. Cytochem.* **23:**507–528.
Lazarides, E., 1975b, Tropomyosin antibody: The specific localization of tropomyosin in nonmuscle cells, *J. Cell Biol.* **65:**549–561.
Lazarides, E., 1976a, Actin, alpha-actinin, and tropomyosin interaction in the structural organization of actin filaments in nonmuscle cells, *J. Cell Biol.* **68:**202–219.
Lazarides, E., 1976b, Two general classes of actin filaments in tissue culture cells: The role of tropomyosin, *J. Supramol. Struct.* **5:**531–563.
Lazarides, E., and Burridge, K., 1975, Alpha-actinin: Immunofluorescent localization of a muscle structural protein in nonmuscle cells, *Cell* **6:**289–298.
Lazarides, E., and Revel, J. P., 1979, The molecular basis of cell movement, *Sci. Am.* **240:**100–113.
Lazarides, E., and Weber, K., 1974, Actin antibody: The specific visualization of actin filaments in non-muscle cells, *Proc. Natl. Acad. Sci. USA* **71:**2268–2272.
Leavitt, J., Bushar, G., Kakunaga, T., Hamada, H., Hirakawa, T., Goldman, D., and Meuil, C., 1982, Variations in expression of mutant beta-actin accompanying incremental increases in human fibroblast tumorigenicity, *Cell* **28:**259–268.
Lewis, L., Verna, J.-M., Levinstone, D., Sher, S., Marek, L., and Bell, E., 1982, The relationship of fibroblast translocations to cell morphology and stress fibre density, *J. Cell Sci.* **53:**21–36.
Lewis, W. H., and Lewis, M. R., 1924, Behavior of cells in tissue cultures, in: *General Cytology* (E. V. Cowdry, ed.), pp. 385–447, The University of Chicago Press, Chicago.
Marshall, C. J., Humphrye, K. C., and Pollack, R. E., 1978, Microfilament bundles, LETS protein and growth control in somatic-cell hybrids, *J. Cell Sci.* **33:**191–201.
Masayoshi, K., and Sato, A., 1978, Circular distribution of microfilaments in cell spreading *in vitro, Exp. Cell Res.* **113:**222–226.
Maupin, P., and Pollard, T. D., 1983, Improved preservation and staining of HeLa cell actin filaments, clathrin-coated membranes, and other cytoplasmic structures by tannic acid-glutaraldehyde-saponin fixation, *J. Cell Biol.* **96:**51–62.
Maupin-Szamier, P., and Pollard, T. D., 1978, Actin filament destruction by osmium tetroxide, *J. Cell Biol.* **77:**837–852.
Mautner, V., and Hynes, R. D., 1977, Surface distribution of LETS protein in relation to the cytoskeleton of normal and transformed cells, *J. Cell Biol.* **75:**743–768.
McClain, D. A., Maness, P. F., and Edelman, G. M., 1978, Assay for early cytoplasmic effects of the *src* gene product of Rous sarcoma virus, *Proc. Natl. Acad. Sci. USA* **75:**2750–2754.
McNabb, J. D., and Sandborn, E., 1964, Filaments in the microvillus border of intestinal cells, *J. Cell Biol.* **22:**701–704.
McNutt, N. S., Culp, L. A., and Black, P. H., 1973, Contact-inhibited revertant cell lines isolated from SV40-transformed cells. IV. Microfilament distribution and cell shape in untransformed, transformed and revertant Balb/C 3T3 cells, *J. Cell Biol.* **56:**412–428.
Meza, I., Ibbara, G., Sabanero, M., Martinez-Palomo, A., and Cerejido, M., 1980, Occluding junctions and cytoskeletal components in a cultured transporting epithelium, *J. Cell Biol.* **87:**746–754.
Mooseker, M. S., and Tilney, L. G., 1975, Organization of an actin filament-membrane complex: Filament polarity and membrane attachment in the microvilli of intestinal epithelial cells, *J. Cell Biol.* **67:**725–743.
Mukherjee, B. B., Mobry, P. M., and Pena, S. D. J., 1982, Retinoic acid induces anchorage- and density-dependent growth without restoring normal cytoskeleton, EGF binding, fibronectin content and ODC activity in retrovirus-transformed mouse cell line, *Exp. Cell Res.* **138:**95–107.
Nachmias, V. T., 1964, Fibrillary structures in the cytoplasm of *Chaos chaos, J. Cell Biol.* **23:**183–188.
Nagai, R., and Rebhun, L. I., 1966, Cytoplasmic filaments in streaming Nitella cells, *J. Ultrastruct. Res.* **14:**571–589.
Norberg, R., Lidman, K., and Fagraeus, A., 1975, Effects of cytochalasin B on fibroblasts, lymphoid cells, and platelets revealed by human anti-actin antibodies, *Cell* **6:**507–512.
Oliver, J. M., Krawiec, J. A., and Berlin, R. D., 1976, A carbamate herbicide causes microtubule

and microfilament disruption and nuclear fragmentation of fibroblasts, *Exp. Cell Res.* **116**:229–237.

Onozato, H., and Watabe, N., 1979, Studies on fish scale formation and resorption, *Cell Tissue Res.* **201**:409–422.

Osborn, M., and Weber, K., 1975, Simian virus 40 gene A function and maintenance of transformation, *J. Virol.* **15**:636–644.

Osborn, M., and Weber, K., 1980, Dimethylsulfoxide and the ionophore A23187 affect the arrangement of actin and induce nuclear actin paracrystals in PtK2 cells, *Exp. Cell Res.* **129**:103–114.

Osborn, M., Born, T., Koitsch, H. J., and Weber, K., 1978, Stereo immunofluorescence microscopy. I. Three-dimensional arrangement of microfilaments, microtubules and tonofilaments, *Cell* **14**:477–488.

Ostlund, R. E., Pastan, I., and Adelstein, R. S., 1974, Myosin in cultured fibroblasts, *J. Biol. Chem.* **249**:3903–3907.

Overton, J., 1966, Microtubules and microfilaments in morphogenesis of the scale cells of *Ephestia Kühniella, J. Cell Biol.* **29**:293–305.

Overton, J., and Shoup, J., 1964, Fine structure of cell surface specializations in the maturing duodenal mucosa of the chick, *J. Cell Biol.* **21**:75–85.

Owaribe, K., and Masuda, H., 1982, Isolation and characterization of circumferential microfilament bundles from retinal pigmented epithelial cells, *J. Cell Biol.* **95**:310–315.

Owaribe, K., Kodama, R., and Eguchi, G., 1981, Demonstration of contractility of circumferential actin bundles and its morphogentic significance in pigmented epithelium in vitro and in vivo, *J. Cell Biol.* **90**:507–514.

Palay, S. L., and Karlin, L. J., 1959, An electron microscopic study of the intestinal villus. I. The fasting animal, *J. Biophys. Biochem. Cytol.* **5**:363–371.

Paulin, D., Nicolas, J. F., Yaniv, M., Jacob, F., Weber, K., and Osborn, M., 1978, Actin and tubulin in teratocarcinoma cells: Amount and intracellular organization upon cytodifferentiation, *Dev. Biol.* **66**:488–499.

Pearlstein, E., 1976, Plasma membrane glycoprotein which mediates adhesion of fibroblasts to collagen, *Nature (London)* **262**:497–500.

Perdue, J. F., 1973, The distribution, ultrastructure, and chemistry of microfilaments in cultured chick embryo fibroblasts, *J. Cell Biol.* **58**:265–283.

Perry, M. M., Tassin, J., and Courtois, Y., 1981, Fine structure of bovine lens epithelial cells *in vitro* in relation to modifications induced by a retinal extract (EDGF), *Exp. Cell Res.* **136**:379–390.

Pollack, R., and Rifkin, D., 1975a, Actin-containing cables within anchorage-dependent rat embryo cells are dissociated by plasmin and trypsin, *Cell* **6**:495–506.

Pollack, R., and Rifkin, D., 1975b, Actin-containing cables within anchorage-dependent rate embryo cell are dissociated by plasmin and trypsin, *Cell* **6**:495–506.

Pollack, R., Osborn, M., and Weber, K., 1975, Patterns of organization of actin and myosin in normal and transformed cultured cells, *Proc. Natl. Acad. Sci. USA* **72**:994–998.

Pollack, R., Nicholson, N., Alcorta, D., Verderame, M., Smith, K., and Steinberg, B., 1982, Actin organization as an *in vitro* assay for tumorigenicity, in: *Cell and Muscle Motility,* Vol. 2 (R. M. Dowben and J. W. Shay, eds.), pp. 1–13, Plenum Press, New York.

Porter, K. R., 1953, Observations on a submicroscopic basophilic component of cytoplasm, *J. Exp. Med.* **97**:727–750.

Porter, K. R., Claude, A., and Fullam, E. F., 1945, A study of tissue culture cells by electron microscopy, *J. Exp. Med.* **81**:233–244.

Porter, K. R., Byers, H. R., and Ellisman, M. H., 1979, The cytoskeleton, in: *The Neurosciences: Fourth Study Program* (F. O. Schmitt and F. G. Worden, eds.), pp. 703–722, M.I.T. Press, Cambridge, Massachusetts.

Raizada, M. K., Fellows, R. E., and Wu, B., 1981, Cytochalasin B-induced alterations of insulin binding and microfilament organization in cultured fibroblasts, *Exp. Cell Res.* **136**:335–341.

Rathke, P. C., Osborn, M., and Weber, K., 1979, Immunological and ultrastructural characterization of microfilament bundles: Polygonal nets and stress fibers in an established cell line, *Eur. J. Cell Biol.* **19:**40–48.

Reaven, E. P., and Axline, S. G., 1973, Subplasmalemmal microfilaments and microtubules in resting and phagocytizing cultivated macrophages, *J. Cell Biol.* **59:**12–27.

Revel, J. P., and Wolken, K., 1973, Electron microscope investigations of the underside of cells in culture, *Exp. Cell Res.* **78:**1–14.

Revel, J. P., Hoch, P., and Ho, D., 1974, Adhesion of culture cells to their substratum, *Exp. Cell Res.* **84:**207–218.

Rifkin, D., Crowe, R., and Pollack, R., 1979, Tumor promotors induce changes in cytoskeletal organization in chick embryo fibroblasts, *Cell* **18:**361–368.

Robbins, P. W., 1975, Comparison of major cell-surface proteins of normal and transformed cells, *Am. J. Clin. Pathol.* **63:**671–676.

Rogers, K. A., and Kalnins, V. I., 1981, The immunofluorescent visualization of microtubules and microfilaments in endothelial cells fixed *in situ, J. Cell Biol.* **91:**328a.

Röhlich, P., and Olah, I., 1967, Cross-striated fibrils in the endothelium of rat myometral arterioles, *J. Ultrastruct Res.* **18:**667–676.

Rohrlich, S. T., and Porter, K. R., 1972, Fine structural observations relating to the production of color by the iridophores of a lizard, *Anolis carolinensis, J. Cell Biol.* **53:**38–52.

Rohrschneider, L. R., 1980, Adhesion plaques of Rous sarcoma virus transformed cells contain the *src* gene product, *Proc. Natl. Acad. Sci. USA* **77:**3514–3518.

Rose, G. G., and Cattoni, M., 1963, Mosaic patterns of stromal cells in tissue cultures, in: *Cinemicrography in Cell Biology* (G. G. Rose, ed.), pp. 445–469, Academic Press, New York and London.

Ruoslahti, E., and Engvall, E., 1980, Complexing of fibronectin glycosaminoglycans and collagen, *Biochim. Biophys. Acta* **631:**350–358.

Sanger, J. M., and Sanger, J. W., 1980, Banding and polarity of actin filaments in interphase and cleaving cells, *J. Cell Biol.* **86:**568–575.

Sanger, J. W., 1975, Intracellular localization of actin with fluorescently labeled heavy meromyosin, *Cell Tissue Res.* **160:**431–444.

Schlessinger, J., and Geiger, B., 1981, Epidermal growth factor induces redistribution of actin and alpha-actinin in human epidermal carcinoma cells, *Exp. Cell Res.* **134:**273–279.

Schliwa, M., 1981, Proteins associated with cytoplasmic actin, *Cell* **25:**587–590.

Schloss, J. A., and Goldman, R. D., 1979, Isolation of a high molecular weight actin-binding protein from baby hamster kidney (BHK-21) cells, *Proc. Natl. Acad. Sci. USA* **76:**4484–4488.

Schloss, J. A., Milsted, A., and Goldman, R. D., 1977, Myosin subfragment binding for the localization of actin-like microfilaments in cultured cells. A light and electron microscope study, *J. Cell Biol.* **74:**794–815.

Schollmeyer, J. E., Furcht, L. T., Goll, D. E., Robson, R. M., and Stromer, M. M., 1976, Localization of contractile proteins in smooth muscle cells and in normal and transformed fibroblasts, in: *Cell Motility* (R. Goldman, T. Pollard, and J. Rosenbaum, eds.), pp. 361–388, Cold Spring Harbor Laboratory, Cold Spring Harbor, New York.

Schroeder, T. E., 1973, Actin in dividing cells: Contractile ring filaments bind heavy meromyosin, *Proc. Natl. Acad. Sci. USA* **70:**1688–1692.

Shin, S., Freedman, V. H., Risser, R., and Pollack, R., 1975, Tumorigenicity of virus-transformed cells in nude mice is correlated specifically with anchorage independent growth *in vitro, Proc. Natl. Acad. Sci. USA* **72:**4435–4439.

Shriver, K., and Rohrschneider, L., 1981, Organization of $pp60^{src}$ and selected cytoskeletal proteins within adhesion plaques and junctions of Rous sarcoma virus-transformed rat cells, *J. Cell Biol.* **89:**525–535.

Singer, I. I., 1979a, The fibronexus: A transmembrane association of fibronectin-containing fibers and bundles of 5nm microfilaments in hamster and human fibroblasts, *Cell* **16:**675–685.

Singer, I. I., 1979b, Microfilament bundles and the control of pinocytotic vesicle distribution at the surfaces of normal and transformed fibroblasts, *Exp. Cell Res.* **122:**251–264.

Singer, I. I., 1982, Association of fibronectin and vinculin with focal contacts and stress fibers in stationary hamster fibroblasts, *J. Cell Biol.* **92:**398–408.

Small, J. V., and Celis, J. E., 1978, Filament arrangements in negatively stained cultured cells: The organization of actin, *Cytobiologie* **16:**308–325.

Small, J. V., and Langanger, G., 1981, Organization of actin in the leading edge of cultured cells: Influence of osmium tetroxide and dehydration on the ultrastructure of actin meshworks, *J. Cell Biol.* **91:**695–705.

Soranno, T., and Bell, E., 1982, Cytostructural dynamics of spreading and translocating cells, *J. Cell Biol.* **95:**127–136.

Spooner, B. S. Yamada, K. M., and Wesselss, N. K., 1971, Microfilaments and cell locomotion, *J. Cell Biol.* **49:**595–613.

Spooner, B. S., Ash, J. F., Wrenn, J. T., Frater, R. B., and Wessells, N. K., 1973, Heavy meromyosin binding to microfilaments involved in cell and morphogenetic movements, *Tissue Cell* **5:**37–46.

Strahs, K. R., and Berns, M. W., 1979, Laser microirradiation of stress fibers and intermediate filaments in non-muscle cells from cultured rat heart, *Exp. Cell Res.* **119:**31–45.

Strenli, C. H., Patel, B., and Critchley, D. R., 1981, The cholera toxin receptor ganglioside GM1 remains associated with Triton X-100 cytoskeletons of Balb/c-3T3 cells, *Exp. Cell Res.* **136:**247–254.

Taylor, D. L., 1976, Motile model systems of amoeboid movements, in: *Cell Motility* (R. Goldman, T. Pollard, and J. Rosenbaum, eds.), pp. 797–821, Cold Spring Harbor Laboratory, Cold Spring Harbor, New York.

Taylor, D. L., and Condeelis, J. S., 1979, Cytoplasmic structure and contractility in amoeboid cells, *Int. Rev. Cytol.* **56:**57–144.

Taylor, D. L., and Wang, Y.-L., 1980, Fluorescently labeled molecules as probes of the structure and function of living cells, *Nature (London)* **284:**405–410.

Tilney, L. G., DeRosier, D. J., and Mulroy, M. J., 1980, The organization of actin filaments in the stereocilia of cochlear hair cells, *J. Cell Biol.* **86:**244–259.

Toh, B. H., Gallichio, H. A., Jeffrey, P. L., Livett, B. G., Muller, H. K., Canchi, M. N., and Clarke, F. M., 1976, Anti-actin stains synapses, *Nature (London)* **264:**648–650.

Toh, B. H., Yildiz, A., Sotelo, J., Osung, O., Holborrow, E. J., and Fairfax, A., 1979, Distribution of actin and myosin in muscle and non-muscle cells, *Cell Tissue Res.* **199:**117–126.

Tomasek, J. J., Hay, E. D., and Fujiwara, K., 1982, Collagen modulates cell shape and cytoskeleton of embryonic corneal and fibroblasts: Distribution of actin, alpha-actinin, and myosin, *Dev. Biol.* **92:**107–122.

Trencheu, P., Sneyd, P., and Holborow, E. J., 1974, Immunofluorescent tracing of smooth muscle contractile protein antigens in tissues other than smooth muscle, *Clin. Exp. Immunol.* **16:**125–136.

Tucker, R. W., Sanford, K. K., and Frankel, F. R., 1978, Tubulin and actin in paired nonneoplastic and spontaneously transformed neoplastic cell lines *in vitro:* Fluorescent antibody studies, *Cell* **13:**629–642.

Verderame, A. D., Egnor, M., Smith, K., and Pollack, R., 1980, Cytoskeletal F-actin patterns quantitated with fluorescein isothiocyanate-phalloidin in normal and transformed cells, *Proc. Natl. Acad. Sci. USA* **77:**6624–6628.

Wang, E., and Goldberg, A. R., 1978, Binding of deoxyribonuclease I to actin: A new way to visualize microfilament bundles in nonmuscle cells, *J. Histochem. Cytochem.* **26:** 745–749.

Wang, K., and Singer, S. J., 1977, Interaction of filamin with F-actin in solution, *Proc. Natl. Acad. Sci. U.S.A.* **74:**2021–2025.

Wang, K., Ash, J. F., and Singer, S. J., 1975, Filamin, a new high-molecular-weight protein found in smooth muscle and non-muscle cells, *Proc. Natl. Acad. Sci. USA* **72:**4483–4486.

Watt, F. M., Harris, H., Weber, K., and Osborn, M., 1978, The distribution of actin cables and microtubules in hybrids between malignant and non-malignant cells, and in tumors derived from them, *J. Cell Sci.* **32:**419–432.

Weber, K., and Gröschel-Stewart, U., 1974, Antibody to myosin: The specific visualization of myosin-containing filaments in non-muscle cells, *Proc. Natl. Acad. Sci. USA* **71:**4561–4564.

Weber, K., Rathke, P. C., Osborn, M., and Franke, W. W., 1976, Distribution of actin and tubulin in cells and in glycerinated cell models after treatment with cytochalsin B, *Exp. Cell Res.* **102:**285–297.

Webster, R. E., Henderson, D., Osborn, M., and Weber, K., 1978, Three-dimensional electron microscopic visualization of the cytoskeleton of animal cells: Immunoferritin identification of actin- and tubulin-containing structures, *Proc. Natl. Acad. Sci. USA* **75:**5511–5515.

Wehland, J., and Weber, K., 1980, Distribution of fluorescently labeled actin and tropomyosin after microinjection in living tissue cultured cells as observed with TV image intensification, *Exp. Cell Res.* **127:**397–408.

Wehland, J., Osborn, M., and Weber, K., 1979, Cell to substratum contacts in living cells, a direct correlation between interference reflexion and indirect immunoflurorescence microscopy using antibodies against actin and alpha-actinin, *J. Cell Sci.* **37:**257–273.

Wessells, N. K., Spooner, B. S., and Luduena, M. A., 1973, Surface movements, microfilaments and cell locomotion, in: *Locomotion of Tissue Cells, Ciba Found. Symp.* **14:**53–77, Elsevier, Amsterdam.

Westermark, B., and Porter, K. R., 1982, Hormonally induced changes in the cytoskeleton of human thyroid cells in culture, *J. Cell Biol.* **94:**42–50.

Wharburton, M. J., Head, L. P., and Rudland, P. S., 1981, Redistribution of fibronectin and cytoskeletal proteins during the differentiation of rat mammary tumor cells *in vitro, Exp. Cell Res.* **132:**57–66.

White, G. E., Fijiwara, K., Shefton, E., Dewey, C. F., Jr., and Gimbrone, M. A., Jr., 1982, Fluid shear stress influences cell shape and cytoskeletal organization in cultured vascular endothelium, *Fed. Proc.* **41:**321 (Abstr.).

White, G. E., Gimbrone, M. A., and Fujiwara, K., 1983, Factors influencing the expression of stress fibers in vascular endothelial cells *in situ, J. Cell Biol.*, in press.

Wickus, G., Gruenstein, E., Robbins, P. W., and Rich, A., 1975, Decrease in membrane-associated actin of fibroblasts after transformation by Rous sarcoma virus, *Proc. Natl. Acad. Sci. USA* **72:**746–749.

Wieland, T., 1977, Modifications of actins by phallotoxins, *Naturwissenschaften* **64:**303–309.

Wieland, T., and Faulstich, H., 1978, Amatoxins, phallotoxins, phallolysin, and antamanide: The biologically active components of poisonous Amanita mushrooms, *CRC Crit. Rev. Biochem.* **5:**185–260.

Wigley, C. B., and Summerhayes, I. C., 1979, Loss of LETS protein is not a marker for salivary gland or bladder epithelial cell transformation, *Exp. Cell Res.* **118:**394–398.

Willingham, M. C., and Pastan, I., 1975, Cyclic AMP and cell morphology in cultured fibroblasts: Effects on cell shape, microfilament and microtubule distribution and orientation to substratum, *J. Cell Biol.* **67:**146–159.

Willingham, M. L., Yamada, K. M., Yamada, S. S., Pouyssegur, J., and Pastan, I., 1977, Microfilament bundles and cell shape are related to adhesiveness to substratum and are dissociable from growth control in cultured fibroblasts, *Cell* **10:**375–380.

Willingham, M. C., Yamada, S. S., Davies, P. J. A., Rutherford, A. V., Gallo, M. G., and Pastan, I., 1981, Intracellular localization of actin in cultured fibroblasts by electron microscopic immunocytochemistry, *J. Histochem. Cytochem.* **29:**17–37.

Wohlfarth-Bottermann, K. E., 1964, Cell Structure and their significance for amoeboid movement, *Int. Rev. Cytol.* **16:**61–131.

Wohlfarth-Bottermann, K. E., and Fleischer, M., 1976, Cycling aggregation patterns of cytoplasmic F-actin coordinated with oscillating tension force generation, *Cell Tissue Res.* **165:**327–344.

Wohlman, A., and Allen, R. D., 1968, Structural organization associated with pseudopod extension and contraction during cell locomotion in Difflugia, *J. Cell Sci.* **3:**105–114.

Wolosewick, J. J., and Porter, K. R., 1976, Stereo high-voltage electron microscopy of whole cells of the human diploid line WI-38, *Am. J. Anat.* **147:**303–323.

Wolosewick, J. J., and Porter, K. R., 1979, Microtrabecular lattice of the cytoplasmic ground substance: Artifact or reality, *J. Cell Biol.* **82:** 114–139.

Wong, A. J., Pollard, T. D., and Herman, I. M., 1983, Actin filament stress fibers in vascular endothelial cells *in vivo, Science* **219:**867–869.

Wulf, E., Deboben, A., Bautz, F. A., Faulstich, H., and Wieland, Th., 1979, Fluorescent phallotoxin, a tool for the visualization of cellular actin, *Proc. Natl. Acad. Sci. USA* **76:**4498–4502.

Yang, N. S., Kirkland, W., Jorgensen, T., and Furmanski, P., 1980, Absence of fibronectin and presence of plasminogen activator in both normal and malignant human mammary epithelial cells in culture, *J. Cell Biol.* **84:**120–130.

Zigmond, S. H., Otto, J. J., and Bryan, J., 1979, Organization of myosin in a submembranous sheath in well-spread human fibroblasts, *Exp. Cell Res.* **119:**205–219.

Zylberberg, L., and Nicolas, G., 1982, Ultrastructure of scales in a teleost (*Carassius auratus L.*) after use of rapid freeze-fixation and freeze-substitution, *Cell Tissue Res.* **223:**349–367.

3

The Form and Function of Actin

A PRODUCT OF ITS UNIQUE DESIGN

David J. DeRosier and Lewis G. Tilney

1. Introduction

Even if actin were only found in muscle, it would be an interesting protein, but it is found in abundance in all cells of the body. For example, it is implicated in pseudopod and filopod formation, in ruffling, in the production of the acrosomal process in invertebrate sperm, in cytokinesis, in the construction of the cytoskeleton, and in specialized extensions such as microvilli and sterocilia. These systems can be quite difficult to attack experimentally and consequently much of our knowledge of actin and motility has come from the study of muscle, a highly periodic structure. In striated muscle, actin, together with troponin and tropomyosin, regulates the attachment of myosin cross-bridges, activates the myosin ATPase, and provides part of the mechanical framework necessary to couple forces from neighboring filaments. Myosin is also found in nonmuscle cells but in much lesser amounts than in muscle. Some but not all of the motile events in nonmuscle cells require myosin and, for these events, it is assumed that muscle serves as a model system. Pseudopod formation and cytokinesis are two motile events thought to be driven by an actomyosin contraction.* For other motile events the involvement of myosin has either been ruled out or at least seems improbable: the extension of the acrosomal process in the sperm of *Thyone* and *Limulus*, for example, in-

*For references, see the excellent collection of articles on the cytoskeleton appearing in *Cold Spring Harbor Symp. Quant. Biol.* **81.**

David J. DeRosier • Rosenstiel Basic Medical Sciences Research Center, Brandeis University, Waltham, Massachusetts 02254. ***Lewis G. Tilney*** • Department of Biology, University of Pennsylvania, Philadelphia, Pennsylvania 19104.

volves the polymerization of actin (*Thyone*; Tilney, 1975a) and a change in twist of the actin filaments (*Limulus;* DeRosier *et al.*, 1982). There is no convincing evidence that myosin is present in these sperm and, moreover, the direction of motion is opposite to that produced by myosin. Thus, the function of actin in muscle contraction does not fully describe its role in nonmuscle motility.

Actin from a wide variety of sources is a remarkably conserved molecule both in terms of amino acid sequence, filament structure, and physical properties (Korn, 1982). As a structural protein, actin is capable of forming a number of different structures, some of which are bundles and other mesh works. Which particular actin structure is built, therefore, must be a function of which auxiliary proteins are present rather than there being different actins required for different cellular machines. For example, fascin (Bryan and Kane, 1978) or fimbrin (Bretscher and Weber, 1980; Bretscher, 1981) promote actin bundle formation, while actin gelation protein (Pollard, 1981) and actin-binding protein (Hartwig *et al.*, 1980) produce an irregular mesh work or gel of actin filaments.

Given that the actin filament is a conserved structure and yet one that is used in a variety of different cellular machinery, which structural features make it such a key protein? An obvious answer is that because actin must interact by specific binding with a number of different proteins, it must be evolutionarily conservative since changes in virtually any position would affect one of its many different binding sites. This is surely only part of the answer. We suggest that actin must have unusual mechanical features which make it suitable for inclusion in a wide variety of different cellular machines. We are now beginning to piece together some of the structural and mechanical properties of actin and to relate these to its function.

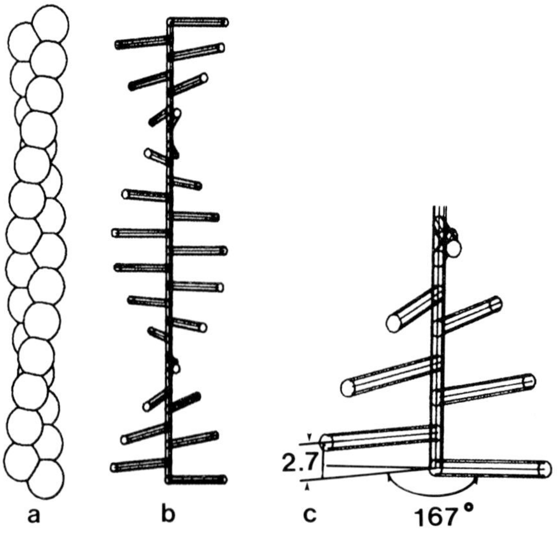

Figure 1. Actin symmetry. (a) Ping-pong ball model of actin showing two repeats of the structure. (b) A stick model depicting the relationship between adjacent subunits. The model consists of a rod running along the helix axis with projecting rods at right angles to it. Each projecting rod would pass through the center of one of the spheres in (a). (c) An enlargement of the model in (b). The symmetry operator for this helix is a rotation of 167° followed by an axial translation of 2.7 nm.

What we have elected to do in this article is first to describe the structure of the actin filament. Recent evidence (Egelman *et al.*, 1982; Egelman *et al.*, 1983; Egelman and DeRosier, 1983a) suggests a structure quite different from the perfectly helical "ping-pong ball" model of actin (Fig. 1a). In fact, the structure is disordered and the disorder, which has been neglected up to now, provides the key to understanding actin bundles. We will then use this new information on the structure of the actin filament to show what kind of bundles can form and thereby derive the "bonding rules" for bundles. We will next consider what happens when a bundle is bent, an important consideration as this gives insight into the rigidity of these bundles and how this rigidity may be modified. Such a consideration is not only important in determining the rigidity of cell extensions and thus how cells can crawl and communicate with each other but also will tell us something about the role of actin bundles in hearing. Finally, we will demonstrate how the structure of actin allows cells to form and maintain bent bundles and how such bundles can produce force without myosin simply by changing the twist of the actin filaments.

2. *The Structure of the Actin Filaments*

Our earliest picture of the anatomy of an actin filament as seen in the electron microscope consisted of a filament whose diameter seemed to be about 6–8 nm (Fig. 2a). From such images, it was concluded that the filament appeared to consist of a helical array of globular subunits (Fig. 1a) with a crossover spacing of 38.2 nm. Since single filaments are not well ordered, investigators, beginning with the pioneering work of Jean Hanson (1967), began to study actin paracrystals (Fig. 2b) because they are extremely well ordered and therefore amenable to image analysis. The object of choice was a paracrystal induced by high concentrations of magnesium as these are easily made from muscle actin. Several groups have used these specimens in order to determine the three-dimensional structure of actin and actin plus tropomyosin. In their reconstructions (Moore *et al.*, 1970; Spudich *et al.*, 1972; Wakabayashi *et al.*, 1975), the actin filament appears to be about 7–8 nm wide. Very recently, Egelman and DeRosier (1983a) have produced a model for actin which is substantially different from that derived from paracrystals. Their model is based on electron micrographs of single filaments rather than paracrystals, and it is the only model that can be reconciled with X-ray diffraction data from muscle. It also accounts for the images observed in the Mg^{2+} paracrystals (Egelman and DeRosier, 1983b; see Figs. 2c and d). Thus, there are two different models of F-actin both of which account for the paracrystal images. This seemingly paradoxical situation revolves around basic assumptions about the arrangement of the filaments in the paracrystal. The earlier workers assumed or concluded that they could successfully extract the image of a single filament from the image of the paracrystal. Thus, they believed the interfilament spacing, 6–7 nm, was approximately equal to the diameter of the filament, 7–8 nm. Egelman *et al.* (1983) conclude that actin is 9.5 nm in

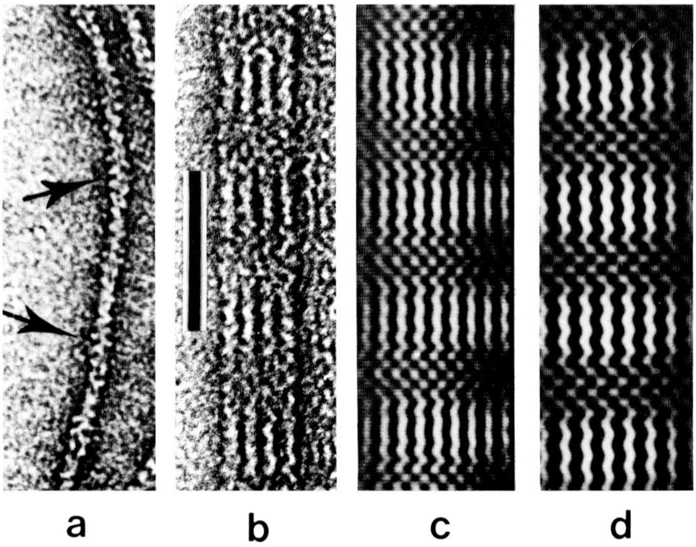

Figure 2. Actin-single filaments and paracrystals. (a) Electron micrograph of a single filament of actin in negative stain. The arrows indicate regions where chevron appearance of actin is clearly visible. (From Vibert and Craig, 1982. Reprinted by permission from Academic Press, Inc.) (b) Electron micrograph of a Mg^{2+} paracrystal of actin in negative stain. (c) A filtered image of a paracrystal. The filtered image is created by averaging together the repeated features in a micrograph of a paracrystal such as the one in (b). (From Egelman and DeRosier, 1983b. Reprinted by permission from Academic Press, Inc.) (d) Computer simulation of a paracrystal. The 9.5 nm wide model of F-actin deduced by Egelman and DeRosier (1983a) and shown in Figure 5 is packed in an antiparallel tetragonal array and the result projected and filtered in order to simulate the image seen in (c). Note that the two images agree with near perfection. (From Egelman and DeRosier, 1983b. Reprinted by permission from Academic Press, Inc.) The bar corresponds to a distance of 50 nm.

diameter and that in the antiparallel arrangement in the paracrystal there is substantial superposition of subunits between adjacent filaments. Thus, they suggested, previous workers by "carving out" a single filament from the paracrystalline array, trimmed off the outer edges of subunits in the selected filament and failed to trim out subunits from neighboring filaments, which are oppositely oriented. By so doing, they have produced a model for actin which is incorrect.

2.1. Filament Symmetry

While controversy still surrounds the question of subunit shape, there is little controversy about filament symmetry. The ideal actin filament consists of identical subunits equivalently arranged on a helical lattice. This lattice determines the symmetry of the filament or, in other words, it describes the spatial relationships between all subunits in the filament. The familiar ping-pong ball model (Fig. 1a) does not reveal the helical parameters in an easy to understand way. We have therefore replaced the ping-pong balls with sticks (Figs. 1b and c) which are radially arranged so that they would poke through the heart of the actin subunits. Two neighboring sticks are related as follows: the second stick is 2.73 nm above the first and is rotated clockwise as viewed from the top by 167°. The relationship between the second and third sticks is the same as that between the first and second. Thus, the symmetry of an ideal filament is described by two parameters, the axial rise per subunit (2.73 nm) and the angle between subunits (167°). There are other ways of describing the symmetry, but this one is the most useful for our purposes.

Do these parameters differ for filaments from different sources? The answer may be "yes" but not by much. In the thin filaments of lobster fast muscle (an invertebrate), the axial rise is 2.76 nm and the angle 167.2° (Wray *et al.*, 1978). In frog sartorius muscle the axial rise is 2.73 nm and the angle is 166.7° (Huxley and Brown, 1967).

Do actin filaments from the same species have different parameters under different conditions? Here, the answer seems to be a definite "yes." Gillis and O'Brien (1975) found two forms of thin filament paracrystals, one having an angle of 166° and one an angle of 167°. The rise per subunit was not reported. In a second system, the sperm of the horseshoe crab, there is an actin bundle which, as part of its physiological function, undergoes a change in the angle from 166° to 167°.

The effect of this change in angle can be seen in Fig. 3. In Fig. 3a the filament is drawn having an angle of 166° and an axial rise of 2.73 nm. In Fig. 3b the filament has the same axial rise (2.73 nm) but a different angle (167°). The most striking difference in these two filaments can be seen in the change in crossover spacing (see arrows). As the angle increases so does the crossover period. Note that this lengthening of crossover spacing does not lengthen the filament since it is due to a change in angle not in axial rise per subunit. Rather, the lengthening of the crossover arises from the change in *twist* of the filaments, that is, a change in angle corresponds to a change in twist. This can be seen by considering the actin filament to be constructed of two strings of beads that wind around each other. Where the two strings lie one behind the other we observe a crossover (see arrows). By changing the angle from 166° to 167° we lessen the twisting of the two strings around each other and the crossovers become less frequent. Thus, variations in angle will be reflected in changes in crossover spacings. A change from 166° to 167° results in a change in crossover from 35.5 to 38.2 nm.

2.2. Angular Disorder

We know that the angle between adjacent subunits is a bit different for different actin filaments and that it can even be different for the same filaments in different conditions. Does the angle also vary within a single filament under a fixed set of conditions? The answer is once again "yes." This variation, which is large, is known as angular disorder (Egelman *et al.*, 1982).

Generally, disorder is the last thing a structural biologist wants to find in a specimen, since it increases the difficulty of analysis, but disorder does not preclude analysis of the structure, nor does it imply artifact. In only a few cases has the form of disorder of a biological specimen, as seen in the electron microscope, been analyzed. Generally, researchers move on to another structure or find some other way to prepare the specimen. In the case of actin, the disorder has proved tractable, and what is more important, the disorder is physiologically interesting. In fact, in retrospect, it was unfortunate that early investigators began the study of actin by studying a very ordered structure, the magnesium paracrystal. By so doing, they neglected one of the most important features of the actin filament, namely, its angular disorder. Once we understand this feature, we have a much greater appreciation about how the actin filament functions in muscle and in actin bundles present in a variety of nonmuscle cells.

There are several ways to detect angular disorder. It affects the diffraction pattern of the structure in a very specific way, which we will discuss later. Its presence also causes variability in the crossover spacings along the filament axis. This latter effect can be easily seen in Fig. 3. On the left-hand side of the figure are two actin filaments having slightly different helical parameters. Both have the same axial rise per subunit, but in the leftmost one, the angle between subunits is 166°, whereas in the center filament, it is 167°. This change in angle or twist of the filament causes a corresponding change in the crossover spacings denoted by arrows. Suppose we now make the angle between a pair of subunits variable so that instead having the same angular rotation, e.g., 166°, between subunits, the angle varies over a range of say 156–176°. Then the angle that exists between neighboring subunits will vary for different pairs of subunits. More importantly, the angle between pair of neighboring subunits does not affect the angle between the next pair. In this situation, the angular deviations become cumulative. To put it differently, if the root mean square (rms) angular deviation between subunits 1 and 2 is 10° then the rms deviation between 1–3 is $10° \sqrt{2}$, and between 1 and n is $10° \sqrt{(n-1)}$.

If we now examine the corresponding filament (Fig. 3b), we see a variation in crossover spacings. Short crossover spacings occur in a region where the angle per subunit is on average less than 166° and longer spacings when it is more. These fluctuations in angle occur randomly and lead to random fluctuations in crossover spacing as shown in Fig. 3. Such variations in crossover spacing were initially noted by Jean Hanson (1967) in her characterization of negatively stained filaments. She measured an average crossover spacing of 38.2 nm with a standard deviation of 2.3 nm. The variation in spacing converts to an rms angular variation of 6° which is very similar to the 10° measured by Egelman *et al.* (1982). The differences between these two values presumably lie in the differences in averaging techniques employed in the two different studies. Hanson measured lengths corresponding to between three and nine crossovers and then divided this by the number of crossovers. This is ideal for measuring crossovers as the average crossover spacing is not affected by the number of crossovers per measurement, but it is not ideal for the measurement of rms angular variation as this measure is affected by the number of crossovers used per measurement.

Egelman *et al.* (1982) used computer diffraction patterns of negatively stained filaments in order to determine the amount of disorder. Diffraction patterns of actin consist of a series of layer lines which are perpendicular to the helix axis (Fig. 4a). The stronger layer lines are the equator ($l=0$), the first ($l=1$), the sixth ($l=6$), and the seventh ($l=7$). Angular disorder affects the

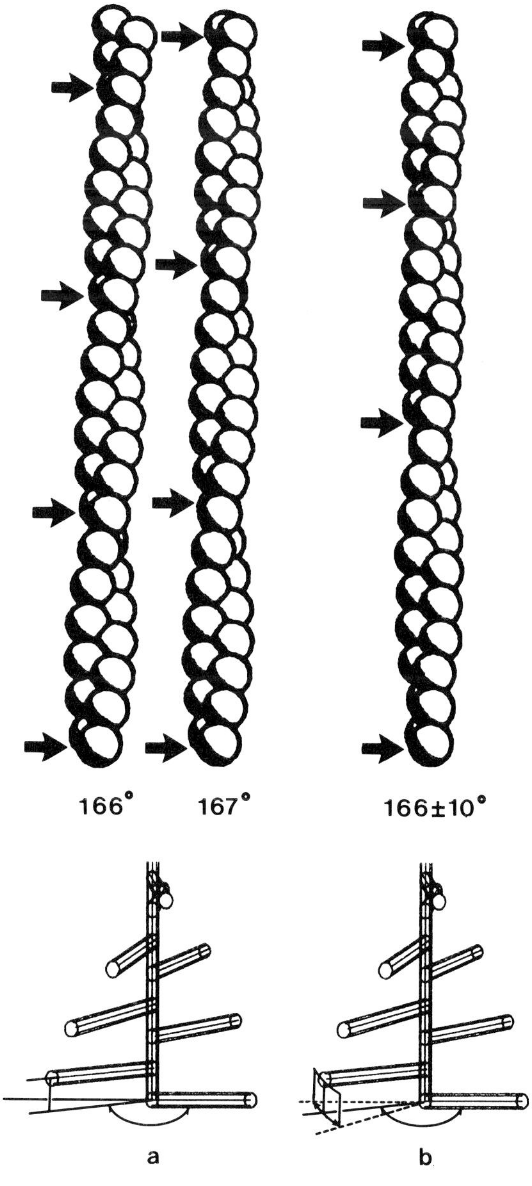

Figure 3. Change in actin twist. (a) Two perfectly helical actin models each of which has a constant axial rise per subunit (2.73 nm) and constant angle between adjacent subunits. In the leftmost filament the angle is 166° and in the filament to its right 167°. Note that the change in angle corresponds to a change in twist. This can be seen in the spacing between crossovers indicated by the arrows (i.e., the region where the twin strands cross over each other). An angle of 166° gives rise to a periodicity of 35.5 nm and a change in twist of 1° per subunit lengthens the crossover to 38.2 nm. This "lengthening" takes place with no change in the axial spacing of subunits. (b) A helix with a random variable twist. In this filament, the expectation value for the angle between subunits is 166° with a standard deviation of 10° between adjacent subunits. The result is that the spacing between crossovers is variable. (Reprinted by permission from Egelman *et al.*, 1982. Copyright © 1982, Macmillan Journals Limited.)

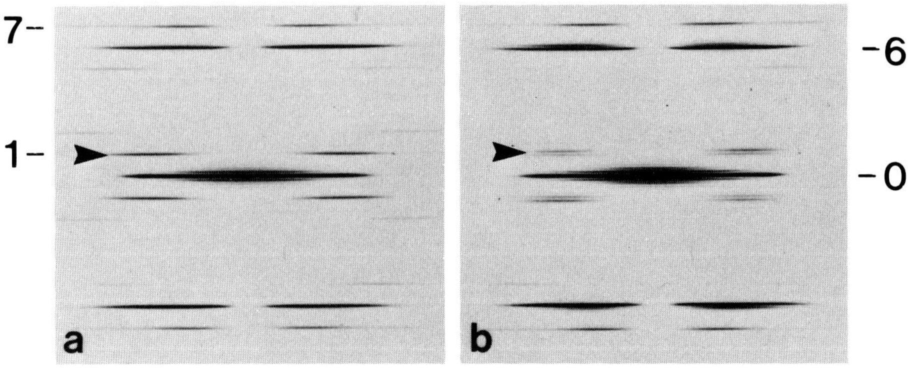

Figure 4. Diffraction patterns of model actin filaments. (a) Diffraction pattern of a perfectly helical (ordered) version of the filament shown in Figure 5. Note that all the layer lines are thin. (b) Diffraction pattern for the same filament but with 10° of angular disorder. Note that the first layer line is weaker relative to the sixth and seventh than it is in (a). Note that it is also split horizontally by the disorder. The equatorial (0), first (1), sixth (6), and seventh (7) layer lines are marked on the left or right margins of the image. The last three correspond to spacings of 1/36, 1/5.9, and 1/5.1 nm^{-1} respectively.

intensities of the layer lines and it affects different layer lines differently. In 1969, Huxley noted the fall-off in diffracted intensity with increasing distance from the meridian and suggested that variability in the twist would produce such an effect. A computer diffraction study of electron micrographs of actin filaments allows one to establish the validity of the angular disorder as well as estimate its magnitude. In particular, Egelman and DeRosier (1982) showed that the relative strengths of layer lines depend on the length of filament diffracted from. If the length is very short, the layer lines will have relative intensities near that of a perfectly helical particle. The problem is that very short lengths have a poor signal to noise ratio. As the length is increased the first layer line falls in intensity relative to the sixth (compare Fig. 4a and b). The effect depends on the square of the angular disorder.

By examining a number of different actin filaments, Egelman and DeRosier (1982) concluded that a value of 10° was about right, that is to say 5° was decidedly too small and 15° too high. Since the effect depends on the square of these numbers, a variation in angle of 5–15° corresponds to roughly an order of magnitude in intensity. The variation in diffraction patterns of the actin images made it difficult to determine the angular disorder more precisely than this.

2.3. Other Forms of Disorder

The reader must now be asking, "What about variations in axial rise of the subunits? Do these occur?" The analysis of electron micrographs shows no detectable variation in the 2.73 nm rise per subunit under conditions where movements resulting from the angular disorder are easily detected. Further, in muscle under full physiological tension, any change must be under .01 nm (Huxley and Brown, 1967; M. Kress, personal communication). Thus, we conclude that the rise per subunit is constant along the length of a filament.

Another form of disorder is filament curvature due to flexibility. All filamentous structures can be characterized by a measure of the length of filament for which the angular orientations of the two ends are correlated, that is, for a short segment there is little or no curvature and the two ends point 180° away from one another. For a very long filament, there is substantial curvature, looping, etc., and one end may point at any direction relative to the other. Put another way, for a very long filament given the axial orientation of one end, there is no way to predict the orientation of the other end without knowledge of the exact form of the loops and curves. The flexibility is characterized by a length, the "persistence length," over which the directionality of the two ends of the filament remain correlated. The longer this length, the less flexible the filament. Actin, interestingly, has a persistence length of 6 μm at 20°C (Oosawa, 1980).

Thus, a description of the geometry of actin needs to be modified to include disorder. Actin is a flexible filament consisting of a subunit having a fixed axial displacement of 2.73 nm relative to its neighbors, but a variable angular displacement of 167° ± 10°. As we shall see, the subunit shape and arrangement appear to dictate the form, flexibility, and angular disorder of the filament. In its turn, the form, flexibility, and angular disorder determine the way in which assemblies of filaments are built.

2.4. Subunit Shape

The flexibility and angular disorder make actin a difficult structure to study in the electron microscope. While it is possible to choose straight segments of filaments, it has not been possible to find segments with no angular disorder. Thus, a long, straight length of filament behaves as a collection of short segments each having a different twist (see Fig. 3b). Since image analysis methods usually depend on averaging together of structural features in different parts of the filament, it is essential that one knows the precise details of the particular variations for an individual filament before averaging can take place. The general characterization of the disorder such as the rms angular deviation of 10° is not sufficient.

In light of this difficulty, Egelman *et al.* (1983) turned to model building where the shape and disorder could be built in and the result compared with results from images of real filaments. In order to carry out this study they had to obtain data which characterized or measured the morphological features of actin. They also had to construct a model with variable parameters which could be fitted to the data.

2.4.1. Characterization of the Features of Actin

Electron micrographs of negatively stained filaments (Fig. 2a) show a chevron-like appearance between crossovers. The detailed appearance is quite variable so that in trying to fit a model to the image, it would be difficult to know which segments to pay attention to and which to ignore. Clearly, what is needed is some way to combine all the segments into a more easily characterized pattern. This is done by producing a diffraction pattern of the micrograph.

Figure 4a shows the diffraction pattern of a model actin filament having no angular disorder. The strongest features are the equator and the first, sixth, and seventh layer lines. These four layer lines would seem to be good ones from which to extract data. For diffraction patterns from negatively stained real filaments, however, the lack of cylindrical symmetry of the stain together with the angular disorder render some of these layer lines less useful. More specifically, the equatorial layer line is affected by the distribution of stain as well as by filament structure, but is not affected by angular disorder. Because it would be necessary to model the general distribution of stain as well as the subunit shape, the equatorial layer line was not used. The remaining three layer lines do not suffer from this defect, but they are all sensitive to the angular disorder. While the angular disorder in the filaments does lead to considerable variability in the transforms, several general features are consistently found. The sixth layer line is always the strongest feature and exhibits a single broad reflection on peak at 0.09 nm^{-1}. The seventh layer line is weaker, having a reflection at 0.08 nm^{-1} and often a second peak of equal intensity to the first at 0.24 nm^{-1}. The first layer is weak, its intensity being at most half that of the sixth even after the disorder is taken into account. The positions and relative intensities of the layer line reflections are enough information from which to determine values for parameters of the model.

2.4.2. *A Model for Actin*

Several groups have been working on actin-containing crystals. Among them are Suck *et al.* (1981) on actin-DNase I and Aebi *et al.* (1981) on the Gd^{3+} sheets. Both find a bilobed actin subunit which is elongated but the sizes of the lobes and the exact dimensions differ somewhat between the two maps. Making use of this information, Egelman and DeRosier (1983a) chose a model consisting of two spheres each of diameter 3.8 nm; the two spheres were separated by 3.0 nm giving rise to a bilobed subunit having a dimension of 6.8 × 3.8 × 3.8 nm. They systematically varied the position and orientation of the subunit and found that only one orientation of the model fits the characteristics of the transform (Fig. 5a); actually, one narrow range of orientations fits the data.

Before describing the result, it is important to qualify it in terms of the relationship between the model and the real filaments. Perhaps the best place to begin is to consider why, for example, the relative radii of the spheres were not variables in the model. There are three reasons for this. First, the number of models to compute increases (to astronomical proportions) as the number of variables increases. To systematically search for the correct orientation requires generating about 200 models. If five radii were tried for each of the two spheres, 5000 models would need to be considered. Second, since the data

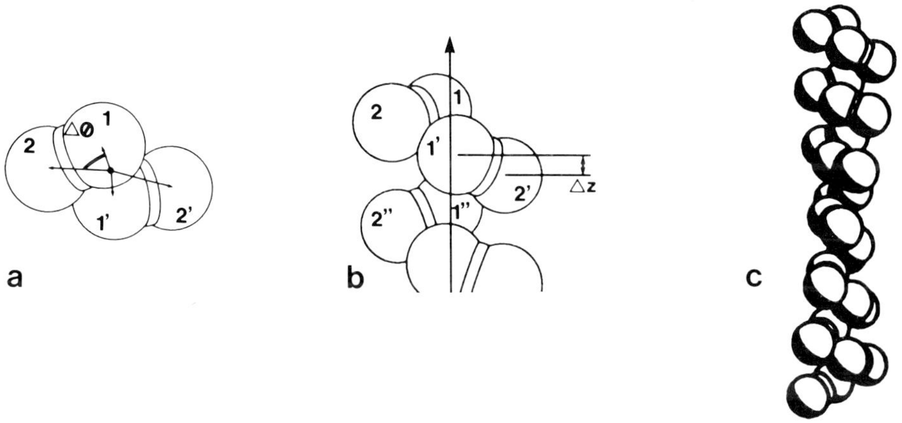

Figure 5. Model for the actin filament. (a) Top view of the model. The subunit consists of two spheres each of diameter 3.8 nm and separated by 3.0 nm. The orientation of the subunit is such that the distance of the inner sphere (#1) from the axis is 1.2 nm and that of the outer sphere (#2) is 3.0 nm. The angle $\Delta\phi$, between the centers, is 70°. (b) Side view of the model. The axial displacement, Δz, of the outer sphere with respect to the inner appears variable lying between 0.5 and 1.5 nm. Only adjacent subunits make contact. Sphere 1 touches sphere 1′, while sphere 2 is only slightly off contacting 1′. In a real filament, it is possible that both domains 1 and 2 contact domain 1′. This would depend on the exact shape of the subunit. (c) A shadowed side view of the model. (From Egelman and DeRosier, 1983a. Reprinted by permission from Academic Press, Inc.)

describe a peak position and relative strength rather than the exact distribution along the layer line, they are probably insufficient to pick out the subtler aspects of the shape such as relative domain size. Third, it may be that departures of domain shape from spherical symmetry may contribute as importantly to the transforms as relative domain size so that one would be finding a best fit for the wrong model.

Now let us consider their result. Actin is best described by a subunit consisting of two 3.8 nm globular masses, one at a radius of about 1.2 nm and one at a radius of about 3.0 nm. The long axis of the subunit is tilted by about 20° off perpendicular to the helix axis to give a chevron appearance to the filament (Fig. 5c). This subunit axis does not pass through the helix axis but is skew to it. In the fitting it was found that different tilts, i.e., 10° to 30°, fit different images and we think that variable tilt of the subunit is a real feature of the filament structure. We will return to this feature in our discussion of bundles.

One of the notable features of the model is the large outer diameter, about 9.5 nm, of the filament. The exact diameter, however, depends on the detailed shape of the subunit, but what is certain is that this is inconsistent with a filament outer diameter in the range 6–8 nm, as has been quoted in the literature. The outer diameter of the filament is a difficult quantity to measure. The reason is that the distribution of negative stain tends to obscure the outer edges of the particle making it appear narrower. If one examines the images of actin filaments carefully, one finds wispy extensions protruding some 4.5–5.0 nm from the axis. These would suggest a much wider filament, but the subjective interpretation required for the measurement makes the value obtained for the radius questionable. An objective measure of width can be obtained from the data contained on the sixth and seventh layer lines of the actin transforms. Each layer line contains information about the radial distribution of density and can provide an objective estimate of the particle diameter. Egelman and DeRosier (1983a) used these techniques to show that the outer diameter of the single actin filaments seen in negative stain is, in fact, approximately 9.5 nm.

Thus, the picture of actin is, we think, substantially different from the rigid ping-pong ball model usually seen in textbooks. We now ask what role these features play in the structure and function of actin assemblies.

3. The Relationship of Structure to Properties of the Filament

Dr. E. Egelman (personal communication) suggests that the tilting of the actin subunit in the filament can be interpreted in terms of its two domains. The inner domain, in the model, is the one that is bonded to produce the filament. The simplest interpretation of the tilt would be an axial movement of the outer domain relative to the inner presumably as a result of some hinge between the two.

In the model, the intrafilament bonds lie close to the filament axis and are only between adjacent subunits. This has implications for models of polymerization, since we are suggesting only one class of interactin bond rather than two. This structure differs appreciably from a uniform rod having the same mass per unit length. In particular, E. Egelman (personal communication) has calculated that actin filaments are substantially more flexible (about 65 times) and have much less torsional rigidity than would be expected for the corresponding uniform rod of protein. This conclusion agrees with our intuition about the mechanical properties of our model filament. Since the intersubunit contacts occur only between nearest neighbors, the bondings all follow a single helical path rather like a wire spring. This is made clearer perhaps by a comparison to the microtubule. In the latter, the bonding between subunits is in at least two directions (rather than one as in the actin model), giving greater rigidity. That is, a microtubule is rather like a *sheet* of subunits rolled into a tube whereas actin is more like a *string* of subunits coiled into a helix. The differences in mechanical properties of these two kinds of filaments are displayed in their organization into cellular assemblies.

4. Actin Bundles

In many nonmuscle cells, extensions protrude from the cell surfaces. Within these extensions is a core composed of a bundle of actin filaments which appears to be responsible for maintenance of its shape. Most spectacular are the brush border of intestinal epithelial cells (Mooseker and Tilney,

1975), the microvilli of sea urchin eggs (Spudich and Amos, 1979), the stereocilia sensory tranducers on the hair cells of the vertebrate ear (Flock and Cheung, 1977; Tilney *et al.*, 1980), and the acrosomal process, a long microvillar-like structure produced by a sperm (Tilney *et al.*, 1973).

The naturally occurring actin bundles in the microvillus, the acrosomal process, and the stereocilium consist of parallel actin filaments cross-bridged by additional proteins. These paracrystalline arrays differ from Mg^{2+} paracrystals in two important ways. First, Mg^{2+} paracrystals are bipolar while naturally occurring paracrystals are polar, and second, the naturally occurring paracrystalline bundles require a cross-bridging protein. In cross section, the filaments in the naturally occurring bundles can be found in some cases to be packed in a hexagonal lattice, for example, in the stereocilia in the chicken (DeRosier and Tilney, 1982; Fig. 6b) or the acrosomal process in *Limulus* sperm (Tilney, 1975b). In other cases, the transverse packing of filaments is liquidlike (the stereocilium of the lizard; Fig. 6d).

What are the bonding rules for building these bundles? This question is dealt with in some detail in DeRosier and Tilney (1982), DeRosier *et al.*, (1980a), and DeRosier *et al.* (1977), and can be summed up as follows: the cross-bridging proteins are known to bond to the actin filaments at specific sites. Furthermore, the actin cross-bridge must have two distinct actin-binding sites (call them A and B) in order to cross-bridge two filaments. Correspondingly, the actin subunits must have two sites, A′ and B′, for the cross-bridge. In a bundle, all cross-bridges make the same two actin bonds, i.e., A-A′ and B-B′.

From studies on the packing of actin filaments in a number of different biological situations, both *in vivo* and *in vitro* (DeRosier and Tilney, 1982), we know that in all cases the crossover points of adjacent actin helices are in register (Figs. 6a and 6c). It turns out that this rule for cross-bridging has the special property that it allows complete freedom in the lateral arrangement of filaments, that is, the filaments can be arranged on some regular lattice or can have a random, liquid packing. This is due to the fact that subunits on a filament are arranged in a helical pattern. Thus, the geometry (helical symmetry) of the actin filament itself is used to its best advantage when the crossover points of adjacent actin filaments lie in register. To see why this rule produces this result, consider the packing of actin filaments in a hexagonal lattice. We know that there are approximately 13 subunits per crossover or each crossover contains 13 subunits pointing outward at equal angles from the axis. Thus, actin is not hexagonally symmetric. This means that to produce a hexagonally packed bundle of filaments that does not have a 6-fold symmetry, there has to be either distortion of the cross-bridging protein, or some movement of the actin subunits. Of course, a similar problem exists with any other packing, for example, tetragonal (square) packing. What is exciting and eminently satisfying to a structural biologist is that the actin filament is designed to accommodate any packing arrangement simply by making use of its angular disorder.

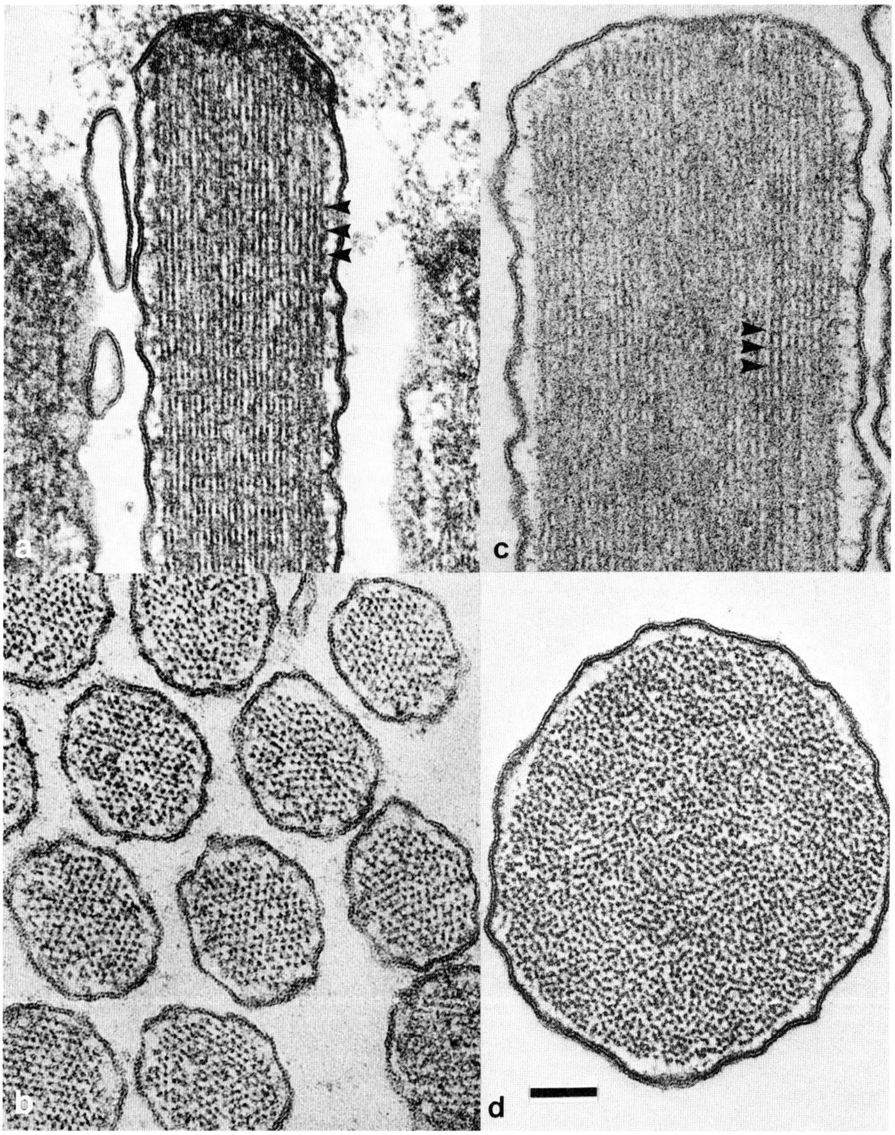

Figure 6. Actin bundles in stereocilia. (a,b) Longitudinal and transverse sections of stereocilia from the hair cells in the chicken. The core of the stereocilia consists of a bundle of actin filaments. The arrows in (a) indicate the crossover points, which are in register. This fixes the bonding rule (i.e., the spatial relationship between any two filaments that permits cross-bridging). In the transverse section in (b), the filament arrangement is essentially hexagonal but there is a good bit of irregularity. (c,d) Longitudinal and transverse of stereocilia from the hair cells of the alligator lizard. In the longitudinal section, the crossover points are in register and, therefore, this bundle has the same bonding rule as that in chicken. In the transverse section in (c), the arrangement of filament is liquid. Therefore, the same bonding rule is consistent with more than one filament packing scheme. Bar corresponds to a length of 100 nm. (Reproduced from Tilney *et al.*, 1983, and Tilney *et al.*, 1980, by copyright permission of the Rockefeller University Press.)

4.1. Hexagonal Bundles

Let us consider the case of hexagonal packing first.

For hexagonal packing, the subunits all must lie close to (if angular disorder is taken into account), or on one of the six hexagonal directions 0°, 60°, 120°, 180°, 240°, and 300°. Table 1 shows the axial and angular positions of the first 14 subunits in an actin filament. Note that subunits 0, 4, 9, and 13 (the subunits which in naturally occurring bundles form bonds) are all within 10° of lying along one of the six hexagonal directions (0°, 60°, 120°, 180°, 240°, and 300°), that is, subunit 0 is at 0°, subunit 4 is 5° off the 300° direction, subunit 9 is 5° off 60°, and subunit 13 (a repeat) is again at 0°.

Let us assume that two filaments, A and B, can be cross-bridged at subunit 0 (Figs. 7a and b). The cross-bridges are illustrated as rods. The angular disorder of filaments tells us that if subunit 0 is fixed at 0°, then subunit 4 will be found at 305° ± 10° $\sqrt{4}$ or between 285–325°. It therefore is capable (as are subunits 3 and 5, incidentally) of forming a cross-bridge to a third filament C lying along the 300° lattice line. With subunits 0 and 4 fixed at 0° and 300°, the remaining subunits will presumably partition themselves to take up the strain roughly uniformly (see column B in the table). The same reasoning can be applied to subunit 9 so that filament A can be cross-bridged to filament D (Figs. 7a and 7b). If we simply repeat this pattern, it will build a hexagonal bundle (Figs. 7c and d).

Table 1. Angular Positions of Actin Subunits

Subunit number	Subunit axial position (Å)	A Subunit angular position (°)	B Distorted angular positions for hexagonal bundle (°)	C Distorted angular positions for tetragonal bundle (°)
0	0	0	0	0
1	27.3	166	165	167
2	54.6	332	330	334
3	81.9	138	135	141
4	109.2	305	300	309
5	136.5	111	108	116
6	163.8	277	276	283
7	191.1	83	84	90
8	218.4	249	252	261
9	245.7	55	60	64
10	273.0	222	225	228
11	300.3	28	30	32
12	327.6	194	195	196
13	354.9	0	0	0

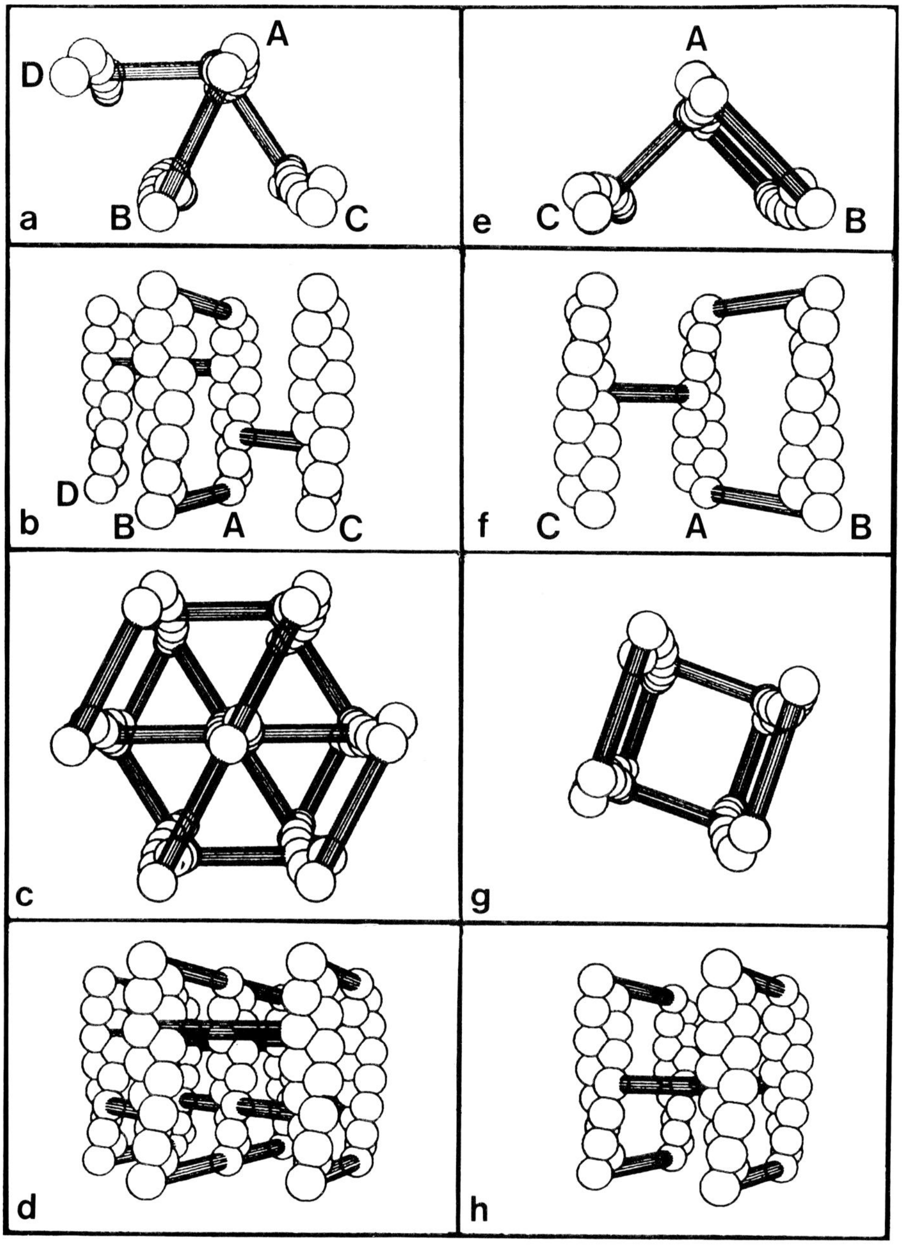

Figure 7. The bonding in actin bundles. (a–d) Hexagonal bundles. (e–h) Tetragonal bundles. (a) Top view of a partial hexagonal bundle. The actin symmetry and angular disorder make it possible to connect filament A to filaments B, C, and D with the same bonding rule. The cross-bridge, indicated by a rod, is arbitrarily positioned between the outermost surface of the bottom subunit (#0) in A and the innermost surface of subunit #0 in B. The exact same cross-bridge is made at subunits 4 and 9 between filament A and filaments C and D, respectively. (b) Side view of (a). (c) Top view of extended hexagonal bundle. The same bonds in (a) are extended to show how

4.2. Tetragonal and Liquid Bundles

Bundles in which the filaments have tetragonal order are equally possible, although none has been found. The difference is that the angular relationships are confined to multiples of 90° rather than 60°. For example, it is possible to cross-bridge three filaments, A, B, and C, whose axes lie on a square lattice, i.e., are related by multiples of 90°. As before, filament A can be cross-bridged to B at subunit 0 (Figs. 7e and 7f). Note in column A of the table that subunit 7 is at 83°, just 7° off 90°. As shown in column C, the filament can be angularly distorted such that subunit 7 is at 90°, so that filament A can be cross-bridged to C. Thus, the bonding shown in Fig. 7a will also produce a tetragonal arrangement.

Since both the tetragonal and hexagonal packings are possible given the *same* bonding rule, it is an easy step to imagine a packing which is a mixture of these two angles, 60° and 90°, and in fact, a mixture of all angles since bonding can take place at any position. Such a mixture of angles generates transverse liquid filament packing as has been observed in the stereocilia of lizards. In the lizard bundle, the average number of neighbors is about five, halfway between four for tetragonal, and six for hexagonal.

If the bonding rules are the same for tetragonal, liquid, and hexagonal packing of filaments, what factors influence which packing will in fact occur. One difference is the number of cross-bridges. Note that in the hexagonal case there are three cross-bridges (at subunits 0, 5, and 9) per repeat, while in the tetragonal case there are only two (at 0 and 7). The liquid-like bundle, which is intermediate, presumably has on average 2.5 cross-bridges per repeat. Thus, the hexagonal bundle corresponds to the maximal extent of cross-bridging. On the other hand, there will be a favorable entropic energy associated with the disorder in a liquid bundle. It is also possible that a liquid bundle may be a metastable form and may correspond to a local energy minimum. Evidence for this was found by Kane (1976), who noted that although bundle formation was very rapid (minutes), the appearance of highly ordered hexagonal from the disordered ones occurring initially occurred only after hours of incubation. One interpretation is that in the rapidly formed bundles there is liquid order, but with time, the cross-bridges can rearrange to form a hexagonally packed bundle.

In summary, the combination of the helical symmetry and angular disorder endow the actin filament with the ability to form paracrystalline bundles with variable transverse order. Such bundles, of course, have the property that they are much stiffer than a single filament, or even a bundle of uncross-bridged filaments.

---

a complete hexagonal bundle is constructed. (d) Side view of (c). (e) Top view of partial tetragonal bundle. Using the same crossbridge contacts as in (a), filament A is cross-bridged to B. Filament C is at 90° with respect to B (i.e., on a square lattice) and can be cross-bridged to A at subunit #7. (f) Side view of (e). (g) Top view of extended tetragonal bundle. This shows how four or any number of filaments can be cross-bridged. Tetragonal bundles, however, have not yet been observed. (h) Side view of (g).

4.3. Comparison with Bundles of Microtubules

This situation can be contrasted with that of microtubules, which lack the high degree of angular disorder of actin. More specifically, we know that the microtubule consists of 13 protofilaments which, like actin, point out radially at equal angles of $360/13 \simeq 28°$. Cross-bridged bundles of microtubules exhibit much greater polymorphism than actin in that they are not confined to liquid or hexagonally close-packed arrangement. Some microtubules are hexagonally close packed, such as that found in *Phaeodaria* (Cachon and Cachon, 1974). In this case the microtubules sit on 6-fold axes. Others are hexagonally packed but on a more open lattice with the microtubule sitting on the 3-fold axes of the lattice, as in *Acantharia,* on the 2-fold axes, *Raphidiophrys,* or on both the 3-fold and 2-fold axes, *Nassellaria* (Cachon and Cachon, 1974; Tilney, 1971). Another favored packing is in the form of sheets which are in turn cross-bridged to other sheets. One particularly striking arrangement is the double spiraled array in *Actinosphærium* (Tilney, 1968). These variations in packing, unlike those of actin, do not derive solely from the helical symmetry and disorder (if any) of the microtubule but rather must be specified by the cross-bridging proteins and/or other auxiliary protein such as MAPs.

5. The Bending of Bundles

From the preceding discussion, the next question that one might ask is: how rigid is a bundle of actin filaments and what happens when a cross-linked bundle of filaments is bent? Biologically, these are important questions because, in most cases, the actin bundles are present in cells or cell extensions to provide support. The actin bundle provides rigidity so that the acrosomal process of the sperm can penetrate the jelly surrounding the egg (see Section 5.2). A particularly interesting example of a bundle supporting a cell extension is found in the stereocilia that extend from the sensory hair cells of the ear. These structures must be extremely rigid as they are capable of detecting displacements of only an angstrom (0.1 nm) at threshold, yet at the same time they are not broken when they are displaced a micron or more, as during stimulation with loud noise.

5.1. Bending Associated with Stereocilia

Recently, we have studied the actin filament bundle in the stereocilia that extend from the hair cells of a bird cochlea (Tilney *et al.*, 1983). The organization of the filaments in this paracrystalline bundle is easy to analyze as the filaments lie on a hexagonal lattice. Thus, in longitudinal sections, the cross-bridges on neighboring filaments all "line up" producing transverse stripes every 120 Å. Besides the striping due to the bridges, the crossover points themselves are visible in longitudinal section (see Fig. 6a). Thus, when one closely examines the actin filament bundle in stereocilia that is standing perfectly upright, we see a set of horizontal stripes crossing the bundle. When the stereocilium is bent, two different things could happen, as illustrated in the model depicted in Figs. 8a–c. First, the filaments could compress slightly on the inside of the bend, and stretch slightly on the outside of the bend. If this were to occur in the bent region, the stripes (due to the crossover points) would remain perpendicular to the axis of the bundle (Fig. 8b), and would lie at an angle with respect to the stripes in the unbent region. The tilt of angle of the former stripes off horizontal will be equal to the amount of bending that has taken place. Alternatively, what might happen is that the filaments are not compressible or extensible. In this case, if the bundle is bent, the filaments would tend to slide relative to each other. In this case, as is suggested in Fig. 8c, the stripes in the bend remain parallel to the stripes in the unbent portion of the stereocilium. From electron micrographs of bent stereocilia (Fig. 8d),

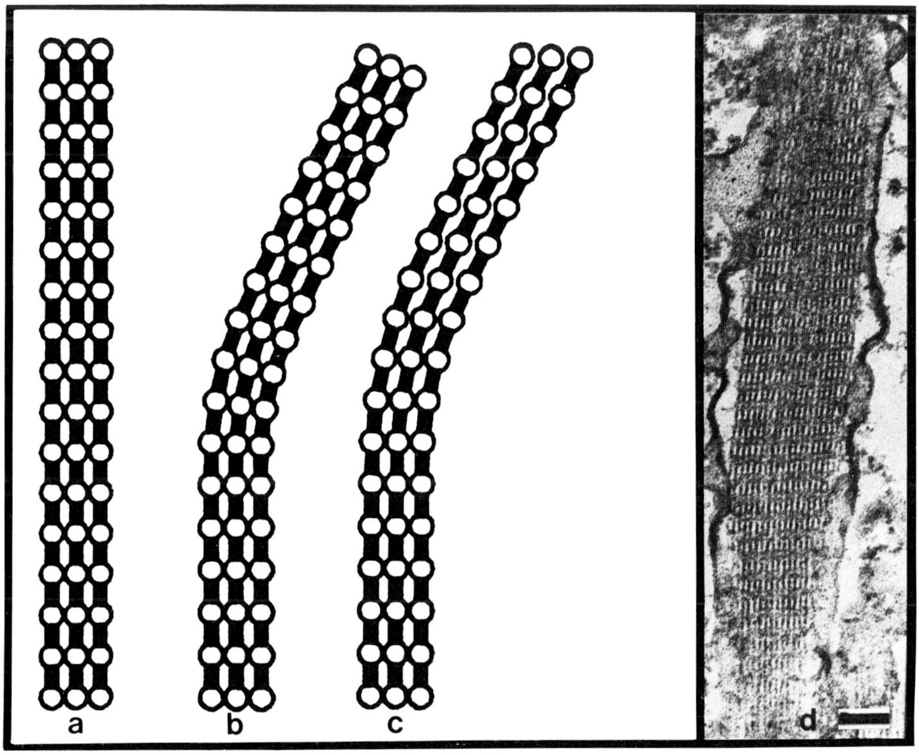

Figure 8. Bent actin bundles. (a–c) Model showing the two extreme possibilities for a bent bundle. When the bundle in (a) is bent and there is exclusively filament stretching and compression, the rows of cross-bridges, indicated by the circles, remain perpendicular to the bundle axis. If there is no filament stretching and compression, but rather filament slippage as in (c), the rows of cross-bridges are all parallel to each other and not necessarily perpendicular to the bundle axis. (d) Image of a bent chicken stereocilium. The bend is about ¼ the way up the bundle. Note that the rows of cross-bridges are all parallel and, therefore, that the bend is more akin to the case in (c) than that in (b). The bar corresponds to a distance of 100 nm.

we see that the situation depicted in Fig. 8c occurs biologically, namely that the filaments slide past one another in the bent region. The stereocilium is an interestingly constructed bundle that is built to bend but not in the middle as shown in Fig. 8d, a bend that is probably a preparative artifact. Instead, it is constructed to move rather like a lever, that is, to pivot at its base. The bundle in the stereocilium is tapered sharply where it enters the main body of the cell. The bulk of the structure consists of hundreds or thousands of filaments but tapers down to 20 at the pivot point, where the stereocilium joins the apical end of the cell (Figs. 9a and 10a). It is, therefore, designed to bend at its base with the remainder of the bundle being more or less rigid. When the stereocilium is stimulated by sound, it rocks to and fro bending at the base. What happens to the bundle is surprising. In Fig. 9b, the stereocilium axis is per-

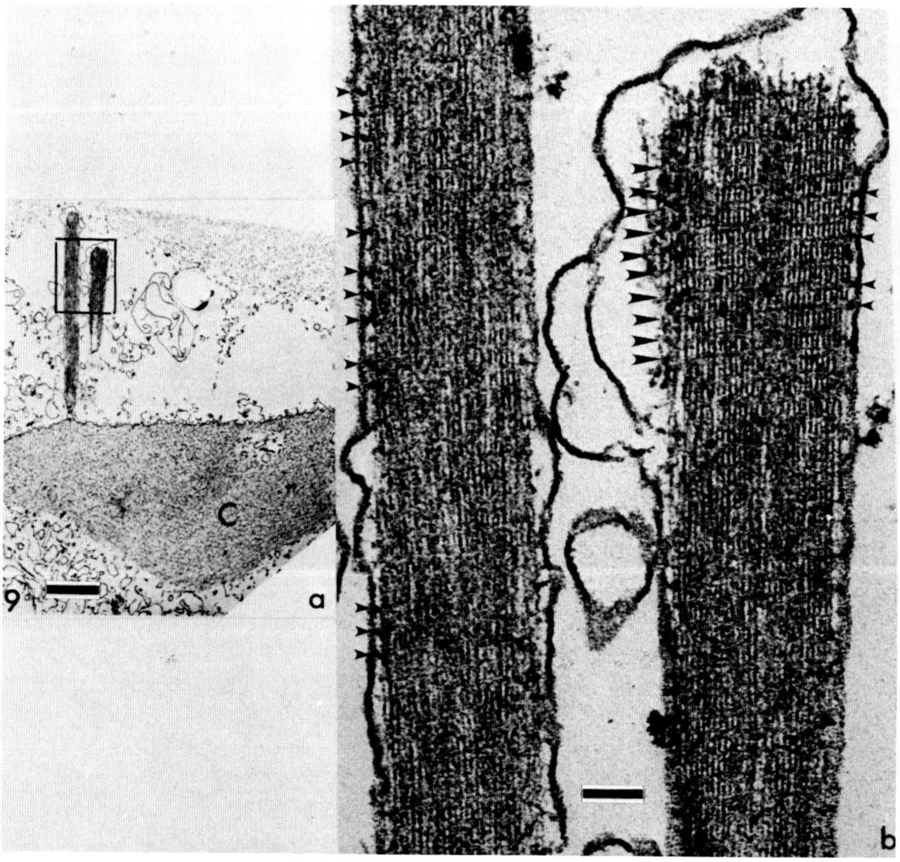

Figure 9. The relationship of the rows of cross-bridges to the apical surface of the cell in an untilted stereocilium. (a) A low magnification picture of a hair cell in which the stereocilia are perpendicular to the apical surface. The bar is 1 μm. (b) A higher magnification of the area shown in the box in (a). Note that the rows of crossover points are all parallel to the apical surface of the cell. The bar corresponds to a distance of 100 nm. (Reproduced from Tilney *et al.*, 1983, by copyright permission of the Rockefeller University Press.)

pendicular to the cell surface and the transverse bands of cross-bridges are parallel to the cell surface. When the stereocilium is tilted off perpendicular (Fig. 10a), the cross-bridges all tilt to remain parallel to the cell surface (Fig. 10b). This means that *all* the cross-bridges in the stereocilium tilt, even those that aren't connecting the 20 filaments, which are actually bent. As a result, the act of pivoting or bending the stereocilium at its narrow base produces a tilting of all cross-bridges throughout the stereocilium. The restoring force, that is, the force which will return the stereocilium to its upright position, contains two components: one, that arises from the 20 bent filaments which is constant for all stereocilia, and a second arising from the tilting of cross-bridges, which is a function of the number of cross-bridges, i.e., the volume of the stereocilium. By changing the volume of the bundle, the cell can adjust the mechanical response of the stereocilium to sound. As stated in Tilney *et al.* (1983), these considerations may be important biologically in the "tuning" of the stereocilia to different sound frequencies.

This interesting property of bundles derives from the properties of the component filaments. We noted earlier that while the actin filament has substantial angular disorder and while the subunits appear to tilt or change angle,

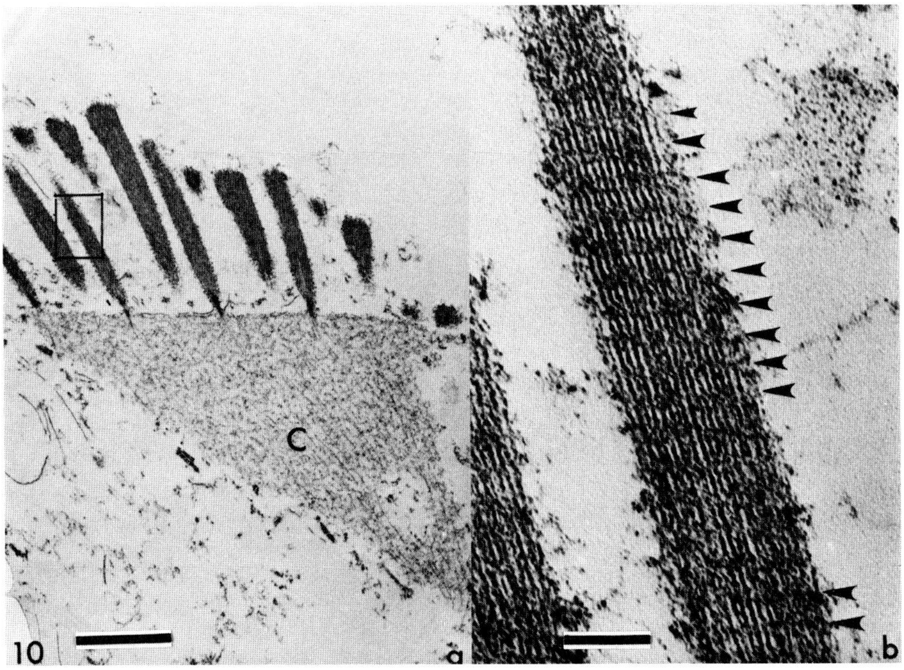

Figure 10. The relationship of the rows of cross-bridges to the apical surface of the cell in a tilted stereocilium. (a) A low magnification of a hair cell. The bar corresponds to a length of 1 μm. (b) Higher magnification of the area in the box. Note that the rows of cross-bridges are tilted to remain parallel to the apical surface of the cell. Bar corresponds to a distance of 100 nm. (Reproduced from Tilney *et al.*, 1983, by copyright permission of the Rockefeller University Press.)

there was little, if any, axial disorder, that is, the axial spacing of subunits remained constant. Put another way, there is little spontaneous stretching or compression of filaments relative to the angular disorder or the variable tilt of subunits. Thus, the filament structure would favor a tilt of the subunit to accommodate the tilted cross-bridges, instead of favoring a stretching or compression of subunit spacing.

We can define more precisely the extent to which filaments can slip past one another as a result of a bend. If the bend is small (~10° off straight) the amount of interfilament slippage is about 20 Å for a typical bundle. This amount of bending is found in stereocilia, and the interfilament cross-bridges remain attached, presumably because actin subunits can tilt by this amount. Recall that this agrees with the 10° of variable tilt found for the outer domain of the actin subunit. If the amount of sliding is larger, say 50 Å, then one supposes the cross-bridges must be broken. In fact, in acoustic trauma, where the stereocilia are subjected to a large amplitude motion, cross-bridge destruction and filament breakage are seen and concomitantly there is a loss in rigidity (Tilney *et al.*, 1982). The breaking of cross-bridges also occurs in the *Limulus* sperm, where changes in bending of 26° off straight result in a breaking and reforming of cross-bridges such that the bend becomes built into the bundle.

5.2. *How to Build a Bend into a Bundle*

In most cells, the bundles of actin filaments present in their cytoplasm are straight. In a few cells, however, most notably the sperm of the horseshoe crab, *Limulus,* the bundle is not straight but is bent. Before applying our rules on the structure of actin filaments to an analysis of this naturally bent bundle, let us briefly tell something about the biology of the sperm as it is essential for understanding this chapter and the next.

When a *Limulus* sperm is activated, such as occurs when it makes contact with the outer coats that surround the egg, an extraordinary thing happens. Following exocytosis of the acrosomal vacuole, a 60 μm long process (true discharge) grows out of the anterior end of the sperm in about 4 sec. (Fig. 11c) (Andre, 1965; Tilney, 1975b). As this process elongates, it rotates, a design used to penetrate (screw) through the tough extracellular layers that surround the *Limulus* egg. Running from the tip of the process to the base of the nucleus is a paracrystalline bundle of actin filaments. In unreacted sperm, this bundle, all 60 μm of it, is neatly coiled up just posterior to the nucleus (Fig. 11b). Thus, the extension of the bundle during activation is the result of its uncoiling. Uncoiling can also occur in which the basal end of the bundle extends out the posterior end of the sperm, a reaction called the false discharge (Fig. 11a). This process is not straight like the true discharge, but is shaped like a corkscrew.

The paracrystalline bundle can be isolated from the sperm in three states: as the false discharge, the coil, or the true discharge (Figs. 12a–c). All three states are stable *in vitro,* thus, the sperm does not maintain these states by

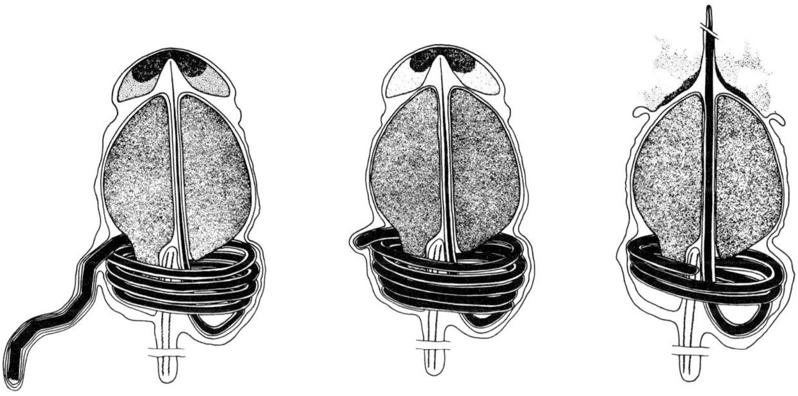

Figure 11. Drawings of the sperm from *Limulus*. (a) The false discharge. In the false discharge reaction the coiled actin bundle extends laterally from the posterior end of the sperm. The form of the false discharge is a left-handed corkscrew having a pitch of 2.8 μm. (b) The unfired sperm. The 60 μm long bundle is coiled about the base of the sperm. (c) The true discharge. In this reaction the bundle extends outward directed anteriorly by the nuclear canal. The elbows or bends built into the bundle are "melted out" to yield a straight, extended filament. (Reproduced from DeRosier *et al.*, 1982, by copyright permission of the Rockefeller University Press.)

external force but rather the form of the bundle is a property of structure itself, e.g., the actin filaments and its cross-bridges. SDS gels of the false discharge and the coil reveal that the bundle consists of actin, and in equal molar ratio a second, 55-kd protein known as scruin. A third protein having a mass of 95-kd is present in ¼ the molar amount. The filaments in the bundle have the appearance of actin but seem more substantial than pure actin, presumably due to the additional proteins. As expected, the filaments are hexagonally packed with the crossover points in register. The most unusual feature of the bundle is the manner in which it is curved. In the coil and in the false discharge one might have expected to find the bundle smoothly bent. What is seen (Figs. 12a and 12b) is that the bundle is really polygonal consisting of straight regions about 0.7 μm in length and then sharply bent vertices (elbows) making angles of 154°, i.e., 26° off straight (DeRosier *et al.*, 1980b). How can structures such as the elbows be built into a bundle? As before, the answer lies in the helical symmetry and angular disorder of the actin filaments (DeRosier and Tilney, unpublished results).

Recall that when a bend is produced in the middle of the bundle, the filaments are not stretched or compressed but rather there is tilting of the cross-bridges so that the filaments slip relative to each other. Examine the pair of filaments in the bend shown in Fig. 13. The right side filament is on the inside track and the left side filament on the outside track. The latter has a longer path to follow than the former and will fall out of being in perfect transverse register, that is, the two filaments will slip relative to each other. To

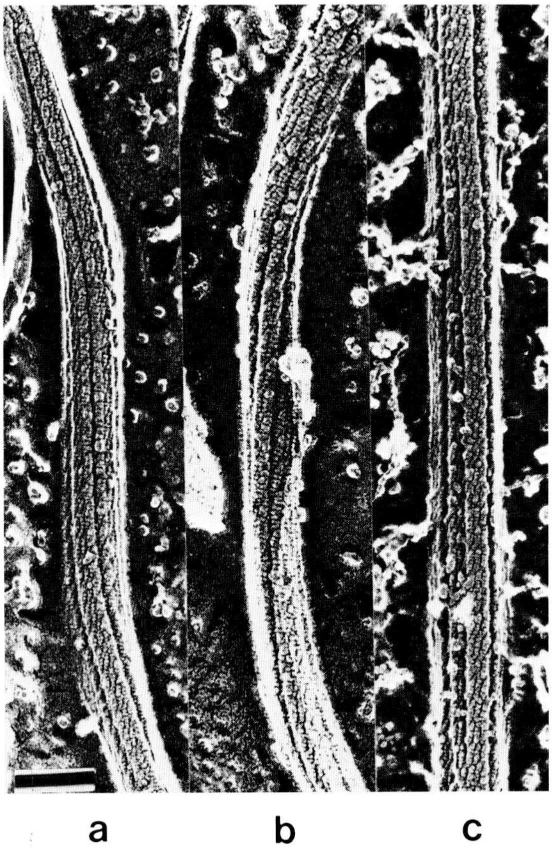

Figure 12. The actin bundle in the sperm of *Limulus*. (a) The false discharge. The notable features are the elbows (154° bends), the 0.7 μm, and the left-handed supercoiling of the filaments (i.e., the filaments wind around the bundle from lower right to upper left). (b) The coil. Again, note the same 0.7 μm arms and 154° elbows but the filaments are supercoiled in a right-handed manner. (c) The true discharge. There are neither elbows nor supercoiling. The bar corresponds to a distance of 100 nm. (Reproduced from DeRosier *et al.*, 1982, by copyright permission of the Rockefeller University Press.)

see this, examine the bottom halves of the filaments and note that the corresponding subunits are in register. Recall that if the filaments are in register, they can be cross-bridged just as in Fig. 7. In Fig. 13, the cross-bridge is from subunit 5 on the left filament to the back of subunit 5 on the right filament (see regions A and B in Fig. 13). In the bend, there is a slippage between filaments due to the difference in path length. Subunit 5 on the left is no longer in register with subunit 6 on the right in region C so that the cross-bridge is tilted. When the bend is past, the full slippage between filaments is realized. If the slippage is equal to 54.6 Å, the distance between two subunits, then it is possible to rebond the two filaments by shifting the cross-bridge to the nearest actin subunit and thereby removing the tilt of the subunit and its cross-bridge. In this case, subunit 7 on the left filament occupies the axial position of subunit 5. Thus, a cross-bridge can be made from subunit 7 on the left filament to the back of subunit 5 on the right (regions D and E in Fig. 13). There is a difficulty, however, because, while subunit 7 is in the correct axial position, it is not at the right angle. The situation can be clarified by examining column A in the table. Note that if one compares the angular positions of

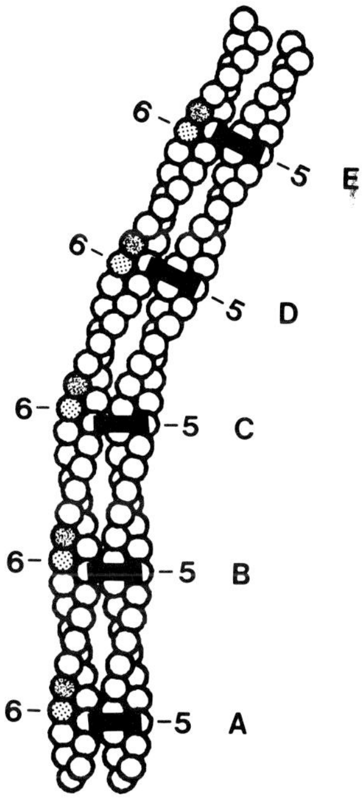

Figure 13. A bend built into a bundle. The two filaments are in register and are cross-bridged at subunit 5 in regions A and B. The cross-bridges are indicated by black bars. The left-side filament is on a longer, outer track in the bend relative to the right-side filament. As a result, the two filaments slip out of register. If the bend is 154° the amount of slippage is just enough to allow rebonding, in this case subunit 7 on the left can be cross-bridged to 5 on the right (in regions D and E). Subunits 6 and 8 have been stippled in order to highlight the shift in crossbridge position. Note that in regions A, B, and C the cross-bridge is below the pair of stippled units, but in C and D it lies between them.

subunits 5 and 7, there is an angular difference of about 28°. In the unbent filament, let us suppose that a cross-bridge would be made at position 5, but as a result of bending, subunit 7 on the outer filaments is now nearer to subunit 5 on the first filament. There would be considerable axial strain on the cross-bridge. If we change the cross-bridge from subunit 5 on the outer filament to subunit 7 we remove the axial strain. However, as can be noted by looking at the angular orientations of subunit 7, angular strain must take its place, that is, subunit 7 is 28° off the correct angular position. This amount of angular strain is somewhat larger than that cited in the cross-bridging of hexagonal bundles but, depending on the details of the interfilament bonding, this strain may be spread out over two or three subunits, bringing it within the 10° per subunit.

The key feature is that re-cross-bridging the filament bundle after a bend locks in the difference in path length and therefore builds a bend into the bundle. Thus, the symmetry of actin coupled with its angular disorder not only allows the cross-bridging of filaments into a bundle but also provides the potential for building in curvature.

6. Actin Generates Motion without Myosin

In the preceding section we have described the construction of the bundle which accounts for its form, rigidity, and curvature. Now we wish to consider the mechanism of its extension. It is fortunate for us that there are two extended forms, and their comparison provides the key to understanding the mechanism of extension. Often when sperm are placed in sea water they undergo a reversible reaction known as the false discharge. In the reaction, the bundle extends out laterally from the posterior end of the cell. There are a number of interesting differences and similarities between the true and false discharge. The true discharge corresponds to an extension of the bundle starting at the thin end of the bundle and the false to an extension starting at the thick end. Both structures rotate as they extend, but the false discharge comes out in the form of a corkscrew, while the true discharge is perfectly straight. In addition the false discharge is a reversible reaction, while the true discharge is irreversible. By analyzing the structures of these two forms relative to that of the coil, it is possible to arrive at a mechanism that accounts for both the two reactions.

The false discharge is the key (DeRosier *et al.*, 1982). Electron micrographs of the false discharge show that it consists of arms and elbows having the same lengths and angles as those of the coil (see Fig. 12b). What is different is that in the coil, the filaments are not straight but are supercoiled about the bundle in a right-handed sense. In the false discharge, the filaments are also superhelical but the hand is left handed. Thus, the extension involves a change in the supercoiling of the bundle. The mechanism is as follows: in the coil all the elbows lie in a single plane so that each turn of the coil is in effect a polygon (see Figs. 14a and c). If the supercoiling were to change, the planes of neighboring elbows must rotate relative to one another, in this case by 90° (see Fig. 14b). The resulting structure in which the plane of each successive elbow is rotated 90° with respect to its neighbors would be a corkscrew-like structure having exactly the hand and period observed in the false discharge. This is shown in Figs. 14c–g. Thus, the change in supercoiling of the bundle causes the transformation from a compact coil to an extended corkscrew.

The next question is what changes the supercoiling of the bundle? The change in supercoiling of the bundle must be accompanied by a change in the symmetry or twist of its component filaments. This is a topological necessity in a bundle of cross-linked filaments (see Fig. 15). Thus, the 90° change in supercoiling per arm must be accompanied by a corresponding 90° change in the twist of each component filament, or the angular relation between subunits in the actin helix must change. More specifically, in the coil, the pitch or twist of each actin filament must be tighter than in those in the false discharge. The twist or pitch of an actin filament can be measured accurately on diffraction patterns and what we find is that, as expected, there is a "loosening" or unwinding of the actin filament in going from the coiled state to the false discharge. Thus, it should be no surprise that the amount of this unwinding turns out to be exactly the amount needed to allow a 90° change in supercoiling. Since the bundle is stable without any external restraining force, e.g., an external or internal scaffold, we conclude that the twist of the actin filament itself controls the form of the bundle. Thus, changes in the twist interconvert the coil and false discharge. This explains the reversibility of this reaction, namely, that the false discharge can return to the coiled state. If the condi-

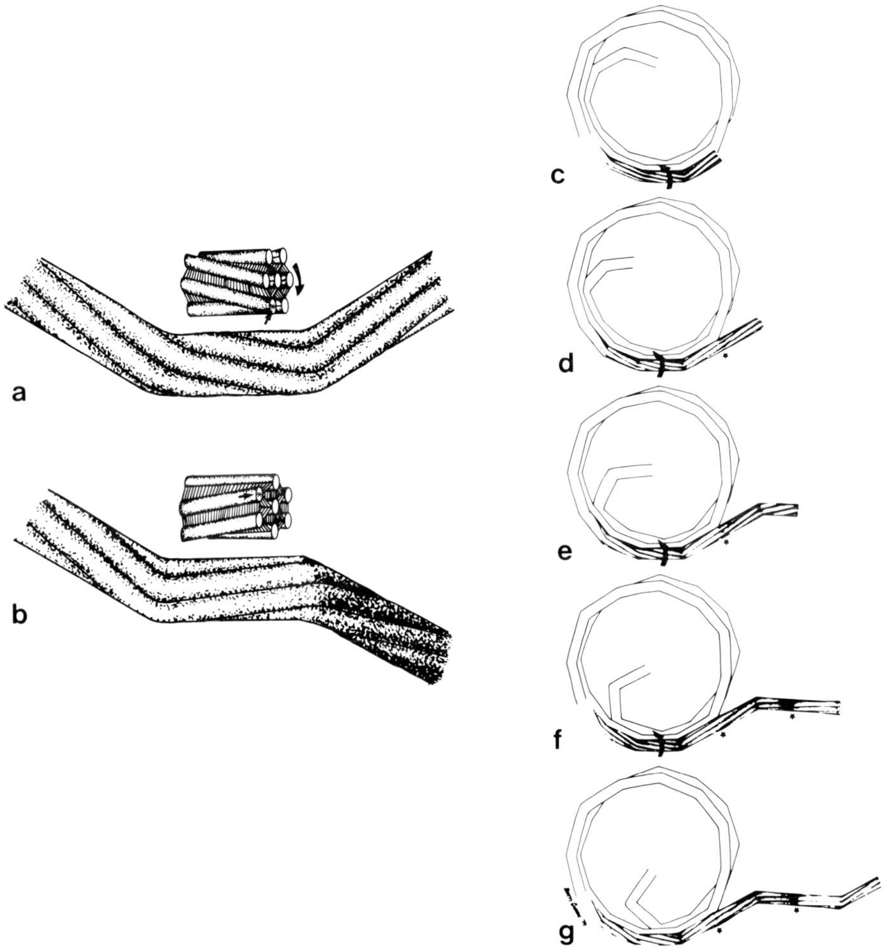

Figure 14. The extension of the false discharge. (a–b) In the upper inset, the change in supercoiling is seen to result in a rotation of the right end of the bundle relative to the left. In the larger figure the effect of this change of supercoiling is to rotate the plane of the right-side elbow relative to that of the left-side elbow. (c–g) The change in supercoiling begins at one end of the bundle and moves progressively around the bundle. As it does so, the plane of each elbow is rotated by 90° relative to its neighbor and as a result the bundle switches from a compact coil to an extended helix. The coil, of course, rotates to feed the growing false discharge. (Reproduced from DeRosier *et al.*, 1982, by copyright permission of the Rockefeller University Press.)

tions are right, the filament bundle will switch from left-handed to right-handed supercoiling and as a result the false discharge will be drawn back into the coil and, thus, back into the sperm.

This unusual cell illuminates a property of actin that has been exploited to produce motion, namely, the state of twist of the filaments. This reaction presumably requires accessory proteins. These accessory proteins are almost certainly involved in the cross-bridging of filaments into a bundle. It also

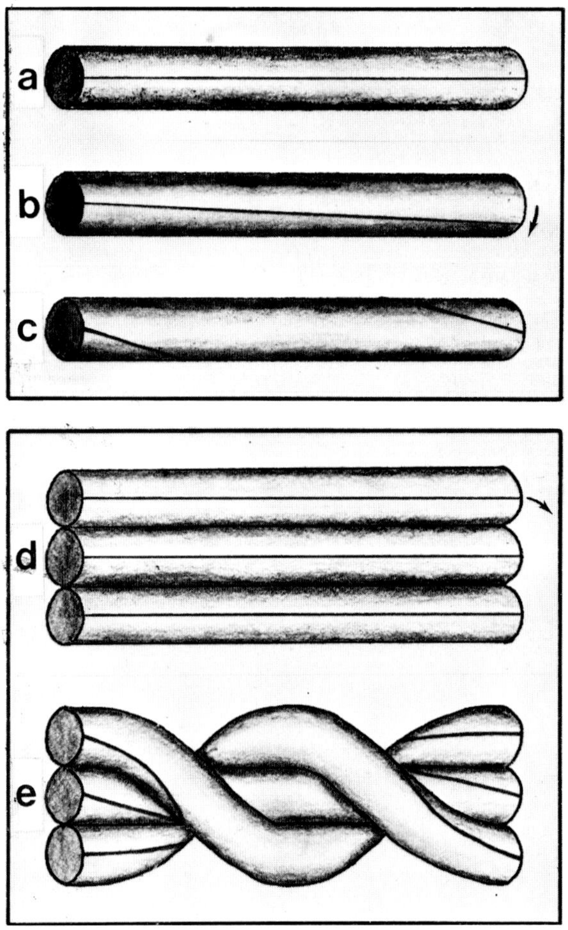

Figure 15. The coupling of the change in twist of filaments to the change in supercoiling of the bundle. (a–c) Shows a change in twist of the filament (rod) achieved by rotating one end of a filament relative to the other. The change in twist is depicted by the straight line in (a) which is transformed into a helix in (c). (d–e) If the right-hand end of a bundle is twisted by 360° relative to the left, not only is 360° of supercoiling introduced but each filament is also twisted by 360°. This is a necessary consequence in a cross-linked bundle. (From DeRosier *et al.*, 1980b. Reprinted by permission from Academic Press, Inc.)

seems reasonable to expect that they are involved in the regulation of the twist of the filaments. Exactly how the cross-bridges determine the twist of the filaments has not yet been determined, but it may be helpful to the reader to put forward a working hypothesis. Our idea takes cognizance of the following facts. First, we have proved that there is one scruin molecule associated with each actin subunit and from our three-dimensional reconstruction we know that these molecules are located on the surface of each subunit. Second, when the filaments are present in the coiled state, the actin filaments are more twisted than in the false discharge. What we propose is that in the coiled state the adjacent scruin molecules on the *same* filament bind to each other, thereby maintaining the more twisted state of the actin subunits. If the scruin molecules bind to each other, they can take advantage of the angular freedom in the subunits in essence by freezing all the subunits to one side of their equilibrium positions. This in turn will, of course, "tighten" or decrease the twist of the filament. To induce the false discharge then, all that has to happen is for the scruin–scruin interaction on the same filament to be broken. Now the subunits will move to their equilibrium position. Since there is little space in the bundle we imagine that this loss of scruin–scruin interaction may be brought about by a small molecule or ion which can easily diffuse into this tightly packed bundle, and by binding to the accessory proteins, alter the twist of the filaments. Release of the substance could result in the formation of the false discharge, or even the true discharge, which is really only a change in the supercoiling of the filaments at the apical end rather than the basal end. If the cell restores its concentration of this agent, the bundle would return to its coiled state. The true discharge would not return to a coiled state, however, since the elbows are missing (DeRosier *et al.*, 1982). Although this model is speculative, it has the property of incorporating the known structural properties of the actin filament.

7. Conclusions

In this article, we have tried to present a new view of the actin filament, its structure, and its properties which make it the protein of evolutionary choice in a variety of structures, both motile as well as skeletal. More specifically, we recognize that actin associates to form compact cross-linked bundles which are endowed with mechanical strength. At the same time, because of the angular disorder of its subunits, bundles with variability in their transverse order, e.g., hexagonal or liquid-type packing, are possible. This feature allows the bundles to be formed extremely quickly, or if formed more slowly can convey on the bundle maximal rigidity. The variable tilt of the subunit dictates the mechanical properties of the bundle when it is bent. Perhaps even more incredible is that this same filament structure allows a stable bend to be a built-in feature, and at least in one cell type, the bundle itself is capable of exerting force, simply by changing the twist of the filaments which make up the bundle.

ACKNOWLEDGMENTS. The authors wish to acknowledge the advice and criticism of Dr. Ed Egelman, the help of Louise Seidel and Molly Tilney in the preparation of the manuscript, and the assistance of Vicki Ragan in preparation of the figures. The authors also acknowledge grant support from the National Institutes of Health Grants GM21189 and GM26357 to D.J.D. and HD14474 to L.G.T.

References

Aebi, U., Fowler, W. E., Isenberg, G., Pollard, T. D., and Smith, P. R., 1981, Crystalline actin sheets: Their structure and polymorphism, *J. Cell Biol.* **91:**340–351.

Andre, J., 1965, A propos d'une lecon sur la Limule, *Ann. Fac. Sci. Clermont* **26:**27–38.

Bretscher, A., 1981, Fimbrin is a cytoskeletal protein that crosslinks F-actin *in vitro*, *Proc. Natl. Acad. Sci. USA* **78:**6849–6853.

Bretscher, A., and Weber, K., 1980, Fimbrin, a new microfilament-associated protein present in microvilli and other cell surface structures, *J. Cell Biol.* **86:**335–340.

Bryan, J., and Kane, R. E., 1978, Separation and interaction of the major components of sea urchin actin gel, *J. Mol. Biol.* **125:**207–224.

Cachon, J., and Cachon, M., 1974, Les systemes axopodiaux, *Annee Biologique* **13:**523–560.

DeRosier, D. J., and Tilney, L. G., 1982, How actin filaments pack into bundles, *Cold Spring Harbor Symp. Quant. Biol.* **81:**525–540.

DeRosier, D. J., Mandelkow, E., Silliman, A., Tilney, L. G., and Kane, R., 1977, The structure of actin-containing filaments from two types of non-muscle cells, *J. Mol. Biol.* **113:**679–695.

DeRosier, D. J., Tilney, L. G., and Egelman, E. H., 1980a, Actin in the inner ear: The remarkable structure of the stereocilium, *Nature* **287:**291–296.

DeRosier, D., Tilney, L., and Flicker, P., 1980b, A change in the twist of the actin-containing filaments occurs during the extension of the acrosomal process in *Limulus* sperm, *J. Mol. Biol.* **137:**375–389.

DeRosier, D. J., Tilney, L. G., Bonder, E. M., and Frankl, P., 1982, A change in twist of actin provides the force for the extension of the acrosomal process in *Limulus* sperm: the false-discharge reaction, *J. Cell Biol.* **93:**324–337.

Egelman, E. H., and DeRosier, D. J., 1982, The Fourier transform of actin and other helical systems with cumulative random angular disorder, *Acta Cryst.* **A38:**796–799.

Egelman, E. H., and DeRosier, D. J., 1983a, Appendix: A model for F-actin derived from image analysis of isolated filaments, *J. Mol. Biol.* **166:**623–629.

Egelman, E. H., and DeRosier, D. J., 1983b, Structural studies of F-actin, in: *Actin: Structure and Function in Muscle and Non-muscle Cells* (C. dosRemedios and J. Barden, eds.), pp. 17–24, Academic Press, Sydney.

Egelman, E. H., Francis, N., and DeRosier, D. J., 1982, F-actin is a helix with a random variable twist, *Nature* **298:**131–135.

Egelman, E. H., Francis, N., and DeRosier, D. J., 1983, Helical disorder and the filament structure of F-actin are elucidated by the angle-layered aggregate, *J. Mol. Biol.* **166:**605–629.

Flock, A., and Cheung, H., 1977, Actin filaments in sensory hairs of the inner ear receptor cells, *J. Cell Biol.* **75:**337–343.

Gillis, J. M., and O'Brien, E. J., 1975, The effect of calcium ions on the structure of reconstituted muscle thin filaments, *J. Mol. Biol.* **99:**445–459.

Hanson, J., 1967, Axial period of actin filaments, *Nature* **213:**353–356.

Hartwig, J. H., Tyler, J., and Stossel, T. P., 1980, Actin-binding protein promotes the bipolar and perpendicular branching of actin filaments, *J. Cell Biol.* **87:**841–848.

Huxley, H., 1969, The mechanism of muscular contraction, *Science* **164:**1356–1366.

Huxley, H., and Brown, W., 1967, The low-angle X-ray diagram of vertebrate striated muscle and its behaviour during contraction and rigor, *J. Mol. Biol.* **30:**383–434.

Kane, R. E., 1976, Actin polymerization and interaction with other proteins in temperature-induced gelation of sea urchin egg extracts, *J. Cell Biol.* **71:**704–714.

Korn, E., 1982, Actin polymerization and its regulation by proteins from non-muscle cells, *Physiol. Rev.* **62:**672–737.

Moore, P. B., Huxley, H. E., and DeRosier, D. J., 1970, Three-dimensional reconstruction of F-actin, thin filaments and decorated thin filaments, *J. Mol. Biol.* **50:**279–295.

Mooseker, M. S., and Tilney, L. G., 1975, Organization of an actin filament-membrane complex. Filament polarity and membrane attachment in the microvilli of intestinal epithelial cells, *J. Cell Biol.* **67:**725–743.

Oosawa, F., 1980, The flexibility of F-actin, *Biophys. Chem.* **11:**443–446.

Pollard, T. D., 1981, Purification of a calcium-sensitive actin gelation protein from *Acanthamoeba*, *J. Biol. Chem.* **256:**7666–7670.

Spudich, J. A., and Amos, L. A., 1979, Structure of actin filament bundles from microvilli of sea urchin eggs, *J. Mol. Biol.* **129:**319–331.

Spudich, J. A., Huxley, H. E., and Finch, J. T., 1972, Regulation of skeletal muscle contraction.

II. Structural studies on the interaction of the tropomyosin-troponin complex with actin, *J. Mol. Biol.* **72:**619–632.

Suck, D., Kabsch, W., and Mannherz, H. G., 1981, Three-dimensional structure of skeletal muscle actin and bovine pancreatic deoxyribonuclease I at 6 Å resolution, *Proc. Natl. Acad. Sci. USA* **78:**4319–4323.

Tilney, L. G., 1968, The assembly of microtubules and their role in the development of cell form, *Dev. Biol. Suppl.* **2:**63–102.

Tilney, L. G., 1971, How microtubule patterns are generated. The relative importance of nucleation and bridging of microtubules in the formation of the axoneme of *Rapidiophrys*, *J. Cell Biol.* **51:**837–854.

Tilney, L. G., 1975a, The role of actin in non-muscle cell motility, in: *Molecules and Cell Movement* (S. Inoue and R. E. Stephens, eds.), pp. 339–388, Raven Press, New York.

Tilney, L. G., 1975b, Actin filaments in the acrosomal reaction of *Limulus* sperm, *J. Cell Biol.* **64:**289–310.

Tilney, L. G., Hatano, S., Ishikawa, H., and Mooseker, M. S., 1973, The polymerization of actin: Its role in the generation of the acrosomal process of certain echinoderm sperm, *J. Cell Biol.* **59:**109–126.

Tilney, L. G., DeRosier, D. J., and Mulroy, M. J., 1980, The organization of actin filaments in the stereocilia of cochlear hair cells, *J. Cell Biol.* **86:**244–259.

Tilney, L. G., Saunders, J. C., Egelman, E., and DeRosier, D. J., 1982, Changes in the organization of actin filaments in the stereocilia of noise damaged lizard cochlea, *Hearing Res.* **7:**181–197.

Tilney, L. G., Egelman, E. H., DeRosier, D. J., and Saunders, J. C., 1983, Actin filaments, stereocilia, and hair cells of the bird cochlea II, *J. Cell Biol.* **96:**822–834.

Vibert, P., and Craig, R., 1982, Three-dimensional reconstruction of thin filaments decorated with a Ca^{2+}-regulated myosin, *J. Mol. Biol.* **157:**299–319.

Wakabayashi, T., Huxley, H. E., Amos, L. A., and Klug, A., 1975, Three-dimensional image reconstruction of actin-tropomyosin complex and actin-tropomyosin-troponin T-troponin I complex, *J. Mol. Biol.* **93:**477–497.

Wray, J. S., Vibert, P. J., and Cohen, C., 1978, Actin filaments in muscle, *J. Mol. Biol.* **124:**501–521.

4

Changes in Actin during Cell Differentiation

Kazuhiro Nagata and Yasuo Ichikawa

1. Introduction

Since the pioneering work by Loewy and Hoffman-Berling, the actomyosin system has been believed to have an important function not only in muscle cells but in nonmuscle cells (reviewed by Pollard and Weihing, 1974). Nonmuscle cell actin was first purified from *Physarum* by Hatano and Oosawa (1966a,b). Success in identifying actin filaments as arrowhead figures *in situ* (Ishikawa *et al.*, 1969), methods devised to purify nonmuscle cell actin (Gordon *et al.*, 1976a), and the establishment of a DNase I inhibition assay that measures actin contents (Blikstad *et al.*, 1978) show the great progress made in this field. In addition, there has been increased interest in various nonmuscle cell actins, functional and chemical differences among actin molecules from different sources, as well as differences in their control mechanisms.

One difference between skeletal muscle and nonmuscle cell actins is that whereas most actin molecules in skeletal muscle are polymerized and well organized in concrete structures called the sarcomeres, nonmuscle actins are in unstable and dynamic equilibrium between the monomer (G) and polymer (F) forms; this is known as G–F transformation. Bray and Thomas (1976) and Abramowitz *et al.* (1975) classified nonmuscle actins into two groups, easily polymerizable and difficult to polymerize. When actins are highly purified, however, no marked difference in polymerization was found even when they come from different sources (Gordon *et al.*, 1977). The high content of G-actin in nonmuscle cells probably results from the presence of the modulator

Kazuhiro Nagata and ***Yasuo Ichikawa*** • Department of Cytochemistry, Chest Disease Research Institute, Kyoto University, Kyoto, Japan.

proteins profilin (Carlsson *et al.*, 1977; Markey *et al.*, 1978), DNase I (Hitchcock *et al.*, 1976; Lazarides and Lindberg, 1974), fragmin (Hasegawa *et al.*, 1980), 17K protein in starfish eggs (Mabuchi, 1981), 65K and 62K protein in human polymorphonuclear leukocytes (Southwick and Stossel, 1981), 65K cytochalasin-like protein in platelets (Grumet and Lin, 1980), and 19K protein in chick embryo brain (Bamburg *et al.*, 1980).

Another difference between nonmuscle and skeletal muscle cells is that in comparison to the myosin content, the actin content is much higher in nonmuscle cells. This excess actin is assumed to contribute to the cytoskeletal structures abundant in nonmuscle cells (Pollard, 1981).

Actin is a very conservative protein; even in actins of different species, only limited numbers of amino acids are exchanged in the whole peptide chain (Vandekerckhove and Weber, 1978). Two isomers are discernible in SDS-gradient polyacrylamide gel electrophoresis that contains urea (Storti and Rich, 1976; Storti *et al.*, 1976). Isoelectric focusing divides the actin of vertebrate cells into three isomers: α, β, and γ (Garrels and Gibson, 1976; Rubenstein and Spudich, 1977). Skeletal muscle actin consists of α-actin, chicken gizzard smooth muscle actin mainly of γ-actin. Most nonmuscle actins consist of β- and γ-actins. All three actin isomers are contained in the myoblasts of the chicken embryo, but after cell fusion that forms the myotube, α-actin is dominant and becomes the single constituent (Rubenstein and Spudich, 1977). A similar phenomenon was quantitatively determined in chicken embryos at different ages after fertilization (Shimizu and Obinata, 1980).

As to the difference in the nature of the actin of skeletal muscle and nonmuscle cells, Korn and his colleagues showed not only the presence of different isoactins, but different critical concentrations for polymerization and differences in the apparent K_m value for the activation of myosin Mg^{2+}-ATPase (Gordon *et al.*, 1976a,b, 1977). Although changes in actin content during ontogenic development have seldom been investigated, Mabuchi *et al.* (1980) have reported a marked increase in actin content (from 0.4–3%) immediately after the fertilization of sea urchin eggs. In a study of newborn rat brain by Schmitt *et al.* (1977), the total actin content reached a maximum 5 days after birth then decreased gradually as the rate of actin polymerization increased.

When nerve growth factor was added to a culture of sympathetic neuron, the induced outgrowth of axon was accompanied by a large increase in the actin and tubulin contents (Fine and Bray, 1971). Morphological changes induced by removing serum from a culture of a neuroblastoma cell line, however, were not accompanied by an increase in actin contents (Schmitt, 1976).

Actin contents also have been compared in malignant transformed cells and normal cells. In normal 3T3 cells, both the actin content and synthesis were higher than in SV40-transformed 3T3 cells (Fine and Taylor, 1976). When cells transformed by a temperature-sensitive mutant of Rous sarcoma virus were transferred to a nonpermissive temperature, the membrane-bound actin decreased although there were no changes in the total actin content (Wickus *et al.*, 1975). Leavitt *et al.* (1980) compared PHA-stimulated normal T lymphocytes and T-lymphoma Molt 4 line cells. In normal T cells, actin synthesis is 3-fold that in leukemic cells, and the former cells have a higher ratio of β-actin in contrast to the almost equal amounts of β- and γ-actins in leukemic cells. Although most cultured cells have both β- and γ-isoactins, Sakiyama *et al.* (1981) have reported a peculiar phenomenon, the absence of γ-actin expression in a mouse fibroblast cell line, L.

Relatively few reports have dealt with differences in contractile proteins in undifferentiated and differentiated cells or in malignant and normal cells. This appears to be due to the difficulty in controlling the induction of cell differentiation and in obtaining a sufficient number of cells in both stages. Generally, undifferentiated or malignant cells proliferate rapidly *in vitro* and *in vivo*, but normal cells and differentiated cells do not.

The M1 cell line, described in the following sections, proliferates as undifferentiated leukemic cells and can be induced to differentiate to nondividing macrophages with such normal functions of mature cells as phagocytosis and cell motility. A large number of cells (more than 1×10^{10} cells, corresponding to about an 8-ml packed volume) are easily obtainable before and after differentiation. This led us to undertake both a quantitative and qualitative comparison of the nature of contractile proteins before and after differentiation.

2. *Differentiation of a Myeloid Leukemia Cell Line*

2.1. *Origin and Nature of the M1 Cell Line*

A cell line, M1, isolated from a spontaneous myeloid leukemia from an SL strain mouse has been cultured in our laboratory since 1969 (Ichikawa, 1969, 1970). Morphologically, the M1 line cells are myeloblasts, grow in suspension with a doubling time of about 20 hr, and develop leukemia when transplanted into isologous strain mice. The most remarkable feature of the cell line is that although the cells continue to grow in conventional culture medium, they differentiate to mature macrophages or to neutrophils and cease to proliferate when cultured with a variety of differentiation-stimulating factors. Differentiation also can be induced in a diffusion chamber inserted into the peritoneal cavities of isologous mice (Honma *et al.*, 1978b; Lotem and Sachs, 1978). Many of the subclones isolated in Sach's laboratory showed different inducibility for differentiation (Sachs, 1978).

We have used three clones derived from our original M1 line cells. One clone is readily induced to differentiate mainly to macrophages by contact with conditioned medium (CM) from rat embryo fibroblasts and is referred to simply as M1 cells. A D^- subline, for which differentiation is not inducible, continues to grow even in CM with no signs of differentiation. And, Mm-1 line cells isolated from differentiated M1 line cells grow while retaining such differentiated functions as phagocytosis and motility (Maeda and Ichikawa, 1973).

2.2. *Differentiation-Stimulating Factors*

Differentiation-stimulating factors for the M1 cell line are separable into proteinous "D factors" and nonproteinous factors. As D factors, a simple protein of molecular weight 68,000 (MGI) (Guez and Sachs, 1973) and a glycoprotein of molecular weight 40,000–50,000 (Maeda *et al.*, 1977) have been isolated from the conditioned medium of embryonic fibroblast cultures. Recently, Sachs divided MGI into two categories: MGI-1, previously a colony-stimulating factor (CSF), and MGI-2 which corresponds to D factor (Sachs, 1982). These active molecules that induce the differentiation of M1 cells, D factor and MGI-2, differ from the CSF for normal hemopoietic cells (Hozumi, 1982; Sachs, 1982).

D factor activity is also present in physiological body fluid, peritoneal fluid (Hozumi *et al.*, 1974), amniotic fluid (Nagata *et al.*, 1977), saliva, urine (Nakayasu *et al.*, 1978), and the serum of mice injected with bacterial endotoxin (Sachs, 1978).

Some glucocorticoids (Honma *et al.*, 1977), poly (I) (Tomida *et al.*, 1978), lipopolysaccharide (Weiss and Sachs, 1978), prostaglandin E, poly(ADP-ribose), and other compounds (Hozumi, 1982; Sachs, 1978) have been reported to be nonproteinous factors.

2.3. *Markers of Differentiation*

After differentiation, M1 cells lose both mitotic activity and the spherical shape characteristic of immature myeloblasts (Fig. 1). The nucleo-cytoplasm ratio decreases and cells become adherent to the bottom of the plate. In a soft agar culture, undifferentiated M1 cells produce only compact-type colonies because of the lack of migration activity. But, when CM is added to its agar medium, colony-forming cells begin to migrate into the surrounding agar because cell motility is induced in the process of differentiation. This results in a diffuse colony type rather than the compact type found with undifferentiated cells (Fig. 1) (Ichikawa, 1969, 1970; Ichikawa *et al.*, 1975).

Differentiated cells also gain phagocytic activity. Fig. 1 shows nonimmune phagocytosis for polystyrene latex particles; immune phagocytosis for IgG-coated red cells has also been found (Sachs, 1978). This suggests the presence of an Fc receptor on the differentiated cell surface (Lotem and Sachs, 1976).

Differentiation is accompanied by an increase in the lysosomal enzyme activities of acid protease, acid phosphatase (Nagata *et al.*, 1976), β-glucuronidase, and lysozyme (Kasukabe *et al.*, 1977). Synthesis of prostaglandin (Hozumi, 1982) and the cytochrome oxidase and glucose-6-phosphatase activities (Hirai *et al.*, 1979) also increase after differentiation. Interestingly, differentiated cells produce CSF for normal hemopoietic cells (Maeda and Ichikawa, 1980) and "DSF," a glycoprotein with a molecular weight of

Figure 1. Tests for cell morphology (a,b), motility (c,d), phagocytosis (e,f), and the Fc receptor (g,h). Untreated Ml cells (a, c, e, and g), CM-treated or differentiated cells (b, d, f, and h).

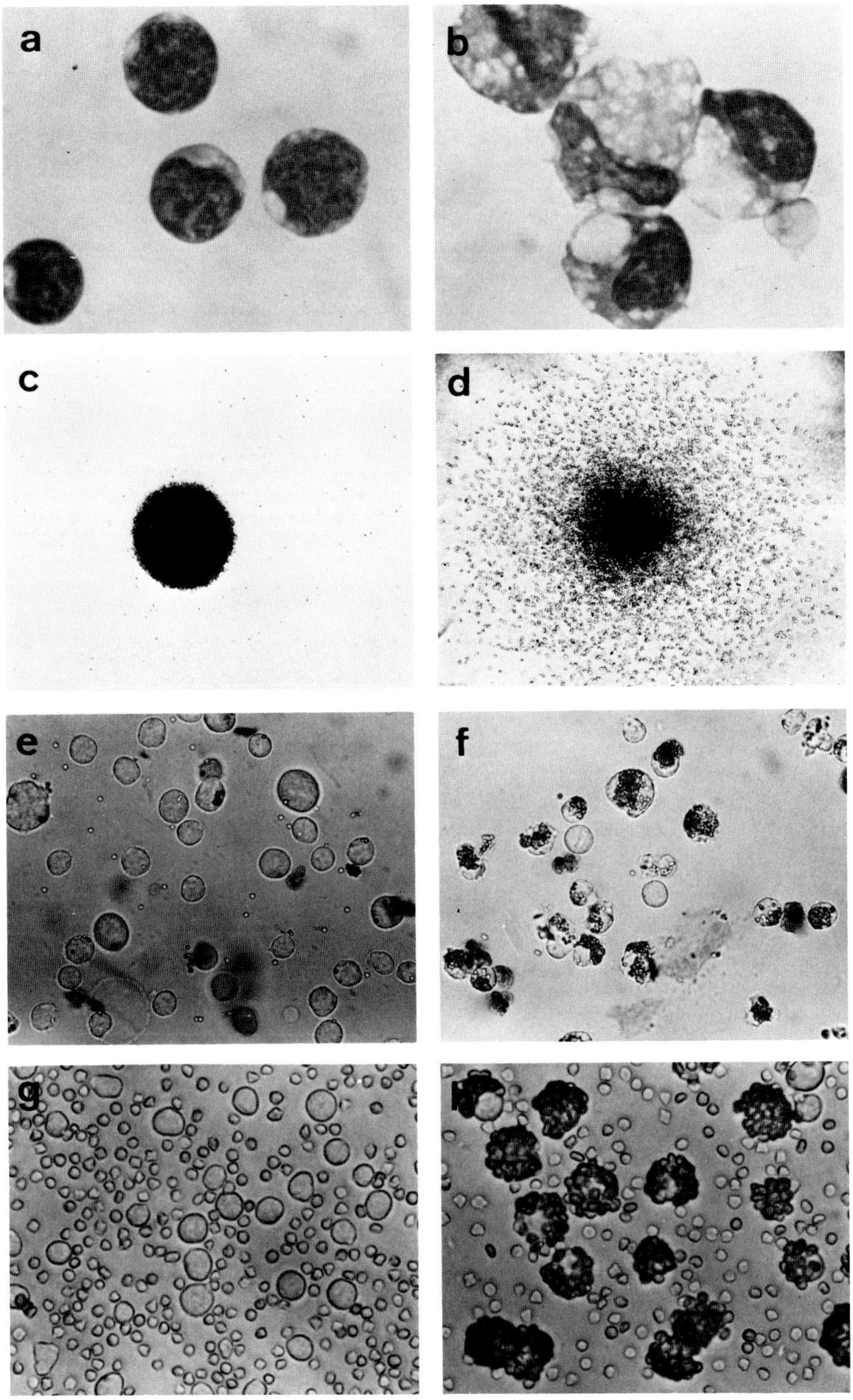
a
b
c
d
e
f
g
h

20,000–40,000, which stimulates differentiation in M1 cells themselves (Honma *et al.*, 1978a).

Changes in the plasma membrane also accompany differentiation: Con A-induced agglutinability after ATP deletion, cap formation by Con A and anti-H2 antibody, desensitization of functional β^2-adrenergic receptors, and ecto-ATPase activity (Sachs, 1978). The appearance of a new glycoprotein, named P180 (Sugiyama *et al.*, 1979), and some changes in the lipid composition of the plasma membrane also have been reported (Hozumi, 1982).

Even in the presence of FUdR and Ara-C, CM induced phagocytic and locomotive activities in the M1 cell line (Ichikawa *et al.*, 1975; Nagata and Ichikawa, 1979), but actinomysin D and puromycin inhibited the induction of differentiation by CM. This means that the differentiation of M1 cells does not require DNA synthesis; it is controlled in the transcriptional stage (Nagata and Ichikawa, 1979). This is also true for the erythro-differentiation of *Friend* leukemia cells (Marks and Rifkind, 1978). In contrast, induction of the Fc receptor, another differentiation-marker of the M1 line, was not inhibited by actinomycin D and puromycin. This suggests that the induction of the Fc receptor and the inductions of phagocytosis and cell motility are controlled by different mechanisms (Nagata and Ichikawa, 1979).

The induction of cell motility, as observed in changes in colony type in agar culture, and the induction of nonimmune phagocytosis for polystyrene latex particles are easily determined and are the most stable signs of differentiation in this cell line. These two functions are known to be controlled by actomyosin and its regulatory proteins (Pollard and Weihing, 1974; Stendahl *et al.*, 1980; Stossel and Hartwig, 1976). Actually, the induction of these functions in the M1 line by CM is reversibly suppressed by cytochalasin B (Ichikawa *et al.*, 1975) which inhibits actin polymerization and gelation (Flanagan and Lin, 1980; Chang *et al.*, 1980; Hartwig and Stossel, 1979; MacLean-Fletcher and Pollard, 1980a). Actomyosin also functions in contractile ring formation during cytokinesis (Pollard and Weihing, 1974). As stated above, during their differentiation from leukemic myeloblasts to macrophages, M1 cells lose mitotic activity and gain phagocytic and locomotive activities. These findings led us to the study that is reviewed hereafter.

3. Quantitative Changes in Actin

3.1. Content and Synthesis

The content and synthesis of the actins of the M1, D^- and Mm-1 cell lines are shown in Table 1 (Nagata *et al.*, 1980). Actin content was determined by a DNase I inhibition test of the cell lysate in buffered solution containing Triton X-100 and guanidine-HC1. Actin synthesis was calculated from the radioactivity present in the actin and total protein after cultured cells had been pulse labeled with ^{35}S-methionine.

After treatment with CM, both the actin content and synthesis increased 1.3 times their values before treatment. This means a 2.5-fold increase for

Table 1. The Actin Content, Rate of Synthesis, and F-Actin Ratio[a]

Cell line	CM-treatment	Phagocytosis (%)	Actin content: Percent total protein	Actin content: mg/10^8 cells	Actin synthesis, percent total protein	F-actin ratio (%)
Ml	−	1.5	4.8	0.83	4.4	48.3
Ml	+	53.0	6.1	1.98	5.7	62.7
D^-	−	0	4.9	0.51	4.4	47.2
D^-	+	0	4.5	0.50	4.3	40.6
Mm-1	−	92.0	7.8	2.56	8.5	—

[a]M1 cells were incubated with or without CM for 3 days. After the test for phagocytosis, actin contents in the cell homogenates were assayed by the DNase I inhibition test. The F-actin ratio was calculated from the difference between the G-actin and total actin contents, before and after incubation with guanidine-HCl. To determine the synthesis rate, cultures were labeled with ^{35}S-methionine for the final 12 hr of the 3-day incubation. The cell lysate was applied to SDS–PAGE, and then the radioactivity of its actin band was divided by the radioactivity of the total protein.

each individual cell because the cellular protein content was twice as great after differentiation. The D^- subline, for which CM treatment did not induce phagocytic activity, showed no changes. For the Mm-1 line, a macrophage line derived from the M1 line, both the values for actin content and synthesis were considerably higher than in newly differentiated M1 cells.

Hoffman-Liebermann and Sachs (1978) reported a 3-fold increase in actin synthesis in differentiated M1 cells compared to the value for untreated M1 cells in the stationary state.

3.2. F-Actin Ratio in Situ

Actin is a simple globular protein with a molecular weight of 42,000. To perform the mechanical processes of extention and contraction, the monomer G-actins have to polymerize and form a filamentous double helix, F-actin. In muscle cells, most actin molecules are present in the F-actin form and make up thin filaments with tropomyosin and troponin. These thin filaments are known to interact in a sliding fashion with thick filaments consisting of myosin molecules. In nonmuscle cells, however, F-actin is only about half, or less, of the total actin present (Blikstad *et al.*, 1978). What remains is either an oligomer bound to submembraneous spectrin (Tilney and Detmers, 1975; Brenner and Korn, 1980) or monomer actin freely suspended in the cytoplasm.

The most dramatic example of G–F interchange is the acrosomal reaction in echinoderm sperm found by Tilney (1976, 1978). Upon contact with the surface substance of an oocyte, the G-actin present in the acrosome of the sperm polymerizes to actin filaments as long as 90 μm within 10–30 sec. There is no myosin present in this acrosomal process. Microvilli of intestinal epithelial cells, in which actin filaments are bundled together, also do not contain myosin (Bretscher and Weber, 1978, 1980).

In cultured fibroblasts, actin filaments are bundled and are distributed in fairly parallel lines that have been called stress fibers (Lazarides and Weber, 1974; Lazarides, 1976). Stress fibers disappear promptly, but may be reorganized depending on such changes in cell conditions as trypsinization on cell transfer, transformation by tumor virus (Bosckek *et al.*, 1981; Carley *et al.*, 1981), or nutritional deficiencies. The deformability of the cell's morphological structure is a prerequisite for cell motility as well as for phagocytosis. Accordingly, G-F transformation is an important process by which nonmuscle cells carry out various cellular functions.

The F-actin ratio in M1 cells was determined by two methods. One makes use of the fact that DNase I is inhibited by G-actin, but not by F-actin. Thus, the actin content was measured by a DNase I inhibition assay before and after treatment with guanidine-HCl which depolymerizes F-actin. In the second methods, a ^{35}S-methionine-labeled cell extract was centrifuged to sediment its F-actin; SDS–PAGE followed. The F-actin ratio was calculated from the radioactivity present in the actin band. These two methods gave the same values although only values from the first are shown in Table 1. Only when phagocytosis was induced did the F-actin ratio show an increase (Nagata *et al.*, 1980).

After ultracentrifugation of the M1 cell homogenate, its sediment was fractionated by sucrose gradient centrifugation in order to isolate the plasma membrane which is identifiable by its Na^+,K^+-ATPase activity. The membrane fraction of differentiated cells contained 3 times as much actin and myosin as before differentiation (Sagara *et al.*, 1982). Thus, during differentiation, large amounts of actin were polymerized to the F-actin and bound to the plasma membrane.

4. *Qualitative Changes in Actin*

4.1. Polymerization of Actin in the Crude Extract

Experiments on G-F transformation were expected to produce important information about the nature of the actin in this cell line. In the experiments made (Nagata *et al.*, 1980), crude extract denotes the supernatant after the cell homogenate had been centrifuged at 100,000*g* for 1 hr in ATP-containing buffer solution. The extract was combined with 0.1 M KCl and 5 mM $MgCl_2$ then incubated at 30°C for 1.5 hr, after which it was ultracentrifuged to check how much actin had completed polymerization under these conditions (Fig. 2).

Although all the crude extracts contained almost the same amount of actin (about 7%), the untreated M1 cell extract showed only a very slight increase in sedimentable actin in contrast to the marked increase seen in CM-treated M1 cells. Varying the concentration of KCl (0–0.2 M) and $MgCl_2$ (0–5 mM) produced no increase in sedimentable actin in the undifferentiated M1 cell extract. With the crude extract of the D^- subline, no increase in sedimen-

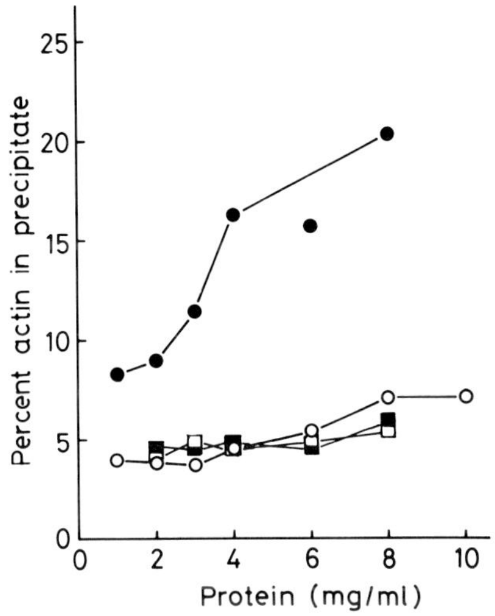

Figure 2. Effects of protein concentration on actin polymerization in crude cell extracts. The ^{35}S-methionine-labeled cell lysate was incubated at 30°C for 1.5 hr in the presence of 0.1 M KCl and 5 mM $MgCl_2$, then it was centrifuged at 100,000 g for 2.5 hr at 25°C. The ordinate shows the count in the actin band/the total count in the pellets put through SDS-PAGE. (○) untreated Ml; (●) CM-treated Ml; (□) untreated D⁻; (■) CM-treated D⁻ cell extracts.

table actin was found whether the cells were treated with CM or not. These results suggest two possibilities: that structural changes are induced in actin molecules during differentiation or that an inhibitor for actin polymerization is present only in undifferentiated cells.

4.2. Polymerization of Purified Actin

4.2.1. Purification

Actin was purified from M1 line cells before and after CM-treatment and from D^- subline cells by the method of Gordon *et al.* (1976a). The crude cell extract was applied to a DEAE-Sephadex column and eluted by a gradient concentration of KCl. The actin fraction, detected by a DNase I inhibition assay, was incubated at 30°C in the presence of 2 mM $MgCl_2$. Polymerized actin was collected by centrifugation, then depolymerized again by dialysis against a low ionic strength buffer, after which it was fractionated on a Sephadex G-150 column.

The actin samples had purities of more than 95%. Their molecular weights, and inhibitory activities for DNase I as well as for the K^+-EDTA-ATPase of myosin did not differ from the values for skeletal muscle actin purified by the method of Spudich and Watt (1971).

4.2.2. Kinetics of Polymerization

Polymerization velocity was measured with a viscometer of the Ostwald type at 25°C in the presence of 2 mM $MgCl_2$ or of 0.1 M KCl (Fig. 3). In the presence of $MgCl_2$, fast polymerization made it difficult to detect differences among the three actin samples, but some differences were evident with KCl

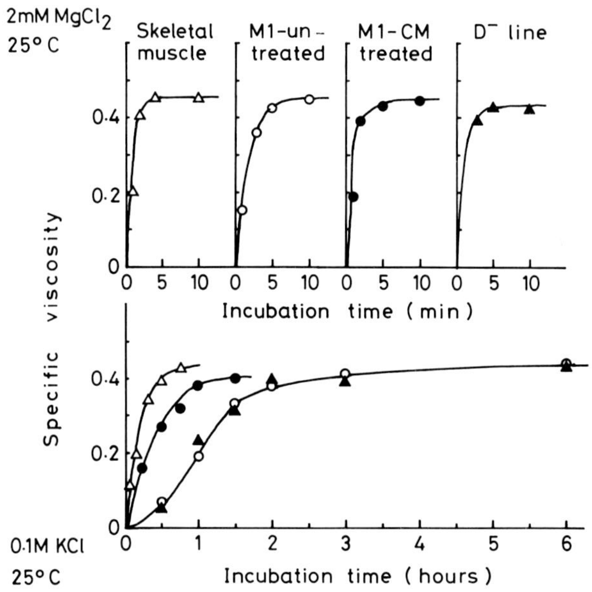

Figure 3. Polymerization velocity of purified actin. The specific viscosity at 25°C was plotted as a function of the length of incubation for four different purified actins. All the samples contained 0.5 mg/ml of actin, 0.2 mM ATP, 0.1 mM $CaCl_2$, 0.75 mM β-mercaptoethanol, 3 mM imidazole-HCl, pH 7.5, 0.02% NaN_3, and 2 mM $MgCl_2$ or 0.1 M KCl. Viscosity was measured in a size-75, Cannon-Manning semimicro viscometer.

that were not present with $MgCl_2$. Polymerization of skeletal muscle actin proceeded the fastest and was complete in approximately 30 min; polymerization of the untreated M1 and D^- actins required 2–3 hr. After CM-treatment, however, polymerization of actin was complete within an hour (Nagata *et al.*, 1982a).

Oosawa and his colleagues (Kasai *et al.*, 1962b; Oosawa and Kasai, 1962) found two steps in actin polymerization: a nucleation step for monomer actin which forms an oligomer of 3–4 actins and an elongation step for the monomer which adds onto the oligomer and elongates the actin filament. The first step is rate-limiting for the whole polymerization process.

Which step is responsible for the delayed polymerization of undifferentiated M1 cell actin? An addition of F-actin fragments to the actin sample greatly promoted polymerization; even D^- actin completed polymerization in 5 min, whereas it would have taken more than 2 hr without the F-actin fragments (Nagata *et al.*, 1982a). This experiment shows that delayed polymerization is caused by a defect in the nucleation step. It also suggests that structural changes take place in the actin molecule during differentiation.

4.2.3. *Critical Concentration*

Oosawa and his colleagues (Oosawa and Kasai, 1962; Kasai *et al.*, 1962a) established that there is an equilibrium between monomer and polymer actin concentrations in all actin samples, such as the condensation phenomenon of water vapor. The concentration of monomer actin in equilibrium with highly polymerized actin is called the critical concentration under that condition. Under physiological conditions (0.1 M KCl, pH 7.0, 25–30°C) the critical concentration of skeletal muscle actin is 0.05 mg/ml or less.

In Korn's laboratory, critical concentrations for the actins of *Acanthamoeba* and several other nonmuscle cells were compared (Gordon *et al.*, 1976a,b, 1977). Although no differences were found under physiological conditions, the critical concentrations at 5°C in 0.1 M KCl were 0.1 mg/ml for skeletal muscle, and about 0.5 mg/ml for human platelets, embryonic chick brain, rat liver, and *Acanthamoeba*.

The critical concentrations of untreated and CM-treated M1 and D⁻

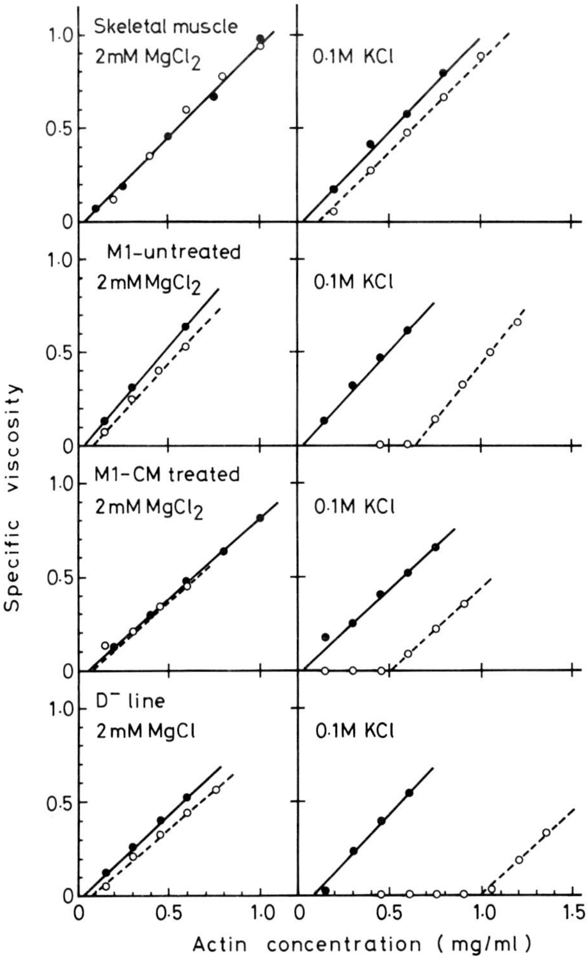

Figure 4. Equilibrium specific viscosity of purified actin. Specific viscosity was plotted as a function of the concentrations of the four purified actins after incubation for 1 day at 25°C (●) or for 2 days at 5°C (○).

actins in the presence of 2 mM $MgCl_2$ were almost the same at 25°C and at 5°C. But when the actin samples were incubated at 5°C for 2 days in the presence of 0.1 M KCl, their critical concentrations differed (Fig. 4; Nagata *et al.*, 1982a). This is evidence that differentiation of the M1 cell line caused the enhanced polymerizability of actin, and that an inhibitor for actin polymerization may be present in the D^- subline extracts because the polymerization of purified D^- actin does not differ from that of CM-treated M1 actin in the presence of $MgCl_2$, in spite of the inability for unpurified D^- actin to polymerize (see Fig. 2).

4.3. Activation of Myosin Mg^{2+}-ATPase

Although it has yet to be proved whether the sliding mechanism between actin and myosin works in cell motility in nonmuscle cells, one would expect that the activation of myosin Mg^{2+}-ATPase by actin supplies the energy required for cell functions that include cell motility and phagocytosis.

Heavy meromyosin (HMM) was prepared from skeletal muscle myosin and was used as the myosin Mg^{2+}-ATPase. Its specific activation by actin was tested with actins from skeletal muscle and from untreated M1 and CM-treated M1 cells. Values were 33.3, 13.3, and 24.3 μM P_i/μM HMM head·sec·mg actin. This means that the specific activity of actin for myosin Mg^{2+}-ATPase-activation was elevated 2-fold during differentiation.

The dose response in the activation of myosin Mg^{2+}-ATPase by actin is shown in a double reciprocal plot (Fig. 5). As reported for various nonmuscle cell actins (Gordon *et al.*, 1976a, 1977; Uyemura *et al.*, 1978), there was no difference in the V_{max} values for the three samples: untreated and CM-treated M1 and skeletal muscle actin. The apparent Km values did differ in the order of untreated M1 > CM-treated M1 > skeletal muscle. This also shows that the affinity of M1 actin for skeletal muscle myosin was elevated during differentiation (Nagata *et al.*, 1982a).

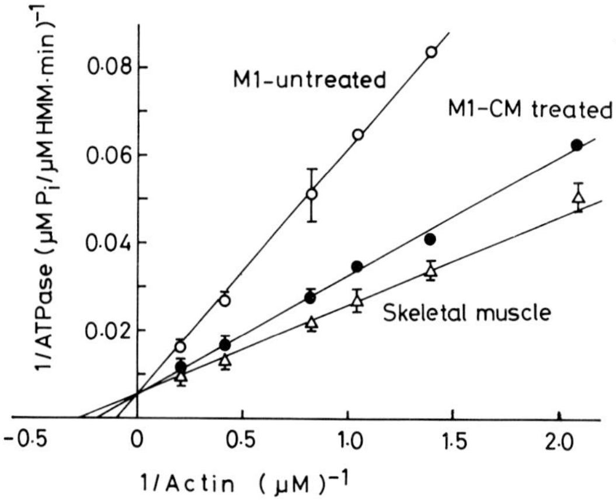

Figure 5. Activation of the Mg^{2+}-ATPase of HMM by actin. HMM was prepared by trypsinization of skeletal muscle myosin. Purified actins were added after polymerization at 25°C for 1 hr in 5 mM $MgCl_2$. Mg^{2+}-ATPase activity was plotted reciprocally against different concentrations of actin.

After differentiation, the characteristics of actin become similar to those of nonmuscle cells as reported previously by others. In other words, the actins of the untreated M1 line and D^- subline cells have exceptional characteristics.

4.4. Chemical Changes in Actin Molecules

Whether the functional changes described in the previous section are accompanied by chemical changes in the actin structure was our next interest. Most nonmuscle cell actin is composed of two isomers with different isoelectric points, β- and γ-isoactins (Garrels and Gibson, 1976; Rubenstein and Spudich, 1977). Actin samples purified from the untreated and CM-treated M1 line and from D^- subline cells were electrophoresed by O'Farrell's (1975) method (Fig. 6). The β/γ ratio was determined densitometrically (Nagata *et al.*, 1982a).

As shown in Table 2, differentiation produces a decrease in the β/γ ratio which seems to be caused by a selective increase in the amount of γ-actin molecule because the total actin content always increases after differentiation (Table 1). The β/γ ratio in most other nonmuscle cells is in the range from 1:1 to 2:1 (Gordon *et al.*, 1977), except as reported by Sakiyama *et al.*, (1981) who detected no γ-actin in their mouse fibroblast cell line, L. Thus, differentiation of the M1 cell line is accompanied by "normalization" of the β/γ ratio of actin, as in polymerization and myosin Mg^{2+}-ATPase activation. D^- subline cells, however, contain only a trace of γ-actin, which results in a very high β/γ ratio.

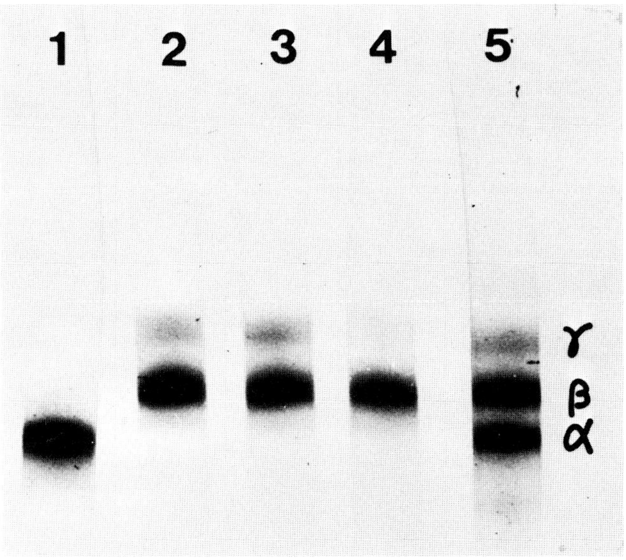

Figure 6. Isoelectric focusing gels of skeletal muscle and Ml cell actins. Purified actins were isoelectrofocused by O'Farrell's method (1975) with 5% pharmalyte (pH 4–6.5). (1) Rabbit skeletal muscle, (2) untreated Ml, (3) CM-treated Ml, (4) D^- subline, and (5) a mixture of rabbit skeletal muscle and untreated Ml cell actins.

Table 2. The Ratio of β- to γ-Actin Isomers[a]

	β/γ Ratio	
	Lot 1	Lot 2
Untreated Ml	4.1 ± 0.5	5.6 ± 1.2
CM-treated Ml	2.6 ± 0.6	3.0 ± 0.1
D^- line	7.4 ± 1.2	7.8 ± 1.2

The ratio of β- to γ-actin isomers. The isoelectrofocusing gel stained with Coomassie Blue (Fig. 6) was scanned densitometrically, and its isoactin areas were cut out and weighed. Mean ± standard error.

Since the differentiation of M1 cells is controlled at the transcriptional stage (Nagata and Ichikawa, 1979), the different isoactin ratios indicate a change in gene expression. Using two-dimensional mapping of the tryptic peptides of actin, we identified several different spots before and after differentiation. Although no functional difference between β- and γ-actins is known, these observed differences may affect functional activities such as polymerization, interaction with myosin, and binding to the plasma membrane.

5. Regulatory Protein for Actin Polymerization

A number of proteins which affect actin polymerization have been isolated and purified. Both DNase I (31K; Hitchcock *et al.*, 1976; Lazarides and Lindberg, 1974) and profilin (16K; Carlsson *et al.*, 1977; Markey *et al.*, 1978) bind with actin in the ratio of 1:1 and prevent it from polymerizing. Capping protein (dimers of 28K and 31K) in *Acanthamoeba* (Isenberg *et al.*, 1980) and 65K protein in platelets (Grumet and Lin, 1980) also inhibit actin polymerization by binding the barbed end of actin molecules. An inhibitor (dimers of 62K and 65K) for actin polymerization was isolated from human neutrophils (Southwick and Stossel, 1981). None of these factors are Ca^{2+}-sensitive or active in the depolymerization of F-actin filaments. Active depolymerizing proteins, 92K (Harris and Schwartz, 1981) was isolated from blood plasma and 17K proteins (Mabuchi, 1981) from starfish eggs; neither are Ca^{2+}-sensitive.

Proteins which cut F-actin into short fragments only in the presence of 10^{-7}–10^{-6} M Ca^{2+} have been isolated: gelsolin (subunit, 91K) from macrophages (Yin and Stossel, 1979; Yin *et al.*, 1980), villin (subunit, 95K) from intestinal epithelial cells (Bretscher and Weber, 1980; Craig and Powell, 1980; Glenney *et al.*, 1981), and fragmin (42K) from *Physarum* (Hasegawa *et al.*, 1980). All are Ca^{2+}-sensitive. Villin and fragmin bind to the barbed end of actin and promote nucleation, which results in an increase of short filaments. Villin shows bundling activity for actin filament in the absence of Ca^{2+}.

As stated before, actin in the crude cell extract of the D^- subline did not polymerize even in the presence of $MgCl_2$ and KCl (Fig. 2), but purified actin polymerized normally in the presence of 2 mM $MgCl_2$ (Figs. 3 and 4). This suggests the presence of an inhibitor for actin polymerization in the D^- cell extract. Actually, additions of increasing concentrations of D^- cell extract to skeletal muscle actin caused a slow-down in the polymerization velocity and a decrease in the final viscosity in the presence of 2 mM $MgCl_2$ (Fig. 7; Nagata *et al.*, 1982b).

D^- cell extract was absorbed on a DEAE-Sephadex A-50 column then eluted by a linear KCl gradient. An inhibitor for actin polymerization, eluted just in advance of the actin peak, was fractionated by a 55/80% saturated solution of $(NH_4)_2SO_4$, then chromatographed on Sephadex G-150 and on hydroxylapatite. The molecular weight of the purified protein, named API (actin polymerization-inhibitor) is 71,000 (Nagata *et al.*, 1982b). API inhibits the polymerization of skeletal muscle actin, but does not depolymerize F-actin. Since the effect of API was not changed by an addition of F-actin fragments to the sample, API probably affects the elongation step, not the nucleation step. Like the other polymerization-inhibitors, API has no Ca^{2+}-sensitivity.

Quantitative changes in API before and after the differentiation of M1 cells are to be investigated.

6. *Changes in Actin-Related Gelation*

Pollard and Ito (1970) and Pollard and Korn (1971) observed crude cell extracts of *Amoeba proteus* in the sol, gel, and contracted states. No filaments were found in the sol sample, but a random meshwork of actin filaments was present in the gel, and bundling of actin filaments was found in the contracted state.

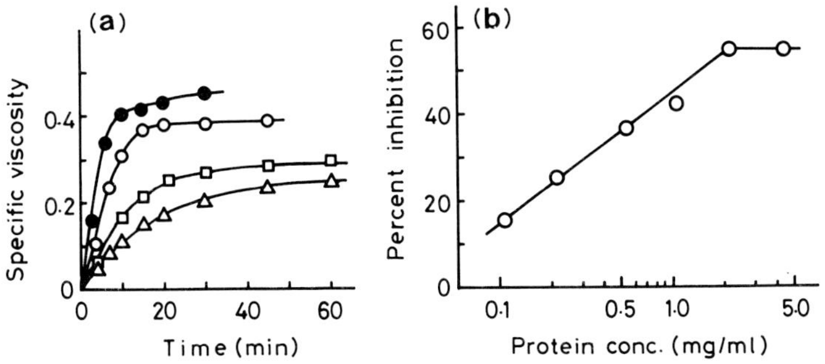

Figure 7. Inhibition of the polymerization of skeletal muscle actin by D^- cell extract. (a) Time course of actin (0.5 mg/ml) polymerization without D^- cell extract (●) and with 0.25 (○), 0.75 (□), or 2.1 (△) mg/ml of D^- cell extract. (b) Percent of inhibition of actin polymerization plotted against the protein concentration in the D^- cell extract.

In 1975, Kane reported that an extract of sea urchin eggs, which had been extracted in a low ionic strength buffer containing glycerol and EGTA, gelled when it was warmed in the presence of ATP and KCl (Kane, 1975, 1976). He also found networks of actin filaments in the gel. A relation between actin-related gelation and cell motility is suggested by the facts that anterior and posterior portions of amoeba have different visco-elasticities (Helleweell and Taylor, 1979; Taylor and Condeelis, 1979) and the cell peripheries of macrophages that adhere firmly to plate bottoms contain larger amounts of actin, myosin, and actin-binding protein (ABP) than other parts of the cytoplasm (Stendahl *et al.*, 1980).

Most gelation is controlled by the physiological concentration of calcium. Gelation-related factors can be divided into three classes: gelation factors which themselves are Ca^{2+}-sensitive such as actinogelin (Mimura and Asano, 1979), the α-actinin of nonmuscle cells (Burridge and Feramisco, 1981), and villin (Bretscher and Weber, 1980; Craig and Powell, 1980). Gelation factors that are not Ca^{2+}-sensitive such as filamin (Shizuta *et al.*, 1976; Wang, 1977), ABP (Hartwig and Stossel, 1975), the 23K, 28K, 32K, 38K, and 280K proteins from *Acanthamoeba* (Maruta and Korn, 1977), and the 58K and 220K proteins from sea urchin (Bryan and Kane, 1978). And, molecules that do not cause gelation by themselves but confer Ca^{2+}-sensitivity on some Ca^{2+}-insensitive gelation factors, gelsolin (Yin and Stossel, 1979; Yin *et al.*, 1980) and fragmin (Hasegawa *et al.*, 1980), belong to this last class.

6.1. Gelation of Crude Extracts

6.1.1. Dose Response

Untreated and CM-treated M1 or Mm-1 cells first were homogenized in a low ionic strength buffer containing ATP and EGTA after which they were ultracentrifuged. After the supernatant had been incubated at 25°C for 1 hr in the presence of 2 mM $MgCl_2$, gelation was tested by the falling ball method of MacLean-Fletcher and Pollard (1980b). The Mm-1 extract gelled at protein concentrations of 7–10 mg/ml, whereas extracts from M1 cells did not gel.

The respective concentrations of actin and ABP were 4.9% and 0.1% for M1 cells, and 6.6% and 0.5% for Mm-1 cells. In another series of experiments, 1 mg/ml of skeletal muscle actin was added to test samples to compensate for their low actin concentrations (Fig. 8). The M1 cell extract still did not gel. The untreated M1 extract gelled only after being dialyzed against a buffer solution containing EGTA, ATP, and DTT that was supplemented with actin.

Extracts from CM-treated M1 cells as well as from Mm-1 cells required a greater protein concentration for gelation after dialysis than they did before. The protein concentration needed for gelation after dialysis is in the order for untreated M1 > CM-treated M1 > Mm-1 cells (Nagata *et al.*, 1983). Thus, dialysis reverses the effects of cell extracts on the gelation found before and after differentiation—promotion for undifferentiated cell extracts and suppression for differentiated cell extracts. What, then, is responsible for the

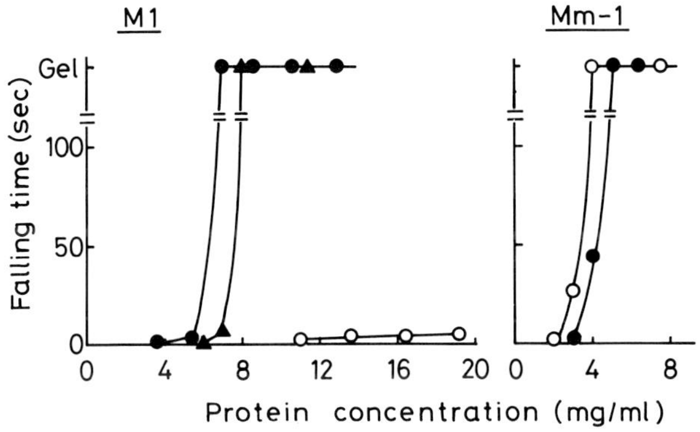

Figure 8. Gelation of crude cell extracts. Crude cell extracts with different protein concentrations were given supplements of rabbit skeletal muscle actin at 1 mg/ml then incubated at 30°C for 1 hr in the presence of 2 mM $MgCl_2$ before (○) and after (●,▲) dialysis against KCl-free buffer. Gelation was checked by the falling ball method of MacLean-Fletcher and Pollard (1980b).

reverse effects? A dialyzable factor would account for the phenomenon in question.

6.1.2. *KCl Effect*

Cell extracts (6–8 mg/ml) supplemented with skeletal muscle actin (1 mg/ml) were incubated at 30°C for 1 hr in the presence of 2 mM $MgCl_2$ and different concentrations of KCl (Fig. 9). Gelation of the CM-treated M1 and Mm-1 extracts was inhibited only when the KCl concentration exceeded 100 mM, whereas gelation of the untreated M1 extract was inhibited by as low a KCl concentration as 25 mM. When the concentration of exogenous actin was decreased to 0.5 mg/ml, 5 mM KCl was sufficient to inhibit gelation of the M1 extract.

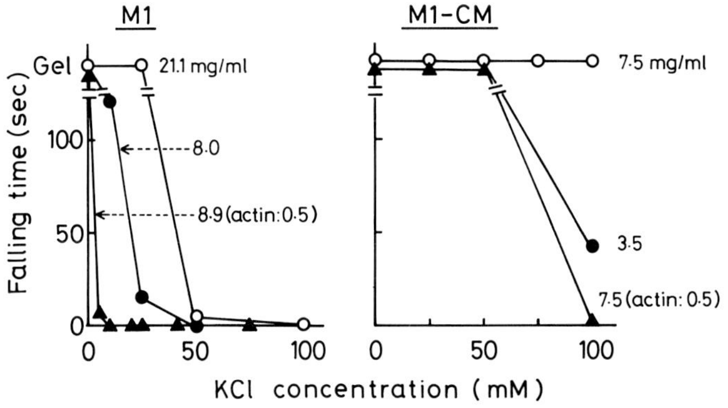

Figure 9. Effect of KCl on gelation of crude cell extracts. Cell extracts were dialyzed against KCl-free buffer, then 1 mg/ml or (0.5 mg/ml as indicated in parentheses) skeletal muscle actin was added. Samples were tested for gelation in the presence of 2 mM $MgCl_2$ and different concentrations of KCl. Figures show protein concentrations in the dialyzed extracts.

With other nonmuscle cell extracts, gelation generally is induced by the presence of KCl alone and is accelerated by the addition of $MgCl_2$. Condeelis and Taylor (1977) reported that 20–60 mM KCl in addition to 1 mM $MgCl_2$ caused maximum gelation of an extract from *Dictyostelium*. Ishiura and Okada (1979) also reported that an Ehrlich ascites tumor cell extract formed its firmest gel when 75 mM KCl was present with 1 mM $MgCl_2$.

Thus, KCl generally promotes gelation of the cell extracts. Gelation of the M1 cell extract, which was inhibited by 5–25 mM KCl, is exceptional. Since inhibition also was caused by NaCl and LiCl, but not by $MgCl_2$ or $MnCl_2$, it must be the effect of monovalent cations.

The gelation of CM-treated M1 and Mm-1 extracts was stimulated by KCl alone and by KCl plus $MgCl_2$ as found for other nonmuscle cell extracts.

6.2. KCl-Sensitive Gelation Factors

What causes the different effects of KCl on the gelations of untreated M1 extract, CM-treated M1, and Mm-1 extracts? We recently isolated three gelation factors from our M1 cell line: a 260K protein which probably is the same as the ABP reported by Hartwig and Stossel (1975), a 105K α-actinin-like protein, and 37–41K protein. We assume that all of their native forms are dimers (unpublished data).

Our preliminary experiments showed that gelation by ABP is KCl-insensitive, whereas gelation by 105K and 37–41K proteins is inhibited by a low concentration of KCl. The different KCl-sensitivity of the gelation of the crude cell extract, therefore, is attributable to one of two mechanisms: quantitative changes in the two factors which change KCl-sensitivity, or changes in the KCl-sensitivities of the factors. The functions of KCl-sensitive factors described here are under investigation.

7. Summary and Future Challenges

The induction of differentiation in a myeloid leukemia cell line, M1, results in the loss of mitotic activity and the appearance of phagocytic and locomotive activities. Actomyosin system is known to have a major function in these activities. As presented, quantitative and qualitative changes were found not only in cellular actin, but in its regulatory proteins [actin polymerization inhibitor (API)] and in several gelation factors, as well.

The following are topics for future investigations:

1. The genetical aspect of gene expression for actin that causes changes in the β/γ isoactin ratio before and after differentiation.
2. Changes in the contents and distribution of API during differentiation, which affect the local density of actin fibers.
3. Changes in the localization of the three gelation factors, and their control mechanism brought about by intracellular Ca^{2+} and K^+ ions.

4. Investigation of an anchor protein in the plasma membrane that binds more actomyosin after differentiation than before.

The results of these investigations should shed new light on the mechanism for the induction of cell motility, as well as on that of "decarcinogenesis" (Sugimura *et al.*, 1972) because leukemic cells are not motile and, once mobility is induced, they are not leukemogenic.

ACKNOWLEDGMENTS. We are grateful to Ms. Patricia Yamada for criticism of the manuscript. This work was supported by a Grant-in-Aid for Cancer Research from the Ministry of Education, Science, and Culture of Japan.

References

Abramowitz, J. W., Stracher, A., and Detwiler, T. C., 1975, A second form of actin: Platelet microfilaments depolymerized by ATP and divalent cations, *Arch. Biochem. Biophys.* **167:**230–237.

Bamburg, J. R., Harris, H. E., and Weeds, A. G., 1980, Partial purification and characterization of an actin depolymerizing factor from brain, *FEBS Lett.* **121:**178–182.

Blikstad, I., Markey, F., Carlsson, L., Persson, T., and Lindberg, U., 1978, Selective assay of monomeric and filamentous actin in cell extracts, using inhibition of deoxyribonuclease I, *Cell* **15:**935–943.

Bosckek, C. B., Jockusch, B. M., Friis, R. R., Back, R., Grundmann, E., and Bauer, H., 1981, Early changes in the distribution and organization of microfilament proteins during cell transformation, *Cell* **24:**175–184.

Bray, D., and Thomas, C., 1976, Unpolymerized actin in fibroblasts and brain, *J. Mol. Biol.* **105:**527–544.

Brenner, S. L., and Korn, E. D., 1980, Spectrin/actin complex isolated from sheep erythrocytes accelerates actin polymerization by simple nucleation, *J. Biol. Chem.* **255:**1670–1676.

Bretscher, A., and Weber, K., 1978, Localization of actin and microfilament-associated proteins in the microvilli and terminal web of the intestinal brush border by immunofluorescence microscopy, *J. Cell Biol.* **79:**839–845.

Bretscher, A., and Weber, K., 1980, Villin is a major protein of the microvillus cytoskeleton which binds both G and F actin in a calcium-dependent manner, *Cell* **20:**839–847.

Bryan, J., and Kane, R. E., 1978, Separation and interaction of the major components of sea urchin actin gel, *J. Mol. Biol.* **125:**207–224.

Burridge, K., and Feramisco, J. R., 1981, Non-muscle α-actinins are calcium-sensitive actin-binding proteins, *Nature* **294:**565–567.

Carley, W. W., Barak, L. S., and Webb, W. W., 1981, F-actin aggregates in transformed cells, *J. Cell Biol.* **90:**797–802.

Carlsson, L., Nyström, L.-E., Sundkvist, I., Markey, F., and Lindberg, U., 1977, Actin polymerizability is influenced by profilin, a low molecular weight protein in non-muscle cells, *J. Mol. Biol.* **115:**465–483.

Chang, D., Tobin, K. D., Grumet, M., and Lin, S., 1980, Cytochalasins inhibit nuclei-induced actin polymerization by blocking filament elongation, *J. Cell Biol.* **84:**455–460.

Condeelis, J. S., and Taylor, D. L., 1977, The contractile bases of amoeboid movement. V. The control of gelation, solation and contraction in extracts from Dictyostelium discoideum, *J. Cell Biol.* **74:**901–927.

Craig, S. W., and Powell, L., 1980, Regulation of actin polymerization by villin, a 95,000 dalton cytoskeletal component of intestinal brush borders, *Cell* **22:**739–746.

Fine, R. E., and Bray, D., 1971, Actin in growing nerve cells, *Nature New Biol.* **234:**115–118.

Fine, R. E., and Taylor, L., 1976, Decreased actin and tubulin synthesis in 3T3 cells after transformation by SV40 virus, *Exp. Cell Res.* **102:**162–168.

Flanagan, M. D., and Lin, S., 1980, Cytochalasins block actin filament elongation by binding to high affinity sites associated with F-actin, *J. Biol. Chem.* **255:**835–838.

Garrels, J. I., and Gibson, W., 1976, Identification and characterization of multiple forms of actin, *Cell* **9:**793–805.

Glenney, J. R., Jr., Kaulfus, P., and Weber, K., 1981, F actin assembly modulated by villin; Ca^{++}-dependent nucleation and capping of the barbed end, *Cell* **24:**471–480.

Gordon, D. J., Eisenberg, E., and Korn, E. D., 1976a, Characterization of cytoplasmic actin isolated from Acanthamoeba castellanii by a new method, *J. Biol. Chem.* **251:**4778–4786.

Gordon, D. J., Yang, Y.-Z. and Korn, E. D., 1976b, Polymerization of Acanthamoeba actin. Kinetics, thermodynamics, and co-polymerization with muscle actin, *J. Biol. Chem.* **251:**7474–7479.

Gordon, D. J., Boyer, J. L., and Korn, E. D., 1977, Comparative biochemistry of non-muscle actins, *J. Biol. Chem.* **252:**8300–8309.

Grumet, M., and Lin, S., 1980, A platelet inhibitor protein with cytochalasin-like activity against actin polymerization in vitro, *Cell* **21:**439–444.

Guez, M., and Sachs, L., 1973, Purification of the protein that induces cell differentiation to macrophages and granulocytes, *FEBS Lett.* **37:**149–154.

Harris, D. A., and Schwartz, J. H., 1981, Characterization of brevin, a serum protein that shortens actin filaments, *Proc. Natl. Acad. Sci. USA* **78:**6798–6802.

Hartwig, J. H., and Stossel, T. P., 1975, Isolation and properties of actin, myosin, and a new actin-binding protein in rabbit alveolar macrophages, *J. Biol. Chem.* **250:**5696–5705.

Hartwig, J. H., and Stossel, T. P., 1979, Cytochalasin B and the structure of actin gels, *J. Mol. Biol.* **134:**539–553.

Hasegawa, T., Takahashi, S., Hayashi, H., and Hatano, S., 1980, Fragmin: A calcium ion sensitive regulatory factor on the formation of actin filaments, *Biochemistry* **19:**2677–2683.

Hatano, S., and Oosawa, F., 1966a, Extraction of an actin-like protein from the plasmodium of a myxomycete and its interaction with myosin A, *J. Cell Physiol.* **68:**197–202.

Hatano, S., and Oosawa, F., 1966b, Isolation and characterization of plasmodium actin, *Biochim. Biophys. Acta* **127:**488–498.

Helleweell, S. B., and Taylor, D. L., 1979, The contractile bases of amoeboid movement. VI. The solation-contraction coupling hypothesis, *J. Cell Biol.* **83:**633–648.

Hirai, K., Nagata, K., Maeda, M., and Ichikawa, Y., 1979, Changes in ultrastructures and enzyme activities during differentiation of myeloid leukemia cells to normal macrophages, *Exp. Cell Res.* **124:**269–283.

Hitchcock, S. E., Carlsson, L., and Lindberg, U., 1976, Depolymerization of F-actin by deoxyribonuclease I, *Cell* **7:**531–542.

Hoffman-Liebermann, B., and Sachs, L., 1978, Regulation of actin and other proteins in the differentiation of myeloid leukemia cells, *Cell* **14:**825–834.

Honma, Y., Kasukabe, T., and Hozumi, M., 1977, Structure requirements and affinity of steroids to bind with receptor for induction of differentiation of cultured mouse myeloid leukemia cells, *Gann* **68:**405–412.

Honma, Y., Kasukabe, T., and Hozumi, M., 1978a, Production of differentiation-stimulating factor in cultured mouse myeloid leukemia cells treated by glucocorticoids, *Exp. Cell Res.* **111:**261–267.

Honma, Y., Kasukabe, T., and Hozumi, M., 1978b, Relationships between leukemogenicity and in vivo inducibility of normal differentiation in mouse myeloid leukemia cells, *J. Natl. Cancer Inst.* **61:**837–841.

Hozumi, M., 1982, A new approach to chemotherapy of myeloid leukemia: Control of leukemogenicity of myeloid leukemia cells by inducer of normal differentiation, in: *Cancer Biology Reviews*, Vol. 3 (J. J. Marchalonis and M. G. Hanna Jr., eds.), pp. 153–211, Marcel Dekker, New York.

Hozumi, M., Sugiyama, K., Mura, M., Takizawa, H., Sugimura, T., Matsushima, T., and Ichikawa, Y., 1974, Factor(s) stimulating differentiation of mouse myeloid leukemia cells found in ascitic fluid, in: *Differentiation and Control of Malignancy of Tumor Cells* (W. Nakahara, T. Ono, T. Sugimura, and H. Sugano, eds.), pp. 471–483, University of Tokyo Press, Tokyo.

Ichikawa, Y., 1969, Differentiation of a cell line of myeloid leukemia, *J. Cell Physiol.* **74:**223–234.

Ichikawa, Y., 1970, Further studies on the differentiation of a cell line of myeloid leukemia, *J. Cell Physiol.* **76:**175–184.

Ichikawa, Y., Maeda, M., and Horiuchi, M., 1975, Induction of differentiated functions which are reversibly suppressed by cytochalasin B, *Exp. Cell Res.* **90:**20–30.

Isenberg, G., Aebi, U., and Pollard, T. D., 1980, An actin-binding protein from Acanthamoeba regulates actin filament polymerization and interactions, *Nature* **288:**455–459.

Ishikawa, H., Bischoff, R., and Holtzer, H., 1969, Formation of arrowhead complexes with heavy meromyosin in a variety of cell types, *J. Cell Biol.* **43:**312–328.

Ishiura, M., and Okada, Y., 1979, The role of actin in temperature-dependent gel-sol transformation of extracts of Ehrlich ascites tumor cells, *J. Cell Biol.* **80:**465–480.

Kane, R. E., 1975, Preparation and purification of polymerized actin from sea urchin egg extracts, *J. Cell Biol.* **66:**305–315.

Kane, R. E., 1976, Actin polymerization and interaction with other proteins in temperature-induced gelation of sea urchin egg extracts, *J. Cell Biol.* **71:**704–714.

Kasai, M., Asakura, S., and Oosawa, F., 1962a, The G-F equilibrium in actin solutions, under various conditions, *Biochim. Biophys. Acta* **57:**13–21.

Kasai, M., Asakura, S., and Oosawa, F., 1962b, The cooperative nature of G-F transformation of actin, *Biochim. Biophys. Acta* **57:**22–31.

Kasukabe, T., Honma, Y., and Hozumi, M., 1977, Induction of lysosomal enzyme activities with glucocorticoids during differentiation of cultured mouse myeloid leukemia cells, *Gann* **68:**765–773.

Lazarides, E., 1976, Actin, α-actinin, and tropomyosin interaction in the structural organization of actin filaments in nonmuscle cells, *J. Cell Biol.* **68:**202–219.

Lazarides, E., and Lindberg, U., 1974, Actin is the naturally occurring inhibitor of deoxyribonuclease I, *Proc. Natl. Acad. Sci. USA* **71:**4742–4746.

Lazarides, E., and Weber, K., 1974, Actin antibody: The specific visualization of actin filaments in non-muscle cells, *Proc. Natl. Acad. Sci. USA* **71:**2268–2272.

Leavitt, J., Leavitt, A., and Attallah, A. M., 1980, Dissimilar modes of expression of β- and γ-actin in normal and leukemic human T lymphocytes, *J. Biol. Chem.* **255:**4984–4987.

Lotem, J., and Sachs, L., 1976, Control of Fc and C3 receptors on myeloid leukemic cells, *J. Immunol.* **116:**580–586.

Lotem, J., and Sachs, L., 1978, In vivo induction of normal differentiation in myeloid leukemia cells, *Proc. Natl. Acad. Sci. USA* **75:**3781–3785.

Mabuchi, I., 1981, Purification from starfish eggs of a protein that depolymerizes actin, *J. Biochem.* **89:**1341–1344.

Mabuchi, I., Hosoya, H., and Sakai, H., 1980, Actin in the cortical layer of the sea urchin egg. Changes in its content during and after fertilization, *Biomed. Res.* **1:**417–426.

MacLean-Fletcher, S. D., and Pollard, T. D., 1980a, Mechanism of action of cytochalasin B on actin, *Cell* **20:**329–341.

MacLean-Fletcher, S. D., and Pollard, T. D., 1980b, Viscometric analysis of the gelation of Acanthamoeba extracts and purification of two gelation factors, *J. Cell Biol.* **85:**414–428.

Maeda, M., and Ichikawa, Y., 1973, Spontaneous development of macrophage-like cells in a culture of myeloid leukemia cells, *Gann* **64:**265–271.

Maeda, M., and Ichikawa, Y., 1980, Production of a colony-stimulating factor following differentiation of leukemic myeloblasts to macrophages, *J. Cell. Physiol.* **102:**323–331.

Maeda, M., Horiuchi, M., Numa, S., and Ichikawa, Y., 1977, Characterization of a differentiation-stimulating factor for mouse meyloid leukemia cells, *Gann* **68:**435–447.

Markey, F., Lindberg, U., and Eriksson, L., 1978, Human platelets contain profilin, a potential regulator of actin polymerisability, *FEBS Lett.* **88:**75–79.

Marks, P., and Rifkind, R. A., 1978, Erythroleukemic differentiation, *Annu. Rev. Biochem.* **47:**419–448.

Maruta, H., and Korn, E. D., 1977, Purification from Acanthamoeba castellanii of proteins that induce gelation and syneresis of F-actin, *J. Biol. Chem.* **252**:399–402.

Mimura, N., and Asano, A., 1979, Ca^{2+}-sensitive gelation of actin filaments by a new protein factor, *Nature* **282**:44–48.

Nagata, K., and Ichikawa, Y., 1979, Requirements for RNA and protein synthesis in the induction of several differentiation-markers in a myeloid leukemia cell line, *J. Cell. Physiol.* **98**:167–176.

Nagata, K., Takahashi, E., Saito, M., Ono, J., Kuboyama, M., and Ogasa, K., 1976, Differentiation of a cell line of mouse myeloid leukemia. I. Simultaneous induction of lysosomal enzyme activities and phagocytosis, *Exp. Cell Res.* **100**:322–328.

Nagata, K., Ooguro, K., Saito, M., Kuboyama, M., and Ogasa, K., 1977, A factor inducing differentiation of mouse myeloid leukemia cells in human amniotic fluid, *Gann* **68**:757–764.

Nagata, K., Sagara, J., and Ichikawa, Y., 1980, Changes in contractile proteins during differentiation of myeloid leukemia cells. I. Polymerization of actin, *J. Cell Biol.* **85**:273–282.

Nagata, K., Sagara, J., and Ichikawa, Y., 1982a, Changes in contractile proteins during differentiation of myeloid leukemia cells. II. Purification and characterization of actin, *J. Cell Biol.* **93**:470–478.

Nagata, K., Sagara, J., and Ichikawa, Y., 1982b, A new protein factor inhibiting actin-polymerization in leukemic myeloblasts, *Cell Struct. Funct.* **7**:1–7.

Nagata, K., Sagara, J., and Ichikawa, Y., 1983, Changes in actin-related gelation of crude cell extracts during differentiation of myeloid leukemia cells, *Cell Struct. Funct.* **8**:171–183.

Nakayasu, M., Shimamura, S., Takeuchi, T., Sato, S., and Sugimura, T., 1978, A factor in human saliva that induces differentiation of mouse myeloid leukemia cells, *Cancer Res.* **38**:103–109.

O'Farrell, P. H., 1975, High resolution two-dimensional electrophoresis of proteins, *J. Biol. Chem.* **250**:4007–4021.

Oosawa, F., and Kasai, M., 1962, A theory of linear and helical aggregation of macromolecules, *J. Mol. Biol.* **4**:10–21.

Pollard, T. D., 1981, Cytoplasmic contractile proteins, *J. Cell Biol.* **91**:156s–165s.

Pollard, T. D., and Ito, S., 1970, Cytoplasmic filaments of Amoeba proteus. I. The role of filaments in consistency changes and movement, *J. Cell Biol.* **46**:267–289.

Pollard, T. D., and Korn, E. D., 1971, Filaments of Amoeba proteus. II. Binding of heavy meromyosin by thin filaments of motile cytoplasmic extracts, *J. Cell Biol.* **48**:216–219.

Pollard, T. D., and Weihing, R. R., 1974, Actin and myosin and cell movement, *CRC Crit. Rev. Biochem.* **2**:1–65.

Rubenstein, P. A., and Spudich, J. A., 1977, Actin microheterogeneity in chick embryo fibroblasts, *Proc. Natl. Acad. Sci. USA* **74**:120–123.

Sachs, L., 1978, Control of normal cell differentiation and the phenotypic reversion of malignancy in myeloid leukemia, *Nature* **274**:535–539.

Sachs, L., 1982, Normal development programs in myeloid leukaemia: Regulatory proteins in the control of growth and differentiation, *Cancer Surv.* **1**:321–342.

Sagara, J., Nagata, K., and Ichikawa, Y., 1982, Changes in myosin during differentiation of myeloid leukemia cells, *J. Biochem.* **91**:1363–1372.

Sakiyama, S., Fujimura, S., and Sakiyama, H., 1981, Absence of γ-actin expression in the mouse fibroblast cell line, L, *J. Biol. Chem.* **256**:31–33.

Schmitt, H., 1976, Control of tubulin and actin synthesis and assembly during differentiation of neuroblastoma cells, *Brain Res.* **115**:165–173.

Schmitt, H., Gozes, I., and Littauer, U. Z., 1977, Decrease in levels and rates of synthesis of tubulin and actin in developing rat brain, *Brain Res.* **121**:327–342.

Shimizu, N., and Obinata, T., 1980, Presence of three actin types in skeletal muscle of chick embryos, *Dev. Growth Diff.* **22**:789–796.

Shizuta, Y., Shizuta, H., Gallo, M., Davies, P., and Pastan, I., 1976, Purification and properties of filamin, an actin binding protein from chicken gizzard, *J. Biol. Chem.* **251**:6562–6567.

Southwick, F. S., and Stossel, T. P., 1981, Isolation of an inhibitor of actin polymerization from human polymorphonuclear leukocytes, *J. Biol. Chem.* **256**:3030–3036.

Spudich, J. A., and Watt, S., 1971, The regulation of rabbit skeletal muscle contraction. I. Biochemical studies of the interaction of the tropomyosin-troponin complex with actin and the proteolytic fragments of myosin, *J. Biol. Chem.* **246:**4866–4871.

Stendahl, O. I., Hartwig, J. H., Brotschi, E. A., and Stossel, T. P., 1980, Distribution of actin-binding protein and myosin in macrophages during spreading and phagocytosis, *J. Cell Biol.* **84:**215–224.

Storti, R. V., and Rich, A., 1976, Chick cytoplasmic actin and muscle actin have different structural genes, *Proc. Natl. Acad. Sci. USA* **73:**2346–2350.

Storti, R. V., Coen, D. M., and Rich, A., 1976, Tissue-specific forms of actin in the developing chick, *Cell* **8:**521–527.

Stossel, T. P., and Hartwig, J. H., 1976, Interaction of actin, myosin, and a new actin-binding protein of rabbit pulmonary macrophages. II. Role in cytoplasmic movement and phagocytosis, *J. Cell Biol.* **68:**602–619.

Sugimura, T., Matsushima, T., Kawachi, T., Kogure, K., Tanaka, N., Miyake, S., Hozumi, M., Sato, S., and Sato, H., 1972, Disdifferentiation and carcinogenesis, *Gann Monogr.* **13:**31–45.

Sugiyama, K., Tomida, M., and Hozumi, M., 1979, Differentiation-associated changes in membrane proteins of mouse myeloid leukemia cells, *Biochim. Biophys. Acta* **587:**169–179.

Taylor, D. L., and Condeelis, J. S., 1979, Cytoplasmic structure and contractility in amoeboid cells, *Int. Rev. Cytol.* **56:**57–144.

Tilney, L. G., 1976, The polymerization of actin. III. Aggregates of nonfilamentous actin and its associated proteins: A storage form of actin, *J. Cell Biol.* **69:**73–89.

Tilney, L. G., 1978, Polymerization of actin. V. A new organelle, the actomere, that initiates the assembly of actin filaments in Thyone sperm, *J. Cell Biol.* **77:**551–564.

Tilney, L. G., and Detmers, P., 1975, Actin in erythrocyte ghosts and its association with spectrin. Evidence for a nonfilamentous form of these two molecules in situ, *J. Cell Biol.* **66:**508–520.

Tomida, M., Yamamoto, Y., and Hozumi, M., 1978, Induction by synthetic polyribonucleotide poly(I) of differentiation of cultured mouse myeloid leukemic cells, *Cell Diff.* **7:**305–312.

Uyemura, D. G., Brown, S. S., and Spudich, J. A., 1978, Biochemical and structural characterization of actin from Dictyostelium discoideum, *J. Biol. Chem.* **253:**9088–9096.

Vandekerckhove, J., and Weber, K., 1978, At least six different actins are expressed in a higher mammal: An analysis based on the amino acid sequence of the amino-terminal tryptic peptide, *J. Mol. Biol.* **126:**782–802.

Wang, K., 1977, Filamin, a new high-molecular-weight protein found in smooth muscle and nonmuscle cells. Purification and properties of chicken gizzard filamin, *Biochemistry* **16:**1857–1865.

Weiss, B., and Sachs, L., 1978, Indirect induction of differentiation in myeloid leukemic cells by lipid A, *Proc. Natl. Acad. Sci. USA* **75:**1374–1378.

Wickus, G., Gruenstein, E., Robbins, P. W., and Rich, A., 1975, Decrease in membrane-associated actin of fibroblasts after transformation by Rous sarcoma virus, *Proc. Natl. Acad. Sci. USA* **72:**746–749.

Yin, H. L., and Stossel, T. P., 1979, Control of cytoplasmic actin gel-sol transformation by gelsolin, a calcium-dependent regulatory protein, *Nature* **281:**583–586.

Yin, H. L., Zaner, K. S., and Stossel, T. P., 1980, Ca^{2+} control of actin gelation. Interaction of gelsolin with actin filaments and regulation of actin gelation, *J. Biol. Chem.* **255:**9494–9500.

5

The Dynamics of Cytoskeletal Organization in Areas of Cell Contact

Benjamin Geiger, Zafrira Avnur, Thomas E. Kreis, and Joseph Schlessinger

1. Membrane–Cytoskeleton Interactions

The progress made in recent years in cell biology has focused much attention on the structure and mechanical properties of the cytoskeleton. Electron microscopic (EM) examinations using the "classic" sections of plastic-embedded tissues, the recently developed high voltage EM of whole cells, or the quick-freezing deep-etching technique revealed a wealth of densely interwoven cytoplasmic filaments (Pollard and Weihing, 1974; Buckley, 1975; Buckley and Porter, 1975; Heuser and Kirschner, 1980). These structures collectively termed "cytoskeletal networks" retained their complex filamentous appearance after extraction of "soluble" cytoplasmic components with nonionic detergents and preserved the overall shape of the cells (Brown *et al.*, 1976; Ben Zeev *et al.*, 1979; Schliwa and van Blerkom, 1980; Schliwa *et al.*, 1981; Fulton *et al.*, 1980; Cervera *et al.*, 1981; Fulton *et al.*, 1981; Penman *et al.*, 1982).

Through the introduction of immunocytochemistry to this field of research about a decade ago some of the molecular properties of the various cytoskeletal filaments were revealed. Microfilaments were shown to contain actin and to be associated with different actin-binding proteins (α-actinin, myosin, tropomyosin, filamin, vinculin, and many others) in a rather complex manner (Gröschel-Stewart, 1980; Lazarides and Weber, 1974; Lazarides and

Benjamin Geiger, Zafrira Avnur, Thomas E. Kreis, and ***Joseph Schlessinger*** • Department of Chemical Immunology, The Weizmann Institute of Science, Rehovot 76100, Israel.

Burridge, 1979; Lazarides, 1976; Wang and Singer, 1977; Geiger, 1979). Microtubules consist of tubulin and several microtubule associated proteins (MAPs). Intermediate filaments, the least soluble cytoskeletal network, were shown to consist of different tissue-specific classes of protein subunits including desmin (muscle cells), vimentin (mesenchyme), glial filament protein (astrocytes), neurofilaments (nerve cells), and prekeratins (epithelia).*

Early attempts to assign specific functions to individual cytoskeletal filaments or to components associated with them are often premature and over simplified. There is no doubt that microtubules in the mitotic spindles are somehow involved with cell division, but due to the complexity of the system it does not seem possible to define the nature of the force-generating system for chromosome movement by characterizing only the properties of tubulin. The microfilament system appears to be involved in cell motility employing different mechanisms including gel–sol transitions, actomyosin sliding, polymerization–depolymerization events, reversible severing of filaments, formation of filament bundles, etc. (Pollard and Weihing, 1974; Tilney, 1975; Stossel and Hartwig, 1976; Condeelis and Taylor, 1977; Korn, 1978; Lindberg *et al.*, 1979; Weatherbee, 1981; Condeelis, 1981; Craig and Pollard, 1982; Geiger, 1983). The prosperity of possible models emphasizes the lack of clarity with respect to the exact involvement of specific cytoskeletal proteins in cellular physiological processes.

The picture is further complicated by the possible interactions between different cytoskeletal networks or between filaments and cellular organelles. Thus, though each of the three major filament classes mentioned above forms its specific network, they are probably all interconnected. It has been shown that the actin-associated system can interact with the cell membrane (see below) and possibly with certain intermediate filaments, for example, the dense plaques and dense bodies of smooth muscle which seem to associate also with desmin (Small and Sobieszek, 1980; Ashton *et al.*, 1975; Stephens, 1977). Microtubules were shown to be closely related to several types of intermediate filaments (desmin, vimentin, and neurofilaments) as well as to mitochondria or pigment granules (Heggeness *et al.*, 1978b; Griffith and Pollard, 1978; Schliwa, 1979; Geiger and Singer, 1980; Schliwa and van Blerkom, 1981; Shelanski *et al.*, 1981; Ball and Singer, 1982).

The mechanical stability and surface properties of cells largely depend on the presence of cytoskeletal filaments, predominantly those which form contacts with the cell membrane. A few examples of the mechanical activities of cytoskeletal meshworks will be mentioned briefly. It appears that the biconcave shape, rigidity, and restriction of surface dynamics of mammalian erythrocytes are maintained or stabilized by the submembraneous mesh of

*Further discussion of the general structure and molecular properties of cytoskeletal elements will be beyond the scope of this chapter and the reader is referred to several review articles (e.g., Olmsted and Borisy, 1973; Stephens and Edds, 1976; Kirschner, 1978; Schlee and Borisy, 1979; Timasheff and Grisham, 1980; Lazarides, 1980; Lazarides, 1981; Anderton, 1981; Franke *et al.*, 1982; Steinert *et al.*, 1982; Weber and Osborn, 1982).

spectrin-actin band 4.1 ankyrin (Steck, 1974; Kirkpatrick, 1976; Marchesi, 1979; Lux, 1979; Branton *et al.*, 1981; Gratzer, 1981; Branton, 1982). In another system, the absorptive intestinal epithelium, the tightly organized structure of microvilli is largely retained by the bundle of actin filaments which run along the core of microvilli from the tip to the rootlets (Brunser and Luft, 1970; Mukherjee and Staehelin, 1979; Mooseker, 1976; Hull and Staehelin, 1979; Matsudaira and Burgess, 1982). In the inner ear, actin bundles appear to stabilize the stereocilia (Hirokawa and Tilney, 1982; DeRosier and Tilney, 1982). Membrane-bound fibers may also be related to the formation of intercellular junctions and cell attachment to extracellular matrices as will be discussed at some length later (for additional details see Geiger, 1983).

Ultrastructural observations, biochemical studies, and recently immunocytochemistry, indicate that cytoskeleton-membrane associations may display enormous variability with respect to the class of interacting filaments, mode of association, specific interconnecting proteins, and the physiological significance of the particular linkage to the membrane. Microtubules are usually located in the medullary cytoplasm of interphase cells, at some distance from the plasma membrane, though in specialized structures such as cilia or flagella contacts with the membrane were reported (Dentler, 1981). Intermediate filaments, mostly those of the prekeratin type, often form regular associations with the cell membrane, through desmosomes and hemidesmosomes. In these structures, prekeratin-containing tonofilaments (and in the exceptional case of cardiac intercalated disk, desmin filaments) are attached to the membrane-bound desmosomal plaque. In smooth muscle, intermediate filaments (desmin) are enriched immediately under the plasma membrane near the dense plaques though the nature of these interactions is still rather unclear.

The microfilament system undoubtedly displays the most extensive and diversified forms of membrane associations. Many of these were described in some detail in recent review articles (Weatherbee, 1981; Geiger, 1983) in which the reader may find complementary information on the membrane-anchorage of actin in different systems.

In only a few cases is it known how specific cytoskeletal filaments are linked to integral elements of the membrane. In mammalian erythrocyte, for example, it is known that the submembranal cytoskeletal network consists predominantly of spectrin filaments which are cross-linked into three-dimensional mesh by oligomeric actin and bands 4.1 protein. The entire meshwork is bound through ankyrin to a population of integral membrane proteins of band 3 (Branton *et al.*, 1981; Branton, 1982). Interestingly, these cytoskeletal proteins, which were considered for many years to be unique to erythrocytes, were recently described in a large variety of nonerythroid cells. This includes ankyrin (Bennett, 1979; Bennett and Davis, 1981, 1982), band 4.1 protein (Cohen *et al.*, 1982), and a variety of spectrin-like molecules notably fodrin, terminal web protein (TW 260/240), and molecules which antigenically cross react with α-spectrin (Levine and Willard, 1981; Goodman *et al.*, 1981; Bennett *et al.*, 1982; Repasky *et al.*, 1982; Glenney *et al.*, 1982a,b; Lazarides and Nelson, 1982; Mangeat *et al.*, 1982).

A different system where the molecular architecture of membrane-microfilament association that has attracted much interest lately is intestinal microvilli. These highly-ordered finger-like projections which increase enormously the absorptive surface of the intestine contain several cytoskeletal proteins. The most abundant is actin which is packed into bundles of filaments with a uniform polarity, attached at their barbed ends to the tip of microvilli, and extend into the rootlets at the terminal web (Hull and Staehelin, 1979; Mooseker and Tilney, 1975; Mooseker, 1976; Matsudaira and Burgess, 1979, 1982). Two major proteins are found within these bundles, fimbrin (molecular weight 68,000) and villin (molecular weight 95,000), both apparently form a highly ordered ternary complex with actin. The latter protein appears to be a Ca^{2+}-dependent protein capable of bundling or severing F-actin filaments (Bretscher and Weber, 1979; Bretscher and Weber, 1980a,b; Craig and Powell, 1980; Glenney *et al.*, 1980; Mooseker *et al.*, 1980; Glenney *et al.*, 1981a,b; Bretscher, 1981). The core microfilament bundle is laterally associated to the microvillar membrane via another cytoskeletal component, the 110,000 molecular weight protein which is also associated with calmodulin (Matsudaira and Burgess, 1979; Coudrier *et al.*, 1981; Glenney *et al.*, 1982c). It has been recently shown that an integral membrane protein with molecular weight of 140,000 and a related 200,000-dalton peptide are tightly associated with the cytoskeletal matrix left after detergent extraction and suggested that they might be involved in linking the actin bundle (through the lateral arms of the 110,000-dalton protein) to the lateral membranes of the microvilli (Coudrier *et al.*, 1981). The molecular nature of the end-on attachment of actin to the tip of microvilli is unknown at the present time.

These are only two examples of systems in which cytoskeletal filaments, related to actin, bind to membranes. There are obviously numerous examples of such interactions throughout many biological systems, though in most cases, the molecular topology of such interactions is far less characterized. These include systems such as the contractile ring of dividing cells, transmembrane associations of clustered membrane molecules with actin filaments, the leading edge and membrane ruffles of motile cells, membrane and cytoskeletal rearrangements in activated platelets, etc. (for review see Geiger, 1983).

In this chapter, we would like to concentrate on one major system and focus mainly on one aspect. We will describe the association of microfilaments with the cell membrane in specific sites of cell contacts either with neighboring cells or with extracellular matrices which we will refer to as "adherens-type junctions." We will discuss below in some detail the dynamic properties of formation, maintenance, and reversal of these cellular junctions.

2. Cytoskeletal Interactions with Specific Cellular Junctions

Electron microscopic studies of cultured cells and intact tissues revealed several types of cell contacts which appeared to be linked to cytoskeletal filaments (for review see Farquhar and Palade, 1963; Staehelin, 1974; Hull and

Staehelin, 1979). We will distinguish here between the adherens junctions, those associated with actin, and desmosomal junctions which are bound to intermediate filaments. Adherens junctions, as we will argue here, are a group of structurally diverse, though molecularly similar, contacts. The cell–cell contacts consist of membrane-bound "plaque" material to which actin filaments are bound forming mirror image symmetry in the two neighboring cells. Their overall shape may, however, vary from one system to the other. In polarized epithelium, such as that of intestine or kidney, the adherens junctions form a continuous ring (zonula) at the periphery of the terminal web. In cardiac myocytes, the junctions appear as distinct patches along the intercalated disks (fascia adhaerens; Fawcett, 1981), while in various locations along basolateral borders of epithelial cells and other cell types, small contact regions can be detected (for discussion see Geiger *et al.*, 1983). Similar junctions with extracellular matrix or artificial subtrates of culture dishes were often noted. The membrane-bound dense plaques of smooth muscle which will be discussed later (see Fig. 5A), are such sites where actin bundles are apparently linked to the membrane. Cultured fibroblasts show a specific association between actin and the cell membrane in regions where the membrane is in contact with extracellular matrix (CM) or with the surface of the culture dish. Contacts of the latter type were extensively studied during the last several years and will be discussed in detail below.

Two major techniques used to visualize cell–substrate contacts are interference-reflection microscopy (IRM) and electron microscopy (EM). IRM (Fig. 1A) is based on the interference of reflected light in the narrow gap between the cell membrane and the substrate in the contact region (for evaluation of the technique see Abercrombie and Dunn, 1975; Izzard and Lochner, 1976, 1980; Gingell, 1981). EM (Fig. 1B) involves fixation of cultured cells *in situ* followed by embedding and preparation of ultrathin cross sections. Both approaches revealed the existence of adhesion plaques or focal contacts in which the membrane runs very close to the plane of the substrate, with an apparent gap of only 100–150 Å. One of the early observations and most important for our discussion here was that focal contacts are associated at their cytoplasmic aspect with bundles of microfilaments (Abercrombie *et al.*, 1971; Heath and Dunn, 1978; Wehland *et al.*, 1979).

What was the basis for the classification of all the cell junctions mentioned above as one class of cell contacts, namely, the adherens junctions? What are the common features of the zonula and fascia adherens, the dense plaque of smooth muscle and the focal contacts? And last, is it justified to use focal contacts as an experimental model system to study the structure, dynamics, and biogenesis of cellular junctions? Some answers to the questions will be given in the sections below.

3. Vinculin: A Cytoskeletal Component of Adherens Junctions

A few years ago we had isolated a new cytoskeletal protein from chick smooth muscle tissue. This protein, later named vinculin, has an apparent molecular

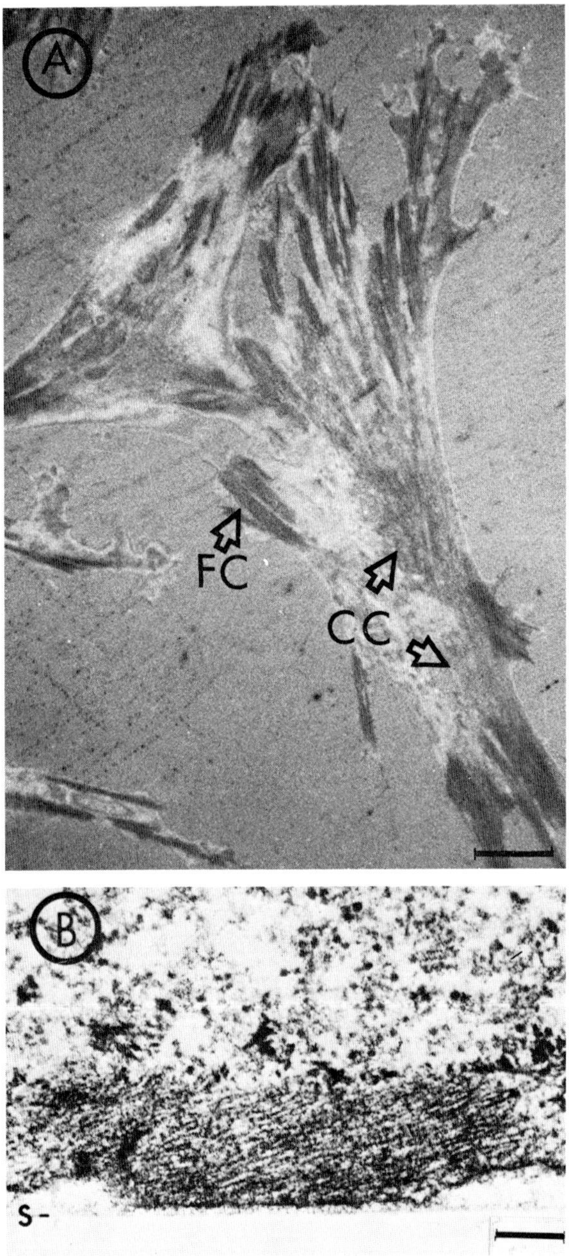

Figure 1. Cell-to-substrate focal contacts of cultured cells as visualized by interference-reflection microscopy (IRM) (A), and transmission EM (B). (A) IRM of chick gizzard fibroblasts reveal many dark patches identified as focal contacts (FC) and gray areas which represent less tight close contacts (CC). (B) Transmission EM of Epon-embedded cultured BMGE (bovine mammary gland epithelium) cells (Geiger *et al.*, 1983) sectioned perpendicularly to the plane of the substrate (S). Notice that the membrane in the focal contact runs parallel to the substrate and that numerous microfilaments appear to emerge from the contact area. Bars in (A) and (B) represent 10 μm and 0.2 μm, respectively.

weight of 130,000 and can be easily extracted at low ionic strength and slightly basic pH (Geiger, 1979). Chick vinculin appeared to be highly immunogenic in rabbits and guinea pigs and the antibodies were used to visualize the antigen in various cells and tissues (Geiger, 1979, 1981a; Geiger *et al.*, 1980, 1981; Burridge and Feramisco, 1980; Bloch and Geiger, 1980). Figure 2A shows the typical pattern of distribution of vinculin in fixed and permeabilized chick fibroblast. It consists of plaques of various sizes located near the ventral cell membrane (arrows) as well as of diffuse cytoplasmic labeling. When the same cells were observed by IRM, the ventral patches were found to coincide with cell–substrate focal contacts (Fig. 2B). In these areas, vinculin appeared to be strictly intracellular and it was our conclusion that vinculin is a cytoplasmic cytoskeletal protein which is peripherally associated with the focal contact membrane.

Double fluorescent labeling of the same cells for both vinculin and actin (Figs. 3A and B, respectively) clearly demonstrated that, in the site of contact,

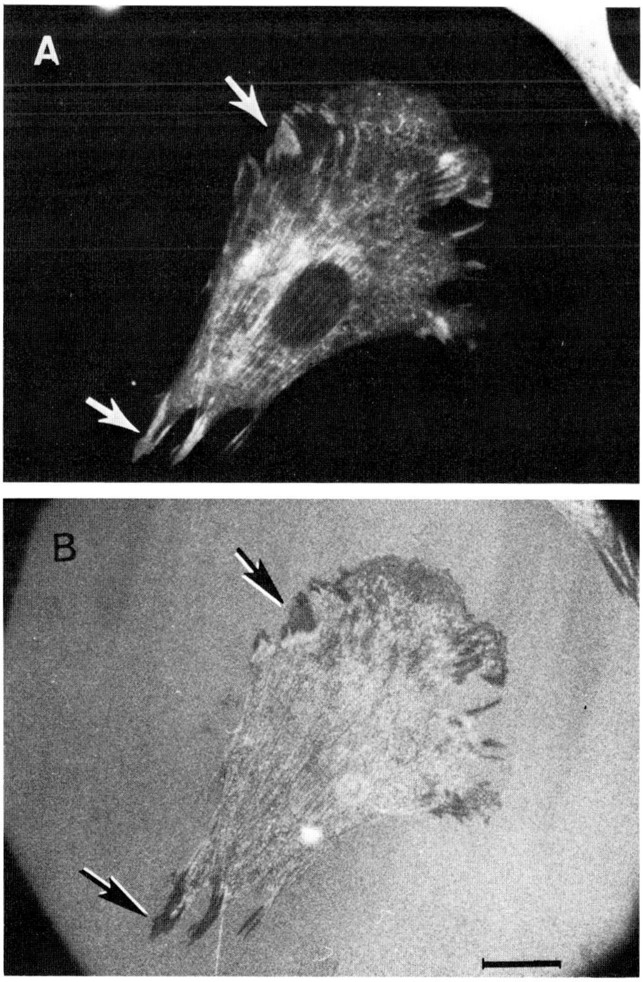

Figure 2. Cultured chick gizzard fibroblast immunolabeled for vinculin (A) or examined by IRM (B). Notice that vinculin is associated with many ventral patches which coincide with IRM-dark focal contact (arrows). In addition, vinculin displays diffuse cytoplasmic staining. The latter can be abolished by preextraction with detergent (see Fig. 3). Bar represents 10 μm.

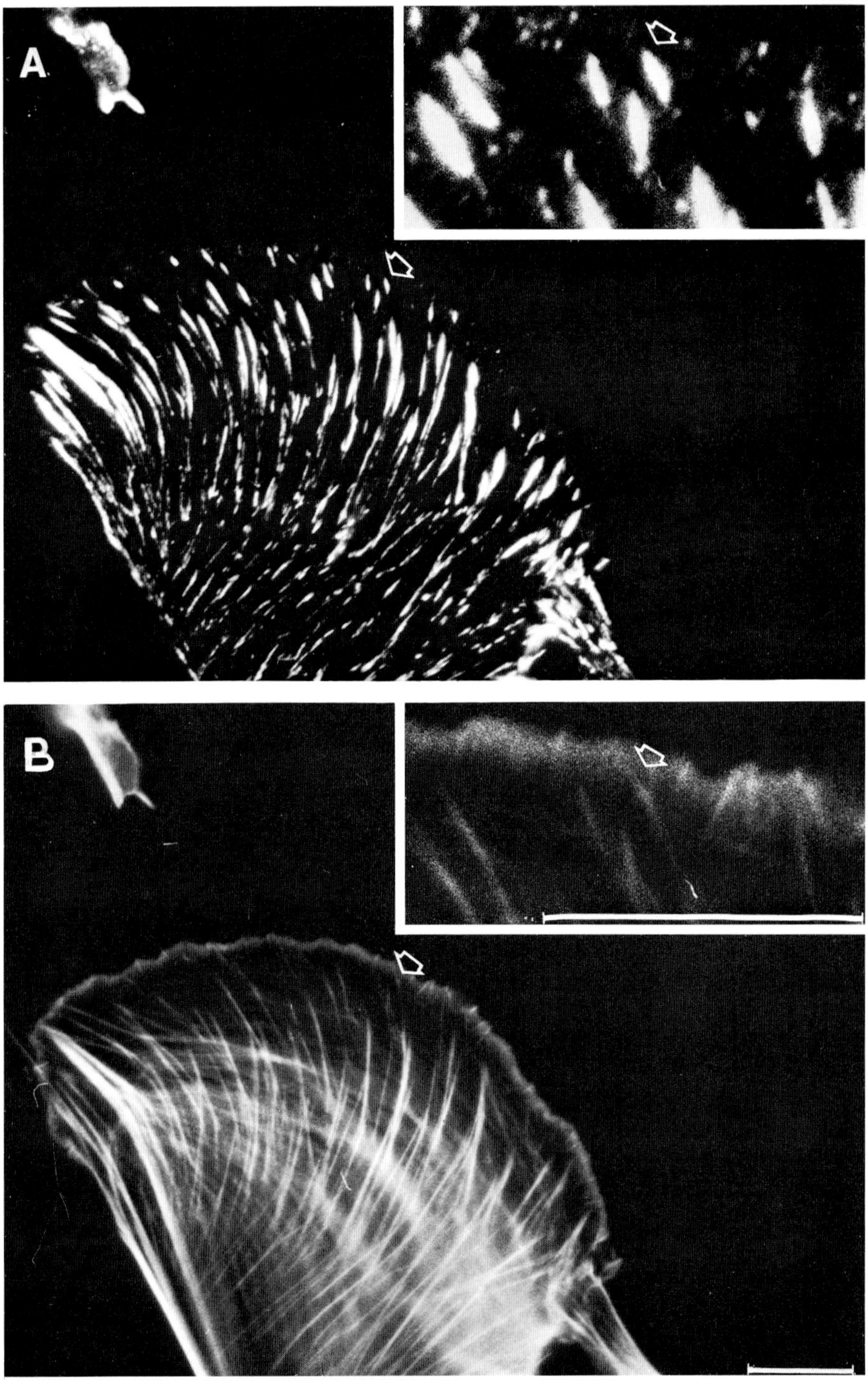

Figure 3. Double fluorescent labeling of chick gizzard cells for vinculin (A) and actin (B). The cells were permeabilized with Trition X-100, fixed and labeled indirectly for vinculin with specific

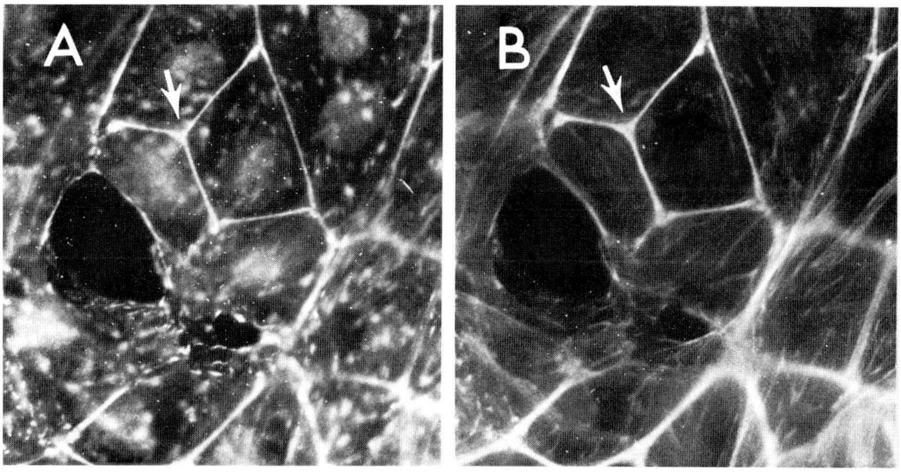

Figure 4. A dense sheet of cultured MDBK cells (bovine kidney epithelium) fluorescently labeled for both vinculin (A) and actin (B) as described in Fig. 3. In these micrographs, the microscope was focused on the subapical region. In that area, both vinculin and actin are associated with the intercellular adherens junction (arrows). At the ventral focal plane (not in focus in these micrographs), vinculin plaques may be seen corresponding to focal contacts and ventral stress fibers. (For additional details see Geiger *et al.*, 1983). Bar represents 10 μm.

vinculin was apparently associated with the termini of actin bundles. These relationships between contact regions, vinculin, and actin were not restricted to cell–substrate focal contacts. Even under culture conditions, cells organized vinculin also in regions of cell–cell contact. This was especially prominent in cultures of polarized epithelial cells (MDBK or MDCK) in which vinculin and actin run along the entire subapical borders in addition to their ventral association with focal contact (Figs. 4A and 4B). Immunofluorescent and immuno-EM labeling of a variety of tissues indicated that vinculin is a rather common component found very close to the membrane of cell junctions such as zonula adherens, in intestine, fascia adherens, in cardiac muscle, and dense plaque in smooth muscle, respectively (Geiger *et al.*, 1981). An example of vinculin localization in chick smooth muscle is shown in Fig. 5. Figure 5A shows by conventional EM (Epon embedded tissue) smooth muscle cells with numerous cytoplasmic filaments. In these cells, cytoplasmic dense bodies (DB) can be detected as well as membrane-bound dense plaques (DP). Double immuno-EM (Fig. 5B) for vinculin and tropomyosin using ferritin and imposil (iron dextran complex) indicated that vinculin was associated only with the membrane-bound dense plaques from which tropomyosin was exluded (for details see Geiger *et al.*, 1981). In fact, vinculin, as judged from the

---

rabbit antibodies followed by fluorescein-conjugated goat anti-rabbit IgG as well as with rhodamine phalloidin. The inserts are enlargements of a region along the leading edge of the cell (all arrows point to the same site). Notice that the prominent vinculin-containing patches coincide with the termini of actin bundles. Bars represent 10 μm.

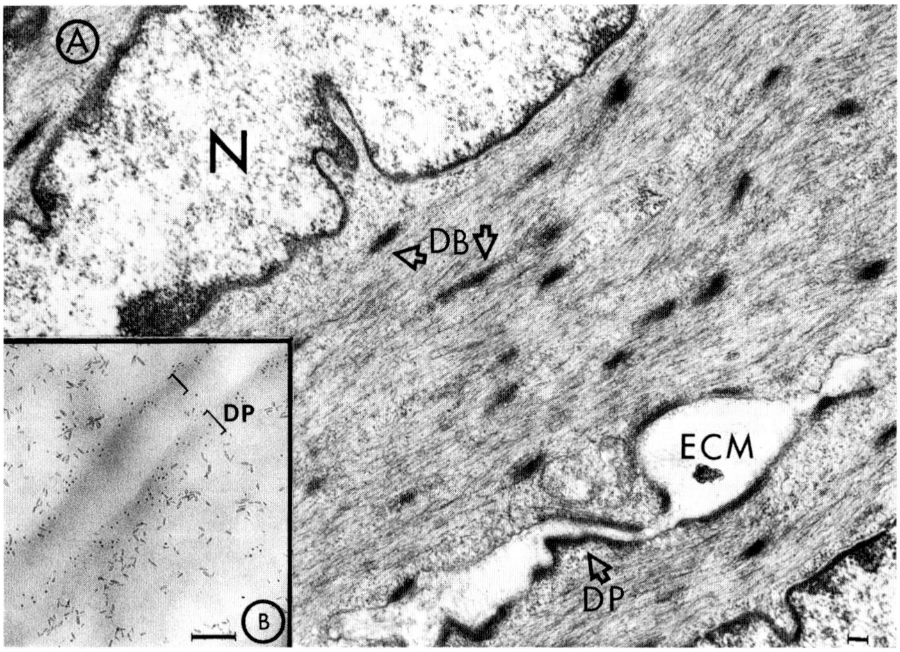

Figure 5. Transmission EM of Epon-embedded (A), as well as frozen-sectioned and immunolabeled (B) chick gizzard smooth muscle. Within the smooth muscle cells, numerous filaments can be detected as well as cytoplasmic dense bodies (DB) and membrane-bound dense plaques (DP). In between individual cells, extracellular matrix and collagen fibers (ECM) can be seen: (N) nucleus. (B) Ultrathin frozen sections of chicken gizzard were double immunolabeled indirectly for both vinculin (with ferritin conjugated antibodies) and for tropomyosin (using Imposil-bound antibodies; for details see Geiger *et al.*, 1981). Notice the specific association of vinculin with the membrane-bound dense plaques and specific exclusion of tropomyosin from these areas. Bar represents 0.1 μm.

distribution of the immunoferritin (or imposil) used for labeling, was significantly closer to the membrane than other actin-associated proteins, including α-actinin in all adherens junctions tested. In general, α-actinin was enriched in regions immediately next to the "vinculin zone" while tropomyosin was characteristically excluded from these areas. These studies could not prove that vinculin was bound to either actin or to structure on the membrane itself, nevertheless, they strongly suggested that vinculin is particularly enriched in the junction area possibly displaying or participating in anchoring of actin. Support for this notion was obtained from a series of studies which will be described in the next section.

4. *Vinculin Can Bind Independently Either to Actin Filaments or to the Plasma Membrane*

A step beyond the structural studies described above involved a characterization of the molecular properties of vinculin and, in particular, its mode of interaction with actin and with the cell membrane. Some of the early biochemical studies mentioned briefly above, suggested that vinculin is not an integral membrane protein; it could be easily extracted in aqueous buffers and did not require detergent for either extraction or solubility. Vinculin has a molecular weight of 130,000 and appears to be monomeric in solution. An interesting recent observation is the presence, in smooth muscle, of a protein with higher molecular weight (150,000-dalton) which is antigenically and structurally related to vinculin (Feramisco *et al.*, 1981; Siliciano and Craig, 1982). This protein, which was termed metavinculin could be extracted from the tissue only by treatment with high salt and detergent, and the possibility was put forward (though still not proven) that it may in fact be an integral membrane protein (Siliciano and Craig, 1982). It still remains to be tested whether metavinculin is present in all cells and to what extent it is truly embedded in the membrane.

The capacity of vinculin to interact with actin and the effect of this interaction on actin organization was investigated in several laboratories. Jockusch and Isenberg (1981, 1982) have shown that addition of gizzard vinculin to polymerizing actin *in vitro* does effect nucleation of polymerization but dramatically suppresses the further increase in viscosity. This effect was detected at relatively low vinculin concentrations (molar ratios of 1:60–1:80), was not temperature-sensitive, and was not affected by Ca^{2+} ions. As for the latter point, it appears that the source of vinculin used for the experiments is of critical importance. Burridge and Feramisco (1982) have shown that, while vinculin from chick smooth muscle reduces the low-shear viscosity in a Ca^{2+}-independent manner, the activity of the same protein, isolated from HeLa cells was highly Ca^{2+}-dependent, and suppression of the apparent viscosity *in vitro* occurred in the presence of 10^{-4} M Ca^{2+} or higher. This raises the possibility that the interaction of vinculin with actin, at least in some cells, may be controlled by Ca^{2+} ions, and hence suggests a regulatory mechanism for a critical step in the transmembrane linkage in contact areas.

Several possible mechanisms could account for the effect of vinculin on the viscosity of polymerizing F-actin. One possibility is that vinculin binds to the barbed ends of F-actin nuclei and caps it in a cytochalasin-like fashion. Such activity, reported by Lin and co-workers (Wilkins and Lin, 1982; Lin *et al.*, 1982), was sensitive to buffer conditions and was practically abolished at high (>4 mM) $MgCl_2$ concentrations. Jockusch and Isenberg (1981, 1982) on the other hand, came forward with a different suggestion based on electron-microscopic observations. They have reported that upon addition of vinculin to slowly polymerizing actin, tightly packed bundles were formed. It has been proposed that vinculin and α-actinin have very different effects on actin filaments, the former decreasing actin viscosity and inducing bundle formation and the latter increasing actin viscosity and forming a cross-linked three-dimensional meshwork. The precise site of vinculin binding along F-actin is not yet known. Isenberg *et al.* (1982) have shown that the protein may bind laterally to actin (an unexpected result if we consider its localization *in vivo*), though it is not clear whether this is the biologically significant site of interaction of vinculin. The results of Wilkins and Lin (1982), showing that one high-affinity vinculin binding site is present for every 1500–2000 actin monomers (see also Burridge and Feramisco, 1982) raises the possibility that the apparent lateral associations seen by Isenberg and Jockusch are with highly abundant sites, not directly related to the bundling activity.

These results indicated that vinculin can directly interact with actin and, at least under certain conditions, induce the formation of actin bundles. It may be premature to use these results and speculate on the similarity of the physiological actin–vinculin interaction. In living cells, many actin-binding proteins are present which could (and probably do) effect the mode of binding of actin to vinculin. Hopefully, additional studies will shed light on this aspect.

Is vinculin associated with the termini of actin bundles only or does it have an independent association with the membrane? To approach this problem we have developed procedures for the selective removal of actin from the cytoplasmic faces of the ventral cell membranes. Cultured cells were briefly incubated with 1 mM $ZnCl_2$ in 50 mM MES buffer containing 3 mM EGTA and 5 mM $MgCl_2$ and then exposed to rigorous pipetting with phosphate-buffered saline. As a result of this treatment most of the cells were opened up and their dorsal parts (including the nuclei and most of the organelles) removed. The residual substrate-attached membranes retained their focal contacts and the associated actin, vinculin, and other actin associated proteins, as shown previously in detail (Avnur and Geiger, 1981a).

These preparations proved to be very useful for the characterization of focal contacts and their interaction with the cytoskeleton. An experiment relevant to the primary site of vinculin binding was performed with such ventral membranes using the Ca^{2+}-dependent actin-severing protein fragmin (Hasegawa *et al.*, 1980; Hinssen, 1981) isolated from *Physarum* by H. Hinssen (for details see Avnur *et al.*, 1983). As mentioned above, the membranes were initially associated with actin and actin-associated proteins. Upon treatment with fragmin, actin was rapidly removed from the membranes along with most of the associated proteins, including α-actinin, tropomyosin, myosin, and filamin. Vinculin was virtually unaffected by this treatment, suggesting that its association with focal contacts is at least to a large degree actin-independent.

In summary, it appears that vinculin is specifically associated with a widespread type of cellular junction: the adherens junction. This type of contact region may be defined and identified by the presence of vinculin and actin. The presumptive role of an actin-membrane linker (for which vinculin got its name; Geiger *et al.*, 1980) is suggested not only from the special location of vinculin on the cell membrane but also from its capacity to bind to pure actin on the one hand and to the membrane (or membrane-associated proteins) on the other.

The experiments described so far were mostly static in nature. They dealt with the location of vinculin in cells and tissues as well as with some of its biochemical properties. In the next sections, we would like to turn to the dynamic features of membrane and cytoskeletal components associated with cellular adherens contacts.

5. *Formation and Maintenance of Adherens Junctions: A Working Model*

It is conceivable that adherens junctions in living cells are dynamic structures. This implies that various components of the junction can move, and the entire structure can be induced to develop and grow or to fade and disappear. In the case of cultured fibroblasts, it has been shown that they might move along the substrate, forming new contacts near the leading edge and releasing the contacts at the trailing edge (Abercrombie *et al.*, 1970, 1971; Harris, 1973; Abercrombie and Dunn, 1975; Izzard and Lochner, 1976, 1980; Chen, 1981a,b). In the case of intact tissues such as the intestine, epithelial cells in large numbers probably exfoliate and new ones move into the gap, a process which requires the continuous reformation of new junctions. Cellular contacts are also progressively formed and reorganized in the embryo during the development of its various tissues.

To study the biogenesis of adherens junctions we have selected as our model system the focal contacts between cultured cells and the substrate. Our working hypothesis was based on the assumption that the timing and location of adherens-junction formation as well as the association of the junctional membrane with actin filaments inside cells are triggered by initial contacts with extracellular structures. The working model (Geiger, 1982a) which we used to define specific experimental approaches are shown in a schematic form in Fig. 6.

We assume that there are different membrane molecules (receptors) which can mediate the attachment of the cell to various extracellular structures (ECM, membrane proteins of neighboring cells, artificial substrate, etc.). Initially, when the cells are suspended, these receptors are mobile and randomly distributed within the plane of the membrane. The microfilament system of suspended cells consists of actin meshworks that are abundant in the cortical cytoplasm. Bundles of actin are not detected in the suspended cells and the association of actin with the membrane occurs at random sites. When plated on a solid substrate the cells may form various types of associations which are mediated through the receptors mentioned above (triangles and flags in Fig. 6A). Some of these associations may be made with components of the ECM as shown schematically in the figure. Subsequently, the "contact receptors" specific for focal contacts (the "flags") aggregate laterally around the initial point of contact, in a positively cooperative fashion, to form an "adhesive patch" (Fig. 6B). We propose that as a result of this lateral aggregation, some changes occur at the cytoplasmic faces of the membrane in that region, forming new binding sites for vinculin. This is followed by accretion of vinculin from a soluble cytoplasmic pool to the membrane. At its new location on the cell membrane, vinculin might then organize filamentous actin into oriented bundles with uniform porarity. These primordial focal contacts may further grow as new membrane receptors of the particular type(s) laterally move into the contact region, eventually forming a "mature" focal contact (Fig. 6C). As the cell moves around, changes may occur in the organization of the focal contact such as those depicted in Fig. 6D. Part of the contacts may be

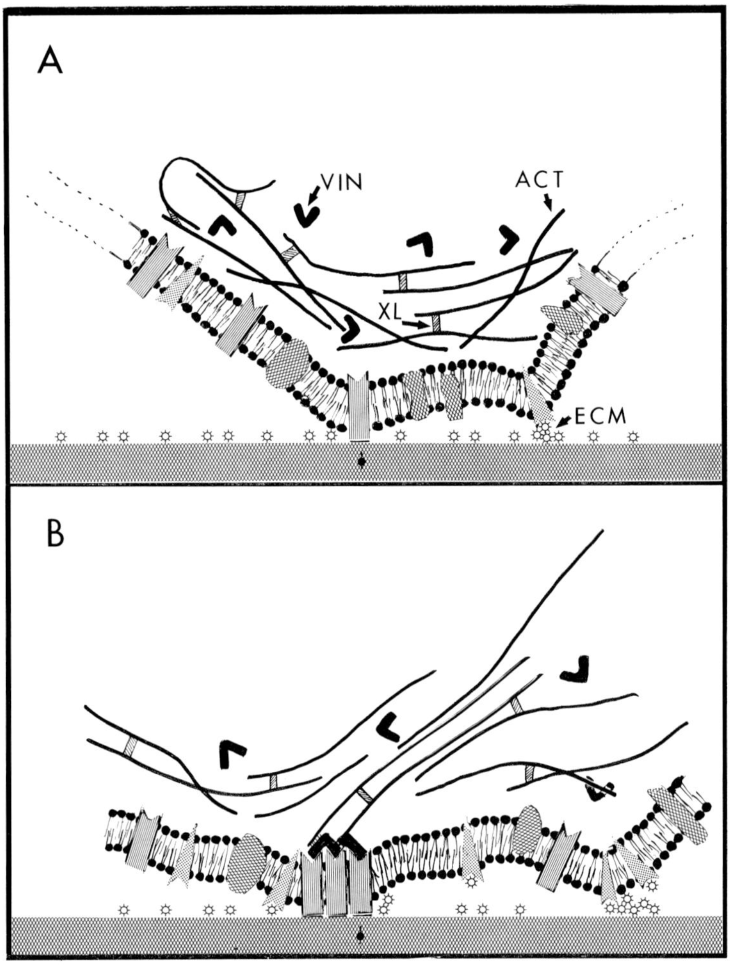

Figure 6. A hypothetical working model depicting steps in the formation of focal contacts. (A) Formation of initial contacts with the substrate mediated through different receptors (flags and triangles). The latter binds to substrate-attached extracellular matrix (ECM) components. Most of the "receptors" at that stage are free to diffuse laterally. Most of the vinculin (VIN) is soluble within the cytoplasm, and actin (ACT) forms three-dimensional meshworks which also contain actin cross-linking proteins (XL) such as α-actinin and filamin. (B) Formation of an adhesive patch around the initial point of contact. In this area, specific receptor(s) (represented by the flags) are laterally clustered. Upon this clustering, binding sites for vinculin become available on

reversed while others are newly formed, etc. There are additional details in this model such as the exclusion of certain ECM components from focal contacts, and the possibility that vinculin forms a multimolecular layer on the membrane which emerged from recent studies that will be described and discussed later in this chapter.

Some predictions suggested by this model which seem to be relevant for the formation of junctions are:

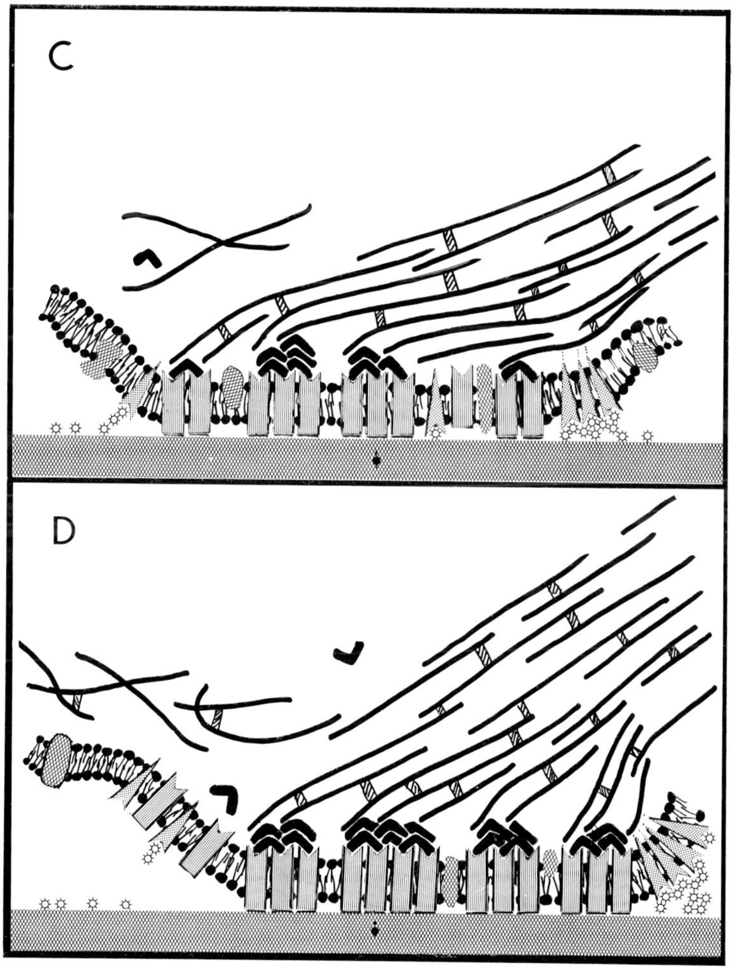

the membrane. Vinculin binds to these sites and links to them actin filaments. ECM receptors (triangles) may also cluster at other sites (see text). (C) Formation of a mature focal contact. Most of the vinculin is bound to the membrane in one or more layers (see text). Actin is primarily packed into tight bundles. In this scheme, ECM is excluded from the focal contact. (D) Rearrangement of focal contact. The attachment at the left side is released, allowing the receptors (flags) and actin to move away from the contact area. At the right side, on the other hand, new associations are formed. An arrow in (A–D) marks a fixed position on the substrate.

1. The adherens junctions should be specialized domains in the cell membrane in which some components are immobilized while others are free to move. This aspect is especially important for the formation and reversal of junctions.
2. There should be a dynamic equilibrium between cytoplasmic pool of free (soluble) cytoskeletal components (vinculin, actin, etc.) and a pool of anchored components on the junctional membrane.

3. The model predicts that the association of vinculin with the membrane of nascent contacts precedes the association of actin with those sites.
4. The presumed primary association of vinculin with the membrane components also implies that its binding to focal contacts is at least partially or transiently actin independent.

The notion that adherens junctions are dynamic in nature also suggests that if mobile junctional components interact directly with ECM they may either become immobilized or alternatively rearrange the matrix. This is not an obligatory requirement from the model; nevertheless, as will be shown later, such dynamic interactions with fibronectin do exist.

These are only some of the aspects of junction formation which we could challenge experimentally. Additional studies on the involvement of ECM in the induction of junction formation will be discussed separately later.

6. Biophysical Approaches for the Measurement of Molecular Dynamics in Cells

The working model presented in the previous section deals with the dynamic properties of the junction. Ironically, however, it is based predominantly on results derived from observations of fixed cells and tissues. In order to test directly the various dynamic aspects of the junction as outlined above, we have used the technique of fluorescence photobleaching recovery (FPR) and combined it, in several cases with image-intensified microscopy and microinjection of fluorescently labeled cytoskeletal elements into living cells. For the benefit of the reader, we would like to turn here to these experimental approaches and discuss briefly their conceptual and technological aspects.

In FPR, the lateral motion of molecules on cell membranes is determined by measuring the rate of local changes in the concentration of fluorescent markers of the cell-surface molecules. A small region of the cell surface which is labeled with fluorescent molecules is illuminated with a focused laser beam (area 3 μm^2). A brief and intense pulse of light causes an irreversible photobleaching of some of the fluorophores in the illuminated region. The diffusion coefficients are determined from the rate of recovery of fresh fluorophores into the bleached area, resulting from the lateral motion of labeled molecules in the plane of the membrane. The fluorescence from the bleached area is excited by an attenuated laser beam in order to avoid photobleaching during the course of fluorescence recovery.

Several geometries can be used in FPR experiments (for review see Schlessinger and Elson, 1982). The most common approach utilizes a Gaussian focused laser beam both for bleaching and after attenuation (of 1000–10,000-fold) for monitoring diffusion of fluorescent molecules into the bleached area.

The theory for FPR assumes that the photobleaching of the fluorophore to a nonfluorescent species is a simple irreversible first order reaction with a rate constant $\alpha I(r)$, $I(r)$ is the bleaching intensity. Then the concentration of the unbleached fluorophore $C(r,t)$ at position r and time t on the absence of diffusion can be calculated from

$$\frac{dC(r,t)}{dt} = \alpha \mathrm{I}(r) \cdot C(r,t) \tag{1}$$

For a short bleaching pulse (T), the fluorescence concentration profile at the beginning of the recovery phase (t=0) is given by

$$C_{(r,o)} = C_o \cdot \exp\ (\alpha \cdot T \cdot I(r)) \tag{2}$$

where Co is the initial uniform fluorophore concentration. The "amount" of bleaching induced in time T is expressed by a parameter K.

$$K = \alpha \cdot T \cdot I(o) \tag{3}$$

For a Gaussian intensity profile, where w is the half width at e^{-2} height and for K<<1 the fluorescence recovery curve can be represented in simple form

$$F(t) = F(o)\ (1 + \tau_D)^{-1} \tag{4}$$

where $\tau_D = w^2/4D$.

Theory was also developed for uniform flow at velocity Vo, or for simultaneous diffusion and flow (Axelrod *et al.*, 1976). More complicated expressions are required for arbitrary K values.

Usually, the diffusion coefficient is determined from $\tau½$, the time required for half of the observed fluorescence recovery to occur. Then,

$$D = (w^2/4\tau½)\ \gamma D, \tag{5}$$

$\gamma D = \tau½/\tau_D$ is a constant which depends on K.

If the fluorophore is attached to a variety of molecules with a range of diffusion coefficients, the recovery curve reflects a weighted average D that depends on the distribution of individual diffusion coefficients. If some of the fluorophores are immobile, the recovery of fluorescence intensity is incomplete and the fraction recovered at long times gives the mobile fraction of fluorophores (fractional recovery). Values of D and the fractional recovery are determined by curve-fitting procedures described elsewhere (Axelrod *et al.*, 1976).

The FPR method was originally designed to determine lateral mobilities on cell surfaces. However, recently, FPR was applied to determine the diffusion coefficient of fluorescently labeled molecules introduced into the cytoplasm of living cells (Wojcieszyn *et al.*, 1981; Kreis *et al.*, 1982). As will be described later, we have shown that it is possible to use high numerical aperture objectives for the measurements of cytoplasmic motion of fluorescently labeled microinjected proteins in living cells (Kreis *et al.*, 1982). By the use of an image plane pinhole for recording the fluorescence recovery curves (Koppel *et al.*, 1976) it is possible to effectively restrict the measured fluorescence intensity from a volume extending no more than approximately 10 μm above and below the focal plane. Since over this range the laser beam expands only by 10%, it is possible to use the two-dimensional treatment for which *D* should not exceed 20% of the value obtained when using the beam size at the focal plane (Kreis *et al.*, 1982).

The favorable features of FPR for membrane and cytoplasmic mobility measurements are as follows (Schlessinger and Elson, 1982):

1. The use of a fluorescence microscope enables a fairly precise localization of the observation region relative to visible cellular structures.
2. A wide range of mobilities is accessible in the range of 10^{-12}–10^{-6} cm^2/sec.
3. The focused laser beam allows mobility measurement in different regions of an individual cell.
4. The FPR method is capable of distinguishing among various mechanisms of transport, e.g., diffusion and flow.
5. Mobility can be measured on individual living cells enabling a correlation of physiological events, morphology, and dynamic properties.
6. Mobility of specific fluorescently labeled components is measured. Moreover, by using two different fluorescent markers it is possible to measure the effect of one cellular molecule on the mobility of a second nearby molecule.

Several reports evaluate the possibility that the bleaching pulse in FPR experiments may damage the cell or the plasma membrane (for review see Schlessinger and Elson, 1982). The conclusion from these studies is that under the usual experimental conditions it is either impossible to detect any cellular damage or that the existing damage is negligible.

The second approach which we combine here with FPR for studying the dynamics of fluorescently labeled cellular proteins is the method of image intensification microscopy. Intensified microscopy overcomes two major shortcomings of conventional fluorescence microscopy. First, in conventional microscopy, the fluorescent dye attached to the cell can be seen only for relatively short periods, because the illumination required to observe and photograph the cells cause rapid photobleaching of the fluorescence intensity. Second, the number of certain cellular components is very low. Image intensification microscopy, being a sensitive means of monitoring low levels of light overcomes these problems.

An FPR system and image intensification microscopy were combined in a single apparatus (Levi *et al.*, 1980), thus allowing the determination of mobility and distribution of various fluorescently labeled markers on a single viable cell under various physiological conditions.

7. Restricted Mobility of Membrane Components in Cell–Substrate Focal Contacts

The approaches outlined in the previous section were first used to characterize the dynamic properties of membrane proteins and lipids in the contact region. In order to perform such experiments on single membranes which are directly associated with the substrate (thus avoiding fluorescent signals from the dorsal cell membrane) we have developed a technique for the isolation of these membranes without affecting their substrate contacts (for details see Avnur and Geiger, 1981a). In brief, cells were incubated for 2 min in buffer A (50 mM MES buffer, 5 mM $MgCl_2$, 3 mM EGTA 1 mM $ZnCl_2$, pH 6.0), then washed and sheared with a stream of phosphate-buffered saline (PBS). Analysis of these preparations with IRM and scanning EM indicated that most of the cells were opened up, leaving attached their ventral membranes and the associated cytoskeleton (Avnur and Geiger, 1981a).

Fluorescent labeling of cells, prior to the preparation of ventral membranes are performed using either a lipid probe WW 591 (5-[3-γ-sodium sulfopropyl-6,7-benzo-2-(3H)-benzoxazolylidene-butinyldene]-1,3-dibutyl-2-thiobarbituric acid; Hillman and Schlessinger, 1982) or with the amino-group reactive, nonpenetrating fluorophore lissamine-rhodamine B sulfonyl chloride (RB200SC). With respect to the latter, it was verified that it does not significantly label membrane lipids or cytoplasmic proteins. Moreover, to prevent extensive labeling of extracellular matrix proteins, the cells were reacted with RB200SC in suspension, and then replated on substrate. After they attached to the glass and spread, their ventral membranes were isolated by the $ZnCl_2$ method.

The substrate-attached membranes, labeled with either WW591 or with RB 200 SC retained their focal contacts as shown in Fig. 7. We were able thus to use IRM to define attached and unattached regions over the membrane (a and b, respectively, in Fig. 7) and measure the lateral diffusion of the lipid probe and of rhodamine-labeled proteins in the two membrane domains by FPR. The values obtained for the diffusion coefficients in the two domains indicated that the lateral diffusion of both lipids and proteins in focal contacts was reduced by a factor of 1.5–2 as compared to those measured in unattached regions (Table 1). The relative retardation of lateral mobility in contact areas was similar for both WW591 and the rhodamine-labeled proteins, and could be attributed to the possible reduction in the volume available for free diffusion in that membrane region or to transient associations with immobile elements in or on the membrane.

An important difference between the lipid probe and the labeled proteins was the extent of fluorescence recovery after photobleaching. The lipid probe exhibited essentially complete recovery while over 50% of the proteins were essentially immobile ($D < 3 \times 10^{-12}$ cm^2/sec). This does not imply that similar proportion of the membrane proteins is directly involved in the contact with the substrate. It is possible for example that membrane components

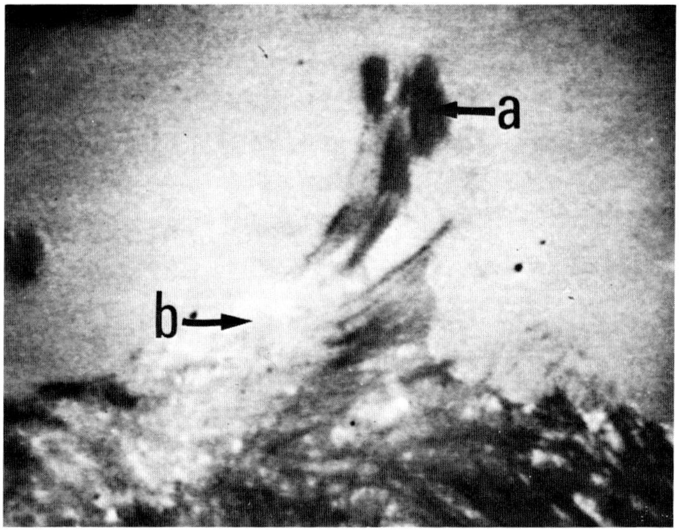

Figure 7. FPR measurements of a lipid probe (WW 591) and rhodamine labeled surface proteins in the ventral membrane of chick gizzard cells. Membranes were prepared by the $ZnCl_2$ method and visualized by IRM. The region marked by (a) represents a site of focal contact and (b) represents an unattached area. Immobile molecules in these experiments have lateral diffusion lower than 3×10^{-12} cm²/sec (for details see Geiger *et al.*, 1981).

which do not interact directly with the substrate or with the cytoskeleton become trapped in the focal contacts and thus appear immobile. It has been shown a few years ago that the immobile acetylcholine receptors of rat myotubes are concentrated in areas of broad close contact, although they seem not to be directly associated with individual contact points but rather to be localized between them (Bloch and Geiger, 1980). It is notable that the values obtained here are based on the residual fluorescence of labeled membrane proteins a few hours after labeling, and do not reflect actual amounts of protein molecules. Nevertheless, it clearly indicates that protein molecules in focal contacts are heterogeneous with respect to their dynamics and fall into at least two categories: those proteins that are free to diffuse into and away from the contact area and others which are immobile. Such organization of focal

Table 1. Fluorescence Photobleaching Recovery of a Lipid Probe (WWS91) and Rhodamine-Labeled Surface Proteins in Focal Contacts and in Loosely Attached Areas

Fluorescent probe	Membrane domain	Diffusion coefficient $D(cm^2/sec) \times 10^{-9}$	Fractional recovery (percent)
WW591	a	7.7 ± 0.8	> 90
WW591	b	12.8 ± 2.0	> 90
RB200SC	a	0.8 ± 0.18	47 ± 9
RB200SC	b	1.4 ± 0.21	90 ± 4

contacts allows for rather rapid exchange of components, addition or loss of contact receptors which may be of significance for the regulation of focal contact formation, maintenance, and reversal. Such an approach, when applied to defined intercellular junctions may shed light on the dynamics of junction biogenesis and maintenance.

8. Dynamic Properties of Cytoskeletal Proteins in Living Cells

Another prediction of the model shown above is that cytoskeletal proteins such as actin and vinculin maintain a dynamic equilibrium between a soluble pool and a membrane-bound pool. To study the kinetic parameters of this process we have prepared fluorescent derivatives of three relevant cytoskeletal proteins, namely actin, α-actinin, and vinculin, injected them into living cells and measured their mobilities by the FPR technique (for details see Kreis *et al.*, 1982). All three proteins became incorporated shortly after injection into specific cellular structures in the injected cells (Fig. 8). Actin was associated with filament bundles which often terminated in distinct focal contacts (Figs. 8A–C). α-Actinin was associated with areas of focal contact as well as with regularly spaced striations along the stress fibers (Fig. 8D) and vinculin was specifically bound to focal contacts (Fig. 8E). The amounts of microinjected proteins were usually quite low and did not exceed 5% of the respective endogenous protein. Moreover, microinjected cells showed normal patterns of attachment spreading, mobility, and division. Fluorescent labeling of injected cells with antibodies directed against the particular injected proteins coupled to different fluorophores indicated that the two fluorescence patterns were essentially identical. We were thus capable of defining some structural domains with which the fluorescent proteins were associated. Actin, for example, was present in focal contacts, stress fibers, leading edge, and interfibrillary spaces (see Kreis *et al.*, 1982). α-Actinin was present in the same areas, while vinculin was detected only in focal contacts and diffusely between the actin bundles.

To determine the dynamic properties of these proteins in the different cellular regions of living cells we have used the FPR technique. The results obtained in such experiments (with rhodamine labeled actin, α-actinin, and vinculin) are summarized in Table 2. Control proteins (rhodamine labeled BSA and IgG) injected at different concentrations into cells were highly mobile (> 90% fluorescence recovery regardless of whether the measurement was performed at the cell periphery or in the perinuclear area, $D = 6 \times 10^{-9}$ cm²/sec). The mobilities of actin, α-actinin, and vinculin were significantly lower (3×10^{-9} cm²/sec). For those proteins, the mobile fraction was lower than that measured for the control proteins. Actin interfibrillary regions had a mobile fraction of about 70% and in focal contacts and stress fibers the mobile fraction was smaller than 20%. Similarly, significant proportion of vinculin and α-actinin molecules (mostly those in focal contacts regions) did not diffuse freely (40% recovery while the cytoplasmic protein was mostly

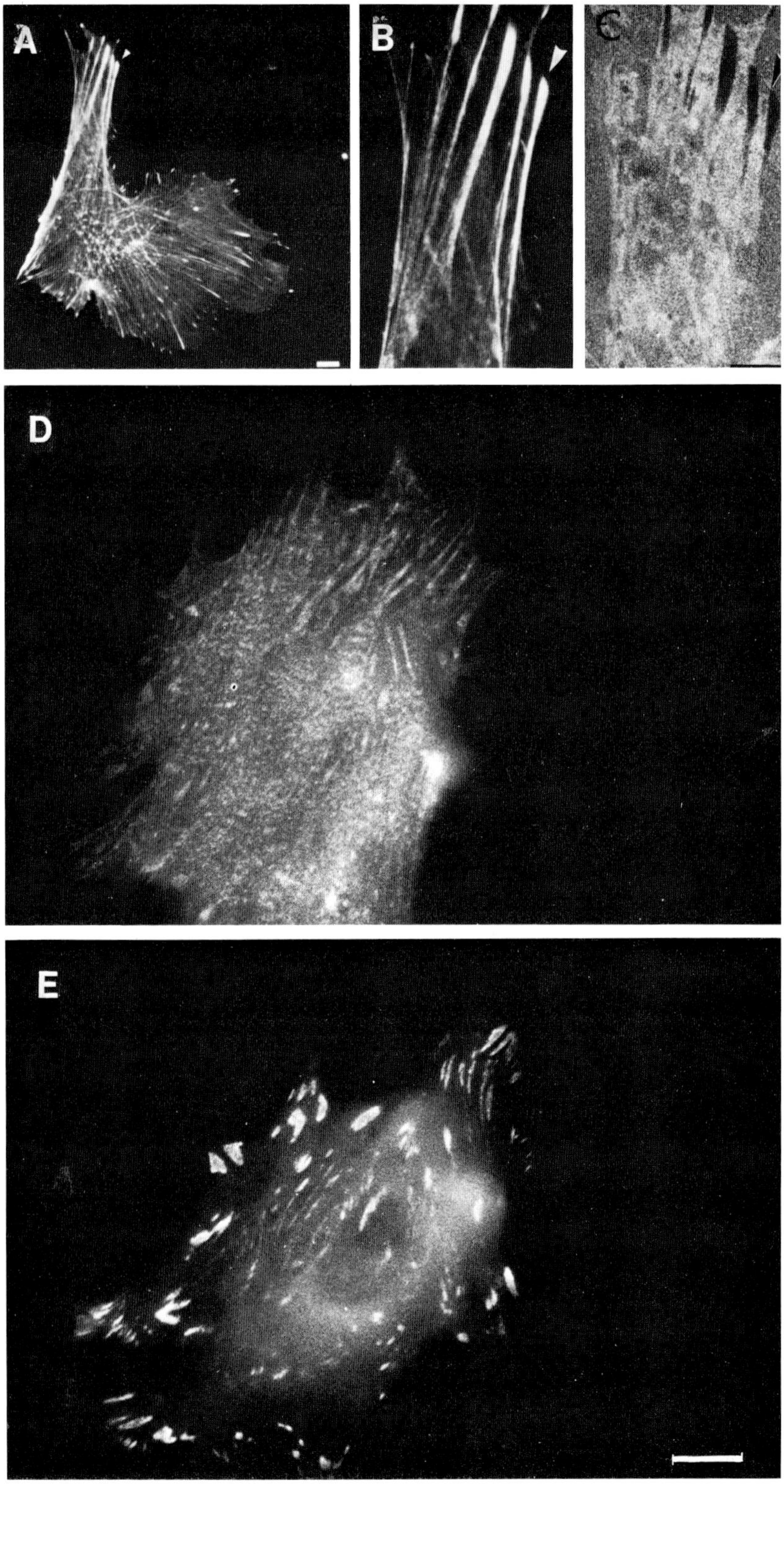

A
B
C
D
E

Table 2. Mobility of Rhodamine-Labeled Actin, α-Actinin, and Vinculin in Microinjected Chicken Gizzard Fibroblasts as Measured by Fluorescence Photobleaching Recovery

Protein	Cellular domain	Diffusion coefficient $D(cm^2/sec) \times 10^{-9}$	Fractional recovery (percent)	Rate of slow recovery τ/2
Actin	Focal contact	3.1 ± 1.1 (30)[b]	18 ± 7	4.1 ± 2.8 (23)
α-Actinin	Focal contact	2.8 ± 1.0 (10)	40 ± 10	2.7 ± 1.2 (6)
Vinculin	Focal contact	3.5 ± 1.2 (20)	43 ± 8	2.1 ± 0.9 (15)
Actin	Interfibrillary	3.2 ± 1.2 (35)	65 ± 13	N.D.[a]
α-Actinin	Interfibrillary	2.5 ± 1.4 (12)	76 ± 5	N.D.
Vinculin	Interfibrillary	2.9 ± 1.1 (30)	~80	N.D.
Bovine serum albumin	Border area	6.2 ± 0.9 (14)	>90	
	Perinuclear	6.3 ± 0.9 (7)	>90	
	Border area	6.0 ± 1.3 (15)	>90	
Goat immunoglobulin	Perinuclear	6.6 ± 0.7 (12)	>90	

[a]Not determined.
[b]Number of determination.

mobile, 80%). It was, nevertheless, possible to detect a slow rate of fluorescence recovery of these proteins to photobleached areas. The rate of recovery of the diffuse components was dependent on the square of the beam radius. However, the rate of slow recovery in focal contacts was independent of the size of the bleaching spot (Kreis *et al.*, 1982; Kreis, Schlessinger, and Geiger, unpublished results).

Extensive measurements with rh-actin in focal contacts or stress fibers indicated that the half time (τ½) of the slow recovery was within the range of 1–11 min (with an average of 4.1; see Table 2). The structural and molecular basis for this wide variability is not yet clear. Vinculin and α-actinin appeared to be more uniform in their dynamic behavior and had slow recovery with τ½ of about 2–3 min. Additional experiments indicated the slow recovery was due to exchange of components between the soluble cytoplasmic pool and the "cytoskeletal" pool, rather than rearrangement of components within the cytoskeleton (Kreis *et al.*, 1982). This was shown for actin by bleaching segments of variable size on stress fibers and measuring the rate of fluorescence recovery in the center of the bleached area. The results indicated that the length of the bleached segment did not alter the rate of slow recovery. These results strongly suggested that there is an exchange between the immobile and

Figure 8. Cytoplasmic distribution in living chicken gizzard cells of microinjected fluorescently labeled cytoskeletal proteins. (A) Microinjected rhodamine actin. Notice the association with stress fibers and perinuclear polygonal baskets. (B and C) Part of the same cell shown in (A) in which the fluorescence is compared with the IRM image (B and C, respectively). Notice the specific association of actin with focal contacts (arrowhead). (D) The distribution of microinjected rhodamine α-actinin. The protein is associated with focal contacts and with striations along stress fibers. (E) The distribution of microinjected rhodamine labeled vinculin. Notice the association with focal contacts and the diffuse labeling in the cytoplasm. Bar represents 10 μm.

the soluble pools in living cells, as predicted in the model outlined above.

A complementary set of experiments which provided interesting hints on the relationships between the soluble and immobile pools of the vinculin involved the addition of fluorescently labeled vinculin to permeabilized cells or to substrate-attached membranes of cultured cells. Upon addition of rhodamine-labeled vinculin to these preparations the labeled protein became specifically associated with the exposed cytoplasmic faces of focal contacts and restricted regions of cell–cell contacts (open arrows and solid arrows in Fig. 9). This indicated that soluble vinculin can indeed bind to preexisting focal contacts. Moreover, we have recently shown that removal of most of the membrane-bound actin using fragmin (Avnur *et al.*, 1983) did not hamper the binding of vinculin to isolated ventral membranes. This again pointed to the actin-independent nature of vinculin association with focal contact as discussed above. An unexpected result emerged from control experiments in which we tried to inhibit the binding of fluorescently labeled vinculin by excess of the unlabeled protein. It was our previous experience that such inhibition was very efficient in the case of α-actinin (Geiger, 1981b). However, with vinculin, the inhibition was quite weak. This indicated that the binding of vinculin is not a simple saturable process and suggested that vinculin might undergo self-aggregation in the contact area. This suggestion might be related to the finding that several isoforms of vinculin are present in cells and that different isovinculins may have distinct subcellular distribution (Geiger, 1982b). If the "self aggregation" indeed occurs, it implies that vinculin is involved in several types of interactions: the first with some elements in or on

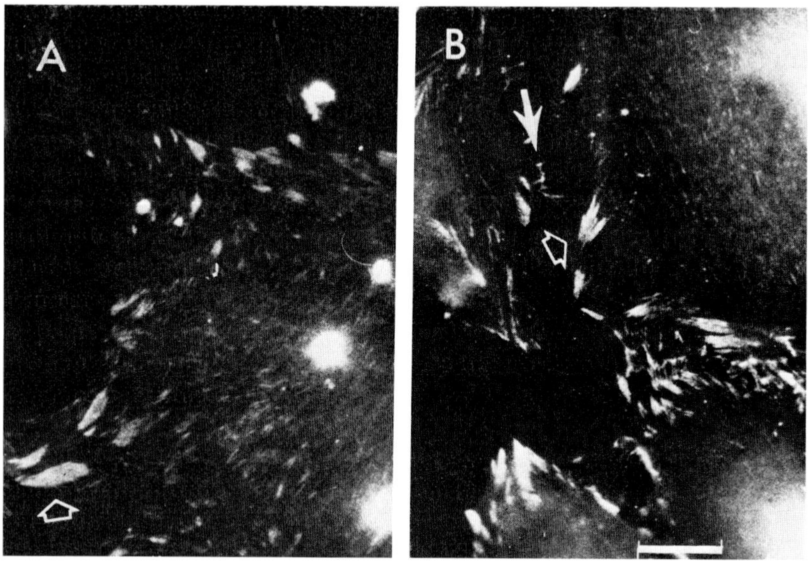

Figure 9. Association of rhodamine labeled vinculin with Triton permeabilized chick fibroblasts (A) or PtK_2 cells (B). The labeled protein associates with focal contacts (empty arrows) as well as with some cell–cell contacts (solid arrow). Bar represents 10 μm.

the membrane, the second with itself, and a third one with actin. This speculation is now open for experimental manipulation which hopefully will soon provide us with a better view of the different dynamic interactions of vinculin with the different cellular elements.

To summarize this section on the membrane and cytoskeletal rearrangements in adherens junctions (represented here by the focal contacts), it appears that although the junction is a special structural domain over the membrane, which appears to be partly immobile, many of its components appear to be capable of movement at one rate or another. This property might be of significance for the establishment of cell contacts. If we refer to the plasma membrane as a plane of reference, we may argue that some of the membrane proteins and most of the membrane lipids can move laterally into, away from, and through the junction. Other proteins may become transiently immobile due to "vertical" interactions with the substrate, on the outside, or possibly with cytoskeletal proteins such as vinculin at the interior. The different components of stress fibers appear to maintain a dynamic equilibrium between their soluble and the junction-bound forms. Though no information is yet available concerning the primary site for vinculin binding in the junction, we suggest that the regulatory mechanisms involved in this binding are of great biological significance. Turning on or turning off of such control mechanisms may shift the dynamic equilibrium described above in one direction or another, bringing about the formation, reversal, or rearrangement of junctions.

The molecular movements which were discussed in this section were those of membrane components and membrane-associated cytoskeletal proteins. The results imply, however, that if "contact receptors" move laterally, they may mobilize extracellular matrix constituents to which they are bound. Some studies along this line will be discussed in the next section.

9. Rearrangement of Extracellular Fibronectin in Areas of Cell Contact

It is now generally believed that cell adhesion is promoted by, and in some cases depends on, "adhesive proteins" that compose the ECM. In this complex network, several proteins and proteoglycans were identified though their individual contribution to cell contact formation is not at all clear. Several studies have directly demonstrated that one of the major components of the ECM, namely, fibronectin (for review see Yamada and Olden, 1978; Grinnell, 1978; Hynes, 1981) might be associated across the membrane with actin filaments. This notion was also supported by several indirect observations. As discussed above, the microfilament system is involved in many cytodynamic processes which are also effected by extracellular fibronectin. The latter protein from either cultured cells or plasma promotes cell attachment, spreading, and in some cases, motility. Interestingly, cultured cells transformed with oncogenic viruses exhibit markedly reduced levels of fibronectin concomitantly with the appearance of disorganized actin bundles and loose substrate attachments. Agents which disrupt the extracellular fibronectin matrix affect, indirectly, the organization of actin and, reciprocally, treatment of cells with microfilament-disrupting drugs such as cytochalasin B, were reported to cause a release of fibronectin from the surface of cultured cells (Yamada and Olden, 1978; Hynes, 1973; Vaheri and Ruoslahti, 1974; Ruoslahti and Vaheri, 1975; Yamada *et al.*, 1976; Ali *et al.*, 1977; Grinnell and Feld, 1979). What are the molecular interrelationships between the extracellular matrix and the cytoskeletal microfilaments?

The spatial transmembrane relationships between fibronectin and actin were demonstrated some years ago, by the use of double fluorescent labeling for the two components. Heggeness *et al.* (1978a) and Hynes and Destree (1978) demonstrated that part of the membrane-bound fibronectin is apparently associated, across the membrane, with actin filaments. These interrelationships were most obvious a few hours to 1 day following plating. The immunofluorescent experiments provided a clear overall picture of the transmembrane relationships of fibronectin and actin but the resolution of the technique was rather limited. A high resolution analysis of these interactions was performed by Singer (1979) who showed by electron microscopy that microfilament bundles are often colinear with extracellular fibronectin filaments at a specialized membrane site denoted "fibronexus." It is, however, not entirely clear what the relationships are between the fibronexus and different membrane areas where both fibronectin and actin can be detected by the double immunolabeling technique. Especially controversial were the relationships between cell–substrate focal contacts and fibronectin. Intuitively, it was expected that fibronectin, being an "adhesive" protein, may directly mediate cell–substrate attachment in these areas. However, experimental results indicated that the relationships are considerably more complex.

In fixed and permeabilized chick-fibroblasts, immunolabeling for fibronectin was apparently missing from most focal contacts and enriched at the periphery of these areas (Grinnell, 1980; Birchmeier *et al.*, 1980; Geiger, 1981a; Avnur and Geiger, 1981b). This rather unexpected observation could, however, be attributed not to an actual absence of this protein but rather to reduced accessibility of these sites to the immunochemical reagents. Indeed, it has been directly shown that penetration of antibodies into the extracellular gap in focal contacts is restricted (Grinnell, 1980; Avnur and Geiger, 1981b). To circumvent this problem, Chen and Singer (1980) prepared ultrathin frozen sections of cultured cells, double immunolabeled them for fibronectin and vinculin, and examined them in the electron microscope. They found that fibronectin was largely absent from vinculin-rich focal contacts. These sites could, however, be efficiently labeled with antibodies to intrinsic concanavalin A-binding membrane proteins of these cells, thus excluding the possibility that antibodies could not penetrate to these areas. Recently, Chen and Singer (1982) demonstrated, using the same technique, that both fibronectin and vinculin can be found in contact regions with ECM in which the membrane does not come close to the plane of the substrate.

Another approach which we used to follow the dynamic interrelationships between fibronectin and cell contacts was to plate cells on substrates coated with fluorescently labeled fibronectin. In these systems, it was possible to continuously follow the fate of the underlying fibronectin in viable cells without fixation or permeabilization. It was found that the fibroblastic cells form contacts with the fibronectin carpet, but subsequently, remove it from many, though not all, the contact regions (Avnur and Geiger, 1981b). Some of the rearranged fibronectin forms extracellular cables similar to those found in normal cultures. Figure 10 shows the patterns of substrate-attached rhodamine fibronectin at different time points after plating of chicken gizzard cells. Initial displacement of the ECM (A) and endocytosis (B) leads, subsequently, to removal of the matrix protein from large areas as shown in (C). Some of the fine fibrillary fibronectin under or on the cell may coincide with intracellular vinculin (for details see Avnur and Geiger, 1981b). In a recent study with melanoma cell lines (of the K 1735 series) which display distinct metastatic properties, we have noticed that the high metastatic cells move rapidly on the substrate, forming small and transient contacts without perturbing underlying ECM carpet. The low metastatic subclones formed large focal contacts, showed restricted motility, and rearranged the matrix into distinct cables (Geiger, Volk, and Raz, unpublished results). Studies with other cell types suggested that fibronectin may be more commonly found in association with vinculin-rich structures coinciding with close substrate contact (Burridge and Feramisco, 1980; Hynes, 1981; Hynes *et al.*, 1982; Singer and Paradiso, 1981). These apparently inconsistent results on the transmembrane-anchorage of fibronectin to the microfilament system in contact regions may be related to intrinsic differences between the cell types used in the various experiments or to the growth state of the cells (Singer and Paradiso, 1981). Another interpretation may be related to our previous results on the rearrangement of an exogenously supplied fibronectin matrix (Avnur and Geiger, 1981b), and those of Chen and Singer (1982). It is possible that vinculin may bind to the membrane in contact regions regardless of the particular ECM elements present. Cellular forces generated in that area may mobilize the external membrane-bound fibronectin by lateral displacement or by endocytosis (Avnur and Geiger, 1981b) and then form extensive focal contacts with the more solid and less deformable substrate. On the other hand, if the cell encounters mechanical difficulties to rearrange and displace the matrix protein (for example when the matrix is locally thick or rigid), it may form stable contacts in that region, mediated by the ECM (Chen and Singer, 1982). These may not be typical focal contacts, as defined by electron microscopy (100–150 Å distance from the substrate), but may otherwise contain all the typical components of an adherens junction. In fact, in certain junctions of intact tissues (for example, dense plaques of smooth muscle shown in Fig. 5 above) the vinculin-rich membrane-bound plaques are usually found in association with cell-ECM contact and not with cell–cell junctions.

In conclusion, it is possible that two or more types of interactions between

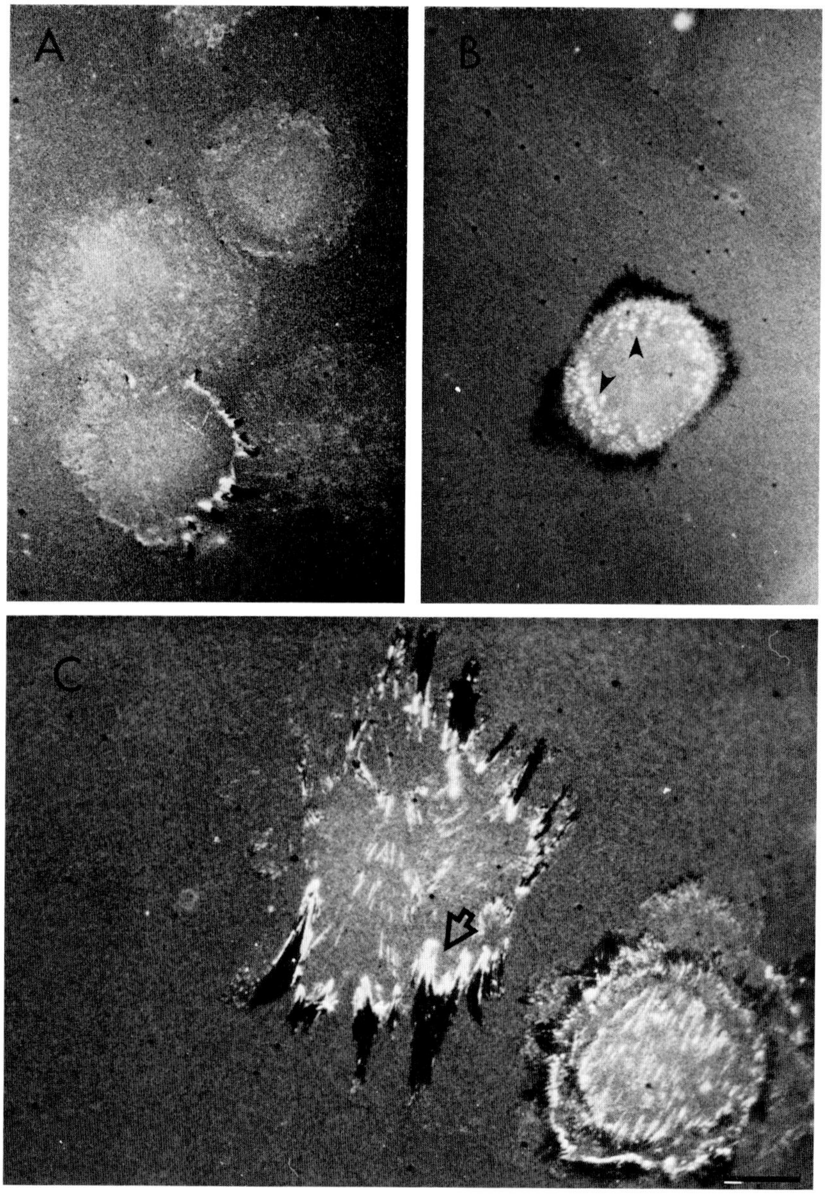

Figure 10. Removal of substrate-attached rhodamine labeled fibronectin from areas of cell contact. (A) Initial steps of fibronectin removal 30 min after plating of chicken gizzard cells. Substrate-bound rhodamine fibronectin is displaced towards the cell center. (B) Endocytosis of rhodamine fibronectin by the cells 30 min after plating. The fluorescent fibronectin in this cell (arrowheads) is localized inside the cell. (C) A more advanced step (60 min after plating) in fibronectin removal from significant regions over the cell periphery (arrow). Some of the fibronectin under central regions of this cell is associated, across the membrane, with vinculin (for additional details see Avnur and Geiger, 1980).

fibronectin in cells lead to distinct transmembrane associations with actin. It seems that fibronectin binds to membrane receptors and, being di- or multivalent, may induce patching in a mechanism analogous to antibody- or lectin-mediated clustering (for review see Geiger, 1983). These clusters may then form transmembrane association with stress fibers along their length. Fibronectin may thus increase the rate and extent of initial contacts with the substrate. When focal contacts are subsequently formed, the complexes of fibronectin and its receptor(s) if and where they retain sufficient freedom of movement, may be excluded from the focal contacts, either by lateral displacement as depicted in Fig. 6, or by endocytosis. The fibronectin which is actively reorganized at the cell exterior may, in turn, effect the spreading and locomotion properties of neighboring cells in the culture, or possibly in analogous systems *in vivo*.

10. Contact Formation, Cytoskeletal Rearrangement, and Growth Regulation: Are They Interrelated and How?

The formation of intercellular junctions or cell substrate contacts has some obvious direct effects on the mechanical properties of cells and tissues as well as on cell polarity and motility. The various studies described above attempted to reveal some of the mechanisms underlaying the various steps from the establishment of the first cell contact to the formation of the fully organized junction. The time scale of such changes ranges from minutes to few hours and, as pointed out, the entire process appears to be rather complex and involve several sequential steps most of which are still poorly understood. Another set of physiological activities which appear to depend on cell contact ("anchorage") includes the regulation of growth and differentiation (Hay and Meier, 1976; Folkman and Moscona, 1978; Gospodarowicz *et al.*, 1978, Ben Zeev *et al.*, 1980). Many of the anchorage-dependent activities of the latter type are manifested at a relatively late stage (hours to days) after establishment of cell contacts. The molecular mechanisms responsible for contact-induced differentiation and growth control are not clear. Moreover, it should be clearly stated that by no means are formation of specific cell contacts and consequent shape changes expected to be the only essential factors. It has been extensively documented that a variety of soluble factors, some of which have been molecularly defined in recent years, have critical effects on cell behavior and growth.

Are the generalized effects of cell adhesion on cellular physiology related to (or even dependent upon) the establishment of structurally and molecularly defined cell junctions? It is clear that if such relationships could be elucidated experimentally it would provide us with an extremely valuable model system for studies on the molecular mechanisms and cellular topology of growth regulation. At the present time, the idea of close relationships between a specific type of cell contact and the pattern of growth is based on indirect indications and correlations which will be discussed below. It will probably remain for future studies to characterize the physiologically relevant signals transmitted across the membrane in adherens junctions.

Indirect relationships between adherens junctions and state of growth emerges from studies on the abnormal organization of the cytoskeleton and the modification of vinculin in transformed cells. It is well established that transformed cells (cultured cells infected with Rous sarcoma virus (RSV) or others) exhibit reduced adhesiveness and display deteriorated cytoskeletal organization (Pollack *et al.*, 1975; Edelman and Yahara, 1976; Ash *et al.*, 1976; Wang and Goldberg, 1976; Weber *et al.*, 1977; Brinkley *et al.*, 1980).

Immunofluorescent labeling for actin indicated that in the RSV transformed cells there were only a few or no detectable stress fibers and that actin and its associated proteins were often organized in the form of "rosettes" (David-Pfuety and Singer, 1980; Rohrschneider, 1980; Rohrschneider *et al.*, 1982). Immunofluorescent labeling of similar cells for vinculin indicated that it was predominantly associated with the actin-containing rosettes of the transformed cells as shown in Fig. 11 (see also David-Pfuety and Singer, 1980). In a different system (mouse melanoma K1735 and fibrosarcoma 2237), we have shown that disorganization of actin and vinculin may be related to the metastatic potential of these cells, rather than to their tumorigenicity (Raz and Geiger, 1982).

In the RSV-transformed cells it has been shown that the abnormal, so-called "transformed phenotype" was related to the product of the src oncogene (Hanafusa, 1977). This product appears to be a tyrosine-specific protein kinase denoted $pp60^{src}$ capable of phosphorylating itself and several additional cellular proteins. Immunoelectron microscopy indicated that $pp60^{src}$ is membrane-bound and immunofluorescent labeling suggested that it was particularly enriched in focal contacts (Willingham *et al.*, 1979; Rohrschneider, 1980; Collett *et al.*, 1980; Hunter and Sefton, 1980; Kreuger *et al.*, 1980). These findings were compatible with the idea that some essential elements of the focal contact might become phosphorylated on tyrosine by the neighboring $pp60^{src}$ and change their properties in a way that brings about a drastic alteration in focal contact structure. Such analysis indicated that the major defined cytoskeletal substrate for the $pp60^{src}$ is vinculin (Sefton *et al.*, 1981, 1982). A tempting interpretation of this finding would suggest that the modified (tyrosine phosphorylated) vinculin cannot bridge between actin and the membrane, thus leading to a contact-insensitive cell with distorted cytoskeleton. There is, however, still little evidence to support this speculation. Focal contacts may, and probably do, contain additional components not known to us at present, the phosphorylation of which may easily be undetected. We therefore cannot exclude the possibility that the physiologically relevant substrate is not vinculin but rather one protein or more which are essential for establishment and maintenance of stable contact. To judge to what extent the results on vinculin phosphorylation are related to the transformed phenotype, it should be directly determined whether the modified

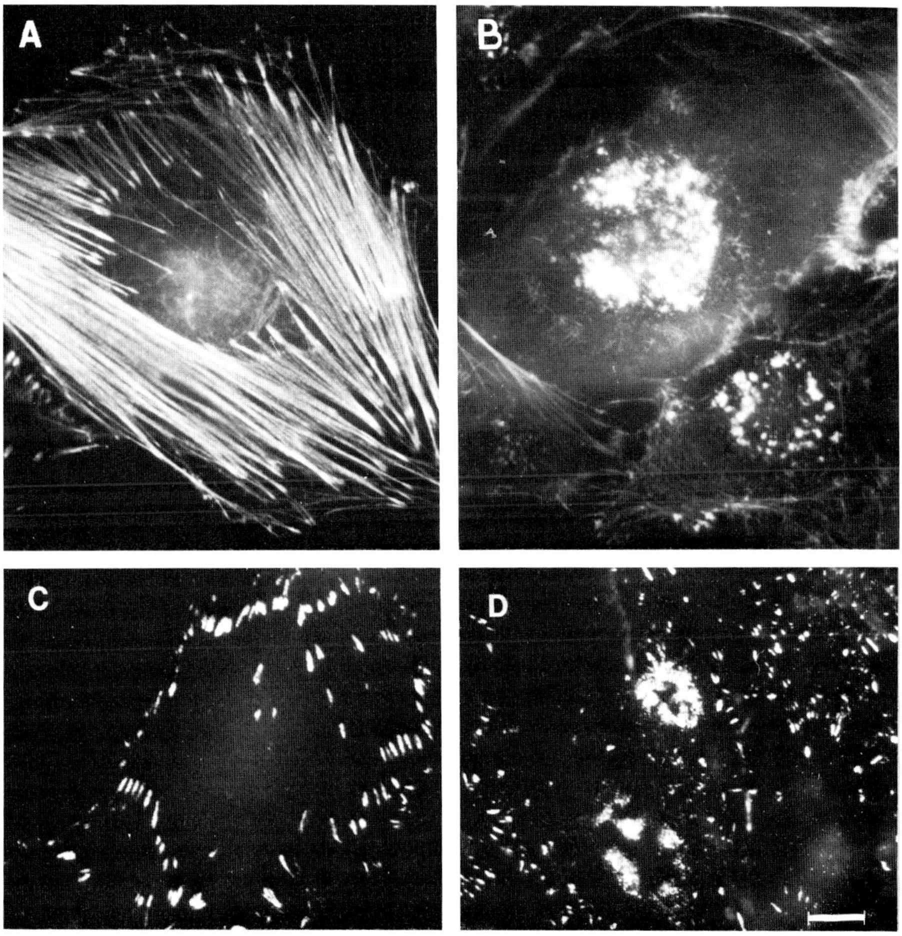

Figure 11. Fluorescent labeling of *ts* LA 23 cells (infected with a temperature-sensitive mutant of RSV) for actin (A and B) or vinculin (C and D). The cells were cultured in nonpermissive temperature, 39°C (A and C), displaying "normal" phenotype with prominent stress fibers and vinculin-rich focal contacts, or in the permissive temperature, 33°C (B and D) displaying a "transformed phenotype." The latter is characterized by disappearance of most of the stress fibers and deterioration of focal contacts. Instead, actin and vinculin become associated with "rosette-like structures." (For details see David-Pfuety and Singer, 1980; Rohrschneider, 1980). Bar represents 10 μm.

protein has unique properties with respect to its binding to actin or to the membrane.

The studies with transformed cells described here pointed to some relationships between formation of specific cell contacts and cytoskeletal rearrangements in focal contacts and the maintenance of a specific pattern of growth. The phosphorylation of vinculin or some of its specific isoforms

(Geiger, 1982b) as well as other focal contact elements may be of major physiological significance or else a fortuitous result of its proximity to membrane-bound kinase serine as well as tyrosine specific.

Where do these results and speculations lead us *vis-à-vis* the contact-dependent mechanism of growth control? One possible approach is to follow the line of studies on vinculin modification to determine directly the effect of phosphorylation at specific sites on its various activities. This would probably require a better understanding of the native physiological molecular interactions of vinculin in the cell. Alternatively, attempts should be made to identify other components of the transmembrane connection between external structures (specific matrix components, for example) and actin filaments. Vinculin may be one of the links in such a transmembrane chain, but certainly not the only one, and failure of cells to normally respond to contact stimuli may be related to an alteration in each of these components.

11. Conclusion

In the previous pages, we have discussed some of the molecular properties and dynamic behavior of a specialized domain in the cell membrane, the adherens junction. We have discussed the criteria for the definition of specific types of junction and proposed that a molecular definition should be used instead of a morphological one (see also Geiger *et al.*, 1983). We propose here that vinculin might be used as such a marker until a more comprehensive picture is available on other adherens junction-specific molecules. The extent of molecular homology between different contact systems is yet to be determined.

Our attempts to elucidate the role of vinculin in the formation of adherens junction suggested that it is involved in the linking of actin to the inner surfaces of the plasma membrane confirming some of our predictions in the working model shown in Fig. 6 above. According to this scheme, vinculin binds first to specific sites on the membrane which are present exclusively in contact areas. Subsequently, it binds to filamentous actin (or directs its polymerization by providing olygomeric actin nuclei) and causes its bundling. Moreover, it was shown that vinculin binds to the membrane independently of the presence of actin.

As for the kinetic properties of the contact region, it was shown that the junction is a specialized membrane domain with rather dynamic properties. Membrane elements (lipids and some of the membrane proteins) were shown to move laterally through focal contacts, though at a reduced speed. Vinculin and actin can probably move "vertically" from the cytoplasm to the junctional membrane and back. This dynamic steady-state equilibrium and its modulation during contact formation and reversal are compatible with our suggestion that local perturbations in the cell membrane of developing junction induce the accretion of vinculin to the membrane and these sites subsequently serve as nucleatin centers for actin bundle assembly. In line with this model are the findings that vinculin can be detected on the membrane before significant amounts of actin are assembled in those sites and that removal of actin does not affect the apparent association of exogenously-added vinculin with the membrane. The studies described here also point to the complex transmembrane relationships of vinculin (and adherens junctions in general) to the ECM protein fibronectin. It has been shown that in some areas, the cell may be attached to exogenous fibronectin through a small adherens-type contact (contains vinculin), nevertheless, fibronectin is usually absent from defined focal contacts probably due to its active removal from these sites. This exclusion also suggests that secretion of cellular fibronectin and possibly other secreted matrix components do not occur at focal contact regions. The process of ECM rearrangement is probably responsible for the formation of fibronectin cables in cultured cells and possibly in intact tissues.

In spite of the progress made in the characterization of adherens junctions, there are many central questions that are as yet unanswered. We would like to know what the extent of molecular homology is between different cellular junctions, cell–substrate and cell–cell. The identity of the membrane constituent(s) specific for adherens junctions (represented by the flags in Fig. 6) should be elucidated, together with the nature of the ECM components to which they can bind (if there is a specific one). The generality and distribution of adherens junctions should, throughout different tissues, be further established by immunocytochemistry and complementary approaches. It will also be necessary to define the particular role of the various actin-binding proteins which are specifically localized near adherens junctions. Among these are α-actinin, fimbrin (Bretscher and Weber, 1980a,b), a new 215,000 dalton protein (Connell and Burridge, 1982), the src-gene product, $pp60^{src}$ (Rohrschneider, 1980), as well as a recently described surface component of focal contacts of chick fibroblasts (Oesch and Birchmeier, 1982).

Future progress in this field and answers to at least some of the major questions listed here will greatly increase our understanding of the transmembrane dynamics and anchorage-dependent processes in living cells.

ACKNOWLEDGMENTS. The studies described in this chapter which were carried out in our laboratories were supported by the Muscular Dystrophy Association (B.G.), the Ch. Revson Foundation (B.G.), the National Institute of Health (J.S.), and the US-Israel Binational Science Foundation (J.S.). We express our gratitude to colleagues whose advice and support were of great help in all the studies described here. The studies on vinculin and the immunoelectron microscopy were initiated in J. Singer's laboratory and carried out with A. Dutton and K. Tokuyasu. Studies on adherens junction and EM of BMGE cells were carried out with E. Schmid and W. Franke. We would like to thank Ms. T. Volberg for excellent technical assistance and labeling of MDBK cells, and Mrs. E. Majerowich for superb secretarial help. Electron microscopy of smooth muscle was carried out by T. Volk. Actin antibodies were provided by V. Small and rhodamine phalloidin by H. Faulstich.

References

Abercrombie, M., and Dunn, G. A., 1975, Adhesions of fibroblasts to substratum during contact inhibition observed by interference reflection microscopy, *Exp. Cell Res.* **92:**57.

Abercrombie, M., Heaysman, J. E. M., and Pegrum, S. M., 1970, The locomotion of fibroblasts in culture. I. Movements of the leading edge, *Exp. Cell Res.* **59:**393.

Abercrombie, M. J., Heaysman, E. M., and Pegrum, S. M., 1971, The locomotion of fibroblasts in culture. IV. Electron microscopy of the leading lamella, *Exp. Cell Res.* **67:**359.

Ali, I. U., Mautner, V., Lanza, R., and Hynes, R. O., 1977, Restoration of normal morphology: Adhesion and cytoskeleton in transformed cells by addition of a transformation-sensitive surface protein, *Cell* **11:**115.

Anderton, B. H., 1981, Intermediate filaments: A family of homologous structures, *J. Muscle Res. Cell Motil.* **2:**141.

Ash, J. F., Vogt, P. K., and Singer, S. J., 1976, Reversion from transformed to normal phenotype by inhibition of protein synthesis in rat kidney cells injected with a temperature sensitive mutant of Rous sarcomavirus, *Proc. Natl. Acad. Sci. USA* **73:**3603.

Ashton, F. T., Somlyo, A. V., and Somlyo, A. B., 1975, The contractile apparatus of vascular smooth muscle: Intermediate high voltage stereo electron microscopy, *J. Mol. Biol.* **98:**17.

Avnur, Z., and Geiger, B., 1981a, Substrate-attached membranes of cultured cells isolation and characterization of ventral cell membranes and the associated cytoskeleton, *J. Mol. Biol.* **153:**361.

Avnur, Z., and Geiger, B., 1981b, The removal of extracellular fibronectin from areas of cell-substrate contact, *Cell* **25:**121.

Avnur, Z., Small, J. V., and Geiger, B., 1983, Actin independent association of vinculin with the cytoplasmic aspects of the plasma membrane in cell-substrate contacts, *J. Cell Biol.* **96:**1622.

Axelrod, D., Koppel, D. E., Schlessinger, J., Elson, E. L., and Webb, W. W., 1976, Mobility measurement by analysis of fluorescence photobleaching recovery kinetics, *Biophys. J.* **16:**1055.

Ball, E. H., and Singer, S. J., 1982, Mitochondria are associated with microtubules and not with intermediate filaments in cultured fibroblasts, *Proc. Natl. Acad. Sci. USA* **79:**123.

Bennett, V., 1979, Immunoreactive forms of human erythrocyte ankyrin are present in diverse cells and tissues, *Nature* **281:**597.

Bennett, V., and Davis, J., 1981, Erythrocyte ankyrin: Immunoreactive analogues are associated with mitotic structures in cultured cells and with microtubules in brain, *Proc. Natl. Acad. Sci. USA* **78:**7550.

Bennett, V., and Davis, J., 1982, Immunoreactive forms of human erythrocyte ankyrin are localized in mitotic structures in cultured cells and are associated with microtubules in brain, *Cold Spring Harbor Symp. Quant. Biol.* **46:**647.

Bennett, V., Davis, J., and Fowler, W. F., 1982, Brain spectrin, a membrane associated protein related in structure and function to erythrocyte spectrin, *Nature* **299:**126.

Ben-Zeev, A., Duerr, A., Solomon, F., and Penman, S., 1979, The outer boundary of the cytoskeleton: A lamina derived from plasma membrane proteins, *Cell* **17:**859.

Ben-Zeev, A., Farmer, S. R., and Penman, S., 1980, Protein synthesis requires cell-surface contact while nuclear events respond to cell-shape in anchorage dependent fibroblasts, *Cell* **21:**365.

Birchmeier, C., Kreis, T. E., Eppenberger, H. M., Winterhalter, K. H., and Birchmeier, W., 1980, Corrugated attachment membrane in WI-38 fibroblasts: Alternating fibronectin fibers and actin-containing focal contacts, *Proc. Natl. Acad. Sci. USA* **77:**4108.

Bloch, R. J., and Geiger, B., 1980, The localization of acetylcholine receptor clusters in areas of cell substrate contact in cultures of rat myotubes, *Cell* **21:**25.

Branton, D., 1982, Membrane cytoskeletal interactions in the human erythrocyte, *Cold Spring Harbor Symp. Quant. Biol.* **46:**1.

Branton, D., Cohen, C. M., and Tyler, J., 1981, Interaction of cytoskeletal proteins on the human erythrocyte membrane, *Cell* **24:**24.

Bretscher, A., 1981, Fimbrin is a cytoskeletal protein that crosslinks F-actin in vitro, *Proc. Natl. Acad. Sci. USA* **78**:6849.

Bretscher, A., and Weber, K., 1979, Villin the major microfilament-associated protein of the intestinal microvillus, *Proc. Natl. Acad. Sci. USA* **76**:2321.

Bretscher, A., and Weber, K., 1980a, Villin: Is a major protein of the microvillus cytoskeleton which binds both G and F actin in a calcium dependent manner, *Cell* **20**:839.

Bretscher, A., and Weber, K., 1980b, Fimbrin, a new microfilament-associated protein present in microvilli and other cell surface structures, *J. Cell Biol.* **86**:335.

Brinkley, B. R., Beall, P. T., Wible, L. J., Mace, M. L., Turner, D. S., and Cailleau, R. M., 1980, Variations in cell form and cytoskeleton in human breast carcinoma cells in vitro, *Cancer Res.* **40**:3118.

Brown, S., Levinson, W., and Spudich, J., 1976, Cytoskeletal elements of chick embryo fibroblasts revealed by detergent extraction, *J. Supramol. Struct.* **5**:119.

Brunser, O., and Luft, J., 1970, Fine structure of the apex of absorptive cells from rat small intestine, *J. Ultrastruct. Res.* **31**:291.

Buckley, I. K., 1975, Three-dimensional fine structure of cultured cells; possible implications for subcellular motility, *Tissue Cell* **7**:51.

Buckley, I. K., and Porter, K. R., 1975, Electron microscopy of critical point dried cultured cells, *J. Microsc.* **104**:107.

Burridge, K., and Feramisco, J. R., 1980, Microinjection and localization of a 130 K protein in living fibroblasts: A relationship to actin and fibronectin, *Cell* **19**:587.

Burridge, K., and Feramisco, J. R., 1982, α-actinin and vinculin from nonmuscle cells: Calcium sensitive interactions with actin, *Cold Spring Harbor Symp. Quant. Biol.* **46**:587.

Cervera, M. G., Dreyfuss, G., and Penman, S., 1981, Messenger RNA is translated when associated with the cytoskeletal framework in normal and VSV injected HeLa cells, *Cell* **23**:113.

Chen, W.-T., 1981a, Mechanism of retraction of the trailing edge during fibroblast movement, *J. Cell Biol.* **90**:187.

Chen, W-T., 1981b, Surface changes during retraction-induced spreading of fibroblasts, *J. Cell Sci.* **49**:1.

Chen, W.-T., and Singer, S. J., 1980, Fibronectin is not present in the focal adhesions formed between normal cultured fibroblasts and their substrate, *Proc. Natl. Acad. Sci. USA* **77**:7318.

Chen, W.-T., and Singer, S. J., 1982, Immunoelectron microscopic studies of the sites of cell–substratum and cell–cell contacts in cultured fibroblasts, *J. Cell Biol.* **95**:205.

Cohen, C. M., Foley, S., and Korsgren, C., 1982, Localization of immunoreactive forms of erythrocyte band 4.1 to fibroblast stress fibers, *J. Cell Biol.* **95**:285a.

Collett, M. S., Purchio, A. F., and Erikson, R. L., 1980, Avian sarcoma virus-transforming protein pp60^{src} shows protein kinase activity specific for tyrosine, *Nature* **285**:167.

Condeelis, J., 1981, Microfilament-membrane interactions in cell shape and surface architecture, in: *International Cell Biology 1980–1981* (H. G. Schweiger, ed.), pp. 306–320, Springer-Verlag, Berlin, Heidelberg, New York.

Condeelis, J., and Taylor, D., 1977, The contractile basis of amoeboid movement, *J. Cell Biol.* **74**:901.

Connell, L., and Burridge, K., 1982, A new protein of the adhesion plaque, *J. Cell Biol.* **95**:299a.

Coudrier, E., Reggio, H., and Louvard, D., 1981, Immunolocalization of the 110 KD molecular weight cytoskeletal protein of intestinal microvillus, *J. Mol. Biol.* **152**:49.

Craig, S. W., and Pollard, T. D., 1982, Actin binding proteins, *Trends Biochem. Res.* **7**:88.

Craig, S. W., and Powell, L. D., 1980, Regulation of actin polymerization by villin, a 95,000 dalton cytoskeletal component of intestinal brush borders, *Cell* **22**:739.

David-Pfuety, T., and Singer, S. J., 1980, Altered distributions of the cytoskeletal proteins, vinculin and α-actinin in cultured fibroblasts transformed by Rous sarcoma virus, *Proc. Natl. Acad. Sci. USA* **77**:6687.

Dentler, W. L., 1981, Microtubule—membrane interactions in cilia and flagella, *Int. Rev. Cytol.* **72.**

DeRosier, D. J., and Tilney, L. G., 1982, How actin filaments pack into bundles, *Cold Spring Harbor Symp. Quant. Biol.* **46**:525.

Edelman, G. M., and Yahara, I., 1976, Temperature-sensitive changes in surface modulating assemblies of fibroblasts transformed by mutants of Rous sarcoma virus, *Proc. Natl. Acad. Sci. USA* **73:**2047.

Farquhar, M. G., and Palade, G. E., 1963, Junctional complexes in various epithelia, *J. Cell Biol.* **17:**375.

Fawcett, D. W., 1981, *The Cell,* 2nd ed., pp. 1–862. W. B. Saunders, Philadelphia.

Feramisco, J. R., Burridge, K., Smart, J. E., and Thomas, G. P., 1981, Co-existence of vinculin and a vinculin-like protein of higher molecular weight in smooth muscle, *J. Cell Biol.* **91:**292a.

Folkman, J., and Moscona, A., 1978, Role of cell shape in growth control, *Nature* **273:**345.

Franke, W. W., Schmid, E., Schiller, D. L., Winter, S., Jarasch, E.-D., Moll, R., Deuk, H., Jackson, B., and Illmensee, K., 1982, Differentiation-related patterns of expression of proteins of intermediate-sized filaments in tissues and cultured cells, *Cold Spring Harbor Symp. Quant. Biol.* **46:**431.

Fulton, A. B., Wan, K., and Penman, S., 1980, The spatial distribution of polyribosomes in 3T3 cells and the associated assembly of proteins into the skeletal framework, *Cell* **20:**849.

Fulton, A. B., Prives, J., Farmer, S. R., and Penman, S., 1981, Developmental reorganization of the skeletal framework and its surface lamina in fusing muscle cells, *J. Cell Biol.* **91:**103.

Geiger, B., 1979, A 130 K protein from chicken gizzard: Its localization at the termini of microfilament bundles in cultured chicken cells, *Cell* **18:**193.

Geiger, B., 1981a, Transmembrane linkage and cell attachment: The role of vinculin, in: *International Cell Biology* (H. G. Schweiger, ed.), pp. 761–773, Springer Verlag, New York.

Geiger, B., 1981b, The association of rhodamine-labelled α-actinin with actin bundles in demembranated cells, *Cell Biol. Int. Rep.* **5:**627.

Geiger, B., 1982a, Involvement of vinculin in contact-induced cytoskeletal interactions, *Cold Spring Harbor Symp. Quant. Biol.* **46:**671.

Geiger, B., 1982b, Microheterogeneity of avian and mammalian vinculin. Distinctive subcellular distribution of different isovinculins, *J. Mol. Biol.* **159:**685.

Geiger, B., 1983, Membrane-cytoskeleton interaction, *Biochim. Biophys. Acta* **737:**305–341.

Geiger, B., and Singer, S. J., 1980, Association of microtubules and intermediate filaments in chicken gizzard cells as detected by double immunofluorescence, *Proc. Natl. Acad. Sci. USA* **77:**4769.

Geiger, B., Tokuyasu, K. T., Dutton, A. H., and Singer, S. J., 1980, Vinculin, an intracellular protein localized at specialized sites where microfilament bundles terminate at cell membranes, *Proc. Natl. Acad. Sci. USA* **77:**4127.

Geiger, B., Dutton, A. H., Tokuyasu, K. T., and Singer, S. J., 1981, Immunoelectron microscopic studies of membrane-microfilament interactions. The distributions of α-actinin, tropomyosin and vinculin in intestinal epithelial brush border and in chicken gizzard smooth muscle, *J. Cell Biol.* **91:**614.

Geiger, B., Avnur, Z., and Schlessinger, J., 1982, Restricted mobility of membrane constituents in cell substrate focal contacts of chicken fibroblasts, *J. Cell Biol.* **93:**495.

Geiger, B., Schmid, E., and Franke, W. W., 1983, Spatial distribution of proteins specific for desmosomes and adherens junctions in epithelial cells demonstrated by double immunofluorescence microscopy, *Differentiation* **23:**189.

Gingell, D., 1981, The interpretation of interference-reflection images of spread cells: Significant contribution from the peripheral cytoplasm, *J. Cell Sci.* **49:**237.

Glenney, J. R., Bretscher, A., and Weber, K., 1980, Calcium control of the intestinal microvillus cytoskeleton; Its implications for the regulation of microfilament organizations, *Proc. Natl. Acad. Sci. USA* **77:**6458.

Glenney, J. R., Kaulfus, Ph., and Weber, K., 1981a, F-actin assembly modulated by villin: Ca^{++}-dependent nucleation and capping of the barbed end, *Cell* **24:**471.

Glenney, J. R., Kaulfus, Ph., Matsudaira, P., and Weber, K., 1981b, F-actin binding and bundling properties of fimbrin, a major cytoskeletal protein of microvillus core filaments, *J. Biol. Chem.* **256:**9283.

Glenney, J. R., Glenney, P., and Weber, K., 1982a, Erythroid spectrin, brain fodrin and intestinal brush border proteins (TW-260/240) are related molecules containing a common calmodulin-binding subunit bound to a variant cell type specific subunit, *Proc. Natl. Acad. Sci. USA* **79:**4002.

Glenney, J. R., Glenney, P., Osborn, M., and Weber, K., 1982b, An F-actin and calmodulin-binding protein from isolated intestinal brush borders has a morphology related to spectrin, *Cell* **28:**843.

Glenney, J. R., Osborn, M., and Weber, K., 1982c, the intracellular localization of the microvillus 110 K protein, a component considered to be involved in side on membrane attachment of F-actin, *Exp. Cell Res.* **138:**199.

Goodman, S. R., Zagon, I. S., and Kulikowski, R. R., 1981, Identification of a spectrin-like protein in nonerythroid cells, *Proc. Natl. Acad. Sci. USA* **78:**7570.

Gospodarowicz, D., Greenburg, G., and Birdwell, C. R., 1978, Determination of cellular shape by the extracellular matrix and its correlation with the control of cellular growth, *Cancer Res.* **38:**4155.

Gratzer, W. B., 1981, The red cell membrane and its cytoskeleton, *Biochem. J.* **198:**1.

Griffith, L. M., and Pollard, T. D., 1978, Evidence for actin filament-microtubule interaction mediated by microtubule-associated proteins, *J. Cell Biol.* **78:**958.

Grinnell, F., 1978, Cellular adhesiveness and extracellular substrata, *Int. Rev. Cytol.* **53:**65.

Grinnell, F., 1980, Visualization of cell substratum adhesion plaques by antibody exclusion, *Cell Biol. Int. Rep.* **4:**1031.

Grinnell, F., and Feld, M. K., 1979, Initial adhesion of human fibroblasts in serum-free medium: Possible role of secreted fibronectin, *Cell* **17:**117.

Gröschel-Stewart, U., 1980, Immunochemistry of cytoplasmic contractile proteins, *Int. Rev. Cytol.* **65:**193.

Hanafusa, H., 1977, Cell transformation by RNA tumor viruses, in: *Comprehensive Virology*, Vol. 10 (Fraenkel-Control and R. R. Wagner, eds.), pp. 401–483, Plenum Press, New York.

Harris, A., 1973, Location of cellular adhesions to solid substrate, *Dev. Biol.* **35:**97.

Hasegawa, T., Takahashi, S., Hayashi, H., and Hatano, S., 1980, Fragmin: A calcium ion sensitive regulatory factor on the formation of actin filaments, *Biochemistry* **19:**2677.

Hay, E. D., and Meier, S., 1976, Stimulation of corneal differentiation by interaction between cell surface and extracellular matrix. II. Further studies on the nature and site of transfilter "induction", *Dev. Biol.* **52:**141.

Heath, J. P., and Dunn, G. A., 1978, Cell to substratum contacts of chick fibroblasts and their relation to the microfilament system. A correlated interference-reflexion and high-voltage electron-microscopy study, *J. Cell Sci.* **29:**197.

Heggeness, M. H., Ash, J. F., and Singer, S. J., 1978a, Transmembrane linkage of fibronectin to intracellular actin-containing filaments in culture human fibroblasts, *Ann. N.Y. Acad. Sci. USA* **312:**414.

Heggeness, M. H., Simon, M., and Singer, S. J., 1978b, Association of mitochondria with microtubules in cultured cells, *Proc. Natl. Acad. Sci. USA* **75:**3863.

Heuser, J. E., and Kirschner, M. W., 1980, Filament organization revealed in platinum replicas of freeze-dried cytoskeletons, *J. Cell Biol.* **86:**212.

Hillman, G., and Schlessinger, J., 1982, The lateral diffusion of epidermal growth factor complexed to its surface receptor does not account for the thermal sensitivity of patch formation and endocytosis, *Biochemistry* **21:**1667.

Hinssen, H., 1981, An actin modulating protein from Physarum polycephalum. II. Ca^{+2} dependence and other properties, *Eur. J. Cell Biol.* **23:**234.

Hirokawa, N., and Tilney, L. G., 1982, Interactions between actin filaments and between actin filaments and membranes in quick frozen and deeply etched hair cells of chick ear, *J. Cell Biol.* **95:**249.

Hull, B. E., and Staehelin, L. A., 1979, The terminal web: A reevaluation of its structure and function, *J. Cell Biol.* **81:**67.

Hunter, T., and Sefton, B. M., 1980, The transforming gene product of Rous sarcoma virus phosphorylates tyrosine, *Proc. Natl. Acad. Sci. USA* **77:**1311.

Hynes, R. O., 1973, Alteration of cell-surface proteins by viral transformation and proteolysis, *Proc. Natl. Acad. Sci. USA* **70:**3170.

Hynes, R. O., 1981, Relationships between fibronectin and the cytoskeleton, *Cell Surf. Rev.* **7:**97.

Hynes, R. O., and Destree, A. T., 1978, Relationships between fibronectin (LETS protein) and actin, *Cell* **15:**875.

Hynms, R. O., Destree, A. T., and Wagner, D. D., 1982, Relationships between microfilaments, cell substratum adhesion, and fibronectin, *Cold Spring Harbor Symp. Quant. Biol.* **46:** 659.

Isenberg, G., Leonard, K., and Jockusch, B. M., 1982, Structural aspects of vinculin-actin interactions, *J. Mol. Biol.* **158:**231.

Izzard, C. S., and Lochner, L. R., 1976, Cell-to-substrate contact in living fibroblasts. An interference reflexion study with an evaluation of the technique, *J. Cell Sci.* **27:**129.

Izzard, C. S., and Lochner, L. R., 1980, Formation of cell-to-substrate contacts during fibroblast motility: An interference-reflection study, *J. Cell Sci.* **41:**81.

Jockusch, B. M., and Isenberg, G., 1981, Interaction of α-actinin and vinculin with actin: Opposite effect on filament network formation, *Proc. Natl. Acad. Sci. USA* **78:**3005.

Jockusch, B. M., and Isenberg, G., 1982, Vinculin and α-actinin: Interaction with actin and effect on microfilament network formation, *Cold Spring Harbor Symp. Quant. Biol.* **46:**613.

Kirkpatrick, F. H., 1976, Spectrin: Current understanding of its physical, biochemical, and functional properties, *Life Sci.* **19:**1.

Kirschner, M. W., 1978, Microtubule assembly and nucleation, *Int. Rev. Cytol.* **54:**1.

Koppel, D. E., Axelrod, D., Schlessinger, J., Elson, E. L., and Webb, W. W., 1976, Dynamics of fluorescence marker concentration as a probe for mobility, *Biophys. J.* **16:**1315.

Korn, E. D., 1978, Biochemistry of actomyosin-dependent cell motility (a review), *Proc. Natl. Acad. Sci. USA* **75:**588.

Kreis, T. E., Geiger, B., and Schlessinger, J., 1982, Mobility of microinjected rhodamine actin within living chicken gizzard cells determined by fluorescence photobleaching recovery, *Cell* **29:**835.

Kreuger, J. G., Wang, E., and Goldberg, A. R., 1980, Evidence that the *src* gene product of Rous sarcoma virus is membrane associated, *Virology* **101:**25.

Lazarides, E., 1976, Actin, α-actinin, and tropomyosin interaction in the structural organization of actin filaments in nonmuscle cells, *J. Cell Biol.* **68:**202.

Lazarides, E., 1980, Intermediate filaments as mechanical integrators of cellular space, *Nature* **283:**249.

Lazarides, E., 1981, Intermediate filaments-chemical heterogeneity in differentiation, *Cell* **23:**649.

Lazarides, E., and Burridge, K., 1979, α-Actinin: Immunofluorescent localization of a muscle structural protein in nonmuscle cells, *Cell* **6:**289.

Lazarides, E., and Nelson, W. J., 1982, Expression of spectrin in nonerythroid cells, *Cell* **31:**505.

Lazarides, E., and Weber, K., 1974, Actin antibody: The specific visualization of actin filaments in non-muscle cells, *Proc. Natl. Acad. Sci. USA* **71:**2268.

Levi, A., Shechter, Y., Newfeld, E. J., and Schlessinger, J., 1980, Mobility, clustering and transport of nerve growth factor in embryonal sensory cells and in a sympathetic neuronal cell line, *Proc. Natl. Acad. Sci. USA* **77:**3465.

Levine, J., and Willard, M., 1981, Fodrin: Axonally transported polypeptides associated with the internal periphery of many cells, *J. Cell Biol.* **90:**631.

Lin, S., Wilkins, J. A., Cribbs, D. H., Grumet, M., and Lin, D. C., 1982, Proteins and complexes that affect actin filament assembly and interactions, *Cold Spring Harbor Symp. Quant. Biol.* **46:**625.

Lindberg, U., Carlsson, L., Markey, F., and Nystrom, L. E., 1979, *Meth. Achiev. Exp. Pathol.* **8:**143.

Lux, S. E., 1979, Spectrin-actin membrane skeleton of normal and abnormal red blood cells, *Sem. Hematol.* **16:**21.

Mangeat, P., Kelly, T., and Burridge, K., 1982, Non-erythrocyte spectrin actin-membrane attachment proteins occurring in many cell types, *J. Cell Biol.* **95:**294a.

Marchesi, V. T., 1979, Spectrin: Present status of a putative cytoskeletal protein of the red cell membrane, *J. Membr. Biol.* **51:**101.

Matsudaira, P. T., and Burgess, D. R., 1979, Identification and organization of the components in the isolated microvillus cytoskeleton, *J. Cell Biol.* **83:**667.

Matsudaira, P. T., and Burgess, D. R., 1982, Structure and function of the brush border cytoskeleton, *Cold Spring Harbor Symp. Quant. Biol.* **46:**845.

Mooseker, M. S., 1976, Actin filament-membrane attachment in microvilli of intestinal epithelial cells, in: *Cell Motility* (R. Goldman, T. Pollard, and J. Rosenbum, eds.), pp. 631–650, Cold Spring Harbor Laboratory.

Mooseker, M. S., and Tilney, L. G., 1975, The organization of an actin filament membrane complex: Filament polarity and membrane attachment in the microvilli of intestinal epithelial cells, *J. Cell Biol.* **67**:725.

Mooseker, M. S., Graves, T. A., Wharton, K. A., Falco, N., and Howe, C. L., 1980, Regulation of microvillus structure: Calcium-dependent solation and cross-linking of actin filaments in the microvilli of intestinal epithelial cells, *J. Cell Biol.* **87**:809.

Mukherjee, T. M., and Staehelin, L. A., 1979, The fine structural organization of the brush border of intestinal epithelial cells, *J. Cell Sci.* **8**:573.

Oesch, B., and Birchmeier, W., 1982, New surface component of fibroblast's focal contacts identified by a monoclonal antibody, *Cell* **31**:671.

Olmsted, J. B., and Borisy, G. G., 1973, Microtubules, *Annu. Rev. Biochem.* **42**:507.

Penman, S., Fulton, A., Capco, D., Ben Zeev, A., Wittelsberger, S., and Tse, C. F., 1982, Cytoplasmic and nuclear architecture in cells and tissue: Form, functions and mode of assembly, *Cold Spring Harbor Symp. Quant. Biol.* **46**:1013.

Pollack, R., Osborn, M., and Weber, K., 1975, Patterns of organization of actin and myosin in normal and transformed cultured cells, *Proc. Natl. Acad. Sci. USA* **72**:994.

Pollard, T. D., and Weihing, R. R., 1974, Actin and myosin end cell movement, *CRC Crit. Rev. Biochem.* **2**:1.

Raz, A., and Geiger, B., 1982, Altered organization of cell-substrate contacts and membrane-associated cytoskeleton in tumor cell variant exhibiting different metastatic capabilities, *Cancer Res.* **42**:5183.

Repasky, E. A., Granger, B. L., and Lazarides, E., 1982, Widespread occurrence of avian spectrin in nonerythroid cells, *Cell* **29**:821.

Rohrschneider, L. R., 1980, Adhesion plaques of Rous sarcoma virus-transformed cells contain the *src* gene product, *Proc. Natl. Acad. Sci. USA* **77**:3514.

Rohrschneider, L., Rosok, M., and Shriver, K., 1982, Mechanism of transformation by Rous sarcoma virus: Events within adhesion plaques, *Cold Spring Harbor Symp. Quant. Biol.* **46**:953.

Rouslahti, E., and Vaheri, A., 1975, Interaction of soluble fibroblast-surface antigen with fibrinogen and fibrin. Identity with cold insoluble globulin of human plasma, *J. Exp. Med.* **141**:497.

Schlee, R. B., and Borisy, G. G., 1979, In vitro assembly of microtubules, in: *Microtubules* (K. Roberts and J. S. Hymas, eds.), pp. 175–254, Academic Press, New York.

Schlessinger, J., and Elson, E. L., 1982, Fluorescence methods for studying membrane dynamics, in :*Methods of Experimental Physics*, Vol. 20 (H. Ehrenstein and H. Lecar, eds.), pp. 197–227, Academic Press, New York.

Schliwa, M., 1979, Stereo high voltage electron microscopy of melanophores, *Exp. Cell Res.* **118**:323.

Schliwa, M., and van Blerkom, J., 1980, Three-dimensional organization and interaction of cytoskeletal structures, *Eur. J. Cell. Biol.* (Abstr.) **22**:352.

Schliwa, M., and van Blerkom, J., 1981, Structural interaction of cytoskeletal components, *J. Cell Biol.* **90**:222.

Schliwa, M., van Blerkom, J., and Porter, K. R., 1981, Stabilization of the cytoplasmic ground substance in detergent opened cells and a structural and biochemical analysis of its composition, *Proc. Natl. Acad. Sci. USA*, **78**:4329–4334.

Sefton, B. M., Hunter, T., Ball, E. H., and Singer, S. J., 1981, Vinculin: A cytoskeletal target of the transforming protein of Rous sarcoma virus, *Cell* **24**:165.

Sefton, B. M., Hunter, T., Nigg, E. A., Singer, S. J., and Walter, G., 1982, Cytoskeletal targets for viral transforming proteins with tyrosine protein kinase activity, *Cold Spring Harbor Symp. Quant. Biol.* **46**:939.

Shelanski, M. L., Liem, R. K. H., Leterrier, J.-F., and Keith, C. H., 1981, The cytoskeleton and neuronal disease, in: *International Cell Biology* (G. H. Schwieger, ed.), pp. 428–439, Springer-Verlag, Berlin, Heidelberg, New York.

Siliciano, J. D., and Craig, S. W., 1982, Meta vinculin: A vinculin related proteins with solubility properties of a membrane protein, *Nature* **300:**533.

Singer, I. I., 1979, The fibronexus: A transmembrane association of fibronectin-containing fibers and bundles of 5 nm microfilaments in hamster and human fibroblasts, *Cell* **16:**675.

Singer, I. I., and Paradiso, P. R., 1981, A transmembrane relationship between fibronectin and vinculin (130 Kd protein): Serum modulation in normal and transformed hamster fibroblasts, *Cell* **24:**481.

Small, J. V., and Sobieszek, A., 1980, The contractile apparatus of smooth muscle, *Int. Rev. Cytol.* **64:**241.

Staehelin, L. A., 1974, Structure and function of intercellular junctions, *Int. Rev. Cytol.* **39:**191.

Steck, T. L., 1974, Organization of proteins in the human red blood cell membrane, *J. Cell Biol.* **62:**1.

Steinert, P., Idler, W., Aynardi-Whitman, M., Zackroff, R., and Goldman, R. D., 1982, Heterogeiety of intermediate filaments assembled in vitro, *Cold Spring Harbor Symp. Quant. Biol.* **46:**465.

Stephens, N. L., (ed.), 1977, Contractile proteins in smooth muscle, in: *The Biochemistry of Smooth Muscle,* pp. 363–551, University Park Press, Baltimore.

Stephens, R. E., and Edds, K. T., 1976, Microtubules: Structure, chemistry and function, *Physiol. Rev.* **56:**709.

Stossel, T., and Hartwig, J., 1976, Interactions of actin, myosis and a new actin binding protein of rabbit pulmonary macrophages. II. Role in cytoplasmic movement and phagocytosis, *J. Cell Biol.* **68:**602.

Tilney, L. G., 1975, The role of actin in non-muscle cell motility, in: *Molecules and Cell Movement* (S. Inoue and R. E. Stephens, eds.), pp. 339–388, Raven Press, New York.

Timasheff, S. N., and Grisham, L. M., 1980, In vitro assembly of cytoplasmic microtubules, *Annu. Rev. Biochem.* **49:**565.

Vaheri, A., and Ruoslahti, E., 1974, Disappearance of a major cell-type specific surface glycoprotein (SF) after transformation of fibroblasts by Rous sarcoma virus, *Int. J. Cancer* **13:**579.

Wang, E., and Goldberg, A. R., 1976, Changes in microfilament organization and surface topography upon transformation of chick embryo fibroblasts with Rous sarcoma virus, *Proc. Natl. Acad. Sci. USA* **73:**4065.

Wang, K., and Singer, S. J., 1977, Interaction of filamin with F actin in solution, *Proc. Natl. Acad. Sci. USA* **74:**2021.

Weatherbee, J. A., 1981, Membranes and cell movement: Interactions of membranes with the proteins of the cytoskeleton, *Int. Rev. Cytol. Suppl.* **12:**113.

Weber, K., and Osborn, M., 1982, Intermediate filaments: Cell type specific markers in differentiation and pathology, *Cell* **31:**303.

Weber, M. J., Hale, A. H., and Losasso, L., 1977, Decreased adherence to the substrate in Rous sarcoma virus-transformed chicken embryo fibroblasts, *Cell* **10:**45.

Wehland, J., Osborn, M., and Weber, K., 1979, Cell-to-substratum contacts in living cells: A direct correlation between interference-refexion and indirect immunofluorescence microscopy using antibodies against actin and α-actinin, *J. Cell Sci.* **37:** 257.

Wilkins, J. A., and Lin, S., 1982, High affinity interaction of vinculin with actin filaments in vitro, *Cell* **28:**83.

Willingham, M. C., Jay, G., and Pastan, I., 1979, Localization of the ASV *src* gene product to the plasma membrane of transformed cells by electron microscopic immunocytochemistry, *Cell* **18:**125.

Wojcieszyn, J. W., Schlegel, R. A., Wu, E.-S., and Jacobson, K., 1981, Diffusion of injected macromolecules within the cytoplasm of living cells, *Proc. Natl. Acad. Sci. USA* **78:**4407.

Yamada, K. M., and Olden, K., 1978, Fibronectins—Adhesive glycoproteins of cell surface and blood, *Nature* **275:**179.

Yamada, K. M., Yamada, S. S., and Pastan, I., 1976, Cell surface protein partially restores morphology, adhesiveness and contact inhibition of movement of transformed fibroblasts, *Proc. Natl. Acad. Sci. USA* **73:**1217.

6

Cell Shape and Membrane Receptor Dynamics

MODULATION BY THE CYTOSKELETON

David F. Albertini and Brian Herman

1. Introduction

1.1. Statement of Problem

The cell surface is thought to be a major site for the transfer of information from the extracellular environment to the cell interior. Glycoprotein receptors, embedded in the plasma membrane, most likely mediate the process of information transfer since they provide a mechanism whereby cells can recognize particular signals in the environment which modulate cellular metabolism. Among the signals affecting cellular metabolism are soluble proteins, such as hormones and toxins, or macromolecular complexes, such as LDL and viruses, each of which bind to specific cell surface receptors to initiate their effects. Subsequent to ligand-receptor binding, many ligands and their receptors are internalized within coated vesicles which form endosomes that travel to specific cytoplasmic locations such as the Golgi-lysosome region where further processing occurs (Pastan and Willingham, 1981). The available evidence indicates that the movement of ligands, and perhaps their receptors, both at the cell surface and within the cytoplasm, is nonrandom and subject to control by the cytoskeleton (Albertini and Anderson, 1977; Edelman, 1976; Herman and Albertini, 1982; Oliver and Berlin, 1982; Rebhun, 1972).

Studies on the human erythrocyte membrane first indicated that the cytoskeleton could modulate receptor topography by forming specific pro-

David F. Albertini and ***Brian Herman*** • Department of Anatomy and Laboratory of Human Reproduction and Reproductive Biology, Harvard Medical School, Boston, Massachusetts 02115.

tein–protein interactions at the inner surface of the lipid bilayer (Branton *et al.*, 1981). Although detailed molecular information is lacking in other systems, there is substantial agreement that similar cytoskeletal–membrane interactions operate in many cell types to control the display of surface receptors, the movement of membrane-bound organelles, and the maintenance of cellular shape (Edelman, 1976; Folkman and Moscona, 1978; Herman and Albertini, 1982; Oliver and Berlin, 1982; Rebhun, 1972; Vasiliev and Gelfand, 1976). The problem addressed in this chapter regards the extent to which one can experimentally separate the functions of the cytoskeleton in these seemingly disparate processes.

1.2. Receptor Display and the Cytoskeleton

Actin microfilaments have been shown to interact with the inner aspect of the plasma membrane in *Acanthamoeba* (Pollard and Korn, 1973). This interaction appears to be a general phenomenon, since the redistribution of surface receptors induced by multivalent ligands such as lectins or antibodies involves spatially correlated changes in the disposition of actin and functionally associated proteins such as myosin, α-actinin, and calmodulin (Albertini and Anderson, 1977; Bourguignon and Singer, 1977; Condeelis, 1979; Salisbury *et al.*, 1981). The topographic disposition of contractile elements during capping may be controlled by microtubules (Albertini *et al.*, 1977). Many pharmacological lines of evidence show that microtubules also modulate receptor mobility and endocytosis at the cell surface perhaps by regulating the distribution of microfilaments (Berlin *et al.*, 1978; Oliver and Berlin, 1982).

1.3. Intracellular Motility and the Cytoskeleton

For most biologically relevant ligands such as hormones or growth factors, subtle receptor movements at the cell surface are coupled with rapid endocytosis of ligands in specialized structures known as coated vesicles. This process is called receptor-mediated endocytosis (Goldstein *et al.*, 1979). Exactly how the cytoskeleton particpates in this process or the subsequent translocation of ligands through the cytoplasm is not understood. Moreover, since many hormones themselves modulate cellular shape and cytoskeletal organization, it becomes difficult to sort out direct actions of the cytoskeleton on receptor display and function from modifications in cytoskeletal function that occur as a result of ligand stimulation.

To better understand the role of the cytoskeleton in modulating ligand and receptor dynamics, we have analyzed the behavior of lectin receptors in a cell type which undergoes characteristic changes in shape and cytoskeletal organization in response to hormones. Primary cultures of rat granulosa cells, responding to the pituitary hormone follicle-stimulating hormone (FSH), rapidly and reversibly round by a cyclic AMP-dependent mechanism (Lawrence *et al.*, 1979). The data presented in this chapter suggest that the cytoskeleton of ovarian granulosa cells regulates the distribution of lectin receptors on the cell surface and plays a major role in orchestrating the nonrandom movement of ligand-bearing endosomes. It is suggested that the hormone-induced reorganization of the cytoskeleton results in alterations in cellular morphology and surface receptor mobility because of modified interactions between the cytoskeleton and plasma membrane.

2. Materials and Methods

2.1. Cell Culture

All experiments were performed with 5 day rat granulosa cell cultures prepared from 21-day-old estrogenized females as described previously (Albertini and Kravit, 1981). Cells were cultured in 35 mm tissue culture dishes or on circular glass coverslips in multiwell plates containing 10% fetal calf serum in McCoy's 5A medium supplemented with glutamine, penicillin, and streptomycin (GIBCO, Grand Island, New York). Rat FSH was obtained from the National Pituitary Agency, NIAMMD, and dibutyryl cyclic AMP (DcAMP), 8-bromo cyclic AMP, adenosine, theophylline, horseradish peroxidase, and colchicine were all purchased from Sigma Chemical Co. (St. Louis, Missouri). Nocodazole was obtained from Aldrich Chemical Co. (Milwaukee, Wisconsin), 7-nitrobenz-2-oxa-1,3-diazole phallicidin (NBD-Ph) was from Molecular Probes (Plano, Texas), and reagents for immunolabeling were obtained from Miles-Yeda Laboratories.

2.2. Rounding Experiments

Rounding experiments were performed by washing cultures twice in Earle's balanced salts buffered with 10 mM Hepes (EBSSH) followed by incubation for 90 min at 37°C in EBSSH containing 5 μg/ml FSH, 0.5 mM DcAMP (±0.1 mM theophylline, a phosphodiesterase inhibitor), 0.5 mM 8-bromo cAMP, or 0.5 mM adenosine. Cultures were then incubated in either EBSSH or complete medium for an additional 90–210 min at 37°C. Samples were fixed with 3.7% formaldehyde in phosphate-buffered saline (PBS) for 10 min at various time intervals for determining the kinetics of hormone-induced rounding (Fig. 3). Some coverslips were mounted in a perfusion chamber through which either FSH, DcAMP, or colchicine were perfused to monitor the rounding process by time lapse video microscopy. Changes in receptor topography were followed by labeling control or rounded cells with 100 μg/ml fluorescein Con A (F-Con A) in PBS for 10 min at 24°C. Cultures were washed in EBSSH and either fixed immediately or at various time intervals after labeling as above.

2.3. Immunofluorescence

To demonstrate actin and tubulin simultaneously, cultures were fixed in 3.7% formaldehyde in a stabilization buffer (SB) composed of 0.1% Triton

X-100, 0.1% aprotinin (Sigma), 0.1M Pipes (pH 6.9), 5.0 mM EGTA, and 2.5 mM $MgCl_2$ in 50% deuterium oxide. Following a 10-min fixation, cultures were washed in SB lacking Triton X-100 and placed in PBS. Coverslips were labeled with a 1/100 dilution of rabbit anti-tubulin in PBS (Fujiwara and Pollard, 1978) for 30 min at 37°C, washed in PBS, and stained with 2ng/ml NBD-Ph in a 1/30 dilution of rhodamine goat anti-rabbit immunoglobulin in PBS for 30 min at 37°C. Tubulin and α-actinin double staining was accomplished as follows. Cells were fixed in Triton-SB as above followed by methanol fixation (−20°C, 10 min). Cells were incubated in anti-tubulin for 30 min at 37°C followed by fluorescein goat anti-rabbit immunoglobulin for 30 min at 37°. These coverslips were then treated with a 1/50 dilution of nonimmune rabbit IgG followed by incubation in a rhodamine conjugated, affinity-purified rabbit anti α-actinin IgG (Fujiwara *et al.*, 1978). Cell surface receptor topography and microtubule organization were simultaneously demonstrated utilizing double label immunofluorescence microscopy. Receptors were visualized utilizing F-Con A and microtubules were demonstrated with anti-tubulin antibodies followed by rhodamine goat anti-rabbit IgG. Cells were exposed to F-Con A and fixed in cold methanol (−20°C for 10 min) after incubation for various times at 37°C, and then incubated with anti-tubulin antibodies. All coverslips were mounted in PBS/glycerol containing 0.01% sodium azide and were viewed in a Zeiss photoscope III equipped with rhodamine and fluorescein (or NBD) filter sets. Photographs were recorded on Tri-X film developed in Acufine.

2.4. Electron Microscopy

Con A-horseradish peroxidase (HRP) staining and all subsequent processing steps for transmission electron microscopy were performed as described previously (Albertini and Anderson, 1977).

3. Results

3.1. Hormone-Induced Changes in Granulosa Cell Shape

Upon exposure to follicle-stimulating hormone (FSH), granulosa cells lose their normal spread polygonal morphology and round up remaining attached to the substrate by a few focal contacts. Time lapse video phase microscopy of living granulosa cells, exposed to 5 μm/ml of FSH, reveals several features of the rounding response exhibited by these cells (Fig. 1). Cultured granulosa cells are usually polygonal in shape but within 20 min after hormone administration, the margins of the cell retract and appear more phase dense. The retraction proceeds along 3–4 cellular processes until, in the fully rounded state, the cell body becomes spherical with several dendritic processes extending from it. Granulosa cells remain rounded unless hormone is removed from the medium which causes respreading onto the

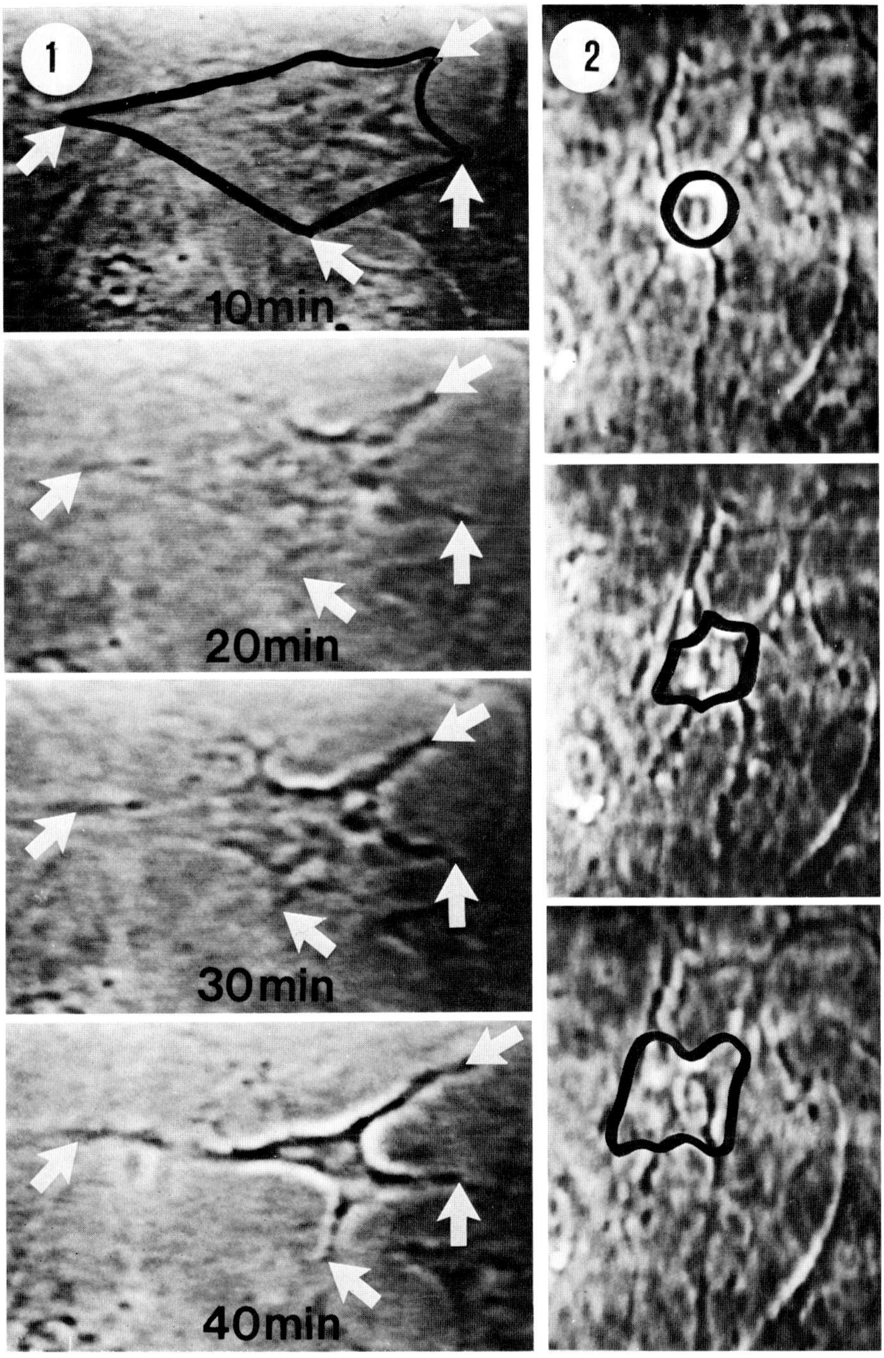

Figures 1 and 2. Time lapse video images of granulosa cell rounding. Figure 1 shows a cell, outlined at the top, which undergoes progressive rounding in response to 1 μg/ml FSH. Note the retraction of the cell body from four points of substrate contact (arrows) which give rise to dendritic processes. Figure 2 illustrates the effects of 1.0 μm colchicine treatment upon an FSH rounded cell. Each frame represents 5 min intervals beginning 15 min after drug addition. Relaxation of the cell body (outlined in black) occurs in response to colchicine.

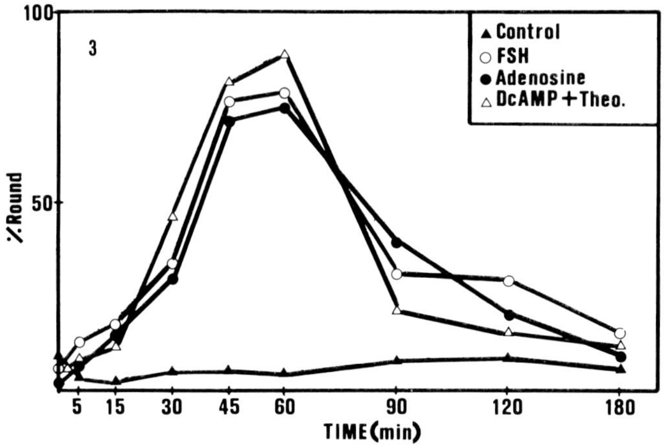

Figure 3. Kinetics of granulosa cell rounding induced by various treatments. Five-day cultures treated with either FSH (○, 1 μg/ml), adenosine (●, 1.0 mM), or dibutyryl cyclic AMP and theophylline (△, 0.1 mM and 1.0 mM, respectively); (▲) Control cultures. Cells undergo complete rounding by 60 min which is followed by respreading within 3 hr of treatment. 150–200 cells per coverslip were counted and scored as flat or rounded.

substratum (see below). Maintenance of the rounded state appears to be microtubule-dependent, since exposure of FSH-treated cells to 1.0 μM colchicine causes flattening of both the cell body and processes (Fig. 2).

The kinetics of the rounding response and its reversibility are illustrated in Fig. 3. Significant rounding is detectable 30 min after FSH treatment. The response is maximal by 60–90 min and removal of hormone results in a gradual respreading of cells after 2 hr of further culturing in either EBSSH or complete medium. Treatment of granulosa cells with DcAMP and theophylline, or adenosine alone, also cause rounding with similar kinetics. Both of these treatments are reversible upon agonist removal. Thus, the rounding response seems to occur by a reversible, cyclic AMP mediated process shortly after ligand addition.

3.2. Consequences of Rounding on Cytoskeletal Organization

To evaluate the effects of hormone-induced rounding on cytoskeletal organization, cultures were fixed at various times after hormone addition and subjected to staining procedures for the demonstration of F-actin, tubulin, and α-actinin. NBD-Ph staining of control spread cultures reveals that F-actin is organized into an extensive network of stress fibers (Fig. 4), as noted in other cultured cells using this staining procedure (Barak *et al.*, 1980). Antitubulin staining in the same cells (Fig. 5) shows a distinctly different fibrillar pattern of microtubules which emanate from a perinuclear organizing center and extend to the cell margins. FSH treatment, within 15 min, causes a progressive shortening and apparent disruption of F-actin stress fibers (Fig. 6). By 30 min, stress fibers are no longer visible and the actin staining is localized

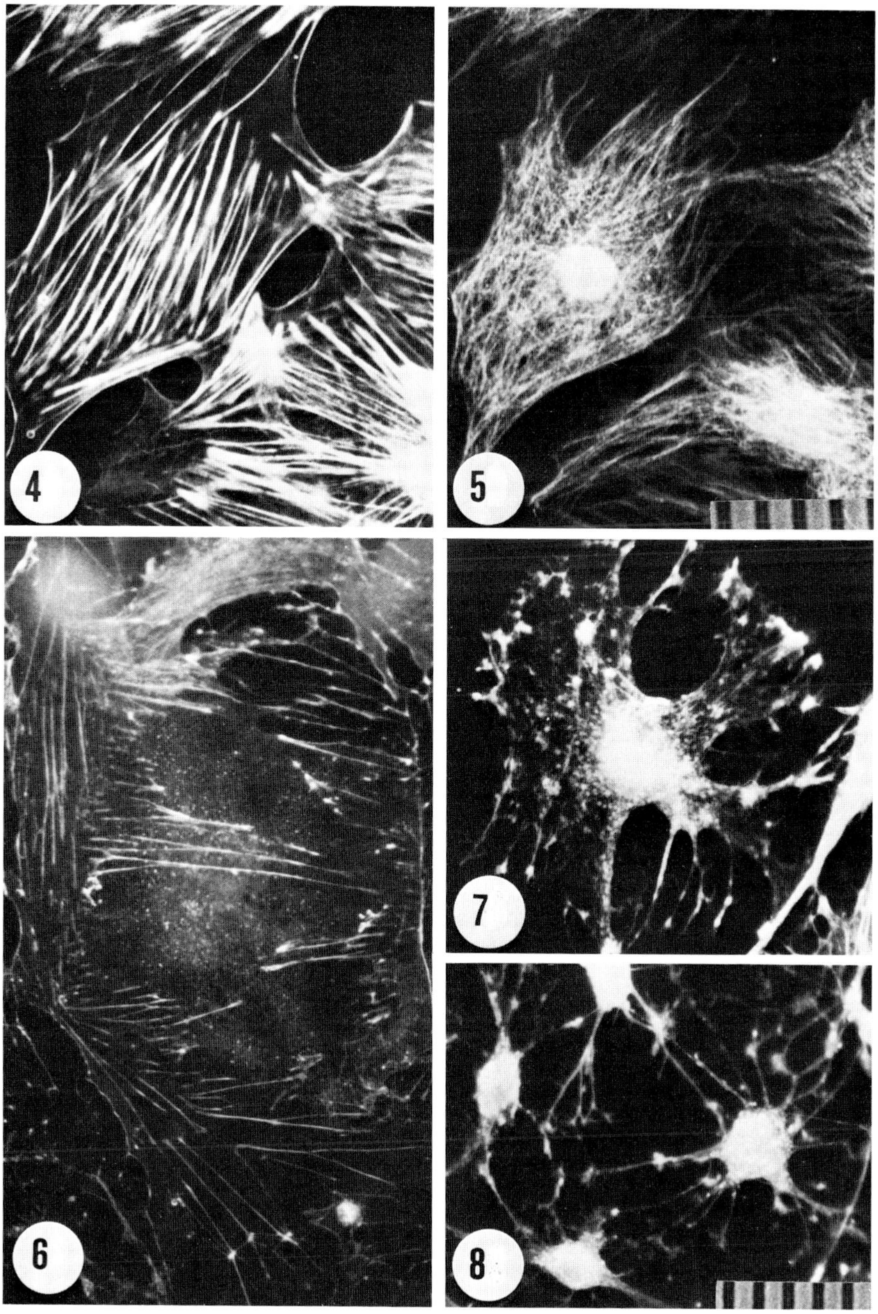

Figures 4–8. NBD-phallicidin staining patterns in 5-day granulosa cell cultures induced to round with 5 μg/ml FSH. Figures 4 and 5 show corresponding NBD-phallicidin and anti-tubulin staining of untreated control cultures in which F-actin is disposed as stress fibers and microtubules radiate to the cell periphery. By 15 min (Fig. 6), stress fibers appear shorter and are concentrated at the cell margins. Complete stress fiber disruption is apparent by 30 (Fig. 7) and 45 (Fig. 8) min after hormone treatment. Note the punctate staining near the cell center and along processes. Scale bars for Figs. 4 and 5 and 6–8 equal 10-μm divisions.

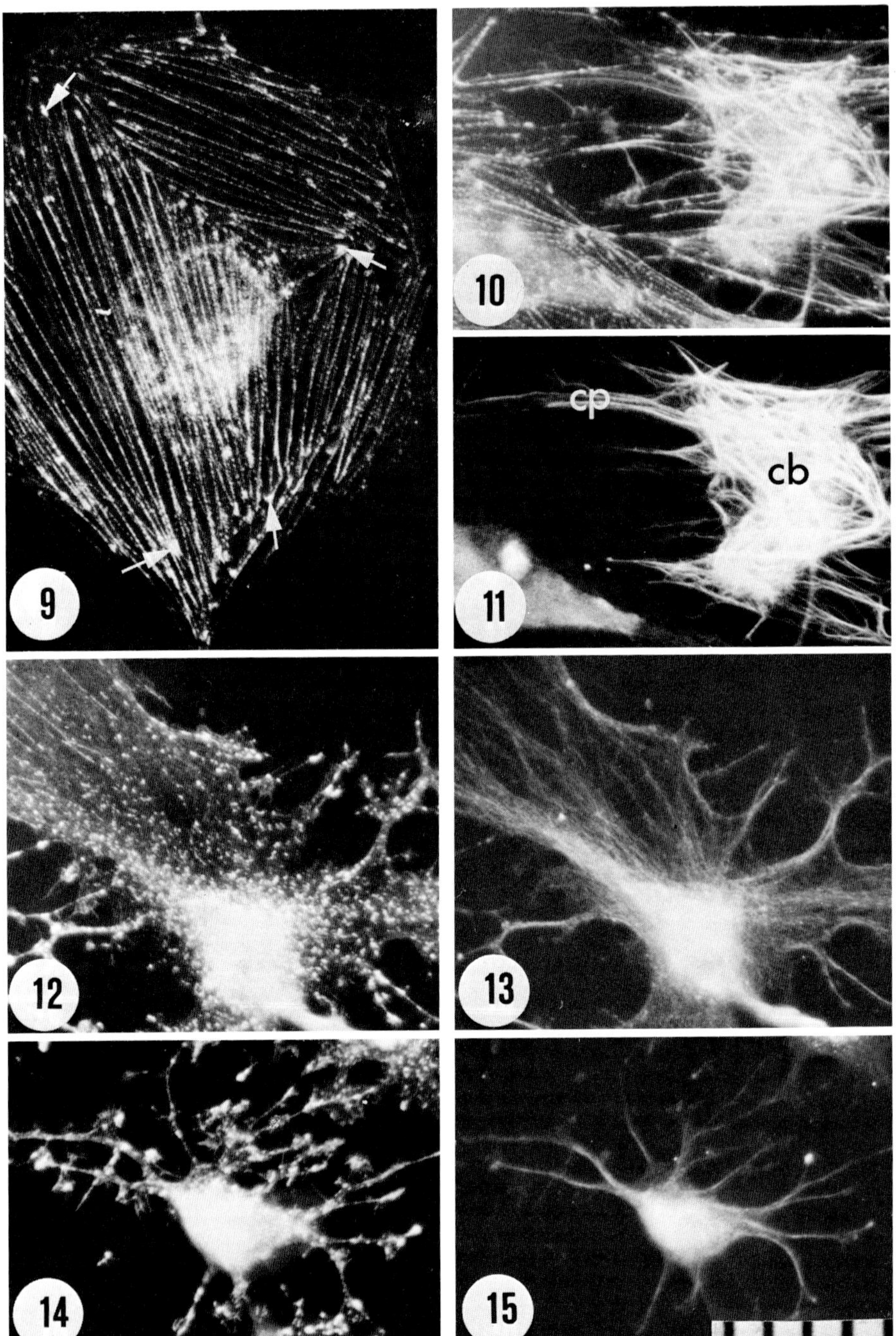

Figures 9–15. Anti α-actinin and anti-tubulin staining patterns of 5-day granulosa cell cultures induced to round with dibutyryl cyclic AMP (0.5 mM) and theophylline (1.0 mM). In control cultures (Fig. 9), α-actinin staining is periodic along the lengths of stress fibers and concentrated

at punctate spots near the cell center and at distal attachment sites where dendritic cellular processes are forming (Fig. 7). This pattern persists in fully rounded cells where most of the NBD-Ph staining is restricted to the cell body (Fig. 8).

Anti α-actinin staining of control granulosa cells is localized to periodic foci along stress fibers (Fig. 9), and as concentrated plaques at the ends of stress fibers which resemble adhesion plaques. This pattern is modified during DcAMP-induced rounding. Initially, continuous stress-fiber staining is observed in forming processes and as disarrayed fibrils near the cell center (Fig. 10). Double labeling with anti-tubulin indicates that this contrinuous fibrillar α-actinin staining coincides with the presence of microtubules located within processes and the cell body (Fig. 11). At later times, linear patterns of α-actinin staining disappear, resulting in the appearance of numerous foci or spots which are most abundant in the cell body but also remain irregularly localized along cellular processes (Figs. 12 and 14). Microtubules undergo a progressive aggregation into bundles which extend from the cell center into the dendritic processes (Figs. 13 and 15). Collectively, these observations indicate that hormone-induced rounding is attended by significant changes in the organization of actin, tubulin, and α-actinin. The consequences of this cytoskeletal reorganization on cell surface receptor dynamics are considered next.

3.3. Regulation of Receptor Topography

Treatment of control spread granulosa cells with F-Con A results in rapid internalization of ligand through coated pits and coated vesicles which give rise to ligand-containing endocytic vesicles within 20 min of F-Con A treatment. Endocytic vesicles subsequently migrate to the Golgi region of the cell (Fig. 16) by a process that previous work has shown requires intact microtubules (Herman and Albertini, 1982). Endocytic vesicles move along bundles of cytoplasmic microtubules in transit to the Golgi complex (Fig. 17). This pattern of ligand uptake and movement of endocytic vesicles is modified when granulosa cells are induced to round prior to F-Con A addition. As shown in Fig. 18, granulosa cells which round upon FSH exposure (see Fig. 7), are unable to endocytose F-Con A when added 60 min after hormone treatment. Both the cell body and dendritic processes bind F-Con A homogeneously. If such cells are subsequently treated with 1.0 μM colchicine, a treatment previously shown to cause flattening of the cell body and processes (Fig. 2), F-Con A redistributes to form large caps on the cell body (Fig. 19). The F-

---

at adhesion plaques (arrows). Within 15 min, periodic fibrillar patterns are disorganized (Fig. 10) and microtubules (Fig. 11) are aggregated in cell processes (cp) and the cell body (cb). By 30 min, α-actinin staining is punctate (Fig. 12) and prominent microtubule bundles (Fig. 13) extend into forming processes. In fully rounded cells, α-actinin exists at multiple foci in the cell body and processes (Fig. 14) and microtubules extend into cell processes (Fig. 15). Scale bar equals 10-μm divisions.

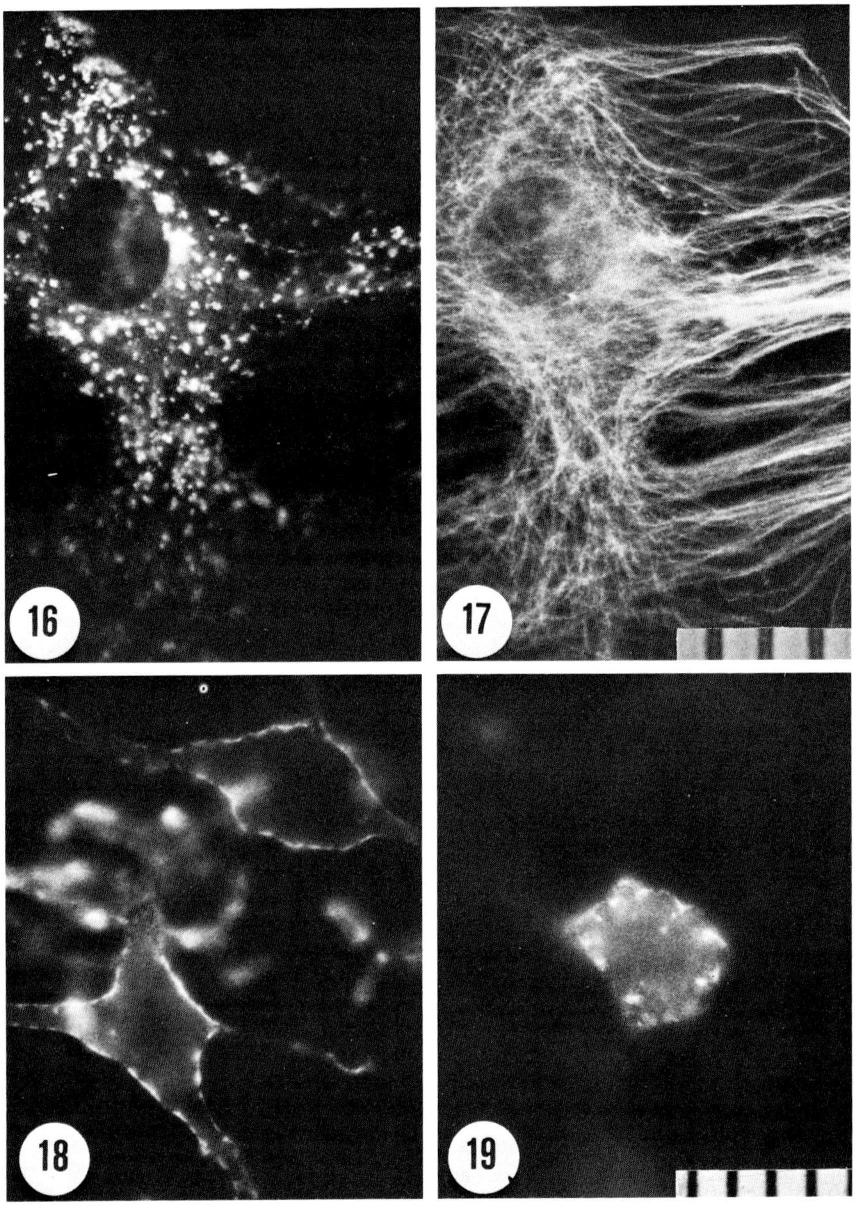

Figures 16–19. F-Con A staining patterns on control (Fig. 16) and FSH rounded (Figs. 18 and 19) 5-day granulosa cell cultures. F-Con A is rapidly internalized in vesicles which are transported to the Golgi area in control cells (Fig. 16). Note the alignment of endocytic vesicles along groups of microtubules (Fig. 17). F-Con A stains the cell bodies and processes on FSH rounded cells (Fig. 18). After colchicine treatment (Fig. 19), Con A receptors redistribute to the cell body forming large surface caps. Scale bars equals 10-μm divisions.

Table 1. Effects of Hormone and Drug Treatments on Con A Receptor Topography

Treatment	Shape	Receptor disposition (%)[a]		
		Endocytosed	Surface/random	Surface/capped
Control	Flat	79	4	17
FSH (μg/ml)	Round	7	83	10
DcAMP (1.0 mM)	Round	10	83	7
Adenosine (1.0 mM)	Round	15	76	9
Colchicine (1.0 μm)	Flat	12	5	83
Colcemid (1.0 μM)	Flat	2	7	91
Nocodazole (3.0 μM)	Flat	20	2	78
FSH nocodazole	Round	14	3	83
DcAMP nocodazole	Round	8	2	90
Adenosine nocodazole	Round	6	2	92

[a]Five-day granulosa cell cultures were treated with various drugs or hormones for 60 min, washed in EBSSH, and labeled with F-Con A for 10 min at 24°C. Cultures were incubated for 50 min at 37°C, fixed, and 150–200 cells per coverslip were scored for F-Con A staining patterns as either punctate (endocytosed; see Fig. 16), diffuse (random surface; see Fig. 18), or aggregated on the cell body (capped surface; see Fig. 19). Percentages of cells in each category are given and the overall morphological state within each group was recorded.

Con A is completely cleared from cellular processes and outlines surface blebs characteristically found on the cell body. Although receptor is able to cap after microtubule disruption, little or no endocytosis of F-Con A is observed.

The effects of various hormones and/or drugs on Con A receptor behavior is summarized in Table 1. The first effect observed was that hormone-induced rounding by FSH, DcAMP, or adenosine-inhibited ligand internalization (lines 1–4). Treatment of granulosa cells with microtubule-disrupting agents did not alter cellular morphology but did result in receptor capping (lines 5–7). When cells were first rounded with hormones and subsequently treated with Nocodazole, a potent microtubule disrupting agent, Con A receptors were cleared from cellular processes and formed caps on the cell body (Fig. 19; lines 8–10). Thus, the ability of granulosa cells to internalize lectin receptors is dependent on both microtubule organization and cellular shape. Rounded cells or cells exposed to antimicrotubule agents are unable to endocytose Con A receptors, however, receptor distribution seems to be regulated by microtubule organization alone since only in the absence of microtubules do lectin receptors cap.

3.4. Cytoskeletal Control of Endosome Movement

As noted previously (Figs. 16 and 17), endosomes induced to form by F-Con A treatment migrate to the cell center along bundles of microtubules. Transmission electron microscopy of Con A-horseradish peroxidase labeled granulosa cells shows that ligand-receptor clusters are first internalized in coated pits and coated vesicles. The coated vesicles pass through a microfilamentous cell cortex, lose their clathrin coats, and subsequently fuse with one

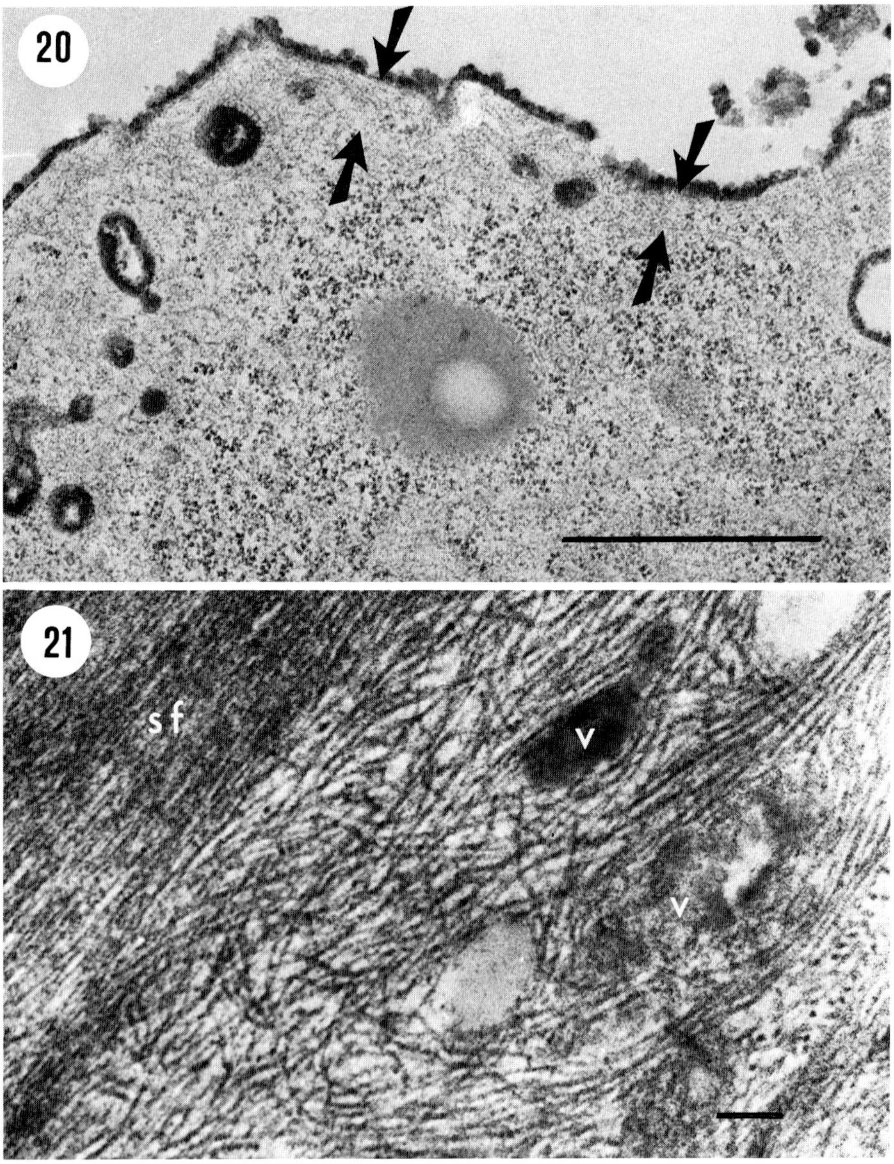

Figures 20 and 21. Electron micrographs of endocytic vesicles containing Con A-horseradish peroxidase. Five minutes after labeling (Fig. 20), Con A is located within coated vesicles in the microfilamentous cell cortex (arrows). Sixty minutes after binding (Fig. 21) ligand is found in vesicles (v) trapped within dense tangles of intermediate filaments. A portion of a stress fiber (sf) is shown on the left. Scale bar equals 1 μm in Fig. 20 and 100 nm in Fig. 21.

another to form endosomes (Fig. 20). Sixty minutes after ligand addition, virtually all endosomes are aggregated around the nucleus and are found within dense tangles of intermediate filaments (Fig. 21) which also become reorganized during endosome translocation (Herman and Albertini, 1982). These data suggest close spatial proximity between cytoskeletal filaments and endosomes. To further explore the possibility that microtubules participate in endosome translocations, we have examined endocytic vesicle migration in granulosa cells treated with nocodazole or taxol, two drugs which perturb microtubule organization by different mechanisms.

If granulosa cells are allowed to form endosomes containing F-Con A and are subsequently treated with 3.0 μM nocodazole for up to 30 min at 37°C, endosomes do not redistribute to the Golgi region as in untreated cells (Fig. 22). That the staining observed is truly intracellular is shown by inaccessibility of the F-Con A to rhodamine tagged rabbit anti-Con A IgG (Fig. 23). A similar inhibition of endosome movement is observed when granulosa cells are pretreated with 1.0 μM taxol for 4 hr before F-Con A labeling. Under these conditions, endosomes form but tend to aggregate in groups throughout the cell (Fig. 24). Since taxol is known to enhance microtubule assembly and bundle formation in cells (Schiff and Horwitz, 1980), it is interesting to note that the endosomes aggregate precisely at sites of microtubule bundle formation (Fig. 25). These data suggest that an intact microtubule system is necessary for the directed translocation of endosomes.

4. Discussion

4.1. Receptor Topography: Two Levels of Cytoskeletal Control

In this chapter, we have presented data indicating that the cytoskeleton can modulate receptor behavior at two levels. The first, and perhaps best studied, suggests that receptor topography in the plasma membrane is controlled by interactions between the cytoskeleton and cell surface receptors. Edelman (1976) first postulated that microtubules impart a constraining or "modulating" effect on receptor mobility. Since these original capping experiments were performed, other examples of microtubular control of cell surface activity have appeared. In differentiated granulosa cells, local rearrangements in microtubules were shown to specify sites of cap formation and endocytosis (Albertini and Clark, 1975). The polarization of the contractile machinery in cells that occurs during capping was shown to be coupled to the degree of microtubule assembly in leukocytes (Albertini *et al.*, 1977), a concept recently extended to phagocytic and dividing cells (Berlin *et al.*, 1978).

Anti-receptor antibodies which cross-link receptors for certain hormones have demonstrated the importance of receptor clustering for the initiation of a biological response (Kahn *et al.*, 1978). The translational movement of plasma membrane receptors which leads to clustering is influenced directly by actin filaments through a contractile mechanism and thus constitutes a deter-

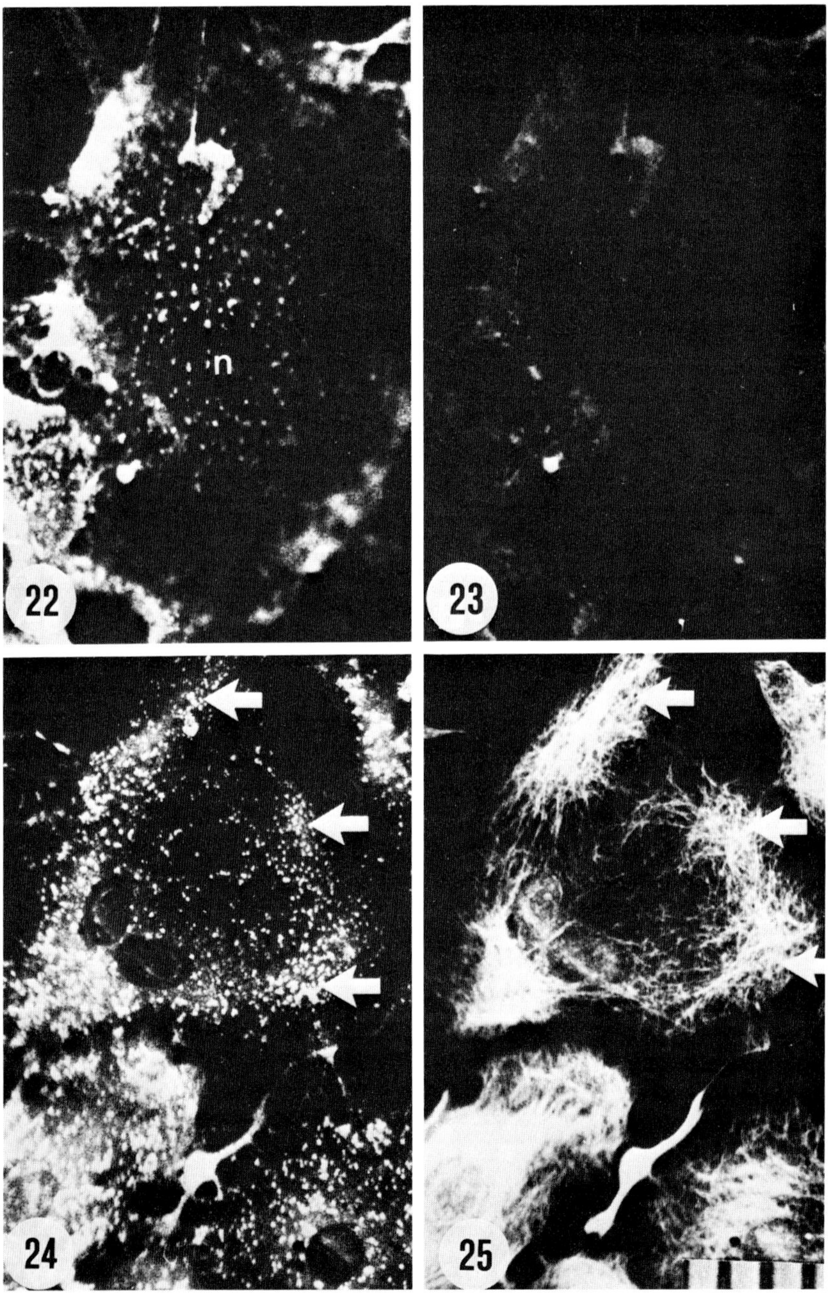

Figures 22–25. Drug treated granulosa cell cultures labeled with F-Con A. Endocytic vesicles are randomly distributed in a cell treated with 3.0 μM. Nocodazole 30 min after F-Con A labeling (Fig. 22). Anti-Con A antibody (Fig. 23) does not stain endocytic vesicles. In cultures treated with 1.0 μM taxol for 4 hr prior to F-Con A labeling, endocytic vesicles form but are unable to redistribute to the Golgi region of the cell (Fig. 24). Note the accumulation of endocytic vesicles at sites of taxol-induced microtubule bundle formation (arrows, Fig. 25). Scale bar equals 10-μm divisions.

minant of receptor topography (Albertini and Anderson, 1977; Bourguignon and Singer, 1977; Condeelis, 1979). More recently, the ubiquitous calmodulin regulatory system (Cheung, 1980) has been shown to participate in both capping and receptor mediated endocytosis (Salisbury *et al.*, 1981). While the molecular basis for microfilament control of receptor aggregation remains unclear, studies on the human erythrocyte membrane have identified a class of polypeptides which link the cytoskeleton to the plasma membrane (Branton *et al.*, 1981). The recent demonstrations of spectrin-like and ankyrin-like molecules in nonerythroid cells suggest that these molecules may generally subserve the function of cytoskeleton–membrane linkers (Levine and Willard, 1983; Bennett *et al.*, 1979; Burridge *et al.*, 1982).

Subsequent to ligand binding, and presumably receptor redistribution, many hormones cause an alteration in cellular morphology similar to the changes in cell shape and the cytoskeleton described in this study (Rapoport and Jones, 1978; Lawrence *et al.*, 1979; Westermark and Porter, 1982). The organization of actin is modified by hormone treatment resulting in the dismantling of the characteristic stress-fiber patterns found in cultured cells (Lawrence *et al.*, 1979; Tramontano *et al.*, 1982) possibly by a divalent cation mediated mechanism (Batten and Anderson, 1981). We have confirmed and extended these observations in granulosa cells by showing that α-actinin and tubulin, in addition to actin, acutely reorganize as a result of hormone stimulation. More pertinent, however, is the finding that under conditions of hormone-induced rounding, granulosa cells are unable to either form receptor clusters or endocytose ligand. The restricted receptor mobility observed in such cells is overcome by microtubule disruption further suggesting that microtubules, in a hormonally activated cell, directly modulate receptor behavior. Recently, Avivi *et al.* (1981) have shown that cAMP can directly modulate thyrotropin receptor topography in thyroid cells. Coupled with earlier observations showing that brain microtubule associated proteins (MAPs) are phosphorylated in a cyclic AMP-dependent manner (Sloboda *et al.*, 1975), these results suggest that microtubular regulation of receptor topography may occur through a cyclic nucleotide mechanism. This idea is further supported by the finding that microtubule depolymerizing agents potentiate the hormone-induced activation of adenylate cyclase in a variety of cells (Hagmann and Fishman, 1980; Rudolph *et al.*, 1977, 1979).

The second level of control exerted on receptor behavior pertains to the fate of ligand-receptor complexes subsequent to their endocytosis. Cells constitutively turn over their plasma membrane by recycling receptors (Steinman *et al.*, 1983). For certain ligands, such as LDL, this provides a possible mechanism for replenishing receptors to the cell surface after ligand is sorted within the cytoplasm (Goldstein *et al.*, 1979). For many of the ligands studied, it appears that coated vesicles mediate the delivery of ligands to lysosomes for their degradation (Pilch *et al.*, 1983; Steinman *et al.*, 1983). The delivery of ligand to lysosomes involves the nonrandom, directed translocation of endocytic vesicles to the Golgi region where lysosomal fusion presumably occurs (Pastan and Willingham, 1981). Since endosomes exhibit saltatory movement in transit to the Golgi, a form of movement exhibited by various intracellular particulates whose regulation seems to involve the cytoskeleton (Rebhun, 1972), it is important to evaluate cytoskeletal regulation of endosome translocations with regard to cytoplasmic organelle interactions.

4.2. *Regulation of Endosome–Lysosome Collisions*

For effective fusion to occur between the endosome and lysosomal compartment of a cell, two criteria must be met: (1) these distinct organelles must become physically approximated in the cytoplasm, and (2) this approximation must involve organelle movements. As noted earlier, endosomes and lysosomes undergo saltatory movement (Pastan and Willingham, 1981). In many systems, particle saltations appear to require microtubules since this type of movement is sensitive to drug-induced microtubule disruption (Rebhun, 1972). How particle saltations occur mechanistically remains unclear and is further complicated by recent views of cytoplasmic organization, since the complex architecture of cytoskeletal filaments may represent a formidable impediment to organelle movement (Wolosewick and Porter, 1979). Thus, mechanisms must exist for locally deforming the cytoskeleton or establishing low-resistance channels through which the rapid, unimpeded translocation of organelles can occur.

Data presented here and elsewhere on the movement of Con A endosomes in granulosa cells indicate that the cytoskeleton figures prominently in effecting endosomal translocation (Herman and Albertini, 1982). Both microtubules and intermediate filaments reorganize during endosome translocation such that microtubules form bundles through which endosomes migrate (see Figs. 16 and 17). The gradual collapse of intermediate filaments around endosomes (Fig. 21) appears to be a consequence of the lectin-induced rearrangement of microtubules and may partly provide a driving force for the centripetal aggregation of endosomes. This prospect receives experimental support from the observation that drug-induced intermediate filament collapse produces net endosome movement (Herman and Albertini, 1982). Linkages between microtubules, intermediate filaments, and endosomes may actively direct this translocation since structural connections of this type have been noted in, for example, axoplasm (Hirokawa, 1982). Additionally, an inherent degree of microtubular plasticity seems to be necessary for endosome movement since taxol or Nocodazole inhibit this directed translocation (Figs. 22–25). These data support the idea that cytoskeletal components participate in the directed translocation of endosomes. The observations that colchicine treatment impairs the degradation of certain ligands which are taken up by receptor-mediated endocytosis (see for example Brown *et al.*, 1980) and that taxol inhibits the mitogenic response of fibroblasts to thrombin and EGF (Crossin and Carney, 1981) bear on this point. Endosome movement is but one aspect of the complex traffic of membranes within cells that ensures homeostatic regulation of endocytic and exocytic processes. Future studies should focus on the influence of cytoplasmic organization on the targeting and movement of veiscles involved in membrane recycling.

5. Conclusions

It is apparent that binding of external signals such as hormones to membrane receptors mediate profound metabolic and structural effects on cells. We have shown that, in ovarian granulosa cells, the consequences of receptor occupancy and activation include acute changes in cytoskeletal organization which lead to modifications in cellular shape and plasma membrane function. The cytoskeleton is emerging as the substrate for the modulation of receptor behavior both at the cell surface and within the cytoplasm. Conditions which predictably deform the cytoskeleton must be more fully understood from a mechanistic and metabolic point of view if the biological importance of cytoskeletal–membrane interactions is to be elucidated.

ACKNOWLEDGMENTS. We would like to thank Drs. Keigi Fujiwara, Mary Porter, and Tom Pollard for providing the various antibodies used in these studies and Dr. John Douros of the National Cancer Institute for the taxol. Portions of this work were supported by NIH grant HD-11769 and a Basil O'Connor Starter Grant from the National Foundation March of Dimes. We thank Ms. Liz Dreesen for her help in the preparation of the manuscript.

References

Albertini, D. F., and Anderson, E., 1977, Microtubule and microfilament rearrangements during capping of concanavalin A receptors on cultured ovarian granulosa cells, *J. Cell Biol.* **73:**111.

Albertini, D. F., and Clark, J. I., 1975, Membrane-microtubule interactions: Concanavalin A capping induced redistribution of cytoplasmic microtubules and colchicine binding proteins, *Proc. Natl. Acad. Sci. USA* **72:**4976.

Albertini, D. F., and Kravit, N. G., 1981, Isolation and biochemical characterization of the ten-nanometer filaments from ovarian granulosa cells, *J. Biol. Chem.* **256:**2484.

Albertini, D. F., Oliver, J. M., and Berlin, R. D. 1977, The mechanism of concanavalin A cap formation is leukocytes, *J. Cell Sci.* **26:**57.

Avivi, A., Tramontano, D., Ambesi-Impionabata, F. S., and Schlessinger, J., 1981, Adenosine 3′:5′-monophosphate modulates thyrotropin receptor clustering and thyrotropin activity in culture, *Science* **214:**1237.

Barak, L. S., Yocum, R. R., Nothnagel, E. A., and Webb, W. W., 1980, Fluorescence staining of the actin cytoskeleton in living cells with 7-nitrobenz-2-oxa-1,3-diazole phallicidin, *Proc. Natl. Acad. Sci. USA* **77:**980.

Batten, B. E., and Anderson, E., 1981, The effects of Ca and Mg deprivation of cell shape in cultured ovarian granulosa cells, *Am. J. Anat.* **161:**101.

Bennet, V., and Davis, J., 1981, Erythrocyte ankyrin: immunoreactive analogues are associated with mitotic structures in cultured cells and with microtubules in brain, *Proc. Natl. Acad. Sci. USA* **78:**7550.

Berlin, R. D., Oliver, J. M., and Walter, R. J., 1978, Surface functions during mitosis I: Phagocytosis, pinocytosis and mobility of surface-bound Con A, *Cell* **15:**327.

Bourguignon, L. W., and Singer, S. J., 1977, Transmembrane interactions and the mechanism of capping of surface receptors by their specific ligands, *Proc. Natl. Acad. Sci. USA* **74:**5031.

Branton, D., Cohen, C. M., and Tyler, J., 1981, Interaction of cytoskeletal proteins on the erythrocyte membrane, *Cell* **24:**24.

Brown, K. D., Friedkin, M., and Rozengurt, E., 1980, Colchicine inhibits epidermal growth factor degradation in 3T3 cells, *Proc. Natl. Acad. Sci. USA* **77:**480.

Burridge, K., Kelly, T., and Mangeat, P., 1982, Nonerythrocyte spectrins: actin-membrane attachment proteins occurring in many cell types, *J. Cell Biol.* **95:**478.

Cheung, W. Y., 1980, Calmodulin plays a pivotal role in cellular regulation, *Science* **207:**19.

Chung, M. P., and Batten, B. E., 1982, The effect of cytoskeletal perturbants on progesterone secretion by rat ovarian granulosa cells, *J. Cell Biol.* **95:**204a.

Condeelis, J., 1979, Isolation of concanavalin A caps during various stages of formation and their association with actin and myosin, *J. Cell Biol.* **80:**751.

Crossin, K. L., and Carney, D. H., 1981, Microtubule stabilization by taxol inhibits initiation of DNA synthesis by thrombin and by epidermal growth factor, *Cell* **27:**341.

Edelman, G. M., 1976, Surface modulation in cell recognition and cell growth, *Science* **192:**218.

Folkman, J., and Moscona, A., 1978, Role of cell shape in growth control, *Nature* **273:**345.

Fujiwara, K., Porter, M. E., and Porter, T. D., 1978, Alpha-actinin localization in the cleavage furrow during cytokinesis, *J. Cell Biol.* **79:**268.

Fujiwara, R., and Pollard, T. D., 1978, Simultaneous localization of myosin and tubulin in human tissue culture cells by double antibody staining, *J. Cell Biol.* **77:**182.

Goldstein, J. L., Anderson, R. G. W., and Brown, M. S., 1979, Coated pits, coated vesicles, and receptor-mediated endocytosis, *Nature* **279:**679.

Hagmann, J., and Fishman, P. H., 1980, Modulation of adenylate cyclase in intact macrophages by microtubules, *J. Biol. Chem.* **255:**2659.

Herman, B., and Albertini, D. F., 1982, The intracellular movement of endocytic vesicles in cultured granulosa cells, *Cell Motil.* **2,** in press.

Herman, B., and Fernandez, S. M., 1982, Dynamics and topographical distribution of concanavalin A receptors during myogenic cell fusion *in vitro, Biochemstry* **21:**3275.

Hirokawa, N., 1982, Cross-linker system between neurofilaments, microtubules, and membranous organelles in frog axons revealed by the quick-freeze, deep-etching method, *J. Cell Biol.* **94:**129.

Kahn, C. R., Baird, K. L., Harrett, D. B., and Flier, J. S., 1978, Direct demonstration that receptor cross-linking or aggregation is important in insulin action, *Proc. Natl. Acad. Sci. USA* **75:**4209.

Lawrence, J. S., Ginzberg, R. D., Gilula, N. B., and Beers, W. H., 1979, Hormonally induced cell shape changes in cultured rat ovarian granulosa cells, *J. Cell Biol.* **80:**21.

Levine, J., and Willard, M., 1983, Redistribution of fodrin (a component of the cortical cytoplasm accompanying capping of cell surface molecules, *Proc. Natl. Acad. Sci. USA* **77:**1561.

Oliver, J., and Berlin, R. D., 1982, Distribution of receptors and functions on cell surfaces: Quantitation of ligand-receptor mobility and a new model for control of plasma membrane topography, *Phil. Trans. R. Soc. London* **B299:**215.

Pastan, I. H., and Willingham, M. C., 1981, Journey to the center of the cell: Role of the receptosome, *Science* **214:**504.

Phaire-Washington, L., Silverstein, S. C., and Wang, E., 1980, Phorbol myristate acetate stimulates microtubule and 10-nm filament extension and lysosome redistribution in mouse macrophages, *J. Cell Biol.* **86:**641.

Pilch, P. F., Shia, M. A., Benson, R. J., and Fine, R. E., 1983, Coated vesicles participate in the receptor-mediated endocytosis of insulin, *J. Cell Biol.* **96:**133.

Pollard, T. D., and Korn, E. D., 1973, Electron microscopic identification of actin associated with isolated plasma membranes, *J. Biol. Chem.* **248:**448.

Rapoport, B., and Jones, A. L., 1978, Acute effects of thyroid-stimulating hormone on cultured thyroid cell morphology, *Endocrinology* **102:**175.

Rebhun, L. I., 1972, Polarized intracellular particle transport: Saltatory movements and cytoplasmic streaming, *Int. Rev. Cytol.* **32**:93.

Rudolph, S. A., Greengard, P., and Malawista, S. E., 1977, Effects of colchicine on cyclic AMP levels in human leukocytes, *Proc. Natl. Acad. Sci. USA* **74**:3404.

Rudolph, S. A., Hegstrand, L. R., Greengard, P., and Malawista, S. E., 1979, The interaction of colchicine with hormone-sensitive adenylate cyclase in human leukocytes, *Mol. Pharmacol.* **16**:805.

Salisbury, J. L., Condeelis, J. S., Maihle, N. J., and Satir, P., 1981, Calmodulin localization during capping and receptor-mediated endocytosis, *Nature* **294**:163.

Sloboda, R. D., Rudolph, S. A., Rosenbaum, J. L., and Greengard, P., 1975, Cyclic AMP-dependent endogenous phosphorylation of a microtubule-associated protein, *Proc. Natl. Acad. Sci. USA* **72**:177.

Solomon, F., 1981, Specification of cell morphology by endogenous determinants, *J. Cell Biol.* **90**:547.

Steinman, R. M., Mellman, I. S., Muller, W. A., and Cohn, Z. A., 1983, Endocytosis and the recycling of plasma membrane, *J. Cell Biol.* **96**:1.

Schiff, P. B., and Horwitz, S. B., 1980, Taxol stabilizes microtubules in mouse fibroblast cells, *Proc. Natl. Acad. Sci. USA* **77**:1561.

Tramontano, D., Avivi, A., Ambesi-Impiombato, F. S., Barak, L., Geiger, B., and Schlessinger, J., 1982, Thyrotropin induced changes in the morphology and the organization of microfilament structures in cultured thyroid cells, *Exp. Cell Res.* **137**:269.

Vasiliev, J. M., and Gelfand, I. M., 1976, Effects of Colcemid on morphogenetic processes and locomotion of fibroblasts, in: *Cell Motility*, Vol. I (R. Goldman, T. Pollard, and J. Rosenbaum, eds.), p. 279, Cold Spring Harbor Laboratory, New York.

Westermark, B., and Porter, K. R., 1982, Hormonally induced changes in the cytoskeleton of human thyroid cells in culture, *J. Cell Biol.* **94**:42.

Wolosewick, J. J., and Porter, K. R., 1979, Microtrabecular lattice of the cytoplasmic ground substance. Artifact or reality, *J. Cell Biol.* **82**:114.

7

The Biochemistry of Microtubules

A REVIEW

Timothy W. McKeithan and Joel L. Rosenbaum

1. Introduction

The whole of the complex field of microtubule structure and function is far too broad to cover in this brief review. The intention has been to give a brief overview of the field and to place emphasis on a few areas of recent significant progress. In particular, the related topics of microtubule "treadmilling," polarity of assembly, and role of nucleotide hydrolysis have been addressed at some length, as has the topic of tubulin heterogeneity. For a more detailed discussion, several recent books can be recommended. *Microtubules* (Roberts and Hyams, eds., 1979) contains excellent reviews on the major areas of microtubule structure, chemistry, mechanisms of assembly, and functions. *Microtubules and Microtubule Inhibitors* (De Brabander and De Mey, eds., 1980) contains briefer articles on many topics of current interest. In *Microtubules* (1978), P. Dustin has authored a remarkably comprehensive review which is especially strong in providing historical perspective, to which this present review is heavily indebted. Also to be recommended is the more recent book *Biological Functions of Microtubules and Related Structures* (Sakai *et al.*, 1982).

2. Microtubule Structure

Microtubules are tubular organelles of indefinite length with an approximate outer diameter of 30 nm composed principally of the protein tubulin, a

Timothy W. McKeithan and ***Joel L. Rosenbaum*** • Department of Biology, Yale University, New Haven, Connecticut 06511. Present address for Dr. McKeithan: Department of Pathology, University of Chicago, Chicago, Illinois 60637.

dimeric molecule with distinct α- and β-subunits, each of which has a molecular weight of approximately 50,000. Microtubules are almost universally present in eukaryotes, most often as single microtubules, but also as doublet microtubules in cilia and flagella and as triplets in centrioles and basal bodies.

Since the diameter of the microtubule is far below the resolving power of the light microscope, a real appreciation of its structure necessarily awaited the development of the electron microscope. Nevertheless, light microscopists observed a number of microtubule organelles and in several cases noted a fine fibrillar ultrastructure, presumably the individual microtubules. Basal bodies, centrioles, and the mitotic apparatus have been observed for over a century. As early as the end of the nineteenth century, neurofibrils were observed and described as a "plexus of fine filaments" (Dustin, 1978). The simple internal structure of the nucleated red cells of nonmammalian vertebrates made these cells a favorable system for detection of microtubules. In these cells, a *marginal band,* now known to be a bundle of microtubules, was noted in 1875 (Ranvier, 1875) and later shown to be formed of tiny parallel fibrils (Jolly, 1920, 1923).

Due to their stability and prominent location, flagella and cilia were the first microtubule organelles observed. As early as 1888, Ballowitz, (Ballowitz, 1888) detected a substructure in the flagellum. Maceration of finch sperm flagella in glycerol revealed as many as 11 constituent fibrils by light microscopy.

These 11 fibrils composing cilia and flagella were visualized in the 1940's as unbranched structures in the electron microscope using whole mount specimens (Grigg and Hodge, 1949, Jakus and Hall, 1946; Schmitt *et al.*, 1943). Manton and Clark described the tubular nature of these fibrils in 1952 and suggested that nine were arranged about a central pair (Manton and Clarke, 1952). Fawcett and Porter directly visualized the 9 + 2 arrangement using thin-section electron microscopy (EM) of ciliated epithelia (Fawcett and Porter, 1954). The term "microtubule" was first proposed in 1961 (Slautterback, 1961, 1963), and was quickly adopted generally. Intracellular microtubules were routinely observed in the cytoplasm only after glutaraldehyde fixation came into general use in 1963 (Sabatini *et al.*, 1963). Earlier workers, however, had noted cytoplasmic microtubules using osmium fixation (Harris, 1962; Roth and Daniels, 1962). Using glutaraldehyde as a prefixative, Ledbetter and Porter observed microtubules in the periphery of plant cells and noted their relationship to the filaments of cilia and flagella (Ledbetter and Porter, 1963). Andre and Thiery (1963) and Pease (1963) detected 4 nm-thick longitudinal protofilaments composing the flagellar microtubules of sperm. Protofilaments were soon also noted in microtubules from salamander blood cells (Gall, 1965) and fibroblast mitotic spindles (Barnicot, 1966). Using rotational enhancement of thin sections of microtubules in juniper root tips, Ledbetter and Porter (1964) showed the cytoplasmic microtubules to contain 13 protofilaments. Tilney *et al.* (1973) showed that cytoplasmic microtubules from a number of sources contain 13 protofilaments. The A-subfiber of outer doublet microtubules of flagella was also shown to contain 13 protofilaments and the B-subfiber to contain ten or 11 protofilaments.

Although most microtubules *in vivo* are composed of 13 protofilaments, a few examples are known of microtubules with different numbers of protofilaments (Burton and Hinkley, 1974; Burton *et al.*, 1975; Nagano and Suzuki, 1975). One of the best studied examples is in the nematode *Caenorhabditis elegans*, in which most cells contain microtubules with 11 protofilaments, but the microtubules of the six touch receptors have 15 protofilaments (Chalfie and Thomson, 1982). In this organism, only the A-subfiber of cilia contains the standard 13 protofilaments. The two classes of microtubules with 11 and 15 protofilaments respond very differently to microtubule-disrupting drugs and are separately affected by various mutations.

Although most microtubules *in vivo* have 13 protofilaments, 14 is the most common number of protofilaments in microtubules which have undergone several cycles of assembly and disassembly *in vitro* (Pierson *et al.*, 1978). The additional protofilament appears to cause a slight tilt in the protofilaments, giving the microtubule a shallow helical pitch of 2.7 μm (Langford, 1980).

Intraprotofilament bonds appear to be stronger than lateral bonds between tubulin subunits, as evidenced by the fact the microtubules may splay out in negative stain into protofilaments. In addition, most aberrant tubulin structures are composed of protofilaments although the arrangement of protofilaments in these structures varies widely (Amos, 1979).

A finer substructure for the microtubule was detected by optical diffraction of negatively stained microtubules. Using this method, Grimstone and Klug (1966) found the longitudinal protofilament to contain a basic subunit 4 nm in diameter. The additional presence of an 8 nm repeat in the diffraction pattern suggested that two morphologically distinct subunits alternated along the length of the protofilament. They suggested that these could correspond to the two monomeric subunits of the tubulin heterodimer. Diffraction studies show that in most microtubules tubulin monomers are arranged in a three-start left-handed helix (Amos and Klug, 1974; Erickson, 1974; Linck and Amos, 1974).

Recent studies using preparations freeze dried, rotary replicated with platinum, and viewed in the transmission electron microscope have also showed a 4-nm periodicity of subunits (Heuser and Kirschner, 1980). This method allows the inner and outer surfaces to be visualized independently. On the outer surface, the longitudinal arrangement into protofilaments is most apparent, but in regions where a microtubule has been broken open to reveal the inner surface, this surface shows clearly the three-start helix.

Using zinc-induced sheets, protofilament structure has been studied with data to 1.5 nm (Baker and Amos, 1978; Amos and Baker, 1979). The α- and β-subunits were shown to be oriented in the same direction and to have an axis pointing at 40–50° from the axis of the protofilament. One side of the protofilament, apparently corresponding to the outer surface in microtubules, extends out as a longitudinal ridge while the other side is rather flat, with grooves imparting a "ladder-like" appearance. The three-dimensional structure predicted for microtubules, based on the structure of protofilaments found in zinc-induced sheets, corresponds well to the appearances of inner and outer surfaces of freeze-dried microtubules seen by Heuser and Kirschner (1980).

Although 24 nm is most often quoted as the diameter of thin-sectioned microtubules, X-ray diffraction studies of oriented hydrated microtubules (Mandelkow *et al.*, 1977) show an inner and an outer diameter of 14 nm and 30 nm. Interprotofilament bonds appear to be greatest at a radius of about 10 nm, somewhat closer to the center of the microtubule than the average radius, with deep grooves between protofilaments on the outside of the wall.

While tubulin monomers are arranged in a three-start left-handed helix, tubulin *dimers* appear to be arranged in a five-start right-handed helix in flagellar axonemal A-subfibers and apparently in brain microtubules assembled *in vitro* with a full complement of accessory proteins. In this lattice α- and β-subunits have monomers of the opposite type as their closest lateral neighbors. In the B-subfiber, in contrast, each monomer has subunits of the same type as closest lateral neighbors (Amos and Klug, 1974; Linck and Amos, 1974). As already noted, microtubules assembled *in vitro* frequently have protofilament numbers other than 13, and in many conditions, microtubules with 14 protofilaments are most common. Neither the A nor the B lattice is compatible with a three-start helix in a microtubule with an even number of protofilaments. There is evidence for a mixed lattice in microtubules *in vitro*, with the B type predominating (McEwen and Edelstein, 1980). There is also evidence that one of the axonemal central pair microtubules has a mixed lattice, and it has been suggested that complex lattices may provide specific sites for attachment of microtubule-associated proteins (Linck and Langevin, 1981).

Since the microtubule lattice, once established, will presumably be maintained throughout the length of the microtubule as it polymerizes, one may speculate that microtubule nucleating sites may specify a specific lattice type for individual microtubules, and this lattice type may affect the functional properties of the microtubule. Microtubule nuclei have in fact been shown to impose their lattice on the microtubules which they nucleate (Sheele *et al.*, 1982).

3. Chemistry of Microtubule Proteins

Colchicine is now only one of many drugs known to bind to tubulin and to block its assembly. However, colchicine has been of crucial historical importance in the understanding of tubulin. As the active principle of the plant *Colchicum automnale*, colchicine has been known since antiquity as a poison and has been used specifically in the treatment of gout since the eighteenth century (Rodnan and Benedek, 1970). As early as 1889, effects on mitosis were noted (Pernice, 1889). However, only in the 1930's was it understood that accumulation of mitotic figures in cells treated with colchicine was due to premetaphase arrest (Bucher, 1939; Ludford, 1936). In 1952, Inoue showed that colchicine caused a reversible loss of birefringence in mitotic spindles and suggested that the drug bound to the subunits of the spindle fibers (Inoue, 1952). Using tritiated colchicine, Taylor (1965) demonstrated colchicine binding to tissue culture cells and suggested that microtubules might be the site of binding. Further work (Borisy and Taylor, 1967a,b) showed that in homogenates, colchicine-binding protein had a sedimentation constant of 6 s. By DEAE chromatography and ammonium sulfate fractionation, Weisenberg *et al.* (1968) purified the colchicine-binding protein and showed that it had a molecular weight of approximately 120,000, bound one mole colchicine, and contained two moles GTP per mole protein. Nearly identical physical and chemical properties were noted in a protein from flagellar central pair microtubules (Shelanski and Taylor, 1967, 1968). From these findings it was concluded that the colchicine-binding protein was tubulin, the subunit of microtubules.

Early electrophoretic analysis of outer doublet tubulin revealed two closely migrating but distinct subunits (Olmsted *et al.*, 1971). Subsequent studies on brain tubulin indicated that microtubules were heteropolymers of two distinct subunits (Bryan and Wilson, 1971). These were named alpha and beta, alpha being the slower migrating polypeptide in most polyacrylamide gel systems. The subunits were almost always found in equal quantities, which suggested that the tubulin dimer contained one α- and one β-subunit. The heterodimeric nature of tubulin has been confirmed by chemical cross-linking of tubulin in solution (Luduena *et al.*, 1977). At low tubulin concentrations, it is possible to detect a rapidly reversible dissociation of the tubulin dimer into its component monomers (Detrich and Williams, 1978). At 4.6°C the dissociation constant is 7.4×10^{-7} M for bovine brain tubulin; colchicine decreases the dissociation constant (Detrich *et al.*, 1982).

In several cases there is evidence that two forms of α- and/or β-tubulin are present in microtubules in approximately equal quantities. For example, Bibring *et al.* (1976) found two forms of α-tubulin in sperm flagella in equimolar quantities. Stephens (1982) used a variety of electrophoretic conditions, including Triton-acid-urea gels and isoelectric focusing in the presence of SDS, to study tubulin heterogeneity. He found, using tubulin from a number of sources, that α- and β-tubulin each split into a pair of equal bands. Similar splitting has been reported by other authors (Berkowitz *et al.*, 1977; Kobayashi, 1982). These results have led to the suggestion that microtubules are assembled from tubulin dimers of two types present in equimolar quantities.

A major advance in the understanding of microtubules occurred in 1972 when Weisenberg experimented with buffer components to find conditions in which microtubule formation from homogenized brain tissue would occur *in vitro* (Weisenberg, 1972). Microtubule assembly occurred at 35°C in the presence of GTP, magnesium, and the calcium chelator EGTA. The microtubules reversibly depolymerized upon cooling. This reversible polymerization was soon used to purify tubulin from crude brain extracts by cycles of temperature-dependent assembly and disassembly (Borisy and Olmsted, 1972).

Early studies of the composition of *in vitro* polymerized microtubules revealed the presence of high molecular weight proteins in addition to tubulin (Dentler *et al.*, 1974; Murphy and Borisy, 1975). These proteins had previously been observed in microtubules isolated intact from brain (Kirkpatrick *et al.*, 1970). Since a constant stoichiometry between tubulin and these high molecular weight proteins was found to be maintained through up to five cycles of *in vitro* polymerization, the term "microtubule-associated proteins" or "MAPs" was coined (Sloboda *et al.*, 1975). Molecular weights of these proteins are about 345,000 (MAP_1) and 271,000 and 286,000 (MAP_2) (Vallee and Borisy, 1977). The proteins were soon shown to greatly reduce the critical concentration and increase the rate of assembly when added to pure 6 s tubulin (Murphy and Borisy, 1975; Sloboda *et al.*, 1976). The decrease in critical concentration was later shown to be due to a reduction in the backward rate constant (Murphy *et al.*, 1977) and MAPs were shown to stabilize microtubules (Sloboda and Rosenbaum, 1977). The proteins apparently are responsible for the filamentous projections seen in microtubules polymerized *in vitro* (Dentler *et al.*, 1975; Murphy and Borisy, 1975) and thus presumably for the same projections seen *in vivo* (Burton and Fernandez, 1973; Smith, 1971; Wuerker and Kirkpatrick, 1972; Wuerker and Palay, 1969; Yamada *et al.*, 1971). The high molecular weight MAPs are arranged periodically along the microtubule in a super-lattice (Amos, 1977), and at saturation bind with an axial periodicity of 32 nm (Kim *et al.*, 1979).

Other microtubule-associated proteins of molecular weight 55,000–70,000, collectively called tau, have also been shown to stimulate the rate and extent of assembly (Lockwood, 1978; Penningroth *et al.*, 1976; Weingarten *et al.*, 1975). Although some have argued that tau is an artifactual proteolytic fragment of high molecular weight MAPs (Sloboda *et al.*, 1976), antibodies to tau and to the high molecular weight MAPs do not cross-react (Connolly *et al.*, 1978). In addition, there appears to be some tissue specificity in the presence of these proteins, for tau is found by immunofluorescence to be a component of microtubules in fibroblasts but not in glial cells, whereas high molecular weight MAPs are detected in glial cells (Connelly *et al.*, 1977, 1978). In neurons, MAP_2 appears to be restricted to the dendritic processes (Matus *et al.*, 1981). Only small quantities of MAP_2 could be detected in HeLa cells by monoclonal antibodies (Weatherbee *et al.*, 1982); a 200,000 molecular weight protein is the principal MAP in these cells (Bulinski and Borisy, 1979). Under certain conditions, MAP_2 and tau promote the formation of different forms of tubulin polymers (Sandoval and Weber, 1981).

Purified tubulin, free of accessory proteins, can polymerize by itself under certain nonphysiological conditions. These include very high tubulin concentration (Dentler *et al.*, 1975), glycerol and high magnesium concentration (Frigon and Timasheff, 1975a,b), high magnesium alone (Herzog and Weber, 1977), and the presence of dimethyl sulfoxide (DMSO) (Himes *et al.*, 1977), of polycations (Erickson and Voter, 1976), and of dextran or polyethylene glycol (Herzog and Weber, 1978).

Early studies (Borisy and Olmsted, 1972) of products of depolymerized microtubules revealed, in addition to 6 s dimers, larger aggregates with sedimentation constants of 20 to 36 s, depending on the conditions (Marcum and Borisy, 1978). The aggregates had the form of rings by negative staining electron microscopy. Depleting microtubule protein of rings was found to delay polymerization; therefore, a role for rings in nucleation was proposed.

MAP proteins promote the formation of rings under depolymerizing conditions (for review see Sheele and Borisy, 1979), and rings contain a higher ratio of high molecular weight MAPs to tubulin (1:5–1:6) than do microtubules (apparent maximum 1:12; Amos, 1977). However, with high magnesium concentrations, tubulin can form rings in the absence of accessory proteins (Frigon *et al.*, 1974). Several models have been proposed for the structure of rings (Sheele and Borisy, 1979). Recent evidence, based on X-ray diffraction analysis, strongly suggests that rings consist of coiled protofilaments (Mandelkow *et al.*, 1983).

4. Mechanism of Tubulin Polymerization

Studying the kinetics of assembly, Johnson and Borisy (1977) showed that when microtubule protein is rapidly warmed to allow polymerization, a pseudo-first-order increase in light scattering follows a lag period in which microtubule nucleation is rate limiting. When microtubule seeds are added, the lag phase is eliminated. The initial rate of polymerization is directly proportional to both subunit concentration and the number concentration of microtubule ends, while the final amount of polymer depends only on the initial subunit concentration. Their data were consistent with a simple nucleated condensation reaction (Oosawa and Asakura, 1975) as a mechanism for *in vitro* assembly. In systems in which this mechanism applies, monomer addition to the polymer (the forward reaction) proceeds at a rate proportional both to free monomer concentration and to polymer number concentration. Monomer loss from the polymer (the backward reaction) is proportional only to polymer number concentration. If the number of nuclei is fixed and small relative to the concentration of monomers and if the concentration of subunits is large enough so that some polymer is present, then steady state occurs at a free monomer concentration called the *critical concentration,* which is independent of initial monomer concentration.

Microtubules polymerized in the absence of MAPs show GTPase activity which correlates with the number of tubulin ends as does the initial rate of depolymerization upon cooling (David-Pfeuty *et al.*, 1978). These data are also consistent with the simple nucleated condensation mechanism.

At very high concentrations of nuclei, the rate of polymerization is no longer proportional to the number of nuclei but approaches a limiting value (Carlier, 1983). These data have been interpreted to imply that a conformational change of tubulin proceed polymerization, and this conformational change is rate limiting at high rates of polymerization.

The experiments of Johnson and Borisy (1977) used microtubule protein depleted of rings. The fact that the initial rate of elongation was directly proportional to the concentration of tubulin dimer implies that tubulin dimer directly adds on to the assembling microtubule without an obligatory oligomeric intermediate.

Under the most commonly used conditions for *in vitro* assembly, ring oligomers appear to have a major role in microtubule nucleation. Although many aspects of the functions of rings and other oligomers remain controversial, a model which explains many of the data is as follows (Sloboda *et al.*, 1976; Kirschner *et al.*, 1975). Early in the polymerization process, rings unroll to form protofilaments which align side-by-side to form sheets. These roll up into short microtubule pieces, which then elongate by addition of dimers. The remaining rings break down into dimers which add on individually (Penningroth *et al.*, 1976; Sloboda *et al.*, 1976; Weisenberg, 1974).

Rings apparently are not an absolute requirement for microtubule nucleation. Bryan (1976) used polyadenylic acid to sequester MAPs. Nucleation of such poly-A inhibited microtubule protein occurred only with the addition of exogenous MAPs, but rings did not form prior to microtubule assembly. On the other hand, under polymerizing conditions (22°C) in which rings break down only slowly, Pantaloni *et al.* (1981) found, using radioactively labeled dimer mixed with unlabeled rings, that in the first 2 min of assembly when 30% of microtubule protein had polymerized, 90% of the incorporated tubulin was derived from rings without exchange with tubulin dimers. Under these conditions, therefore, direct incorporation of rings into microtubules is a significant contribution to net microtubule assembly. These observations suggest that several different pathways to microtubule assembly may exist under different conditions.

5. Polarity of Assembly and Microtubule Treadmilling

Polarity of assembly was early shown to be a property of microtubules (Allen and Borisy, 1974; Binder and Rosenbaum, 1973; Dentler *et al.*, 1974). Microtubules growing off flagellar axonemes and basal bodies grew much more rapidly from the distal end, the end corresponding to the *in vivo* site of assembly (Rosenbaum and Child, 1967; Witman, 1975), than from the proximal end (Allen and Borisy, 1974; Binder *et al.*, 1975; Binder and Rosenbaum, 1973). Microtubules growing from short microtubule pieces coated with DEAE-dextran also showed polarity of growth (Olmsted, *et al.*, 1974). There was evidence from this early work that proximal assembly not only had a slower rate but also a higher critical concentration than did distal assembly. Although not appreciated at the time, such a difference in critical concentration at the two ends necessarily implied that at a subunit concentration intermediate between these two critical concentrations, the microtubule would simultaneously assemble at the distal end and disassemble at the proximal end. Margolis and Wilson (1978) were the first to show that at steady state there is indeed net assembly of microtubules at one end and disassembly at the other, a process called "treadmilling," which had earlier been detected in actin (Wegener, 1976). The efficiency of subunit flux at steady state, *s*, is defined as the ratio of the subunit flux through the polymer to the total number of association events per unit time at the two ends.

The subunit flux rate and the efficiency of subunit flux have been shown to vary with experimental conditions. Using microtubule protein composed of about 75% tubulin and 25% associated proteins, Margolis and Wilson (1978) found a subunit flux rate of 0.3 subunits/sec. Bergen and Borisy (1980), also using microtubule protein containing both tubulin and accessory proteins, measured the individual association and dissociation rate constants for the two ends. The calculated flux rate is approximately 1.6 subunits/sec, and the flux efficiency is 0.07. Cote and Borisy (1981), using tubulin depleted of associated proteins, found a flux rate about 100 times greater (28/sec), and the measured flux efficiency was 0.26. As predicted, flux was negligible in the presence of the nonhydrolyzable GTP analogue GMPPCP. The reported values for flux rate and flux efficiency thus vary widely. Several factors have been determined to affect these values. The rate of treadmilling is affected by the concentration of GTP, being efficient at 0.1 mM GTP, but low at 1 mM (Margolis, 1981); the mechanism for this effect has not been determined.

The flux rate has also been shown to be affected by ATP. In the presence of 1 mM ATP, the rate of flux through microtubules containing associated proteins increases 20-fold (Margolis and Wilson, 1979), and this increased flux rate persists even following removal of the ATP, suggesting that an ATP-dependent protein kinase may be involved. Phosphorylation of MAPs has been shown to increase the overall dissociation rate of tubulin with little effect on the critical concentration (Jameson and Caplow, 1981; Jameson *et al.*, 1980).

These data are consistent with the observation that depletion of ATP in tissue culture cells by metabolic inhibitors leads to inhibition of microtubule disassembly by colcemid or vinblastine (Bershadsky and Gelfand, 1981). Addition of ATP to the cytoskeleton isolated from such cells leads to more rapid microtubule depolymerization.

The kinetics of treadmilling remain highly controversial. Recently, Caplow *et al.* (1982), in a careful study at steady-state conditions, found a treadmilling efficiency *s* of only 0.0005–0.001 with porcine brain tubulin. These authors have evidence that at the ends of microtubules are stretches of tubulin-GTP, at which extremely facile (2500–5000 subunits/sec) association and dissociation can occur. The dissociation rate of the tubulin-GTP in the cap is much greater than that of the tubulin-GDP which makes up the bulk of the microtubule. The authors argue that this fact has resulted in errors in previous calculations of the treadmilling efficiency. Despite a low treadmilling *efficiency*, these workers still found a relatively high treadmilling *rate*, 4.2–6.3 subunits/sec. Other authors (Carlier and Pantaloni, 1981; Carlier, 1983) also have evidence for a cap of tubulin-GTP on microtubule ends at steady state and calculate that the cap may be made up of about 400 tubulin dimers or about 5% of the average microtubule length.

Several functions *in vivo* have been suggested for the difference in critical concentration at the two ends (Kirschner, 1980). First, it was noted that in a situation in which there are both microtubules with both ends free and microtubules with the minus end capped, the capped microtubules will tend to lengthen at the expense of the free microtubules since there will be net dimer addition to the plus ends of both sets of microtubules, but net dimer loss only from the minus ends of the free microtubules. Therefore, only capped microtubules would tend to persist, and spontaneous and random polymerization would be suppressed. It was suggested that the distribution of microtubules in a cell is determined by the anchoring to specific organizing centers of nucleating factors which allow polymerization at only the plus end. Thus, these factors not only would nucleate assembly but would stabilize the microtubules by capping the nonprefered ends. Formation or loss of the nucleating/capping factor would indirectly lead to assembly or disassembly of an entire microtubule.

The hypothesis that anchored microtubules are selectively stabilized serves to explain the observation that, in many cell types, all or most microtubules are bound at one end to discrete structures called microtubule-organizing centers (MTOCs) (Pickett-Heaps, 1969). For example, in the mitotic spindles of most higher animals, polar microtubules originate in amorphous material surrounding the centrioles. In protozoa, MTOCs are frequently complex structures (Tilney, 1971; Tucker *et al.*, 1975). A relatively well-studied example is in *Polytomella agilis* (Brown *et al.*, 1976), in which acting as an MTOC is a "rootlet complex" consisting of four basal bodies at the anterior of the cell and of eight rootlets, partially composed of microtubules, extending radially outward and posteriorly from the region of the basal bodies. Several hundred microtubules originate in amorphous material present along the sides of the rootlets, and these microtubules extend posteriorly just below the plasma membrane. The amorphous material can be extracted from isolated rootlet complexes. While unextracted rootlet complexes can initiate polymerization of exogenous tubulin, extracted complexes cannot. When the extracted material is added back, the amorphous material is again found along the sides of the rootlets and polymerization of exogenous tubulin occurs from the rootlet complexes (Stearns and Brown, 1979, 1981; Stearns *et al.*, 1976). The extracted material itself can initiate microtubule polymerization.

The hypothesis that anchored microtubules are selectively stabilized supposes, as the most straightforward case, that microtubules are anchored at the minus end, since this anchorage would additionally act to stabilize the microtubule. To test this hypothesis, the polarity of the microtubule must be determined. Until recently, few methods were available to determine microtubule polarity. Polarity could be established by comparing the rate of *in vitro* microtubule assembly onto the microtubule ends with the values for control microtubule nuclei. Unfortunately, if nuclei are used which initially are without attached microtubules or which contain microtubules which cannot be distinguished from the stretches of microtubule experimentally assembled from exogenous tubulin, the straightforward application of this method yields results showing only whether plus or minus ends are the predominant site of tubulin assembly and not the intrinsic polarity of the microtubule. This method has yielded evidence that the plus ends are the sites of polymerization both of microtubules associated with centrioles and of kinetochore microtubules (Bergen *et al.*, 1980; Summers and Kirschner, 1979). Using conditions which cause *in vitro* assembly of microtubules with abnormal morphology, Heidemann *et al.* (1980) established that microtubules associated with mitotic poles assemble at their distal ends.

Recently, two techniques have been developed to determine the intrinsic polarity of microtubules *in vivo.* One technique makes use of the fact that, under certain conditions, tubulin will polymerize to form sheets of protofilaments with abnormal wall junctions (Burton and Himes, 1978). Some of these protofilament aggregates show enantiomorphic images allowing determination of the intrinsic polarity of the polymers (Mandelkow and Mandelkow, 1979). Heidemann and McIntosh (1980) found related conditions in which microtubule polymerization, when seeded by exogenous microtubules, results in decoration of these microtubules with C-shaped sheets of protofilaments attached along the wall of the microtubule. In cross section, these sheets appear as hooks whose directions of curvature are determined by the polarity of the microtubule. A second method uses decoration of microtubules with exogenous dynein (Haimo *et al.*, 1979), which binds to microtubules in a specific orientation allowing determination of microtubule polarity in both longitudinal and cross section (Haimo *et al.*, 1979; Telzer and Haimo, 1981). This method is analogous to the determination of actin polarity by decoration with myosin fragments to form "arrowhead complexes" (Ishikawa *et al.*, 1969).

Both techniques have been used in lysed cells and show, for example, that in neuronal axons, the plus ends of microtubules are oriented away from the cell body (Burton and Paige, 1981) and that in the mitotic apparatus of both plants and animals, both polar and kinetochore microtubules have their plus ends oriented toward the equator and away from the poles (Euteneuer *et al.*, 1982; Euteneuer and McIntosh, 1981; Haimo and Telzer, 1981; Telzer and Haimo, 1981). The orientation of polar microtubules is that predicted by the model of selective stabilization of anchored microtubules.

The data suggest a less straightforward model for the polarity of kinetochore microtubules. Isolated chromosomes can initiate polymerization of exogenous tubulin from their kinetochores (McGill and Brinkley, 1975; Pepper and Brinkley, 1979; Telzer *et al.*, 1975). The rate of polymerization is that expected from a microtubule plus end (Bergen *et al.*, 1980; Summers and Kirschner, 1979). This result led to the suggestion that the minus end is bound to the kinetochore and that kinetochore microtubules assemble at their free ends. This interpretation was disproved by direct determination of kinetochore microtubule polarity, by formation of tubulin "hooks," and by dynein binding. Both techniques show that the plus ends of microtubules are bound to the kinetochore. Therefore, if kinetochore microtubule polarity *in vitro* is the same as *in vivo,* tubulin polymerization can occur at the end which is apparently anchored.

Recent evidence suggests that in the very earliest stages of polymerization at the kinetochore, short microtubule pieces are randomly oriented and that only later are they attached to the kinetochore at their plus ends (De Brabander *et al.*, 1980; Euteneuer *et al.*, 1983; Witt *et al.*, 1980). De Brabander (1982) has proposed that in the vicinity of kinetochores and centrioles, the critical concentration for microtubule assembly is lower than that in the rest of the cytoplasm and that microtubules may initially assemble randomly and only later attach to the MTOCs. At later stages of polymerization, kinetochore microtubules may assemble from the anchored ends.

6. Nucleotide Interactions

Simultaneous assembly at one end and disassembly at the other is thermodynamically possible only because GTP is hydrolyzed during polymerization.

Since the initial purification of brain tubulin (Weisenberg *et al.*, 1968), the tubulin dimer has been known to contain two bound guanine nucleotides, one of which rapidly exchanges with exogenous GTP (the E site) and one which is relatively nonexchangeable (the N site). After tubulin polymerization, neither site is exchangeable. GTP was found to be essential for assembly when microtubules were first polymerized *in vitro* (Weisenberg, 1972). The following scheme represents the present consensus (for review see Jacobs, 1979). Polymerizable tubulin contains two moles GTP, and GTP is hydrolyzed during polymerization at the E, but not at the N, site. After depolymerization, tubulin containing one mole GTP and one mole GDP is released. The GDP is either replaced by GTP from the medium or is phosphorylated to GTP by a transphosphorylase (nucleoside diphosphokinase) using ATP (Jacobs, 1975; Jacobs and Huitorel, 1979; Jacobs *et al.*, 1974). Nucleotide present on the E site of tubulin in rings and in microtubules exchanges undetectably slowly with GTP in the medium; bound GDP in rings is directly phosphorylated by the transphosphorylase (Jacobs and Huitorel, 1979).

Polymerization has been shown to occur with the E site occupied by the nonhydrolyzable GTP analogues β,γ-methylene guanosine-5′-triphosphate (GMPPCP) (Lockwood *et al.*, 1975) and 5′-guanylyl imidodiphosphate (GMP-PNP; Arai and Kaziro, 1976). After assembly with GMPPNP, depolymerization by calcium or by dilution is reduced. In addition, the analogue guanyl 5′-methylene diphosphonate (GMPCPP) is much more effective than GTP in promoting assembly, and the microtubules formed are resistant to calcium and relatively resistant to cold (Sandoval *et al.*, 1977). The GDP analogue guanosine 5′-methylene diphosphonate (GMPCP) will also promote tubulin assembly to some extent (Sandoval *et al.*, 1978). Zackroff and Weisenberg (1978) have shown that in the presence of short microtubule pieces GDP can promote limited bidirectional assembly (also see below).The relative resistance of microtubules polymerized in the presence of GMPPNP to depolymerization may mean that GTP hydrolysis, while occurring at the time of assembly, is not necessary for polymerization, but instead promotes the later disassembly of the microtubule (but see below).

Recent work casts doubt on the conclusions derived from work with GTP analogues (Maccioni and Seeds, 1982). These workers have found that conditions used by others to remove GTP from tubulin leave GTP or GDP in a large fraction of the E sites. More stringent extraction conditions produce tubulin which cannot polymerize with either GMPPCP or GMPPNP, but which polymerizes in the presence of GTP. These authors suggest that GTP or GDP in both E and N sites is required for polymerization, and that the effects of GMPPNP and GMPPCP are due to interactions at a third nucleotide-binding site. These results are difficult to reconcile with those of Purich *et al.* (1982), who found polymerization with GMPPNP when exchangeable GDP and GTP were hydrolyzed to guanosine using alkaline phosphatase.

GTP hydrolysis in newly polymerized tubulin appears to occur at a constant rate independent of tubulin concentration so that microtubules polymerized very rapidly, from very concentrated tubulin, may initially contain as much as 70% GTP at their E sites (Carlier and Pantaloni, 1981). These microtubules are more cold sensitive and depolymerize faster in the few minutes after polymerization than they do at later times, after GTP hydrolysis (Bonne and Pantaloni, 1982). These data suggest that GTP hydrolysis acts to *stabilize* microtubules, a conclusion exactly opposite that that suggested by studies with nonhydrolyzable GTP analogues.

Comparison of the polymerization properties of tubulin-GDP with those of tubulin-GTP gives a very complex picture. Analysis of polymerization in the absence of MAPs and presence of high magnesium concentration and glycerol (Carlier and Pantaloni, 1978) showed that while tubulin-GDP did not nucleate, it assembled onto preformed seeds; the critical concentration was only about twice that with tubulin-GTP. The absence of nucleation by tubulin-GDP is consistent with the model of nucleation of Erickson and Pantaloni (1980), in which nucleation is expected only at a concentration at least 3.5–7 times greater than the critical concentration. These authors suggest that facilitation of nucleation by GTP hydrolysis is a possible explanation of the fact that under standard conditions nucleation can occur at much lower concentrations.

In contrast to the straightforward kinetics in the absence of MAPs, GDP has highly unusual effects on polymerization in the presence of MAPs and absence of glycerol. Several authors (Weisenberg *et al.,* 1976; Margolis, 1981; Zeeberg and Caplow, 1981; Zackroff *et al.,* 1980) found that addition of GDP to tubulin polymerized in the presence of GTP arrests further polymerication but results in little net disassembly. The system is left in a *metastable state,* which could be disrupted by various treatments; for example, dilution results in depolymerization of the microtubules.

Partly to explain the unusual effects of GDP on microtubule polymerization, Weisenberg (1980a,b) has proposed a complex model which postulates that polymerization occurs by the obligatory incorporation of tubulin oligomers in rounds of elongation. The first oligomer to add at the beginning of a round of elongation requires the presence of GTP for stabilization. Subsequent addition of oligomers to the remaining protofilaments can take place with tubulin-GDP. According to this model, the metastable state induced by

GDP is the result of blockage of polymerization due to the inability of tubulin-GDP to initiate a round of elongation. Note that in this model oligomeric intermediates play an obligatory role in microtubule elongation. This assertion is in conflict with some experimental data (see Section 4).

In contrast to the results described above, other workers (Engelborghs and Van Houtte, 1981; Carlier and Pantaloni, 1982) found that addition of GDP to microtubules polymerized with GTP and MAPs induced net depolymerization. The apparent critical concentration of tubulin-GDP appeared to be out of the range investigated. Karr *et al.* (1979) found that tubulin-GDP could not nucleate microtubules but would assemble onto preformed seeds; they found only a doubling of the critical concentration with tubulin-GDP. Hamel *et al.* (1983) found that GDP inhibits microtubule assembly maximally at low magnesium concentrations and has relatively little effect at higher magnesium concentrations. Thus, differences in experimental conditions may explain the apparent differences found by various groups in the effect of GDP on tubulin critical concentration.

Lee *et al.* (1982) found that the properties of the microtubule protein depended on the method of isolation. GDP induced a metastable state during polymerization of microtubule protein isolated by the hypotonic solutions customarily used. In contrast, no metastable state occurred with microtubule protein isolated using a sucrose-containing isolation buffer which yields tubulin with a higher complement of MAPs and less contamination with proteins released from membrane-bound organelles.

There is another circumstance in which microtubules appear to be in a metastable state. A significant fraction of polymerized microtubules are cold-stable. Mild shearing leads to considerable cold depolymerization of these microtubules. These and other data suggest that accessory proteins may cap microtubule ends and block depolymerization (Job *et al.*, 1982). It seems possible that a similar mechanism may be responsible for the metastability induced by GDP and that no fundamental modification of the standard model for tubulin polymerization may be necessary.

ATP has many complex effects on microtubule assembly. GDP can be phosphorylated by a transphosphorylase using ATP, as described above. Microtubule assembly and disassembly can also be affected by ATP-dependent phosphorylation of microtubule-associated proteins, also described above. In addition, ATP affects tubulin polymerization even in the absence of accessory proteins. ATP lowers the critical concentration and also increases the rate of nucleation (Zabrecky and Cole, 1982b). ATP has been shown to bind to tubulin (Zabrecky and Cole, 1982a); it apparently is not hydrolyzed during assembly.

7. *Divalent Cations*

One molecule of magnesium is tightly bound to each tubulin dimer and is required for polymerization (Olmsted and Borisy, 1975). The critical concentration for polymerization of pure tubulin is constant from 3–6 mM

$MgCl_2$. The critical concentration for assembly increases with further decrease in magnesium concentration. In addition, there is a lag in polymerization which lengthens with decreasing magnesium concentration (Papaconstantinou and Pantaloni, 1982).

The first *in vitro* polymerization of tubulin in crude brain extracts indicated that calcium was inhibitory at concentrations of 10 μM or higher (Weisenberg, 1972). In later studies with purified brain tubulin, polymerization was inhibited only by much higher concentration (about 1 mM; Olmsted and Borisy, 1975). Recent studies suggest that the discrepancy is due to loss during purification of a factor conferring calcium sensitivity (Nishida and Sakai, 1977). This factor appears to be the ubiquitous protein calmodulin (also known as calcium-dependent regulatory protein or CDR). Calmodulin inhibits and reverses microtubule assembly in the presence of 10 μM calcium (Marcum *et al.*, 1978). Cytoplasmic microtubules in lysed cells depolymerize in the presence of micromolar calcium; this effect is blocked by calmodulin inhibitors such as trifluoperazine (Schliwa, 1980; Schliwa *et al.*, 1981).

Although MAPs by themselves appear to stabilize microtubules against depolymerization by calcium, it appears to be through interaction with MAPs that calmodulin–calcium complex leads to microtubule depolymerization. In the absence of MAPs, calmodulin actually enhances the rate and extent of polymerization in the presence of calcium by sequestering calcium while it increases the inhibitory effects of calcium in the presence of MAPs (Lee and Wolff, 1982). There is evidence for a direct interaction of MAPs with calmodulin (Rebhun *et al.*, 1980), and especially for the interaction of tau factors with calmodulin (Sobue *et al.*, 1981; Kakuichi and Sobue, 1981).

8. Chemical Composition of Tubulin

Building on earlier work, the complete amino acid sequence of α- and β-tubulin from porcine brain has recently been established (Ponstingl *et al.*, 1981; Krauhs *et al.*, 1981). Probably the most striking feature of the alpha sequence is the very unusual composition of the C-terminal portion. Of the last 40 positions, 47% are acidic (16 glutamate, 3 asparate). The C-terminal tryptic fragment contains 26 amino acid residues, of which 13 are acidic. Only one basic residue is present in the fragment. The C-terminal portion of the beta subunit is similarly acidic. Microheterogeneity was found in at least six positions. Although 55,000 had been previously accepted as the molecular weight of both α- and β-tubulin, the sequence data implies a molecular weight of about 50,000 for both subunits.

Most of the nucleotide sequences of the mRNA coding for α- and β-tubulin in chick brain have been determined, including the entire translated sequence for the β-mRNA and all but about the 38 amino-terminal amino acids for the alpha sequence (Valenzuela *et al.*, 1981).

An unusual feature of the C-terminus of α-tubulin is the fact that although the predicted amino acid sequence shows tyrosine as the C-terminal amino acid (Valenzuela *et al.*, 1981), a large fraction of α-tubulin lacks this amino acid at its C-terminus. A specific carboxypeptidase has been isolated which removes C-terminal tyrosine from α-tubulin but appears to have no effects on other proteins. The enzyme preferentially acts on polymerized tubulin (Kumar and Flavin, 1981). A separate protein, tubulin-tyrosine ligase, has been isolated from brain and characterized (Arce *et al.*, 1975; Barra *et al.*, 1974; Hallack *et al.*, 1977; Raybin and Flavin, 1975). This protein incorporates tyrosine onto the terminal glutamate in conjunction with ATP hydrolysis. The enzymatic activity has been observed in every rat tissue tested and in sea urchin eggs (Kobayashi and Flavin, 1978), but not in yeast or *Tetrahymena* cells or cilia. Tyrosylation has been shown to occur *in vivo* (Argarana *et al.*, 1977). Surprisingly, it appears to have no effect on rate or extent of *in vitro* polymerization or on sensitivity to calcium, ionic strength, or temperature (Raybin and Flavin, 1977). Tubulin in tissue extracts appears to be a mixture of tyrosylated and nontyrosylated forms, and of the latter only a portion will accept exogenous tyrosine. The proportion of tyrosylated tubulin changes during development in the chick (Rodriguez and Borisy, 1977). In addition, there appear to be changes during the mitotic cycle in the proportion of tubulin which can accept tyrosine after carboxypeptidase treatment to remove tyrosine already present (Nath *et al.*, 1978). Tubulin from synaptic membranes is unusual in being completely nontyrosylated (Nath and Flavin, 1978). *In vivo* incorporation of tyrosine into the carboxyl terminus requires intact microtubules (Thompson *et al.*, 1979), and *in vivo* incorporation into Chinese hamster ovary cells appears to be dependent on the cell morphology (Deanin *et al.*, 1981). Chemotactic factors stimulate *in vivo* tubulin tyrosylation in polymorphonuclear leukocytes (Nath *et al.*, 1982).

While brain alpha chain can be tyrosylated, the beta chain of brain is at least partially phosphorylated. Brain tubulin isolated by DEAE chromatography contains on the average 0.8 moles phosphorus per mole tubulin (Eipper, 1974b). The phosphate is located on a serine residue near the C-terminus of the α-subunit and thus serves, remarkably, to make this very acidic region even more acidic. When brain microtubule portion is incubated *in vitro* with $^{32}PO_4$, some phosphate is incorporated. However, the pattern of phosphorylation does not correspond to that *in vivo*. Several peptide fragments on both alpha and beta are labeled, and the labeled tubulin is mostly in an aggregated state (Eipper, 1974a). These results suggest that only denatured tubulin is phosphorylated *in vitro*.

In contrast to brain tubulin, neuroblastoma tubulin is not phosphorylated *in vivo* (Solomon *et al.*, 1976). In HeLa cells, the extent of phosphorylation varies greatly with the cell cycle; the maximum extent of phosphorylation occurs during S and M phases (Piras and Piras, 1975). Based on identification of spots in 2-D gels, the degree of β-tubulin phosphorylation in parturient myometrium decreases on induction of labor while α-tubulin phosphorylation is induced (Joseph *et al.*, 1982). In *Chlamydomonas* flagella, β-tubulin is not phosphorylated, however, about one third of the protein migrating with α-tubulin in SDS gels is phosphorylated *in vivo* (Piperno and Luck, 1976). This protein has not been rigorously identified as a tubulin, and little or no phosphorylated tubulin is seen after two-dimensional electrophoresis of *Chlamydomonas* proteins (McKeithan, unpublished data).

A third form of post-translational modification of tubulin is oxidation or blocking of sulfhydryl groups. Both have been shown to prevent microtubule assembly *in vitro* (Himes *et al.*, 1976; Kuriyama and Sakai, 1974; Mellon and Rebhun, 1976). Agents which oxidize sulfhydryls *in vitro*, diamide and butyl hydroperoxide, cause depolymerization of microtubules and oxidation of glutathione, one of the principal intracellular reducing agents (Burchill *et al.*, 1978; Oliver *et al.*, 1976; Rebhun *et al.*, 1976). In addition, in some cases in which methylxanthines cause microtubule depolymerization *in vivo*, the effect may be due to inhibition of glutathione reductase, leading to lower levels of reduced glutathione and indirectly to oxidation of sulfhydryls on tubulin (Rebhun *et al.*, 1976). Burchill *et al.* (1978) have shown that during the normal depolymerization of microtubules that occurs at the end of phagocytosis in polymorphonuclear leukocytes, the level of glutathione cross-linked to protein by disulfide linkages dramatically increases, suggesting that formation of such disulfide linkages may cause depolymerization of microtubules *in vivo*. Unfortunately, PMNs, which produce hydrogen peroxide as part of a bacteriocidal mechanism, may be a special case.

9. *Tubulin Genes*

By use of cloned DNA probes, many higher eukaryotes have been found to have multiple tubulin genes. Four genes each for α- and β-tubulin have been found in chicken DNA (Cleveland *et al.*, 1980) and in *Drosophila* DNA (Kalfayan *et al.*, 1981), about 20 genes each in sea urchin DNA and about ten each in human DNA (Cleveland *et al.*, 1980). *Chlamydomonas* appears to have exactly two genes each for α- and β-tubulin (Brunke *et al.*, 1982b; Silflow and Rosenbaum, 1981). Many of the apparent human tubulin genes have been identified as processed pseudogenes (Lee *et al.*, 1983; Wilde *et al.*, 1982). In most systems studied, tubulin genes are dispersed; in contrast, *Trypanosoma* contains a cluster of alternating α- and β-tubulin genes (13–17 copies per haploid genome) (Thomashow *et al.*, 1983).

One of the best studied systems has been *Drosophila*, where four alpha genes are found (Kalfayan *et al.*, 1981). The three ends of these genes were subcloned and three of the four subclones were found not to cross hybridize. These clones were used to detect mRNAs specific for each gene. In adults, α-2 and α-3 hybridized to one RNA, α-1 to one or two, and α-4 to four mRNAs. A second apparent α-2 transcript is found in larvae. Further studies (Kalfayan and Wensink, 1982) suggested that only one of the four apparent α-4 transcripts actually coded for tubulin. This transcript appears to be a maternal mRNA since it is present only in females and very young embryos. There is some evidence that α-2 and possibly α-1 may encode two mRNAs. While the levels of α-1 and α-2 mRNAs appear to change in parallel, the remaining mRNAs appear to be independently regulated.

In other systems, multiple mRNAs and multiple protein products are encoded by a single gene by variation in mRNA initiation, termination, and splicing. Even in the absence of DNA rearrangements, as many as four distinct immunoglobulin heavy chains, i.e., membrane bound or secretory, mu or delta, may be encoded by a single gene (Maki *et al.*, 1981; Moore *et al.*, 1981; Rogers *et al.*, 1980). A single calcitonin gene encodes for two polypeptide products (Amara *et al.*, 1982), one gene encodes two forms of pyruvate kinase subunits (Marie *et al.*, 1981), and distinct mRNAs for cytoplasmic and secreted yeast invertase are encoded by a single gene (Perlman and Halvorson, 1981).

As mentioned above, a single gene encodes two β-tubulins in *Aspergillus* (Sheir-Neiss *et al.*, 1978). It is not known whether the two tubulins are translated from two mRNAs resulting from differential RNA processing or, instead, are formed by post-translational modification of a common protein product.

10. Control of Tubulin Gene Expression

In many types of tissue culture cells from higher animals, the level of tubulin synthesis appears to be controlled by the concentration of unpolymerized tubulin dimer. In many, although not all, of these cell types, tubulin synthesis drops rapidly in response to inhibitors which depolymerize microtubules and increase the dimer pool (Ben-Ze'ev *et al.*, 1979; Cleveland *et al.*, 1981). In contrast, other inhibitors which promote tubulin assembly (taxol) or cause aberrant assembly (vinblastine) produce a slight increase or no change in tubulin synthesis.

One of the most extensively studied systems is the stimulation of tubulin synthesis in conjunction with growth of flagella or cilia. *In vivo* labeling techniques have been used to study the synthesis of tubulin and other flagellar or ciliary proteins after deflagellation or deciliation in sea urchins (Stephens, 1977), *Tetrahymena* (Guttman and Gorovsky, 1979), *Polytomella* (Brown and Rogers, 1978), and *Chlamydomonas* (Lefebvre *et al.*, 1978). Tubulin synthesis is also greatly stimulated during the differentiation of *Naegleria* from the ameboid to the flagellate form (Kowit and Fulton, 1974a; Lai *et al.*, 1979). In *Naegleria, Tetrahymena* (Marcaud and Hayes, 1979), *Chlamydomonas* (Lefebvre *et al.*, 1980), and sea urchins (Merlino *et al.*, 1978), *in vitro* translation of isolated RNA has shown a large increase in translatable tubulin mRNAs accompanying growth of cilia or flagella. In *Chlamydomonas,* polysomes from cells regenerating flagella have been shown to synthesize more tubulin than polysomes from control, unregenerating cells (Weeks and Collis, 1976). The quantity of tubulin mRNA increases within 8 min following deflagellation (Lefebvre *et al.*, 1980), and the level of mRNA at least roughly parallels the rate of tubulin synthesis (Silflow and Rosenbaum, 1981), suggesting that tubulin synthesis may be controlled principally at the transcriptional level. In *Naegleria,* tubulin synthesis also appeared to be roughly proportional to the quantity of mRNA (Lai *et al.*, 1979).

Chlamydomonas can be induced to resorb its flagella in the presence of the phosphodiesterase inhibitor isobutylmethylxanthine or in the presence of media containing calcium-chelating agents (Lefebvre *et al.*, 1978). When this flagellar shortening is reversed, synthesis of tubulin and other flagellar proteins is stimulated, suggesting that the cell monitors flagellar length and responds directly to subnormal length by increasing flagellar protein synthesis. In addition, deflagellation stimulates flagellar protein synthesis even in the presence of agents such as colchicine which block flagellar regeneration (Lefebvre *et al.*, 1978). Presumably, either the deflagellation event itself or the absence of flagella increases flagellar protein synthesis. Since tubulin synthesis is shut off abnormally early following deflagellation in the presence of colchicine, it is possible that tubulin synthesis is also partially controlled by the concentration of unpolymerized tubulin dimers as described in higher organisms.

11. Similarities among Tubulins and Evidence for Heterogeneity

Microtubules from different sources show great variability in stability and solubility properties. Brain microtubules are sensitive to cold, calcium, and colchicine, whereas flagellar axonemes are stable under all these conditions. Sensitivity to agents such as these or sensitivity to dissolution under various conditions of pH, ionic strength, or detergents have permitted the identification of multiple classes of microtubules (Behnke and Forer, 1967). However, at least some of these differences in stability do not seem intrinsic to the tubulin itself, for microtubules assembled *in vitro* from flagellar tubulin are sensitive to cold and to calcium, in contrast to the flagellar axoneme itself (Binder and Rosenbaum, 1978; Kuriyama, 1976). In addition, there is direct evidence that microtubule-associated proteins can stabilize microtubules (Sloboda and Rosenbaum, 1977).

Microtubules from very different sources can copolymerize. For example, tubulins in yeast (Shriver and Byers, 1977; Water and Kleinsmith, 1976) and *Aspergillus* (Davidse, 1975; Sheir-Neiss *et al.*, 1978) have each been identified by copolymerization with brain tubulin. Copolymerization with brain tubulin has also been used to identify *in vitro* translated tubulin from *Chlamydomonas* (Weeks and Collis, 1976) and from *Naegleria* (Lai *et al.*, 1979). In addition, microtubules from one source can initiate the polymerization of microtubules from another (Burns and Starling, 1974). Flagellar axonemes (Allen and Borisy, 1974; Binder and Rosenbaum, 1973; Binder *et al.*, 1975), basal bodies (Snell *et al.*, 1974), yeast spindle pole bodies (Hyams and Borisy, 1978), and chromosomal kinetochores (Telzer *et al.*, 1975) can all initiate polymerization of brain tubulin. Conversely, brain microtubule pieces can initiate assembly of flagellar tubulin (Kuriyama, 1976). There appears to be no example in the literature showing failure of copolymerization in vitro, that is, utilization of only one type of tubulin in assembly when a second type of polymerization-competent tubulin is available as well. Despite the promiscuous behavior of tubulin *in vitro,* there is no direct evidence *in vivo* that tubulin from one microtubule organelle can be used to build another type of organelle. Nevertheless, in some cases, there is circumstantial evidence for a common tubulin pool. For instance, in some cells, cytoplasmic or flagellar microtubules disassemble before assembly of the mitotic apparatus (Bloodgood, 1974).

Immunological comparisons have indicated both similarities (Dales, 1972) and differences among tubulins. For instance, some sera react to tubulins from evolutionarily distant species. A case in point is the reactivity of polyclonal antibody prepared against β-tubulin from *Chlamydomonas* flagella with brain tubulin (Piperno and Luck, 1977). By immunofluorescence, antibody to brain tubulin reacts with flagella, rootlets, and cytoskeletal microtubules in the colorless alga *Polytomella agilis* (Brown *et al.*, 1976). However, even in the cases of cross reactivity there may be variation in the quantity of tubulin required to give a precipitin band. For example, an antibody to *Arbacia punctulata* sperm flagellar tubulin reacted with tubulins from several sea urchins and a sand dollar, but not with that from a starfish (Fulton *et al.*, 1971). The quantity of tubulin required to give a precipitin band varied from species to species and, most notably, from organelle to organelle within the same species.

Drosophila testis contains a form of β-tubulin distinct from that found in all other organs in the insect. Analysis of mutants shows that a single gene encodes this form of β-tubulin (Kemphues *et al.*, 1979, 1980). This form comprises all of the β-tubulin in sperm. Since sperm contains a central pair of singlet microtubules and nine accessory singlet microtubules in addition to the nine outer doublets, a single gene encodes β-tubulin in both singlet and doublet microtubules. Although one cannot rule out the possible presence of distinctive protein modifications which do not affect migration on 2-D gels, these results strongly suggest that a single form of β-tubulin is present in both singlet and doublet microtubules in sperm. In addition, analysis of mutants provides strong evidence that the distinctive β-tubulin is found in the microtubules of the meiotic apparatus as well (Kemphues *et al.*, 1982).

In another case, mutants in the Ben A locus coding for β-tubulin in *Aspergillus* appear to be altered in both nuclear division, involving mitotic microtubules, and in nuclear migration, which depends on cytoplasmic microtubules. Both processes are also affected by a different mutation in a structural gene for α-tubulin (Oakley and Morris, 1981).

Multiple forms of tubulin have been detected in single neurons (Gozes and Sweadner, 1981). Heterogeneity of tubulin in brain increases with development.

The most dramatic immunological evidence for tubulin diversity is in *Naegleria gruberi*, where antibody prepared against outer doublet tubulin was highly specific, failing to react with any other tubulin tested (Fulton and Kowit, 1975). *Naegleria* normally grows as an ameoba, but under certain conditions will differentiate into a flagellate (Fulton, 1977). Before induction of differentiation, amoeba contain only 2–3% of the amount of flagellar tubulin antigen found in flagellates as judged by radioimmunoassay (Kowit and

Fulton, 1974b) despite the apparent presence of abundant tubulin in these cells.

Further evidence for *de novo* synthesis of a distinct flagellar tubulin in this organism derives from radioactive pulse-chase experiments during flagellar assembly (Kowit and Fulton, 1974a). At least 70% of flagellar outer doublet tubulin is synthesized after the induction of differentiation. At the same time, putative cytoplasmic tubulin is not "chased" into the flagella (Fulton and Simpson, 1976).

Stephens (1970, 1975) and Safer (1973) have evidence for differences in tryptic peptide maps among tubulins of central pair, A-subfiber, and B-subfiber of flagella and cilia. Stephens has also shown differences in amino-acid composition among these types of tubulin. Recently, he has shown that peptide maps of cytoplasmic tubulin from sea urchin eggs have a major difference in both α- and β-subunits from the peptide maps of A- and B-subfiber tubulins, and major differences are also found between the peptide maps of cytoplasmic and central pair tubulins (Stephens, 1978).

Two-dimensional analysis of tubulin from *Aspergillus*, identified by copolymerization with brain tubulin, showed multiple tubulin spots (Sheir-Neiss *et al.*, 1978). One major and two minor alpha spots were seen as well as two major beta spots and two minor betas. In several mutants resistant to benzimidazole derivatives, the position of all four betas was changed. This suggested that a single gene for β-tubulin coded for several beta forms, presumably by post-translational modification. However, certain minor beta spots were unaffected by the mutation and may represent one or more separate genes. Certain revertants of temperature-sensitive benzimidazole-resistant mutants are mutant in α-tubulin (Morris *et al.*, 1979). In these mutants, the position of the major species is changed while two minor spots are unaffected. This again suggests the existence of at least two α-tubulin genes. Unfortunately, there is no evidence yet in this species that specific microtubules are assembled from particular forms of tubulin.

The colorless alga *Polytomella agilis* contains several types of microtubule organelles, including four flagella and several hundred cytoskeletal microtubules. Analysis by two-dimensional electrophoresis has shown that the major forms of α-tubulin in the flagella and in the cytoskeletal microtubules are different (McKeithan and Rosenbaum, 1981). In addition, several minor forms of β-tubulin appear to be specific for flagellar microtubules.

12. Post-translational Modification of Tubulin

As described above, in some organisms, tubulin may be modified by phosphorylation or by removal of the C-terminal tyrosine. In other systems, there is indirect evidence for post-translational modification of tubulin although the specific modification has not been determined.

In *Aspergillus*, a single gene specifies two forms of β-tubulin which migrate differently on SDS gels but have identical isoelectric points (Morris *et al.*,

1979). In this case, the modification must be something other than a simple phosphorylation, sulfation, or acetylation. It is not known whether the two forms of β-tubulin have different functions.

In the alga *Polytomella,* the major α-tubulin found in flagella is different from that found in the cytoskeletal microtubules, and deflagellation induces synthesis of tubulin and of tubulin mRNA. The tubulin synthesized by *in vitro* translation of RNA from regenerating cells does not correspond to the major flagellar α-tubulin (α-3), but, indeed, comigrates with the major cytoskeletal α-tubulin (α-1), which has a more basic isoelectric point (McKeithan *et al.,* 1979; McKeithan *et al.,* 1983). These results raise the possibility that α-1 is a precursor to α-3. Alpha-1 tubulin may not only be a component of cytoskeletal microtubules but may also be the precursor of the form of α-tubulin present in flagella.

The α-tubulins in *Chlamydomonas* can be named in conformity with the nomenclature in *Polytomella.* Deflagellation of *Chlamyodomonas* results in a dramatic increase in synthesis of tubulin and of tubulin mRNA. The form of α-tubulin synthesized following deflagellation, both by *in vivo* and *in vitro* translation of isolated RNA, is not identical to α-3, the major form in flagella, but instead corresponds to α-1 (Lefebvre *et al.,* 1980). Pulse-chase experiments in *Chlamydomonas* lend support to the hypothesis that α-1 is a precursor to α-3. When *Chlamydomonas* is deflagellated, pulse labeled, and subjected to a chase, there is a net increase in the quantity of α-3. This α-3 is found in the flagella; α-1 remains the predominant form of α-tubulin in the cell body (L'Hernault and Rosenbaum, 1983). In addition, in both *Polytomella* and *Chlamydomonas,* some conversion of α-1 to α-3 appears to occur even when protein synthesis is inhibited (McKeithan *et al.,* 1983).

The α-tubulin modification is closely coupled to flagellar assembly (L'Hernault and Rosenbaum, 1983; Brunke *et al.,* 1982a). Tubulin modification does not occur when flagellar assembly is blocked by colchicine. Remarkably, the modification is reversed when flagella are resorbed (L'Hernault and Rosenbaum, 1983). The modification is not a simple phosphorylation or sulfation (McKeithan *et al.,* 1979; S. L'Hernault, personal communication).

Chlamydomonas contains only two α-tubulin genes (Brunke *et al.,* 1982b; Silflow and Rosenbaum, 1981). Two α-tubulin mRNAs are coordinately induced upon deflagellation in roughly equal quantities (Silflow and Rosenbaum, 1981). These two mRNAs appear to be transcribed from the two α-tubulin genes (Brunke *et al.,* 1982b; C. Silflow, personal communication). As determined by *in vitro* translation, both mRNAs appear to encode polypeptides comigrating with α-1 (Silflow and Rosenbaum, 1981; Silflow *et al.,* 1981). Thus, this system is a particularly simple one for further studies of tubulin modification.

In summary, in different systems, tubulin heterogeneity can arise either from multiple tubulin genes or from post-translational modification of a common precursor. There is as yet no data to suggest a function for the presence of multiple forms of tubulin in the same cell.

References

Allen, C., and Borisy, G. G., 1974, Structural polarity and directional growth of microtubules of *Chlamydomonas* flagella, *J. Mol. Biol.* **90:**381–402.

Amara, S. G., Jonas, V., Rosenfeld, M. G., Ong, E. S., and Evans, R. M., 1982, Alternative RNA processing in calcitonin gene expression generates mRNAs encoding different polypeptide products, *Nature (London)* **298:**240–244.

Amos, L. A., 1977, Arrangement of high molecular weight associated proteins on purified mammalian brain microtubules, *J. Cell Biol.* **72:**642–654.

Amos, L. A., 1979, Structure of microtubules, in: *Microtubules* (K. Roberts and J. S. Hyams, eds.), pp. 1–64, Academic Press, London.

Amos, L. A., and Baker, T. S., 1979, The three-dimensional structure of tubulin protofilaments, *Nature (London)* **279:**607–612.

Amos, L. A., and Klug, A., 1974, Arrangement of subunits in flagellar microtubules, *J. Cell Sci.* **14:**523–549.

Andre, J., and Thiery, J. P., 1963, Mise en evidence d'une sous-structure fibrillaire dans les filaments axonematique des flagelles, *J. Microscopie* **2:**71–80.

Arai, T., and Kaziro, Y., 1976, Effect of guanine nucleotides on the assembly of brain microtubules: Ability of 5′-guanylyl imidophosphate to replace GTP in promoting the polymerization of microtubules *in vitro*, *Biochem. Biophys. Res. Commun.* **69:**369–376.

Arce, C. A., Rodriguez, J. A., Barra, H. S., and Caputto, R. 1975, Incorporation of L-tyrosine, L-phenylalanine, and L-3,4-dihydroxyphenylalanine as single units into rat brain tubulin, *Eur. J. Biochem.* **59:**145–149.

Argarana, C. E., Arce, C. A., Barra, H. S., and Caputto, R., 1977, *In vivo* incorporation of ^{14}C-tyrosine into the C-terminal position of the alpha subunit of tubulin, *Arch. Biochem. Biophys.* **180:**264–268.

Baker, T. S., and Amos, L. A., 1978, Structure of the tubulin dimer in zinc-induced sheets, *J. Mol. Biol.* **123:**89–106.

Ballowitz, E., 1888, Untersuchungen uber die structur der spermatozoen zuggleich ein beitrag zur lehre vom feineren bau der contraktilen elemente. *Arch. Mikroskop. Anat.* **32:**401–473.

Barnicot, N. A., 1966, A note on the structure of spindle fibers, *J. Cell Sci.* **1:**217–222.

Barra, H. S., Arce, C. A. A., Rodriguez, J. A., and Caputto, R., 1974, Some common properties of the protein that incorporates tyrosine as a single unit and the microtubule proteins, *Biochem. Biophys. Res. Commun.* **60:**1384–1390.

Behnke, O., and Forer, A., 1967, Evidence for four classes of microtubules in individual cells, *J. Cell Sci.* **2:**169–192.

Ben-Ze'ev, A., Farmer, S. R., and Penman, S., 1979, Mechanisms of regulated tubulin synthesis in cultured mammalian cells, *Cell* **17:**319–325.

Bergen, L. G., and Borisy, G. G., 1980, Head-to-tail polymerization of microtubules in vitro: Electron microscope analysis of seeded assembly, *J. Cell Biol.* **84:**141–150.

Bergen, L. G., Kuriyama, R., and Borisy, G. G., 1980, Polarity of microtubules nucleated by centrosomes and chromosomes of Chinese hamster ovary cells in vitro, *J. Cell Biol.* **84:**151–159.

Berkowitz, S. A., Katagiri, J., Binder, H. K., and Williams, R. C., Jr., 1977, Separation and characterization of microtubule proteins from calf brain, *Biochemistry* **16:**5610–5617.

Bershadsky, A. D., and Gelfand, V. I., 1981, ATP-dependent regulation of cytoplasmic microtubule disassembly, *Proc. Natl. Acad. Sci. USA* **78:**3610–3613.

Bibring, T., Baxandall, J., Denslow, S., and Walker, B., 1976, Heterogeneity of the alpha subunit of tubulin and the variability of tubulin within a single organism, *J. Cell Biol.* **69:**301–312.

Binder, L. I., and Rosenbaum, J. L., 1973, Directionality of assembly of chick brain tubulin onto sea urchin flagellar microtubules, *Biol. Bull. (Woods Hole)* **145:**425.

Binder, L. I., and Rosenbaum, J. L., 1978, The *in vitro* assembly of flagellar outer doublet tubulin, *J. Cell Biol.* **79:**500–515.

Binder, L. I., Dentler, W. L., and Rosenbaum, J. L., 1975, Assembly of chick brain tubulin onto flagellar axonemes of *Chlamydomonas* and sea urchin sperm, *Proc. Natl. Acad. Sci. USA* **72:**1122–1126.

Bloodgood, R. A., 1974, Resorption of organelles containing microtubules, *Cytobios* **9:**143–161.

Bonne, D., and Pantoloni, D., 1982, Mechanism of tubulin assembly: Guanosine 5′ triphosphate hydrolysis decreases the rate of microtubule depolymerization, *Biochemistry* **21:**1075–1081.

Borisy, G. G., and Olmsted, J. B., 1972, Nucleated assembly of microtubules in porcine brain extracts, *Science* **177:**1196–1197.

Borisy, G. G., and Taylor, E. W., 1967a, The mechanism of action of colchicine. Binding of colchicine-^{3}-H to cellular protein, *J. Cell Biol.* **34:**525–533.

Borisy, G. G., and Taylor, E. W., 1967b, The mechanism of action of colchicine. Colchicine binding to sea urchin egg and the mitotic apparatus, *J. Cell Biol.* **34:**535–548.

Brown, D. L., and Rogers, K. A., 1978, Hydrostatic pressure-induced internalization of flagellar axonemes, disassembly, and reutilization during flagellar regeneration in *Polytomella, Exp. Cell Res.* **117:**313–324.

Brown, D. L., Massalski, A., and Patenaude, R., 1976, Organization of the flagellar apparatus and associated cytoplasmic microtubules in the quadriflagellate alga *Polytomella agilis, J. Cell Biol.* **69:**106–125.

Brunke, K. J., Collis, P. S., and Weeks, D. P., 1982a, Post-translational modification of tubulin dependent on organelle assembly, *Nature* (*London*) **297:**516–518.

Brunke, K. J., Young, E. E., Buchbinder, B. U., and Weeks, D. P., 1982b, Coordinate regulation of the four tubulin genes of *Chlamydomonas reinhardi, Nucleic Acids Res.* **71:**749–767.

Bryan, J., 1976, A quantitative analysis of microtubule elongation, *J. Cell Biol.* **10:**1295–1310.

Bryan, J., and Wilson, L., 1971, Are cytoplasmic microtubules heteropolymers? *Proc. Natl. Acad. Sci. USA* **68:**1762–1766.

Bucher, P., 1939, Zur kenntnis der mitose. VI. Der einfluss von colchicin und trypaflavin auf den wachstumsrhthmus und auf die zellteilung in fibrocyten-kulturen, Z. Zellforsch. **29:** 283–322.

Bulinski, J. C., and Borisy, G. G., 1979, Self-assembly of microtubules in extracts of cultured HeLa cells and the identification of HeLa microtubule-associated proteins, *Proc. Natl. Acad. Sci. USA* **76:**293–297.

Burchill, B. R., Oliver, J. M., Pearson, C. B., Leinbach, E. D., and Berlin, R. D., 1978, Microtubule dynamics and glutathione metabolism in phagocytizing human polymorphonuclear leukocytes, *J. Cell Biol.* **76:**439–447.

Burns, R. G., and Starling, D., 1974, The *in vitro* assembly of tubulins from sea-urchin eggs and rat brain: Use of heterologous seeds, *J. Cell Sci.* **14:**411–419.

Burton, P. R., and Fernandez, H. L., 1973, Delineation by lanthanum staining of filamentous elements associated with the surfaces of axonal microtubules, *J. Cell Sci.* **12:**567–583.

Burton, P. R., and Himes, R. H., 1978, Electron microscope studies of pH effects on assembly of tubulin free of associated proteins, *J. Cell Biol.* **77:**120–133.

Burton, B. R., and Hinkley, R. E., 1974, Further electron microscopic characterization of axoplasmic microtubules of the ventral cord of the crayfish, *J. Submicrosc. Cytol.* **6:**311–326.

Burton, P. R., and Paige, J. L., 1981, Polarity of axoplasmic microtubules in the olfactory nerve of the frog, *Proc. Natl. Acad. Sci. USA* **78:**3269–3273.

Burton, P. R., Hinkley, R. E., and Pierson, G. B., 1975, Tannic acid-stained microtubules with 12, 13, and 15 protofilaments, *J. Cell Biol.* **65:**227–233.

Caplow, M., Langford, G. M., and Zeeberg, B., 1982, Concerning the efficiency of the treadmilling phenomenon with microtubules, *J. Biol. Chem.* **257:**15012–15021.

Carlier, M. F., 1982, Guanosine-5′-triphosphate hydrolysis and tubulin polymerization, *Mol. Cell Biochem.* **47:**97–113.

Carlier, M. F., 1983, Kinetic evidence for a conformation change of tubulin preceding microtubule assembly, *J. Biol. Chem.* **258:**2315–2420.

Carlier, M. F., and Pantaloni, D., 1978, Kinetic analysis of cooperativity in tubulin polymerization in the presence of guanosine di- or triphosphate nucleotides, *Biochemistry* **17:**1908–1915.

Carlier, M. F., and Pantaloni, D., 1981, Kinetic analysis of guanosine 5′-triphosphate hydrolysis associated with tubulin polymerization, *Biochemistry* **20:**1918–1924.

Carlier, M. F., and Pantaloni, D., 1982, Assembly of microtubule protein: role of guanosine di- and triphospate nucleotides, *Biochem.* **21:**1215–1224.

Chalfie, M., and Thomson, J. N., 1982, Structural and functional diversity in the neuronal microtubules of *Caenorhabditis elegans, J. Cell Biol.* **93:**15–23.

Cleveland, D. W., Lopata, M. A., MacDonald, R. J., Cowan, N. J., Rutter, W. J., and Kirschner, M. W., 1980, Number and evolutionary conservation of α and β tubulin and cytoplasmic β and γ actin genes using specific cloned cDNA probes, *Cell* **20:**95–105.

Cleveland, D. W., Lopata, M. A., Sherline, P., and Kirschner, M. W., 1981, Unpolymerized tubulin modulates the level of tubulin mRNAs, *Cell* **25:**537–546.

Connolly, J. A., Kalnins, V. I., Cleveland, D. W., and Kirschner, M. W., 1977, Immunofluorescent staining of cytoplasmic and spindle microtubules in mouse fibroblasts with antibody to tau protein, *Proc. Natl. Acad. Sci. USA* **74:**2437–2440.

Connolly, J. A., Kalnins, V. I., Cleveland, D. W., and Kirschner, M. W., 1978, Intracellular localization of the high molecular weight microtubule accessory protein by indirect immunofluorescence, *J. Cell Biol.* **76:**R781–786.

Cote, R. H., and Borisy, G. G., 1981, Head-to-tail polymerization of microtubules in vitro, *J. Mol. Biol.* **150:**577–602.

Dales, S., 1972, Concerning the universality of a microtubule antigen in animal cells, *J. Cell Biol.* **52:**748–754.

David-Pfeuty, T., Laport, J., and Pantaloni, D., 1978, GTPase activity at ends of microtubules, *Nature (London)* **272:**282–284.

Davidse, L. C., 1975, Antimitotic activity of methyl benzimidazol-2-ylcarbamate in fungi and its binding to cellular protein, in: *Microtubules and Microtubule Inhibitors* (M. Borgers and M. De Brabander, eds.), pp. 483–495, ASP Biological and Medical Press, New York.

Deanin, G. G., Preston, S. F., and Gordon, M. W., 1981, Carboxyl terminal tryosine metabolism of alpha tubulin and changes in cell shape: Chinese hamster ovary cells, *Biochem. Biophys. Res. Commun.* **100:**1642–1650.

De Brabander, M., 1982, A model for the microtubule organizing activity of the centrosomes and kinetochores in mammalian cells, *Cell Biol. Int. Rep.* **6:**901–915.

De Brabander, M., and De Mey, J. (eds.), 1980, *Microtubules and Microtubule Inhibitors,* Elsevier/North Holland Biomedical Press, Amsterdam.

De Brabander, M., Geuens, G., Nuydens, R., Willebrords, R., and De Mey, J., 1980, The microtubule nucleating and organizing activity of kinetochores and centrosomes in living PTK-2 cells, in: *Microtubules and Microtubule Inhibitors* (M. De Brabander and J. De Mey, eds.), pp. 255–270, Elsevier/North Holland Biomedical Press, Amsterdam.

Dentler, W. L., Granett, S., Witman, G. B., and Rosenbaum, J. L., 1974, Directionality of brain microtubule assembly *in vitro, Proc. Natl. Acad. Sci. USA* **71:**1710–1714.

Dentler, W. L., Granett, S., and Rosenbaum, J. L., 1975, Ultrastructural localization of the high molecular weight proteins associated with *in vitro*-assembled brain microtubules, *J. Cell Biol.* **65:**237–241.

Detrich, H. W., and Williams, R. C., 1978, Reversible dissociation of the αβ dimer of tubulin from bovine brain, *Biochemistry* **17:**3900–3907.

Detrich, H. W., Williams, R. C., and Wilson, L., 1982, Effect of colchicine binding on the reversible dissociation of the tubulin dimer, *Biochemistry* **21:**2392–2400.

Dustin, P., 1978, *Microtubules,* Springer-Verlag, Berlin.

Eipper, B. A., 1974a, Rat brain tubulin and protein kinase activity, *J. Biol. Chem.* **249:**1398–1406.

Eipper, B. A., 1974b, Properties of rat brain tubulin, *J. Biol. Chem.* **249:**1407–1416.

Engelbroghs, Y., and Van Houtte, A., 1981, Temperature jump relaxation study of microtubule elongation in the presence of GTP/GDP mixtures, *Biophys. Chem.* **14:**195–202.

Erickson, H. P., 1974, Microtubule surface lattice and subunit structure and observations on reassembly, *J. Cell Biol.* **60:**153–167.

Erickson, H. P., and Pantaloni, D., 1980, Nucleation of microtubule assembly—a simple model based on thermodynamics, in: *Microtubules and Microtubule Inhibitors* (M. De Brabander and J. De Mey, eds.), pp. 119–132, Elsevier/North Holland Biomedical Press, Amsterdam.

Erickson, H. P., and Voter, W. A., 1976, Polycation-induced assembly of purified tubulin, *Proc. Natl. Acad. Sci. USA* **73:**2813–2817.

Euteneuer, U., and McIntosh, J. R., 1981, Polarity of some motility-related microtubules, *Proc. Natl. Acad. Sci. USA* **78:**372–376.

Euteneuer, U., Jackson, W. T., and McIntosh, J. R., 1982, Polarity of spindle microtubules in *Haemanthus* endosperm, *J. Cell Biol.* **94:**644–653.

Euteneuer, U., Ris, H., and Borisy, G. G., 1983. Polarity of kinetochore microtubules in Chinese hamster ovary cells after recovery from a colcemid block, *J. Cell Biol.* **97:**202–208.

Fawcett, D. W., and Porter, K. R., 1954, A study of the fine structure of ciliated epithelia, *J. Morphol.* **94:**221–228.

Frigon, R. P., and Timasheff, S. H., 1975a, Magnesium-induced self-association of calf brain tubulin. I. Stoichiometry, *Biochemistry* **14:**4559–4566.

Frigon, R. P., and Timasheff, S. H., 1975b, Magnesium-induced self-association of calf brain tubulin. II. Thermodynamics, *Biochemistry* **14:**4567–4573.

Frigon, R. P., Valenzuela, M. S., and Timasheff, S. N., 1974, Structure of a magnesium-induced polymer of calf brain microtubule protein, *Arch. Biochem. Biophys.* **165:**442–443.

Fulton, C., 1977, Cell differentiation in *Naegleria gruberi*, *Annu. Rev. Microbiol.* **31:**597–629.

Fulton, C., and Kowit, J. D., 1975, Programmed synthesis of flagellar tubulin during cell differentiation in *Naegleria*, *Ann. N. Y. Acad. Sci.* **253:**318–332.

Fulton, C., and Simpson, P. A., 1976, Selective synthesis and utilization of flagellar tubulin. The multitubulin hypothesis, in: Cell Motility (R. Goldman, T. Pollard, J. Rosenbaum, eds.), pp. 987–1005, Cold Spring Laboratory.

Fulton, C., Kane, R. E., and Stephens, R. E., 1971, Serological similarity of flagella and mitotic microtubules, *J. Cell Biol.* **50:**762–773.

Gall, J. G., 1965, Fine structure of microtubules (Abstr.) *J. Cell Biol.* **27:**32a.

Gozes, I., and Sweadner, K. J., 1981, Multiple brain tubulin forms are expressed by a single neuron, *Nature* **294:**477–480.

Grigg, G. W., and Hodge, A. J., 1949, Electron microscopic studies of spermatozoa. I. The morphology of the spermatozoan of the common domestic fowl (*Gallus domesticus*), *Austr. J. Sci. Res. Ser. B. Biol. Sci.* **2:**271–286.

Grimstone, A. V., and Klug, A., 1966, Observations on the substructure of flagellar fibers, *J. Cell Sci.* **1:**351–362.

Guttman, S. D., and Gorovsky, M. A., 1979, Cilia regeneration in starved Tetrahymena: An inducible system for studying gene expression and organelle biosynthesis, *Cell* **17:**307–317.

Haimo, L. T., and Telzer, B. R., 1981, Dynein-microtubule interactions: ATP-sensitive dynein binding and the structural polarity of mitotic microtubules, *Cold Spring Harbor Symp. Quant. Biol.* **46:**207–217.

Haimo, L. T., Telzer, B. R., and Rosenbaum, J. L., 1979, Dynein binds to and crossbridges cytoplasmic microtubules, *Proc. Natl. Acad. Sci. USA* **76:**5759–5763.

Hallack, M. E., Rodriguez, J. A., Barra, H. S., and Caputto, R., 1977, Release of tyrosine from tyrosinated tubulin. Some common factors that affect this process and the assembly of tubulin, *FEBS Lett.* **73:**147–150.

Hamel, E., del Campo, A. A., and Lin, C. M., 1983, Microtubule assembly with the guanosine 5'-diphosphate analogue 2',3'-dideoxyguanosine 5'-diphosphate, *Biochem.* **22:**3664–3671.

Harris, P., 1962, Some structural and functional aspects of the mitotic apparatus in sea urchin embryos, *J. Cell Biol.* **14:**475–488.

Heidemann, S. R., and McIntosh, J. R., 1980, Visualization of the structural polarity of microtubules, *Nature (London)* **286:**517–519.

Heidemann, S. R., Zieve, G. G., and McIntosh, J. R., 1980, Evidence for microtubule addition to the distal end of mitotic structures *in vitro*, *J. Cell Biol.* **87:**152–159.

Herzog, W., and Weber, K., 1977, *In vitro* assembly of pure tubulin into microtubules in the absence of microtubule-associated proteins and glycerol, *Proc. Natl. Acad. Sci. USA* **74:**1860–1864.

Herzog, W., and Weber, K., 1978, Microtubule formation by pure brain tubulin in vitro: The influence of dextran and poly(ethylene glycol), *Eur. J. Biochem.* **91:**249–254.

Heuser, J. E., and Kirschner, M. W., 1980, Filament organization revealed in platinum replicas of freeze-dried cytoskeletons, *J. Cell Biol.* **86:**212–234.

Himes, R. H., Kersey, R. N., Ruscha, M., and Houston, L. L., 1976, Cytochalasin A inhibits the *in vitro* polymerization of brain tubulin and muscle actin, *Biochem. Biophys. Res. Commun.* **68:**1362–1370.

Himes, R. H., Burton, P. R., and Gaito, J. M., 1977, Dimethyl sulfoxide-induced self-assembly of tubulin lacking associated proteins, *J. Biol. Chem.* **252:**6222–6228.

Hyams, J. S., and Borisy, G. G., 1978, Nucleation of microtubules *in vitro* by isolated spindle pole bodies of the yeast *Saccharomyces cerevisiae, J. Cell Biol.* **78:**401–414.

Inoue, S., 1952, The effect of colchicine on the microscopic and submicroscopic structure of the mitotic spindle, *Exp. Cell Res. Suppl.* **2:**305–318.

Ishikawa, H., Bischoff, R., and Holtzer, H., 1969, The formation of arrowhead complexes with heavy meromyosin in a variety of cell types, *J. Cell Biol.* **43:**312–328.

Jacobs, M., 1975, Tubulin nucleotide reactions and their role in microtubule assembly and dissociation, *Ann. N.Y. Acad. Sci.* **253:**562–572.

Jacobs, M., 1979, Tubulin and nucleotides, in: *Microtubules* (K. Roberts and J. S. Hyams, eds.), pp. 255–277, Academic Press, London.

Jacobs, M., and Huitorel, P., 1979, Tubulin-associated nucleoside diphosphokinase, *Eur. J. Biochem.* **99:**613–622.

Jacobs, M., Smith, H., and Taylor, E. W., 1974, Tubulin-nucleotide binding and enzymic activity, *J. Mol. Biol.* **89:**455–468.

Jakus, M. A., and Hall, C. E., 1946, Electron microscope observations of the trichocysts and cilia of *Paramecium, Biol. Bull. (Woods Hole)* **91:**141–144.

Jameson, L., and Caplow, M., 1981, Modification of microtubule steady-state dynamics by phosphorylation of the microtubule-associated proteins, *Proc. Natl. Acad. Sci. USA* **78:**3413–3417.

Jameson, L., Frey, T., Zeeberg, B., Dalldorf, F., and Caplow, M., 1980, Inhibition of microtubule assembly by phosphorylation of microtubule-associated proteins, *Biochemistry* **19:**2472–2479.

Job, D., Rauch, C. T., Fischer, E. H., and Margolis, R. L., 1982, Recycling of cold-stable microtubules: Evidence that cold stability is due to substoichiometric polymer blocks, *Biochemistry* **21:**509–515.

Johnson, K. A., and Borisy, G. G., 1977, Kinetic analysis of microtubule self-assembly *in vitro, J. Mol. Biol.* **117:**1–31.

Jolly, J., 1920, Hematies des tylopodes, *C. R. Soc. Biol.* **93:**125–127.

Jolly, J., 1923, *Traite Technique d'Hematologie,* Maloine et Fils, Paris.

Joseph, M. K., Fernstrom, M. A., and Soloff, M. S., 1982, Switching of β- to α-tubulin phosphorylation in uterine smooth muscle of parturient rats, *J. Biol. Chem.* **257:**11728–11733.

Kakuichi, S., and Sobue, K., 1981, Ca^{2+}- and calmodulin-dependent flip-flop mechanism in microtubule assembly-disassembly, *FEBS Lett.* **132:**141–143.

Kalfayan, L., and Wensink, P. C., 1982, Developmental regulation of Drosophila α-tubulin genes, *Cell* **29:**91–98.

Kalfayan, L., Lowenberg, J., and Wensick, P. C., 1981, *Drosophila* α-tubulin genes and their transcription patterns, *Cold Spring Harbor Symp. Quant. Biol.* **46:**183–190.

Karr, T. L., Prodrasky, A. E., and Purich, D. L., 1979, Participation of guanine nucleotides in nucleation and elongation steps of microtubule assembly, *Proc. Natl. Acad. Sci. USA* **76:**5475–5479.

Kemphues, K. J., Raff, R. A., Kaufman, T. C., and Raff, E. C, 1979, Mutation in a structural gene for a beta-tubulin specific to testis in *Drosophila melanogaster, Proc. Natl. Acad. Sci. USA* **76:**3991–3995.

Kemphues, K. J., Raff, E. C., Raff, R. A., and Kaufman, T. C., 1980, Mutation in a testis-specific beta-tubulin in *Drosophila:* Analysis of its effects on meiosis and map location, *Cell* **21:**445–451.

Kemphues, K. J., Kaufman, T. C., Raff, R. A., and Raff, E. C., 1982. The testis-specific β-tubulin in Drosophila melanogaster has multiple functions in spermatogenesis, *Cell* **31:**655–670.

Kim, H., Binder, L. I., and Rosenbaum, J. L., 1979, The periodic association of MAP_2 with brain microtubules *in vitro, J. Cell Biol.* **80:**266–276.

Kirkpatrick, J. B., Hyams, L., Thomas, V. L., and Howley, P. M., 1970, Purification of intact microtubules from brain, *J. Cell Biol.* **47:**389–394.

Kirschner, M. W., 1980, Implication of treadmilling for the stability and polarity of actin and tubulin polymers in vivo, *J. Cell Biol.* **86:**330–334.

Kirschner, M. W., Honig, L. S., and Williams, R. C., 1975, Quantitative electron microscopy of microtubule assembly *in vitro, J. Mol. Biol.* **99:**263–276.

Kobayashi, Y., 1982. Stable microtubules in starfish sperm flagellum: their structure and heterogeneity of tubulin, *J. Biochem.* (*Tokyo*) **92:**1305–1318.

Kobayashi, T., and Flavin, M., 1978, Tubulin-tyrosine ligase: Do invertebrates have it? *J. Cell Biol.* (Abstr.) **79:**285a.

Kowit, J. D., and Fulton, C., 1974a, Programmed synthesis of tubulin for the flagella that develop during cell differentiation in *Naugleria gruberi, Proc. Natl. Acad. Sci. USA* **71:**2877–2881.

Kowit, J. D., and Fulton, C., 1974b, Purification and properties of flagellar outer doublet tubulin from *Naegleria* gruberi and a radioimmune assay for tubulin, *J. Biol. Chem.* **249:**3638–3646.

Krauhs, E., Little, M., Kempf, T., Hofer-Warbinek, R., Ade, W., and Pontingl, H., 1981, Complete amino acid sequence of β-tubulin from porcine brain, *Proc. Natl. Acad. Sci. USA* **78:**4156–4160.

Kumar, N., and Flavin, M., 1981, Preferential action of a brain detyrosinolating carboxypeptidase on polymerized tubulin, *J. Biol. Chem.* **256:**7678–7686.

Kuriyama, R., 1976, *In vitro* polymerization of flagellar and ciliary outer fiber tubulin into microtubules, *J. Biochem.* (*Tokyo*) **80:**153–165.

Kuriyama, R., and Sakai, H., 1974, Role of tubulin -SH groups in polymerization to microtubules. Functional -SH groups in tubulin for polymerization, *J. Biochem.* (*Tokyo*) **76:**651–654.

Lai, E. Y., Walsh, C., Wardell, D., and Fulton, C., 1979, Programmed appearance of translatable flagellar tubulin mRNA during cell differentiation in *Naegleria, Cell* **17:**867–878.

Langford, G. E., 1980, Arrangement of subunits in microtubules with 14 protofilaments, *J. Cell Biol.* **87:**521–526.

Ledbetter, M. C., and Porter, K. R., 1963, A microtubule in plant fine structure, *J. Cell Biol.* **19:**239–250.

Ledbetter, M. C., and Porter, K. R., 1964, The morphology of microtubules of plant cells, *Science* **144:**872–874.

Lee, M. G.-S., Lewis, S. A., Wilde, C. D., and Cowan, N. J., 1983, Evolutionary history of a multigene family: an expressed human β-tubulin gene and three processed pseudogenes, *Cell* **33:**477–487.

Lee, S.-H., Kristofferson, D., and Purich, D. L., 1982, Microtubule interactions with GDP provide evidence that assembly-disassembly properties depend on the method of brain microtubulle protein isolation, *Biochem. Biophys. Res. Commun.* **105:**1605–1610.

Lee, Y. C., and Wolff, J., 1982, Two opposing effects of calmodulin on microtubule assembly depend on the presence of microtubule-associated proteins, *J. Biol. Chem.* **257:**6306–6310.

Lefebvre, P. A., Nordstrom, S. A., Moulder, J. E., and Rosenbaum, J. L., 1978, Flagellar elongation and shortening in *Chlamydomonas.* IV. Effects of flagellar detachment, regeneration, and resorption on the induction of flagellar protein synthesis, *J. Cell Biol.* **78:**8–27.

Lefebvre, P. A., Silflow, C. D., Wieben, E. D., and Rosenbaum, J. L., 1980, Increased levels of mRNAs for tubulin and other flagellar proteins after amputation or shortening of *Chlamyodomonas* flagella, *Cell* **20:**469–477.

L'Hernault, S. W., and Rosenbaum, J. L., 1983, Chlamydomonas α-tubulin is posttranslationally modified in the flagella during flagellar assembly, *J. Cell Biol.* **97:**258–263.

Linck, R. W., and Amos, L. A., 1974, The hands of helical lattices in flagellar doublet microtubules, *J. Cell Sci.* **14:**551–559.

Linck, R. W., and Langevin, G. L., 1981, Reassembly of flagellar B (αβ) tubulin into singlet microtubules. Consequences for cytoplasmic microtubule structure and assembly, *J. Cell Biol.* **89:**323–337.

Lockwood, A. H., 1978, Tubulin assembly protein: Immunochemical and immunofluorescent studies on its function and distribution in microtubules and cultured cells, *Cell* **13:**613–628.

Lockwood, A., Penningroth, S. M., and Kirschner, M. W., 1975, Function of GTP in microtubule formation, *Fed. Proc.* (Abstr.) **34:**540.

Ludford, R. J., 1936, The action of toxic substances upon the division of normal and malignant cells in vitro and in vivo, *Arch. Exp. Zellforsch.* **18:**411–441.

Luduena, R. F., Shooter, E. M., and Wilson, L., 1977, Structure of the tubulin dimer, *J. Biol. Chem.* **252:**7006–7114.

Maccioni, R. B., and Seeds, N. W., 1982, Residual nucleotide and tubulin's ability to polymerize with nucleotide analogues, *J. Biol. Chem.* **257:**3334–3338.

Maki, R., Roeder, W., Traunecker, A., Sidman, C., Wabl, M., Raschke, W., and Tonegawa, S., 1981, The role of DNA rearrangement and alternative RNA processing in the expression of immunoglobulin delta genes, *Cell* **24:**353–365.

Mandelkow, E.-M., and Mandelkow, E., 1979, Junctions between microtubule walls, *J. Mol. Biol.* **129:**135–148.

Mandelkow, E., Mandelkow, E.-M., and Bordas, J., 1983, Structure of tubulin rings studied by x-ray scattering using synchrotron radiation, *J. Mol. Biol.* **167:**179–196.

Mandelkow, E., Thomas, J., and Cohen, C., 1977, Microtubule structure at low resolution by x-ray diffraction, *Proc. Natl. Acad. Sci. USA* **74:**3370–3374.

Manton, I., and Clarke, B., 1952, An electron microscope study of spermatozoid of *Sphagnum*, *J. Exp. Bot.* **3:**265–275.

Marcaud, L., and Hayes, D., 1979, RNA synthesis in starved deciliated *Tetrahymena pyriformis*, *Eur. J. Biochem.* **98:**267–273.

Marcum, J. M., and Borisy, G. G., 1978, Characterization of microtubule protein oligomers by analytical ultracentrifugation, *J. Biol. Chem.* **253:**2825–2833.

Marcum, J. M., Dedman, J. R., Brinkley, B. R., and Means, A. R., 1978, Control of microtubule assembly-disassembly by calcium-dependent regulator protein, *Proc. Natl. Acad. Sci. USA* **75:**3771–3775.

Margolis, R. L., 1981, Role of GTP hydrolysis in microtubule treadmilling and assembly, *Proc. Natl. Acad. Sci. USA* **78:**1586–1590.

Margolis, R. L., and Wilson, L., 1978, Opposite end assembly and disassembly of microtubules at steady state *in vitro*, *Cell* **13:**1–8.

Margolis, R. L., and Wilson, L., 1979, Regulation of the microtubule steady state in vitro by ATP, *Cell* **18:**673–679.

Marie, J., Simon, M.-P., Dreyfus, J.-C., and Kahn, A., 1981, One gene, but two messenger RNAs encode liver L and red cell L′ pyruvate kinase subunits, *Nature (London)* **292:**70–72.

Matus, A., Bernhardt, R., and Hugh-Jones, T., 1981, High-molecular weight microtubule-associated proteins are preferentially associated with dendritic microtubules in brain, *Proc. Natl. Acad. Sci. USA* **78:**3010–3014.

McEwen, B., and Edelstein, S. J., 1980, Evidence for a mixed lattice in microtubules reassembled *in vitro*, *J. Mol. Biol.* **139:**123–145.

McGill, M., and Brinkley, B. R., 1975, Human chromosomes and centrioles as nucleating sites for the *in vitro* assembly of microtubules from bovine brain tubulin, *J. Cell Biol.* **67:**189–199.

McKeithan, T. W., and Rosenbaum, J. L., 1981, Multiple forms of tubulin in the cytoskeletal and flagellar microtubules of *Polytomella*, *J. Cell Biol.* **91:**352–360.

McKeithan, T. W., Lefebvre, P. A., Silflow, C. D., and Rosenbaum, J. L., 1979, Posttranslational modification of flagellar tubulin, *J. Cell Biol.* (Abstr.) **83:**338a.

McKeithan, T., W., Lefebvre, P. A., Silflow, C. D., and Rosenbaum, J. L., 1983, Tubulin heterogeneity in *Polytomella* and *Chlamydomonas*. Evidence for a precursor of flagellar alpha tubulin, *J. Cell Biol.*, **96:**1056–1063

Mellon, M., and Rebhun, L. I., 1976, Sulfhydryls and in vitro polymerization of tubulin, *J. Cell Biol.* **70:**226–238.

Merlino, G. T., Chamberlin, J. P., and Kleinsmith, L. J., 1978, Effects of deciliation on tubulin messenger RNA activity in sea urchin embryos, *J. Biol. Chem.* **253:**7078–7085.

Moore, K. W., Roger, J., Hunkapiller, T., Early, P., Nottemburg, C., Weissman, I., Bazin, H., Wall, R., and Hood, L. E., 1981, Expression of IgD may use both DNA rearrangement and RNA splicing mechanisms, *Proc. Natl. Acad. Sci. USA* **78:**1800–1804.

Morris, N. R., Lai, M. H., and Oakley, C. E., 1979, Identification of a gene for α-tubulin in *Aspergillus nidulans*, *Cell* **16:**437–442.

Murphy, D. B., and Borisy, G. G., 1975, Association of high molecular weight proteins with microtubules and their role in microtubule assembly *in vitro*, *Proc. Natl. Acad. Sci. USA* **72**:2696–2700.

Murphy, D. B., Johnson, K. A., and Borisy, G. G., 1977, Role of tubulin-associated proteins in microtubule nucleation and elongation, *J. Mol. Biol.* **117**:33–52.

Nagano, T., and Suzuki, F., 1975, Microtubules with 15 subunits in cockroach epidermal cells, *J. Cell Biol.* **64**:242–245.

Nath, J., and Flavin, M., 1978, A structural difference between cytoplasmic and membrane-bound tubulin of brain, *FEBS Lett.* **95**:335–338.

Nath, J., Flavin, M., and Gallin, J. I., 1982, Tubulin tyrosinolation in human polymorphonuclear leukocytes: studies in normal subjects and in patients with the Chediak–Higashi syndromes, *J. Cell Biol.* **95**:519–526.

Nath, J., Whitlock, J., and Flavin, M., 1978, Tyrosylation of tubulin in synchronized HeLa cells, *J. Cell Biol.* (Abstr.) **79**:294a.

Nishida, E., and Sakai, H., 1977, Calcium-sensitivity of the microtubule reassembly system. Difference between crude brain extract and purified microtubule proteins, *J. Biochem.* (*Tokyo*) **82**:303–306.

Oakley, B. R., and Morris, N. R., 1981, A beta-tubulin mutation in *Aspergillus nidulans* that blocks microtubule function without blocking assembly, *Cell* **24**:837–845.

Oliver, J. M., Albertini, D. F., and Berlin, R. D., 1976, Effects of glutathione-oxidizing agents on microtubule assembly and microtubule-dependent surface properties of human neutrophils, *J. Cell Biol.* **71**:921–932.

Olmsted, J. B., and Borisy, G. G., 1975, Ionic and nucleotide requirements for microtubule polymerization *in vitro*, *Biochemistry* **14**:2996–3005.

Olmsted, J. B., Carlson, K., Klebe, R., Ruddle, F., and Rosenbaum, J., 1970, Isolation of microtubule protein from cultured mouse neuroblastoma cells, *Proc. Natl. Acad. Sci. USA* **65**:129–136.

Olmsted, J. B., Marcum, J. M., Johnson, K. A., Allen, C., and Borisy, G. G. 1974, Microtubule assembly: Some possible regulatory mechanisms, *J. Supramol. Struct.* **2**:429–450.

Oosawa, F., and Asakura, S., 1975, *Thermodynamics of the Polymerization of Protein* Academic Press, London.

Pantaloni, D., Carlier, M.-F., Simon, C., and Batelier, G., 1981, Mechanism of tubulin assembly: Role of rings in the nucleation process and of associated proteins in the stabilization of microtubules, *Biochemistry* **20**:3709–4716.

Papaconstantinou, E., and Pantaloni, D., 1982, Mechanism of tubulin assembly: Evidence for GTP hydrolysis on microtubule precursors, *Biol. Cell* (Abstr.) **45**:257.

Pease, D. C., 1963, The ultrastructure of flagellar fibrils, *J. Cell Biol.* **18**:313–326.

Penningroth, S. M., Cleveland, D. W., Kirschner, M. W., 1976, *In vitro* studies of the regulation of microtubule assembly, in: *Cell Motility* (R. Goldman, T. Pollard, J. Rosenbaum, eds.), pp. 1233–1258, Cold Spring Harbor Laboratory.

Pepper, D. A., and Brinkley, B. R., 1979, Microtubule initiation at kinetochores and centrosomes in lysed mitotic cells, *J. Cell Biol.* **82**:585–591.

Perlman, D., and Halvorson, H. O., 1981, Distinct repressible mRNAs for cytoplasmic and secreted yeast invertase are encoded by a single gene, *Cell* **25**:526–536.

Pernice, B., 1889, Sulla cariocinesi delle cellule epiteliali e dell'endotelio dei vasi della mucosa dello stomaco e dell'intestino, nelle studio della gastroenterite sperimentale (nell'avvelenamento per colchico), *Sicilia Med.* **1**:265–279.

Pickett-Heaps, J. D., 1969, The evolution of the mitotic apparatus: An attempt at comparative ultrastructural cytology in dividing plant cells, *Cytobios* **1**:257–280.

Pierson, G. B., Burton, P. R., and Himes, R. H., 1978, Alterations in number of protofilaments in microtubules assembled in vitro, *J. Cell Biol.* **76**:223–228.

Piperno, G., and Luck, D. J., 1976, Phosphorylation of axonemal proteins in *Chlamydomonas reinhardtii*, *J. Biol. Chem.* **251**:2161–2167.

Piperno, G., and Luck, D. J., 1977, Microtubule proteins of *Chlamydomonas reinhardtii*. An immunochemical study based on the use of an antibody specific for the β-tubulin subunit, *J. Biol. Chem.* **252**:383–391.

Piras, R., and Piras, M. M., 1975, Changes in microtubule phosphorylation during cell cycle of HeLa cells, *Proc. Natl. Acad. Sci. USA* **72:**1161–1165.

Ponstingl, H., Krauhs, E., Little, M., and Kempf, T., 1981, Complete amino acid sequence of α-tubulin from porcine brain, *Proc. Natl. Acad. Sci. USA* **78:**2757–2761.

Purich, D. L., Terry, B. J., MacNeal, R. K., and Karr, T. L., 1982, Characterization of tubulin and microtubule-associated protein interactions with guanine nucleotides and their non-hydrolyzable analogues, *Methods Enzym.* **85:**416–433.

Ranvier, L., 1875, Recherches sur les elements du sang, *Arch. Physiol.* **2:**1–15.

Raybin, D., and Flavin, M., 1975, An enzyme tyrosylating α-tubulin and its role in microtubule assembly, *Biochem. Biophys. Res. Commun.* **65:**1088–1095.

Raybin, D., and Flavin, M., 1977, Modification of tubulin by tyrosylation in cells and extracts and its effect on assembly *in vitro, J. Cell Biol.* **73:**492–504.

Rebhun, L. I., Miller, M., Schnaitman, T. C., Nath, J., and Mellon, M., 1976, Cyclic nucleotides, thioldisulfide status of proteins, and cellular control processes, *J. Supramol. Struct.* **5:**199–219.

Rebhun, L. I., Jemiolo, D., Keller, T., Burgess, W., and Kretsinger, R., 1980, Calcium, calmodulin, and control of assembly of brain and spindle microtubules, in: *Microtubules and Microtubule Inhibitors* (M. De Brabander and J. De Mey, eds.), pp. 243–252, Elsevier/North Holland Biomedical Press, Amsterdam.

Roberts, K., and Hyams, J. S. (eds.), 1979, *Microtubules,* Academic Press, London.

Rodnan, G. P., and Benedek, T. G., 1970, The early history of antirheumatic drugs, *Arth. Rheum.* **13:**145–165.

Rodriguez, J. A., and Borisy, G. G., 1977, Developmental studies on the tyrosination of chick brain tubulin, *J. Cell Biol.* (Abstr.) **75:**296a.

Rogers, J., Early, P., Carter, C., Calame, K., Bond, M., Hood, L., and Wall, R., 1980, Two mRNAs with different 3′ ends encode membrane-bound and secreted forms of immunoglobulin μ chain, *Cell* **20:**303–312.

Rosenbaum, J. L., and Child, F. M., 1967, Flagellar regeneration in protozoan flagellates, *J. Cell Biol.* **34:**345–364.

Roth, L. E., and Daniels, E. W., 1962, Electron microscope studies of mitosis in amebae. II. The giant ameba *Pelomyxa carolinensis, J. Cell Biol.* **12:**57–78.

Sabatini, D. D., Bensch, K., and Barrnett, R. J., 1963, Cytochemistry and electron microscopy. The preservation of cellular ultrastructure and enzymatic activity by aldehyde fixation, *J. Cell Biol.* **17:**19–58.

Safer, D., 1973, Comparison of ciliary and flagellar microtubule subunits, *J. Cell Biol.* (Abstr.) **59:**299a.

Sakai, H., Mohri, H., and Borisy, G. G., 1982, *Biological Functions of Microtubules and Related Structures,* Academic Press, Tokyo.

Sandoval, I. V., and Weber, K., 1981, Different tubulin polymers are produced by microtubule-associated proteins MAP_2 and tau in the presence of guanosine 5′-(α,β methylene)triphosphate, *J. Biol. Chem.* **255:**8952–8954.

Sandoval, I. V., MacDonald, E., Jameson, L. J., and Cuatrecasas, P., 1977, Role of nucleotide in tubulin polymerization: Effect of guanylyl 5′-methylenediphosphonate, *Proc. Natl. Acad. Sci. USA* **74:**4881–4885.

Sandoval, I. V., Jameson, J. L., Niedel, J., MacDonald, E., and Cuatrecasas, P., 1978, Role of nucleotides in tubulin polymerization. Effect of guanosine 5′-methylene diphosphonate, *Proc. Natl. Acad. Sci. USA* **75:**3178–3182.

Schliwa, M., 1980, Pharmacological evidence for an involvement of calmodulin in calcium-induced microtubule disassembly in lysed tissue culture cells, in: *Microtubules and Microtubule Inhibitors* (M. De Brabander and J. De Mey, eds.), pp. 57–70, Elsevier/North Holland Biomedical Press, Amsterdam.

Schliwa, M., Euteneuer, U., Bulinski, J. C., and Izant, J. G., 1981, Calcium lability of cytoplasmic microtubules and its modulation by microtubule-associated proteins, *Proc. Natl. Acad. Sci. USA* **78:**1037–1041.

Schmitt, F. O., Hall, C. E., and Jakus, M. A., 1943, The ultrastructure of protoplasmic fibrils, *Biol. Symp.* **10:**261–276.
Sheele, R. B., and Borisy, G. G., 1979, *In vitro* assembly of microtubules, in: *Microtubules* (K. Roberts and J. S. Hyams, eds.), pp. 175–254, Academic Press, London.
Sheele, R. B., Bergen, L. G., and Borisy, G. G., 1982, Control of the structural fidelity of microtubules by initiation sites, *J. Mol. Biol.* **154:**485–500.
Sheir-Neiss, G., Lai, M. H., and Morris, N. R., 1978, Identification of a gene for β-tubulin in *Aspergillus nidulans, Cell* **15:**638–647.
Shelanski, M. L., and Taylor, E. W., 1967, Isolation of a protein subunit from microtubules, *J. Cell Biol.* **34:**549–554.
Shelanski, M. L., and Taylor, E. W., 1968, Properties of the protein subunit of central pair and outer-doublet microtubules of sea urchin flagella, *J. Cell Biol.* **38:**304–315.
Shriver, K., and Byers, B., 1977, Yeast microtubules: Constituent proteins and their synthesis, *J. Cell Biol.* (Abstr.) **75:**297a.
Silflow, C. D., and Rosenbaum, J. L., 1981, Multiple alpha- and beta-tubulin genes in *Chlamyodomonas* and regulation of tubulin mRNA levels after deflagellation, *Cell* **24:**81–88.
Silflow, C. D., Lefebvre, P. A., McKeithan, T. W., Schloss, J. A., Keller, L. R., and Rosenbaum, J. L., 1981, Expression of flagellar protein genes during flagellar regeneration in *Chlamydomonas, Cold Spring Harbor Symp. Quant. Biol.* **45:**157–169.
Slautterback, D. B., 1961, A fine tubular component of secretory cells, *Am. Soc. Cell Biol.* (Abstr.) **199.**
Slautterback, D. B., 1963, Cytoplasmic microtubules. I. Hydra, *J. Cell Biol.* **18:**367–388.
Sloboda, R. D., and Rosenbaum, J. L., 1977, Decoration of intact, smooth-walled microtubules with microtubule-associated proteins, *J. Cell Biol.* (Abstr.) **76:**286a.
Sloboda, R. D., Dentler, W. L., Bloodgood, R. A., Telzer, B., Granett, S., and Rosenbaum, J. L., 1976, Microtubule-associated proteins (MAPs) and the assembly of microtubules in vitro, in: *Cell Motility* (R. Goldman, T. Pollard, and J. Rosenbaum, eds.), pp. 987–1005, Cold Spring Harbor Laboratory, New York.
Sloboda, R. D., Rudolph, S. A., Rosenbaum, J. L., and Greengard, P., 1975, Cyclic AMP-dependent endogenous phosphorylation of a microtubule-associated protein, *Proc. Natl. Acad. Sci. USA* **72:**177–181.
Smith, D. S., 1971, On the significance of cross-bridges between microtubules and synaptic vesicles, *Phil Trans. R. Soc. Ser. B. Biol. Sci.* **261:**395.
Snell, W. J., Dentler, W. L., Haimo, L. T., Binder, L. I., and Rosenbaum, J. L., 1974, Assembly of chick brain tubulin onto isolated basal bodies of *Chlamydomonas reinhardii, Science* **185:**357–360.
Sobue, K., Fujita, M., Muramoto, Y., and Kakuichi, S., 1981, The calmodulin-binding protein in microtubules is tau factor, *FEBS Lett.* **132:**137–143.
Solomon, F., Gyson, R., Rentsch, M., and Monard, D., 1976, Purification of tubulin from neuroblastoma cells: Absence of covalently bound phosphate in tubulin from normal and morphologically differentiated cells, *FEBS Lett.* **63:**316–319.
Stearns, M. E., and Brown, D. L., 1979, Purification of cytoplasmic tubulin and microtubule organizing center proteins functioning in microtubule initiation from the alga *Polytomella, Proc. Natl. Acad. Sci. USA* **76:**5745–5749.
Stearns, M. E., and Brown, D. L., 1981, Microtubule organizing centers (MTOCs) of the alga *Polytomella* exert spatial control over microtubule initiation *in vivo* and *in vitro, J. Ultrastruct. Res.* **77:**366–378.
Stearns, M. E., Connolly, J. A., and Brown, D. L., 1976, Cytoplasmic microtubule organizing centers isolated from *Polytomella agilis, Science* **191:**188–191.
Stephens, R. E., 1970, Thermal fractionation of outer doublet microtubules into A- and B-tubulin, *J. Mol. Biol.* **47:**353–363.
Stephens, R. E., 1975, Structural chemistry of the axoneme: Evidence for chemically and functionally unique tubulin dimers in outer fibers, in: *Molecules and Cell Movement* (S. Inoue and R. E. Stephens, eds.), Raven Press, New York.

Stephens, R. E., 1977, Differential protein synthesis and utilization during cilia formation in sea urchin embryos, *Dev. Biol.* **61:**311–329.
Stephens, R. E., 1978, Primary structural differences among tubulin subunits from flagella, cilia, and the cytoplasm, *Biochemistry* **14:**2882–2891.
Stephens, R. E., 1982, Equimolar heterodimers in microtubules, *J. Cell Biol.* **94:**263–270.
Summers, K., and Kirschner, M. W., 1979, Characteristics of the polar assembly and disassembly of microtubules observed in vitro by dark-field light microscopy, *J. Cell Biol.* **83:**205–217.
Taylor, E. W., 1965, The mechanism of colchicine inhibition of mitosis. I. Kinetics of inhibition and the binding of ^{3}H-colchicine, *J. Cell Biol.* **25:**145–160.
Telzer, B. R., and Haimo, L. T., 1981, Decoration of spindle microtubules with dynein: Evidence for uniform polarity, *J. Cell Biol.* **89:**373–378.
Telzer, B. R., Moses, M. J., and Rosenbaum, J. L., 1975, Assembly of microtubules onto kinetochores of isolated mitotic chromosomes of HeLa cells, *Proc. Natl. Acad. Sci. USA* **72:**4023–4027.
Thomashow, L. S., Milhausen, M., Rutter, W. J., and Agabian, N., 1983. Tubulin genes are tandemly linked and clustered in the genome of Trypanosoma brucei, *Cell* **32:**35–43.
Thompson, W. C., Deanin, G. G., and Gordon, M. W., 1979, Intact microtubules are required for rapid turnover of carboxyl-terminal tyrosine of α-tubulin in cell cultures, *Proc. Natl. Acad. Sci. USA* **76:**1318–1322.
Tilney, L. G., 1971, Origin and continuity of microtubules, in: *Origin and Continuity of Cell Organelles* (J. Reinert and H. Ursprung, eds.), pp. 222–260, Springer, Berlin.
Tilney, L. G., Bryan, J., Bush, D. J., Fugiwara, K., Mooseker, M. S., Murphy, D. B., and Snyder, D. H., 1973, Microtubules: Evidence for 13 protofilaments, *J. Cell Biol.* **59:**267–275.
Tucker, J. B., Dunn, M., and Pattesson, J. B., 1975, Control of microtubule pattern during the development of a large organelle in the ciliate *Nassula, Dev. Biol.* **47:**439–453.
Valenzuela, P., Quiroga, M., Zaldivar, J., Rutter, W. J., Kirschner, M. W., and Cleveland, D. W., 1981, Nucleotide and corresponding amino acid sequences encoded by α and β tubulin mRNAs, *Nature (London)* **289:**650–655.
Vallee, R., and Borisy, G. G., 1977, Removal of the projections from cytoplasmic microtubules *in vitro* by digestion with trypsin, *J. Biol. Chem.* **252:**377–382.
Water, R. D., and Kleinsmith, L. J., 1976, Identification of α and β tubulin in yeast, *Biochem. Biophys. Res. Commun.* **70:**704–708.
Weatherbee, J. A., Sherline, P., Mascardo, R. N., Izart, J. G., Luftig, R. B., and Weihing, R. R., 1982, Microtubule-associated proteins of HeLa cells: Heat stability of the 200,000 mol wt HeLa MAPs and detection of the presence of MAP_2 in HeLa cell extracts and cycled microtubules, *J. Cell Biol.* **92:**155–163.
Weeks, D. P., and Collis, P. S., 1976, Induction of microtubule protein synthesis in *Chlamydomonas reinhardii* during flagellar regeneration, *Cell* **9:**15–27.
Wegener, A., 1976, Head to tail polymerization of actin, *J. Mol. Biol.* **108:**139–150.
Weingarten, M. D., Lockwood, A. H., Hwo, S. Y., and Kirschner, M. W., 1975, A protein factor essential for microtubule assembly, *Proc. Natl. Acad. Sci. USA* **72:**1858–1862.
Weisenberg, R. C., 1972, Microtubule formation *in vitro* in solutions containing low calcium concentration, *Science* **177:**1104–1105.
Weisenberg, R. C., 1974, The role of ring aggregates and other structures in the assembly of microtubules, *J. Supramol. Struct.* **2:**451–465.
Weisenberg, R. C., 1980a, A microtubule assembly model, in: *Microtubules and Microtubule Inhibitors* (M. De Brabander and J. De Mey, eds.), pp. 161–172, Elsevier/North Holland Biomedical Press, Amsterdam.
Weisenberg, R. C., 1980b, Role of co-operative interactions, microtubule-associated proteins and guanosine triphosphate in microtubule assembly: A model, *J. Mol. Biol.* **139:**660–667.
Weisenberg, R. C., Borisy, G. G., and Taylor, E. W., 1968, The colchicine binding protein of mammalian brain and its relationship to microtubules, *Biochemistry* **7:**4466–4479.
Weisenberg, R. C., Deery, W. J., and Dickenson, P. J., 1976, Tubulin-nucleotide interaction

during the polymerization and depolymerization of microtubules, *Biochemistry* **15**:4248–4254.

Wilde, C. D., Crowther, C. E., Cripe, T. P., Lee, M. G.-S., and Cowan, N. J., 1982, Evidence that a human β-tubulin gene is derived from its corresponding mRNA, *Nature* **292**:83–84.

Witman, G. B., 1975, The site of *in vivo* assembly of flagellar microtubules, *Ann. N.Y. Acad. Sci.* **253**:178–191.

Witt, P. L., Ris, H., and Borisy, G. G., 1980, Origin of kinetochore microtubules in Chinese hamster ovary cells, *Chromosoma (Berlin)* **81**:483–505.

Wuerker, R. B., and Kirkpatrick, J. B., 1972, Neuronal microtubules, neurofilaments, and microfilaments, *Int. Rev. Cytol.* **33**:45–75.

Wuerker, R. B., and Palay, S. L., 1969, Neurofilaments and microtubules in anterior horn cells of the rat, *Tissue Cell* **1**:387–402.

Yamada, K. M., Spooner, B. S., and Wessells, N. K., 1971, Ultrastructure and function of growth cones and axons of cultured nerve cells, *J. Cell Biol.* **49**:614–635.

Zabrecky, J. R., and Cole, R. D., 1982a, Binding of ATP to tubulin, *Nature* **296**:775–776.

Zabrecky, J. R., and Cole, R. D., 1982b, Effect of ATP on the kinetics of microtubule assembly, *J. Biol. Chem.* **257**:4633–4638.

Zackroff, V., and Weisenberg, R. C., 1978, Microtubule assembly competence of GDP-tubulin, *J. Supramol. Struct.* (Abstr.) **7**:327.

Zackroff, R. V., Weisenberg, R. C., and Deery, W. J., 1980, Equilibrium and kinetic analysis of microtubule assembly in the presence of guanosine diphosphate, *J. Mol. Biol.* **139**:641–677.

Zeeberg, B., and Caplow, M., 1981, An isoenergetic exchange mechanism which accounts for tubulin-GDP stabilization of microtubules, *J. Biol. Chem.* **256**:12051–12057.

8

MAP_2 (Microtubule-Associated Protein 2)

Richard B. Vallee

1. General Background

Microtubules are known to play a role in a wide variety of cellular processes. The major component of these structures is tubulin, a globular protein that makes up the microtubule wall. With the introduction of procedures for purifying microtubules (Weisenberg, 1972) it soon became clear that they contained a number of proteins in addition to tubulin (Borisy *et al.*, 1975; Sloboda *et al.*, 1975; Weingarten *et al.*, 1975). These proteins have been referred to by the acronym MAPs, or microtubule-associated proteins (Sloboda *et al.*, 1975). At least some of these proteins represent fine filamentous projections regularly arranged on the microtubule surface (Murphy and Borisy, 1975; Dentler *et al.*, 1975). This suggests that the MAPs are involved in mediating the interaction of microtubules with other components of the cell, while tubulin itself makes up the structural backbone of the microtubule. In this view, understanding the MAPs may ultimately provide answers to two key questions regarding how cells work. What precisely do microtubules do in cells, and how do they do it?

The MAPs have been found to have an additional property that has received considerable attention. In the absence of the MAPs, tubulin assembles poorly, if at all, under most *in vitro* conditions. The MAPs dramatically promote the assembly of tubulin into microtubules (Murphy and Borisy, 1975; Weingarten *et al.*, 1975). This suggests that, in addition to being centrally involved in microtubule function, the MAPs may also be involved in regulating the assembly state of microtubules in the cell.

Richard B. Vallee • Cell Biology Group, The Worcester Foundation for Experimental Biology, Shrewsbury, Massachusetts 01545.

Numerous nontubulin proteins have been detected in microtubule preparations. These proteins are probably not all MAPs. A simple definition of a MAP would be any protein that has a specific binding site for microtubules. Proof of the existence of such a site is not necessarily a simple undertaking and extensive documentation has been produced for only a limited number of proteins. These are listed in Table 1.

Of these proteins, MAP_2 is the most extensively characterized. It is the most prominent MAP in brain tissue which, in turn, is the richest source for microtubules. It has been possible to obtain MAP_2 in quantities sufficient for structural and enzymological analysis, something that has been more difficult for some other prominent MAPs and impossible for others. While MAP_2 may prove to have some unique properties, it appears to have several important properties that are characteristic of MAPs in general. Therefore, the information that has been obtained on MAP_2 may prove useful not only for understanding this protein, but for understanding other MAPs as well. In this chapter, I will review the existing evidence on the biochemical properties of MAP_2 as well as on its distribution in cells. In addition, I will attempt to interpret the existing evidence in terms of the possible functional roles of MAPs in the cell.

Table 1. Well Characterized Microtubule Associated Proteins

Name	M_r	Principal source	Structural properties	References
MAP_1	350,000	Brain tissue	Projection on microtubule surface	Sloboda *et al.*, 1975; Vallee and Davis, 1983
MAP_2	270,000		Projection on microtubule surface	Herzog and Weber, 1978; Kim *et al.*, 1979
	70,000		Protein associated with MAP_2, function unknown	Vallee *et al.*, 1981
Tau	55,000–62,000		Asymmetric	Cleveland *et al.*, 1977a,b
Type II cAMP-dependent protein kinase	54,000; 39,000		Enzyme associated with MAP_2	Vallee *et al.*, 1981; Theurkauf and Vallee, 1982
LMW MAPs	28,000; 30,000		Light chains of MAP_1	Berkowitz *et al.*, 1977; Vallee and Davis, 1983
	255,000 210,000 125,000	Cultured cells	Asymmetric Asymmetric Asymmetric	Bulinski and Borisy, 1979; Weatherbee *et al.*, 1980; Olmsted and Lyon, 1981; Duerr *et al.*, 1981
	80,000		Unknown	Duerr *et al.*, 1981
	69,000		Unknown	Duerr *et al.*, 1981

2. *Purification*

MAP_2 has been purified using two approaches. In one, microtubules are exposed to elevated temperature (90–100°C) which coagulates tubulin, MAP_1, and a variety of minor proteins (Fellous *et al.*, 1977; Herzog and Weber, 1978; Kim *et al.*, 1979). Centrifugation yields MAP_2 and tau almost quantitatively in the supernate. MAP_2 can be separated from tau by gel filtration chromatography. This approach produces MAP_2 in excellent yield. The protein promotes microtubule assembly and binds to microtubules with apparently normal stoichiometry (see below; Herzog and Weber, 1978; Kim *et al.*, 1979) via the same portion of the molecule as for MAP_2 produced by other means (Vallee, 1980a). MAP_2 produced in this manner is devoid of associated proteins and enzymatic activity (Vallee *et al.*, 1981).

MAP_2 can also be purified by traditional chromatographic means (Vallee *et al.*, 1981). A fraction of the MAP_2 in microtubule preparations exists in aggregated form along with other protein species. Free MAP_2 can be separated from the aggregated material. The free MAP_2 can then be readily purified by further chromatography. This preparation of MAP_2 contains at least three low molecular weight polypeptide species (Vallee *et al.*, 1981), two of which represent the subunits of an associated protein kinase (Theurkauf and Vallee, 1982; see below).

The heat stability of MAP_2 has led to an operational biochemical definition of the protein. This has proven to be quite useful since MAP_2 consists of more than one electrophoretic species (see below) and must be distinguished from several other species of similar electrophoretic mobility. Hence, MAP_2 is a protein of M_r ~270,000 that is capable of binding to microtubules and remains soluble at elevated temperatures.

3. *Physical Properties*

All available evidence indicates that MAP_2 is a large, highly asymmetric molecule. The subunit molecular weight based on polyacrylamide gel electrophoresis in the presence of SDS has been estimated to be from 270,000–300,000 (Borisy *et al.*, 1975; Sloboda *et al.*, 1975). Actually, the protein runs as two bands separated by approximately 15,000 molecular weight. It is possible that the smaller polypeptide is derived from the larger by proteolytic modification. However, both bands are observed even when tissue is solubilized directly in SDS gel sample buffer. In addition, the ratio of intensity of the bands seems to be characteristic of the animal species of origin. This suggests that the smaller polypeptide is not generated as an artifact of purification. Both polypeptides can be phosphorylated *in vitro* and persist after dephosphorylation (Theurkauf and Vallee, unpublished results), indicating that the two species do not simply represent phospho- and dephospho-MAP_2. Existing antibodies to MAP_2 recognize both bands (Bloom and Vallee, 1983, unpublished results) indicating that the two forms of MAP_2 are closely related. The two polypeptides must also be quite similar in size, shape, and charge, since they are not detectably separated by gel filtration or ion exchange chromatography. No evidence exists that they form a heterodimer.

MAP_2 is difficult to observe by negative stain electron microscopy, apparently due to a tendency to collapse under negative staining conditions (Zingsheim *et al.*, 1979). In thin sections of microtubules, MAP_2 was observed in the same study as a fine projection on the microtubule surface up to 300–400 Å in length. In metal shadowed microtubule preparations, a similar length was obtained. In another study in which unfixed MAP_2 was examined and glycerol was used during sample drying, the protein was observed to protrude up to 900 Å from the microtubule wall (Voter and Erickson, 1982). The free MAP_2 molecule was observed in the latter study to have an even greater length, up to 1650–1800 Å. This suggests that the part of MAP_2 that binds to the microtubule wall (see below) could be as much as 750–900 Å in length. Alternatively, some fraction of dimeric MAP_2 could have been present in the preparations examined.

The asymmetry of the protein is also indicated by its hydrodynamic properties. The sedimentation coefficient is approximately 3–4.5 s (Vallee and Borisy, 1978; Bulinski and Borisy, 1980). If MAP_2 were globular, an s value of approximately 12–14 s would be expected. MAP_2 is sufficiently asymmetric to actually decrease the sedimentation rate of ring-shaped tubulin oligomers by 25% (Vallee and Borisy, 1978). The Stokes radius has been reported to be approximately 15 nm (Bulinski and Borisy, 1980), similar to that of myosin, an extremely elongated molecule. Unlike myosin, MAP_2 has a low content of α-helix (Voter and Erickson, 1982). MAP_2 has been reported to exist in solution as a monomer based on sedimentation and gel filtration data (Bulinski and Borisy, 1980). In addition, a sizeable fraction of MAP_2 in any given preparation exists in an aggregated form (Vallee *et al.*, 1981). Whether the larger MAP_2 species if of biological importance is not known.

4. Domain Structure

MAP_2 is extremely susceptible to attack by proteolytic enzymes. Under conditions that do not result in detectable fragmentation of tubulin, the electrophoretic bands corresponding to MAP_2 could be completely destroyed (Vallee and Borisy, 1977). Microtubule assembly was only slightly affected during the period of time required to eliminate the MAP_2 bands. This suggested that the part of MAP_2 that controlled assembly survived the protease treatment.

We subsequently identified a group of MAP_2 fragments of 32,000–39,000 daltons that bound to microtubules (Vallee, 1980a; see Fig. 1). These fragments promoted microtubule assembly and competed with intact MAP_2 for binding to the microtubule surface. Binding was saturable as is the case for intact MAP_2 (Vallee, 1980b). The fragments showed identical peptides when further digested with *S. aureus* V-8 protease (M. DiBartolomeis and R. Vallee,

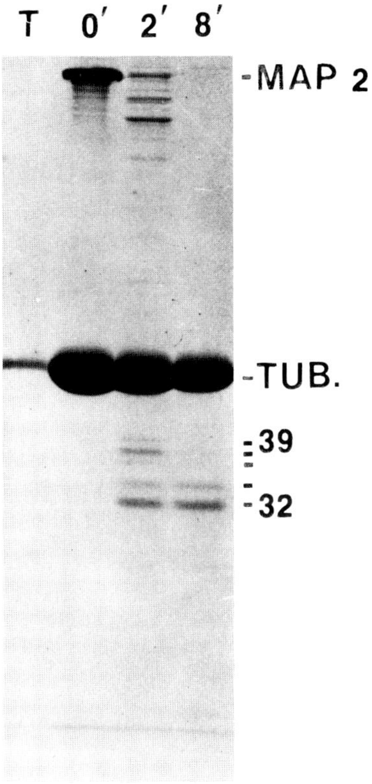

Figure 1. Microtubule-binding fragments of MAP_2. MAP_2 was digested with chymotrypsin for a series of times and combined with purified tubulin. Microtubules were assembled and then sedimented. The figure shows an electrophoretic gel loaded with protein from the microtubule pellets. T, Tubulin alone, microtubule assembly did not occur. 0′, Undigested MAP_2 plus tubulin, extensive microtubule assembly occurred. 2′ and 8′, MAP_2 digested for 2 min or 8 min plus tubulin, extensive microtubule assembly still occurred, and a number of fragments cosedimented with the microtubules. The fragments between 32,000–39,000 molecular weight represent the microtubule binding and assembly promoting domain of the MAP_2 molecule. From Vallee, 1980a.

unpublished results), indicating a common origin in the MAP_2 polypeptide chain. On the basis of these results, it was concluded that the fragments represented the microtubule binding and assembly promoting domain of MAP_2. Presumably, chymotrypsin cleaved the MAP_2 polypeptide chains at several sites within a region of approximately 7000 daltons to generate the observed group of peptides. Upon relatively prolonged exposure to chymotrypsin, the number of microtubule binding fragments was reduced to two of M_r 32,000 and 34,000.

In view of the size of the microtubule-binding domain, it would be expected that most of the MAP_2 polypeptide chain would comprise the part not in direct contact with the microtubule, i.e., the filamentous projection observed on the microtubule surface by electron microscopy. This, in fact, appears to be the case. Upon exposure to either trypsin or chymotrypsin, a large fragment of MAP_2 was generated ($M_r \sim 240{,}000$) that showed no microtubule-binding activity (Vallee and Borisy, 1977; Vallee, 1980a,b). This fragment was approximately 30,000–40,000 daltons smaller than intact MAP_2, and, therefore, was the correct size to account for the remainder of the MAP molecule. Coincident with the disappearance of the MAP_2 bands and the appearance of the M_r 240,000 species was the disappearance of the MAP_2

arms from the microtubule surface, further indicating that the large fragment represents the MAP_2 projection domain. We have purified the projection fragment chromatographically (Vallee *et al.*, 1981), and observed that it has an apparent Stokes radius slightly smaller than that observed for MAP_2.

The projection fragment appeared early in the digestion process, but soon gave way to a number of subfragments (Vallee, 1980a). In addition, low levels of high molecular weight microtubule-binding fragments were observed that, because of their size, must contain part of the projection portion of MAP_2 as well as the microtubule-binding domain. To help in sorting out the disposition of the fragments in the MAP_2 polypeptide chain, the protein was phosphorylated by its associated kinase (see below) and the sites of phosphorylation were used as fixed markers in the amino acid sequence. Based on the available data, we proposed that all of the MAP_2 fragments that were observed could be accounted for assuming parallel attack by chymotrypsin at three sites in the MAP_2 polypeptide chain (Fig. 2, arrows at 1, 2, and 3) with different rates of attack at the three sites. Site 1 would be at the base of the projection, and would represent a stretch of sequence approximately 70 amino acids long to account for the multiple assembly-promoting fragments that were observed. This region may be relatively unstructured to allow a degree of mobility for the projection domain. This may be necessary to permit the molecule to accommodate to the imperfect organization of other cytoplasmic organelles, while at the same time binding to the rigidly organized tubulin dimer lattice. Both the assembly promoting and projection domains of MAP_2 contained sites for cAMP-dependent phosphorylation suggesting a role for phosphorylation in regulating both microtubule assembly and the interaction of microtubules with other cellular structures (see below). The molecule exhibited an interesting pattern of phosphorylation with most phosphate incorporation occurring on two domains of M_r ~35,000. It is tempting to speculate that the M_r 35,000 projection subfragment is the binding site for an, as yet, unspecified cellular organelle.

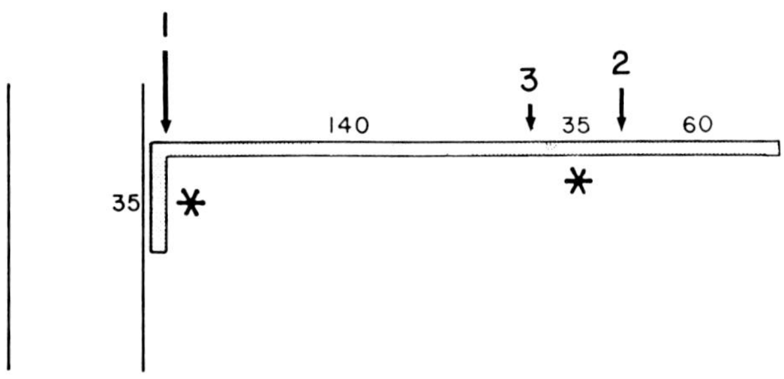

Figure 2. Schematic representation of the MAP_2 digestion process. A microtubule is represented by the two vertical lines at the left in the diagram. The MAP_2 polypeptide chain (shaded area) is shown with the assembly promoting domain bound to the microtubule. Attack at the site indicated by arrow 1 releases the projection domain of the molecule in its entirety. Less rapid cleavage occurs at sites 2 and 3, resulting in the formation of other fragments observed during digestion. From Vallee, 1980a.

5. *Binding of MAP_2 to Microtubules and Promotion of Tubulin Assembly*

MAP_2 has been observed to bind along the entire length of the microtubule (Herzog and Weber, 1978; Kim *et al.*, 1979). It could be dissociated intact from the microtubule (Vallee, 1982) and bound to preformed microtubules (Sloboda and Rosenbaum, 1979), indicating that it associates with the microtubule surface, rather than being intercalated between tubulin subunits. MAP_2 could be displaced from microtubules by the microtubule-binding fragments of the molecule (Vallee, 1980a,b). This indicates that MAP_2 associates reversibly with the microtubule surface. The interaction of MAP_2 with the microtubule appears to involve ionic bonds, since the association was abolished at elevated ionic strength (Vallee, 1982).

The details of the organization of MAP_2 on the microtubule surface are not yet certain. Amos (1977) proposed that MAP projections were distributed in a complex, double helical pattern (see Fig. 3). Since her results are based on work with unfractionated MAPs, which contain two classes of projection protein, MAP_2 and MAP_1 (Vallee and Davis, 1983), it is not certain that her model is valid for MAP_2 alone. According to her model, one MAP binds to the microtubule per each 12 tubulin subunits. Kim *et al.* (1979) found that MAP_2 binding to microtubules was saturable at a level of one MAP_2/nine tubulin dimers. Our laboratory has confirmed this result, but obtained yet a higher value, one MAP_2/seven tubulin dimers (R. Vallee, unpublished results). Whether these numbers are at variance with the Amos model is by no means certain since the stoichiometry values are based on Coomassie Blue staining intensities and rely on SDS gel-determined molecular weights. What does seem certain from these studies is that MAP_2 will bind only to a limited number of sites in the microtubule wall. In support of the model is the average spacing of projections observed by Kim and co-workers, 320 Å, which is consistent with the lattice proposed by Amos.

In addition to binding to tubulin, MAP_2 promotes tubulin assembly. At physiological temperatures, the product of assembly is microtubules (Murphy *et al.*, 1977; Herzog and Weber, 1978; Kim *et al.*, 1979). In the cold, assembly promotion was also observed (Vallee and Borisy, 1978). However, the predominant polymeric species that was formed was a small, ring-shaped oligomer, somewhat larger in diameter than the microtubule (390 Å O. D. for the ring, 250 Å O. D. for the microtubule). A particular form of ring oligomer was induced by MAP_2, a double-layered structure of $s_{20,w} = 30$ s. Formation of a smaller (probably 18 s) oligomer was also stimulated by MAP_2. The structure of this oligomeric species is unknown. Only these cooperative interactions of MAP_2 with tubulin have been described. No evidence exists for a significant interaction of individual tubulin subunits with MAP_2.

In view of these observations, and the fact that MAP_2 saturates the microtubule surface at a molar level considerably lower than the number of potential tubulin binding sites, it seems likely that MAP_2 has an extended micro-

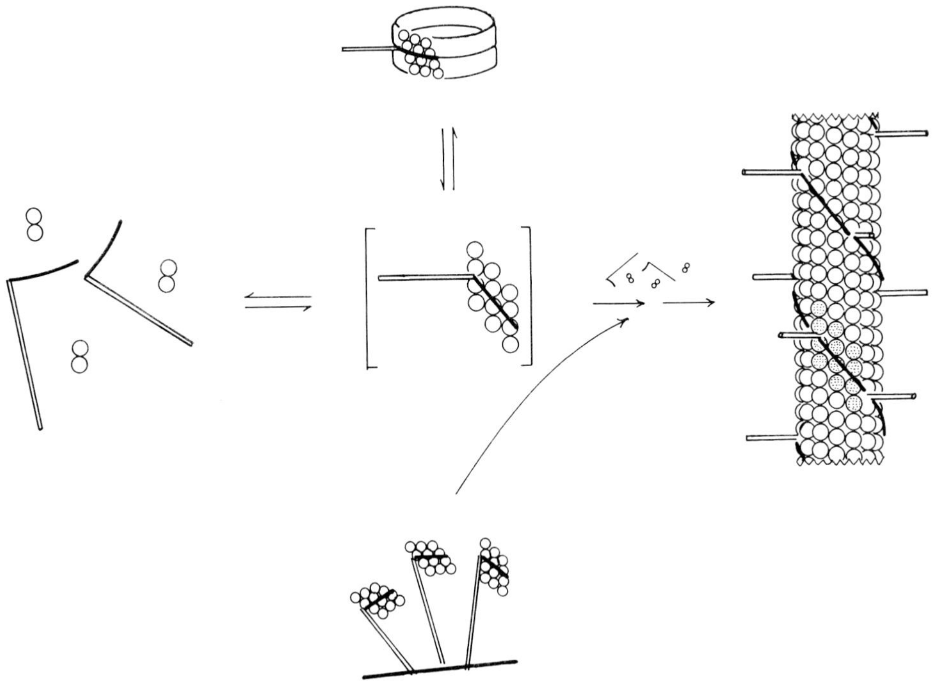

Figure 3. Possible pathway for microtubule assembly with a MAP-tubulin oligomer as an intermediate. Free tubulin dimers (pairs of circles) and MAP_2 molecules (elongated structures) are shown at left. These could associate weakly into an intermediate, such as the structure depicted in the center consisting of several tubulin dimers associated with a MAP molecule (center, in brackets). The organization of the oligomeric intermediate would be preserved in the formation of the ring-shaped particle observed in microtubule preparations (top) as well as in the formation of the microtubule (right). The microtubule binding and assembly-promoting domain of MAP_2 is shown as a elongated structure, possibly spacing adjacent MAP_2 molecules. The distribution of MAP molecules is taken from Amos (1977), as is the tubulin dimer lattice. Shown at bottom is a speculative scheme for how a MAP-tubulin oligomeric intermediate such as that shown in the center of the diagram could be involved in nucleation of microtubule assembly in the cell. The heavy line at bottom represents a microtubule-organizing center, perhaps the pericentriolar material or an adjacent microtubule. Tubulin dimers are shown accreting onto the equivalent of the microtubule-binding sites of MAP-like molecules which are linked to the microtubule-organizing center by the equivalent of MAP projections.

tubule-binding region that contacts multiple tubulin dimers (Fig. 3). As previously pointed out (Vallee and Borisy, 1978), the size of the microtubule-binding domain of MAP_2 (Figs. 1 and 2) is sufficient to span the distance between adjacent MAP_2 molecules distributed according to the model of Amos (1977). The shortest distance between adjacent molecules (~210 Å) would be spanned almost exactly if the microtubule-binding domain of MAP_2 were similar in structure to a molecule such as tropomyosin, an elongated, double α-helix (l ~400 Å, native M_r ~65,000–70,000; see for example Cohen *et al.*, 1972; Hodges *et al.*, 1972). The microtubule-binding domain would cross tubulin protofilaments in such a model, a possibility which has also been raised by Ludueña and co-workers (1981) in interpreting the inhibitory effect of MAP$_2$ on the vinblastine-induced assembly of tubulin protofilaments.

Figure 3 offers a possible scheme for how MAP$_2$ is involved in microtubule assembly, and how MAP$_2$ may be associated with the fully formed microtubule polymer. It seems reasonable, in view of the stabilizing effect of MAP$_2$ (and other MAPs) on the interaction of tubulin subunits, that MAP-stabilized intermediates would be significant in the assembly pathway. Figure 3 shows a possible assembly intermediate, a complex of MAP$_2$ with several tubulin dimers. The organization of MAP$_2$ and tubulin in the intermediate would be related to that in the microtubule and, possibly, as is shown, in the ring oligomer. MAP$_2$ molecules could be self-spacing on the microtubule, as is also shown.

In the cell, microtubules often appear to assemble from an amorphous mass of material surrounding the centriole (see for example Gould and Borisy, 1977), rather than free in solution, as in *in vitro* assembly. In neurons, microtubules are organized in parallel arrays, but the mechanism of assembly of these arrays is not known (see Peters *et al.*, 1976; Chalfie and Thomson, 1979). Presumably, the principles guiding the assembly of microtubules *in vitro* operate under *in vivo* conditions as well. Possible intermediates in *in vivo* assembly are also shown in Fig. 3 to indicate the potential relevance of the *in vitro* studies to understanding assembly in the cell.

6. Phosphorylation

MAP$_2$ was first observed to be a substrate for phosphorylation in preparations of purified microtubules (Sloboda *et al.*, 1975; see Fig. 4), where it was clearly the predominant substrate. It is now clear that MAP$_2$ is a major substrate for phosphorylation even in unfractionated cytosolic extracts of brain tissue (Fig. 5).

Phosphorylation of MAP$_2$ *in vivo* has been more difficult to investigate because of the lack of suitable cultured cell lines containing this protein (see below). By intracranial injection of [^{32}P]orthophosphate into chick or rat brain, it was possible to show incorporation of phosphate into MAP$_2$. In one experiment, microtubules were purified from the labeled tissue (Sloboda *et al.*, 1975). While the procedure took several hours, with considerable time spent at 37°, [^{32}P]phosphate was still found associated with MAP$_2$ in the purified microtubules. Presumably, this phosphate was relatively insensitive to removal by phosphatases which are present in brain cytosolic extracts (Vallee *et al.*, 1981). In another experiment, performed in our laboratory (Vallee, 1980b), labeled brain tissue was homogenized in a boiling water bath. This resulted in the destruction of phosphatase and kinase activities and precipitated most brain proteins. MAP$_2$ was detectable as one of the few soluble proteins and was clearly labeled with [^{32}P]phosphate.

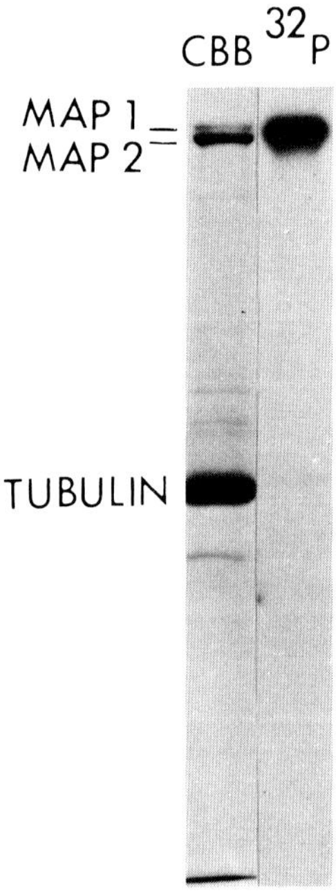

Figure 4. Endogenous phosphorylation of microtubule protein. Microtubules prepared by two cycles of assembly–disassembly purification were exposed to [^{32}P]-ATP in the presence of cAMP. The sample was subjected to electrophoresis, stained with Coomassie brilliant blue (CBB) and then autoradiographed (^{32}P). Most of the radioactivity was incorporated into MAP_2. From Vallee, 1980a.

Neither of those approaches yielded information regarding number or identity of phosphorylation sites. To obtain information on the number of MAP_2 phosphorylation sites, we assayed purified MAP_2 directly for phosphate content (Theurkauf and Vallee, 1983). MAP_2, as isolated from purified microtubules, contained a variable level of phosphate ranging from 7–13 mol phosphate/mol protein. Exposure of MAP_2 to the purified catalytic subunit of cAMP-dependent protein kinase (see below) resulted in a reproducible total of 20–22 mol phosphate/mol MAP_2. Exposure of enzymatically dephosphorylated MAP_2 to the cAMP-dependent catalytic subunit allowed an estimate of the total number of cAMP-dependent phosphorylation sites. This analysis revealed that approximately 13 of the MAP_2 phosphates can be attributed to cAMP-dependent phosphorylation. The remainder are presumably introduced by a kinase of unknown identity. A possible candidate for such a kinase is a neurofilament associated cAMP-independent kinase (Runge *et al.*, 1981). This enzyme was found to phosphorylate the neurofilament peptides but could also use MAP_2 as a substrate. Whether, in fact, this kinase can account for the cAMP-independent phosphates present on MAP_2 as it is isolated is not yet known.

7. *cAMP-Dependent Protein Kinase Associated with MAP_2*

The intrinsic protein kinase activity present in microtubule preparations that was responsible for phosphorylating MAP_2 was observed to persist at

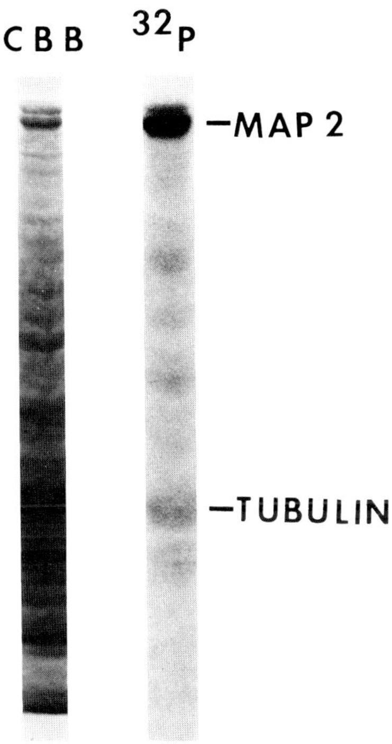

Figure 5. Endogenous phosphorylation of brain cytosolic proteins. A cytosolic extract of calf cerebral cortex prepared as in Vallee *et al.* (1981) and further diluted two-fold was exposed to 1 mM [^{32}P]-ATP for 10 min in the presence of 0.6 mM theophylline, 10 μM cAMP, 1 mM EGTA, 5 mM $MgSO_4$, at 37°C. The sample was analyzed by electrophoresis as described in Fig. 4. MAP_2 was the most prominently phosphorylated protein observed.

constant specific activity through several cycles of assembly–disassembly purification (Sloboda *et al.*, 1975). This observation suggested to us that a protein kinase might be specifically associated with microtubules.

We have since identified a cAMP-dependent protein kinase that is tightly associated with microtubules isolated from brain tissue (Vallee, 1980a; Vallee *et al.*, 1981; Theurkauf and Vallee, 1982; see Fig. 6). The enzyme is quite similar to previously characterized cAMP-dependent kinases, and appears to be virtually identical to the type II kinase isolated from bovine heart (Theurkauf and Vallee, 1982). The microtubule associated enzyme has the same affinity for cAMP as other type II enzymes. Photoaffinity labeling with [^{32}P]-8-azido-cAMP showed that the cAMP-binding subunit of the microtubule associated enzyme had an apparent molecular weight of 54,000, and could be shifted to 57,000 by incubation with ATP. A similar change in electrophoretic mobility has been observed with bovine heart type II cAMP-binding subunit as the result of autophosphorylation by the associated catalytic subunit (Rangel-Aldao *et al.*, 1979). The catalytic subunit of the microtubule associated enzyme was identical to that of other known cAMP-dependent kinases in molecular weight (M_r 39,000) and catalytic properties.

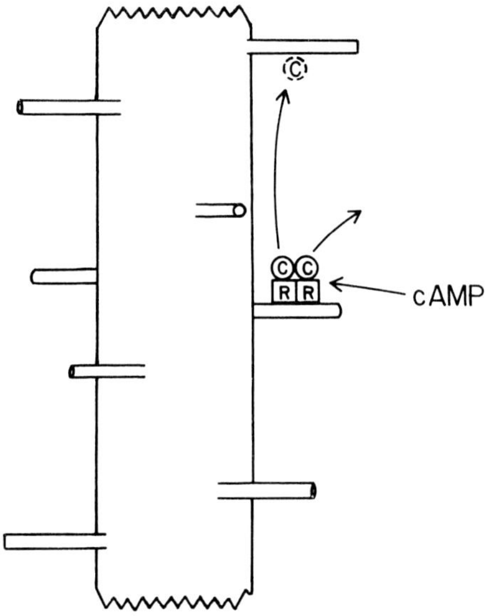

Figure 6. The MAP_2–protein kinase complex. A type II cAMP-dependent protein kinase holoenzyme molecule is shown associated with the projection portion of MAP_2. From Theurkauf and Vallee, 1982.

The enzyme was found to copurify with MAP_2, specifically with the projection domain of the MAP_2 molecule (Vallee *et al.*, 1981). Incubation of microtubule protein with cAMP, which dissociates the regulatory and catalytic subunits of the kinase, resulted in the total release of the catalytic subunit from MAP_2 (Theurkauf and Vallee, 1982). cAMP-binding activity remained associated with MAP_2. This indicates that the enzyme is associated with MAP_2 via its regulatory subunits.

To determine the fraction of brain cytosolic protein kinase that was associated in this manner with MAP_2, the kinase was separated from a cytosolic extract of brain tissue by three independent procedures: (1) a one step microtubule isolation involving the use of taxol (Vallee, 1982), (2) immunoprecipitation with anti-MAP_2 antibody, and (3) gel filtration chromatography. Amount of kinase was assayed by a cAMP-binding assay. The three approaches gave almost identical results; 30–35% of total cytosolic cAMP-dependent protein kinase, approximately one third, was associated with MAP_2. Since the MAP_2 bound kinase was of the type II isozymic form (see above), a somewhat greater fraction of total soluble type II kinase in brain must be associated with MAP_2. Further analysis revealed that only very few (about one in 40) MAP_2 molecules had an associated enzyme molecule. This is not surprising, in view of the high molar concentration of MAP_2 in brain relative to protein kinase. However, since two thirds of the soluble kinase was *not* associated with MAP_2, these results indicate that free MAP_2 and free protein kinase both exist in excess over the complex in the brain. To account for this observation, we suggest that the MAP_2-bound enzyme contains a unique variant isozymic form of the type II regulatory subunit, R_{II}, which differs from the remaining brain R_{II}.

8. Cellular Distribution

MAP_2 is a major component of microtubules isolated from brain tissue. A number of laboratories have prepared antibodies to MAP_2 and used these to determine whether the protein was present in other tissues or cells. The consensus from these studies, described below, seems to be that MAP_2 is most highly concentrated in brain tissue. This is not entirely surprising since tubulin is also most highly concentrated in brain. Thus, it would be useful to know how much MAP_2 was present in a given tissue or cell type on a per tubulin, or, even better, a per microtubule basis.

Immunofluorescence microscopy allows for a qualitative estimate of MAP content on a per microtubule basis. A wide variety of cultured cell lines were found to be negative for MAP_2 by immunofluorescence microscopy using either a polyclonal rabbit antibody to pig brain MAP_2 or a mouse monoclonal antibody to rat brain MAP_2 (Peloquin and Borisy, 1979; Izant and McIntosh, 1980). Only in primary brain cell cultures were strongly MAP_2 positive cells found. Many of these cells had long processes and were assumed to be neurons. In addition, numerous flat cells with fibroblastic or epitheleoid morphology were stained with anti-MAP_2 antibody. The results were interpreted to indicate that MAP_2 was restricted to neurons, though the identity of the flat cells was uncertain.

We have found by double labeling with anti-MAP_2 and antiglial fibrillary acidic protein (anti-GFA) that astroglial cells in primary cultures of rat brain are devoid of MAP_2 (Bloom and Vallee, unpublished results). Cells of neuronal morphology that were MAP_2 positive also bound tetanus toxin, evidence that these cells were, indeed, neurons (Bloom and Vallee, 1983). Flat MAP_2-positive cells were GFA and neurofilament negative and failed to bind tetanus toxin, but were positive for vimentin. We believe that these cells represent neurons at an early stage of differentiation.

Examination of tissue sections has confirmed that MAP_2 is highly concentrated in neurons (Miller *et al.*, 1982; Fig. 7). Somewhat surprising and probably of considerable importance in understanding the function of MAP_2 has been the finding that the protein is preferentially localized in the dendritic processes of neurons as opposed to the axonal processes of these cells. Matus and co-workers (1981) stained sections of brain tissue with antibody prepared against whole high-molecular weight brain MAPs, i.e., MAP_1 and MAP_2. The antibody was found to stain microtubules in dendrites but not in axons. Subsequent work has shown that MAP_2 in particular is concentrated in dendrites; antibody specific for MAP_2 prepared in our laboratory stained dendrites and cell bodies but did not stain axons (Miller *et al.*, 1982; DeCamilli, in preparation; Fig. 7).

We have also used a biochemical approach to address this question (Vallee, 1982). Using taxol to promote assembly, microtubules were purified from calf white matter and compared with those from cerebral cortex, the commonly used source for microtubule protein. Microtubules prepared from white matter contained MAP_2 but at a considerably lower level than those from gray matter. A similar result was obtained by immunoblot analysis of gray and white matter (R. Vallee, unpublished results). It is important to note that these results indicate that MAP_2 is not exclusively limited to dendrites, but is probably present in axons as well, though at a lower level. The important point is that MAP_2 is highly concentrated in dendrites and may be a major cytoskeletal component in dendritic cytoplasm.

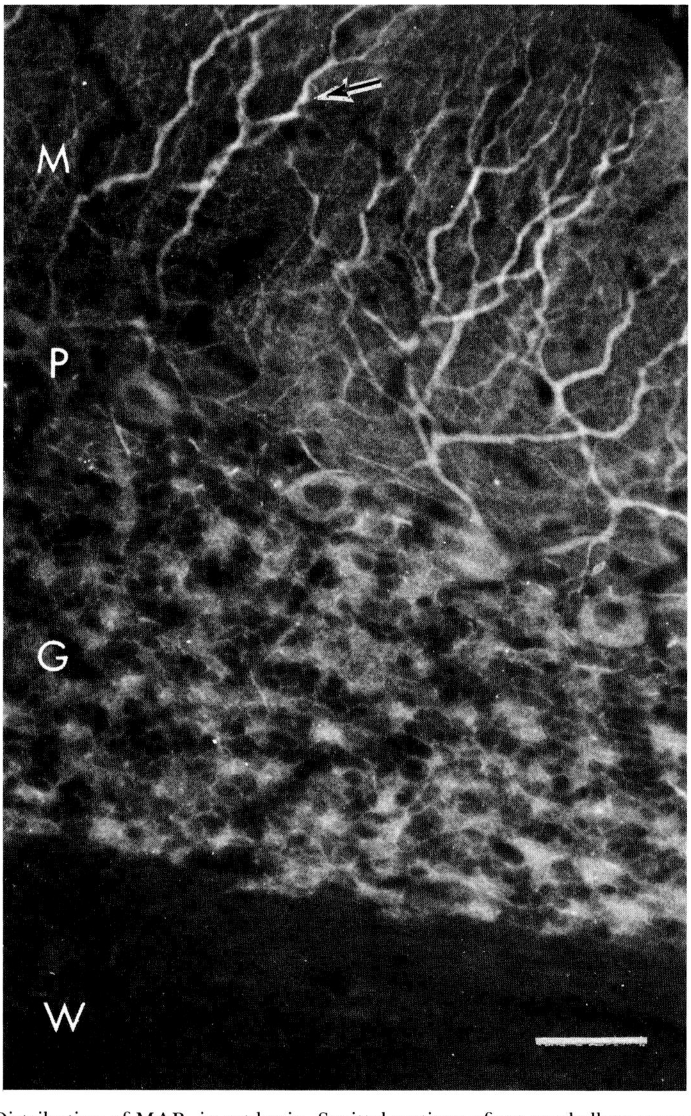

Figure 7. Distribution of MAP_2 in rat brain. Sagittal sections of rat cerebellum were stained with rabbit anti-MAP_2 antibody, followed by fluorescein conjugated goat anti-rabbit IgG. W, white matter; G, granule cell layer; P, Purkinje cell layer; M. molecular layer. The arrow indicates one of the numerous Purkinje cell dendritic processes. Bloom, Schoenfeld and Vallee, unpublished data. Bar: 40 μm.

Some evidence exists that MAP_2 is not exclusively restricted to neurons. MAP_2 has been detected in HeLa cells (Weatherbee *et al.*, 1982), amphibian erythrocytes (Sloboda and Dickersin, 1980), ovarian granulosa cells (Sheterline, 1978, 1980) pig kidney epithelial cells (Valdivia *et al.*, 1982), and bovine trachea epithelial cells (Kuznetsov *et al.*, 1980). The amount of MAP_2 in at least some of these cells (see Weatherbee *et al.*, 1982) was extremely low, and it is not clear whether the protein is functionally significant in those cells. We have recently found MAP_2 in bovine anterior pituitary and cultured GH_3 pituitary cells by immunofluorescence microscopy and by immunoblot analysis of isolated microtubules (Bloom and Vallee, manuscript in preparation). Using a radioimmunoassay, Valdivia *et al.* (1982) have also detected MAP_2 in kidney, lung, spleen, liver, and thyroid tissue. The levels of MAP_2 were approximately 100 to 1000-fold lower in these tissues than in brain. However, it is important to note that tubulin is also present at much lower concentration in these tissues than in brain. Thus, some nonneuronal cells could yet prove to have a significant content of MAP_2 per microtubule.

Thus, it appears that MAP_2 exists in at least some nonneuronal cells as well as in neurons. It seems unlikely that the protein will be universally distributed, however.

9. Function of MAP_2

At least three activities may be envisioned for MAP_2: binding to microtubules, regulating microtubule assembly, and binding to other cytoplasmic organelles, thereby cross-linking microtubules to these structures. The first two activities are well documented *in vitro*. In addition, MAP_2 has been found to be associated with microtubules in the cell by immunofluorescence microscopy. No direct evidence exists as yet for a microtubule assembly promoting role for MAP_2 in cells.

It may be proper to see microtubule binding and promotion of assembly as two aspects of the same phenomenon. The affinity of MAP_2 for microtubules insures that MAP_2 will be associated with microtubules in the cell. The assembly promoting activity of MAP_2 may, in turn, insure that microtubules do not exist without bound MAP_2 (or some other MAP). This would be a reasonable goal for the cell if MAPs are required for microtubules to function in cells.

The third activity of MAP_2 listed above is, at present, the most controversial. What is the function of the long, filamentous projection observed to extend from the microtubule surface? Similar projections have been observed in cells by electron microscopy, and appear to cross-link microtubules with each other, with intermediate filaments, and with membrane bounded organelles such as mitochondria (see Smith *et al.*, 1975; McIntosh, 1974; Hirokawa, 1982). Are these arms MAP_2, do they, in fact, bind to the structures in question, and what is the purpose of this binding?

MAP_2 will bind *in vitro* to a number of structures other than microtubules. These include actin filaments (Griffith and Pollard, 1978, 1982; Sattilaro *et al.*, 1981), secretory granules (Sherline *et al.*, 1977), coated vesicles (Sattilaro *et al.*, 1980), and neurofilaments (Leterrier *et al.*, 1982; Aamodt and Williams, 1983). In addition, microtubules composed of purified tubulin and MAP_2 will gel at high concentrations, indicating an interaction of MAP_2 with more than one microtubule (Griffith and Pollard, 1982; R. Vallee, unpublished observations). While these interactions do occur in the test tube, their specificity has not yet been established. The interaction of MAP_2 with actin was too weak to detect by cosedimentation (Griffith and Pollard, 1978, 1982) except at low ionic strength (Sattilaro *et al.*, 1981), raising the possibility of a nonspecific ionic interaction. The presence of tubulin interfered with the binding of actin to MAP_2 in the latter study, further suggesting that the interaction occurs at least in part through the microtubule-binding domain of MAP_2. The tau MAPs also bound to actin (Griffith and Pollard, 1982) and hence the interaction is not unique to MAP_2. In the case of neurofilaments, binding of MAP_2 was strong, and was specific for neurofilaments rather than glial filaments (Leterrier *et al.*, 1982). However, tubulin and MAP_1 also bound to the neurofilaments, and the stoichiometry of binding was extremely low. Similarly, in the case of secretory granules (Sherline *et al.*, 1977), it appeared that multiple MAP species bound to the granules. In the other cases reported, the data are not yet sufficient to evaluate binding specificity.

It seems clear, therefore, that further work is needed to establish a link between MAP_2 and other organelles. In addition, the available evidence raises the question of whether MAP_2 can bind to such a large variety of organelles. If so, how is binding regulated?

As an alternative to the reconstructive approach to investigating MAP_2 binding sites, we have used an approach designed to perturb cells minimally in the hope of preserving the naturally occurring interactions between MAP_2 and its cellular binding sites (Bloom and Vallee, 1983; Fig. 8). Primary brain cultures were used because of their high content of MAP_2. Flat MAP_2-positive cells were examined because they were particularly well suited for immunofluorescence microscopy. In untreated cells, MAP_2 was observed in a distribution characteristic of microtubules, i.e., the protein was associated with cytoplasmic fibers which emanated radially from the nucleus and extended to the boundary of the cell. After vinblastine treatment, an entirely different distribution was observed. MAP_2 was now found associated with perinuclear cables (Fig. 8). Such structures are characteristic of some classes of intermediate filament suggesting that MAP_2 binds to intermediate filaments in cells. That this was the case was demonstrated directly by double labeling cells with anti-MAP_2 antibody and anti-vimentin, a component of intermediate filaments. MAP_2 also colocalized with tubulin in these cells, both during interphase and during mitosis. This indicated that MAP_2 has at least two binding sites, one for microtubules, and one for intermediate filaments. Presumably, MAP_2 cross-links the two structures.

It is unlikely that this is the only function of MAP_2 in neurons, since the

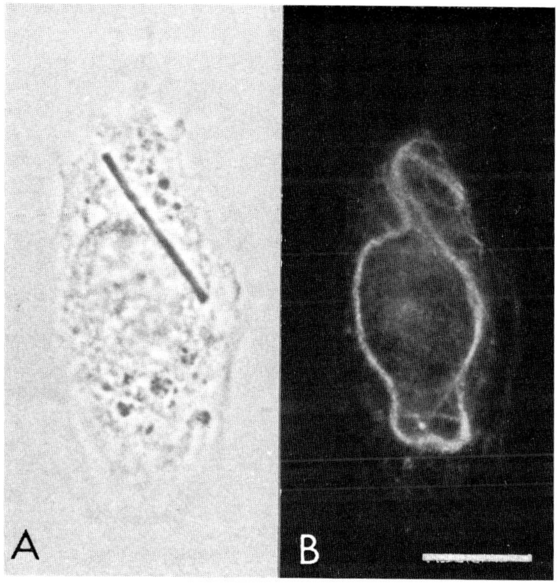

Figure 8. Association of MAP_2 with intermediate filaments. A primary culture of rat brain cells was incubated with vinblastine sulfate (10 μm) for 15 hr to destroy microtubules. The cells were stained with rabbit anti-MAP_2 antibody as described in Fig. 7. (A) Phase contrast, (B) fluorescence. MAP_2 staining is observed in a cable circling the nucleus, characteristic of some classes of intermediate filament. The phase dense bar is a vinblastine-induced paracrystal of tubulin. Bloom and Vallee, unpublished data; see also Bloom and Vallee, 1983. Bar: 10 μm.

protein is most highly concentrated in dendrites. Intermediate filaments are relatively rare in dendrites but are abundant in axons (see Peters *et al.*, 1976). Examination of the cytoplasm of dendrites reveals a highly ordered cytoskeleton with microtubules as the major detectable cytoskeletal component. The microtubules appear to be highly organized and run parallel to the axis of the process. Despite this high degree of organization the microtubules are quite widely spaced. Thus, they must in some way interact, but the distance over which the interaction would occur, several microtubule diameters, seems large. In view of its size and shape, MAP_2 seems an excellent candidate for mediating the interaction of the dendritic microtubules. This might occur in a number of ways (Fig. 9). MAP_2 could have two binding sites for tubulin, the previously identified assembly promoting domain (Figs. 1 and 2; Vallee, 1980a), and an additional site on the projection domain (Fig. 9A). Binding via this second site, if it occurs, must be weak since the MAP_2 projection fragments have shown no apparent affinity for microtubules (Vallee and Borisy, 1977; Vallee, 1980a). Alternatively, the MAP_2 projections could bind to other MAP_2 projections (Fig. 9B). This would allow for extremely long cross-bridges between microtubules. In addition, according to this model, the projection fragment should not bind directly to microtubules.

It is also conceivable that the MAP_2 projections do not bind at all to other microtubules, either directly to tubulin or to another MAP molecule (Fig. 9C). MAP_2 could serve as a semirigid spacer which would serve to organize microtubules in the dendrite by keeping them apart. The dendritic microtubules would be parallel because they would be close packed in actuality, despite the large distance between the walls of adjacent microtubules. This possibility is

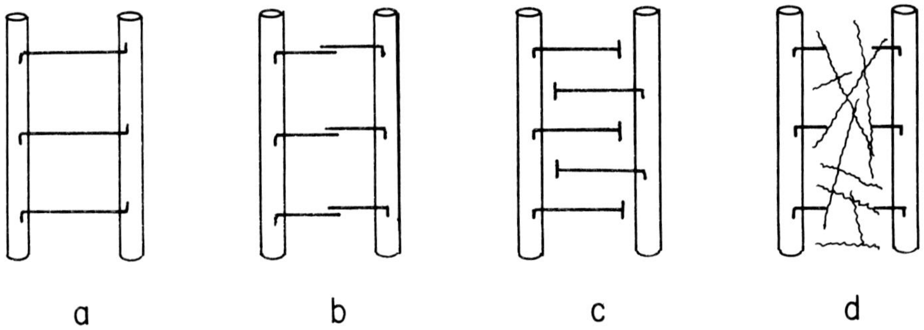

Figure 9. Alternative models for how MAP_2 could organize dendritic microtubules. (a) MAP_2 could have two microtubule binding sites and directly cross-link microtubules, (b) MAP_2 molecules could interact with other MAP_2 molecules, (c) MAP_2 could act as a passive bumper which would sterically interfere with the close approach of microtubule walls, (d) a "microtrabecular" or actin meshwork could cross-link MAP_2 molecules.

supported by observations of purified microtubules. Centrifugation of microtubules bearing projections results in the formation of bundles of parallel microtubules which, like dendritic microtubules, are rather distantly spaced. Microtubules lacking projections pack much more closely (Murphy and Borisy, 1975; Vallee and Borisy, 1977; Kim *et al.*, 1979), resulting in dramatically reduced pellet volumes (R. Vallee, unpublished observation).

A final possibility is that the link between microtubules is indirect, with one or more macromolecular elements being situated between MAP_2 arms on adjacent microtubules (Fig. 9D). For example, an actin filament meshwork could be situated between microtubules. Such a scheme is attractive since it would account for the binding of MAP_2 to actin observed *in vitro*. The filaments in such a scheme need not be actin of course, but could be some other "microtrabecular" element (Wolosewick and Porter, 1976; Ellisman and Porter, 1980). The intervening filaments might well be difficult to observe in traditional thin sections of dendrites, accounting for their apparent absence in published electron micrographs. However, one restriction must be placed on any such model. The intervening structures must themselves be highly organized for order to be transmitted from microtubule to microtubule. Thus, a truly random network of actin (or other) filaments would be difficult to incorporate into any scheme designed to describe the organization of dendritic cytoplasm.

10. *Function of MAP_2 Phosphorylation*

The MAP_2 phosphorylation reaction now appears to be a major phosphorylation reaction in brain tissue in view of the abundance of the protein, the large numbers of phosphorylation sites, and the large fraction of brain cAMP-dependent protein kinase associated with MAP_2. Considering the preferential localization of MAP_2 to neurons it seems likely that MAP_2 phosphorylation is particularly important in these cells. It is not yet certain whether the MAP_2 bound kinase is enriched in dendrites along with MAP_2, but this seems likely. Certainly, more binding sites for cAMP-dependent kinase exist in dendrites than in axons and other cells that make up neuronal tissue (Miller *et al.*, 1982). Thus, MAP_2 phosphorylation is likely to be particularly active in dendrites.

Several pieces of evidence indicate that phosphorylation affects microtubule assembly. Margolis and Wilson (1979) observed that ATP increased the rate of tubulin treadmilling in unfractionated microtubules *in vitro*. This was confirmed by Jameson and co-workers (1980, 1981), who also found that the overall time course of microtubule assembly was depressed by prior incubation of unfractionated MAPs with ATP. Since these results involved the use of unfractionated protein mixtures, it is not clear what the individual effect of MAP_2 phosphorylation may be. In addition to MAP_2, which is clearly the major substrate for phosphorylation of all the microtubule proteins, other minor protein species are phosphorylated *in vitro* (Margolis and Rauch, 1981). These alone appear to affect microtubule assembly and could be in part, or even fully, responsible for the effect of ATP on assembly. In addition, tubulin is a phosphorprotein, containing approximately one mole phosphate per mole dimer as isolated (Eipper, 1974), and a tubulin kinase activity has been detected *in vitro* (Burke and DeLorenzo, 1981). Conceivably, tubulin phosphorylation could affect assembly. Nonetheless, it seems likely that the phosphorylation of MAP_2 itself affects assembly in view of our own result that the microtubule binding and assembly promoting domain of MAP_2 contains sites for cAMP-dependent phosphorylation (Vallee, 1980a; Fig. 2).

We also found phosphorylation sites on the projection portion of MAP_2, suggesting that the interaction of MAP_2 with other cellular structures may also be regulated by this reaction. Some evidence has been reported that is consistent with a role for phosphorylation in regulating the interaction of MAP_2 with actin (Nishida *et al.*, 1981; Selden and Pollard, 1982). The effect of phosphorylation on other possible interactions of MAP_2 has not been explored.

How these results relate to the function of MAP_2 in the cell is not clear, and will probably not be fully clear until more is known about the assembly properties of microtubules in cells and the binding sites of MAP_2 for cellular organelles. However, in view of what has already been learned about the protein, it is possible to begin to speculate about its cellular behavior. As I have indicated above, present evidence suggests that MAP_2 may exist in two functional states in neurons. One would involve an interaction between microtubules and intermediate filaments. This interaction might occur in developing neurons (Fig. 8) and in axons, where MAP_2 probably is present at low levels (Vallee, 1982) and where intermediate filaments (neurofilaments) are abundant. MAP_2 clearly is also present in dendrites where it is unlikely that its sole role would be in mediating the interaction between microtubules and intermediate filaments. Its most likely role would seem to be in organizing microtubules. Whether it does this by directly cross-linking microtubules, by passively interfering with their close approach, or by binding to an intervening cytoskeletal system is not clear (see Fig. 9).

Whatever the case, it seems an attractive hypothesis that phosphorylation defines the functional state of MAP_2. Thus, for example, the neurofilament-associated MAP_2 kinase (Runge *et al.*, 1981) which should be concentrated in axons, could promote binding of MAP_2 to the neurofilaments themselves. The cAMP-dependent kinase, activated by neurotransmitter stimulated cAMP production, could specify localization of MAP_2 in dendrites. Thus, a gradient of some kind (perhaps involving cAMP alone, or a variety of effectors) specified by external stimuli, could define the structural polarity of the neuron, with MAP_2 playing a key role in differentially organizing the cytoplasm of axon and dendrite.

Before any such hypothesis can be fully tested, more information will be needed regarding the localization of MAP_2 in cells (see Fig. 8). Particularly valuable would be more information obtained at the ultrastructural level to help define which organelles interact with microtubules via MAP_2. Coupled with such investigations should be further work on the binding of MAP_2 to cellular organelles *in vitro*. A further understanding of the domain structure of MAP_2 (see Fig. 2) would be useful in these studies to help distinguish between specific and nonspecific interactions of the protein. Thus, for example, it will be of interest to determine whether MAP_2 binds to actin or to intermediate filaments via a unique site on the MAP_2 molecule or via the part of the molecule already known to bind to microtubules. Together, such studies should eventually serve to establish the binding sites of MAP_2 in the cell. What will still remain to be learned, that is of even greater interest, is how the interactions of MAP_2 with other cellular structures are coordinated into the behavior of the cytoskeleton as a whole.

References

Aamodt, E., and Williams, R. C., Jr., 1983, MAPs mediate association of microtubules and neurofilaments *in vitro*, *Biophys. J.* **41**:86a.

Amos, L. A., 1977, Arrangement of high molecular weight associated proteins on purified mammalian brain microtubules, *J. Cell Biol.* **72**:642–654.

Berkowitz, S. A., Katagiri, J., Binder, H.-K., and Williams, R. C., Jr., 1977, Separation and characterization of microtubule proteins from calf brain, *Biochemistry* **16**:5610–5617.

Bloom, G. S., and Vallee, R. B., 1983, Association of MAP_2 with microtubules and intermediate filaments in cultured brain cells, *J. Cell Biol.* **96**:1523–1531.

Borisy, G. G., Marcum, J. M., Olmsted, J. B., Murphy, D. B., and Johnson, K. A., 1975, Purification of tubulin and associated high molecular weight proteins from porcine brain and characterization of microtubule assembly *in vitro*, *Ann. N.Y. Acad. Sci.* **253**:107–132.

Bulinski, J. C., and Borisy, G. G., 1979, Self-assembly of microtubules in extracts of cultured HeLa cells and the identification of HeLa microtubule-associated proteins, *Proc. Natl. Acad. Sci. USA* **76**:293–297.

Bulinski, J. C., and Borisy, G. G., 1980, Microtubule-associated proteins from cultured HeLa cells, *J. Biol. Chem.* **255**:11570–11576.

Burke, B. E., and DeLorenzo, R. J., 1981, Ca^{2+}- and calmodulin-stimulated endogenous phosphorylation of neurotubulin, *Proc. Natl. Acad. Sci. USA* **78:**991–995.
Chalfie, M., and Thomson, J. N., 1979, Organization of neuronal microtubules in the nematode *Caenorhabditis elegans, J. Cell Biol.* **82:**278–289.
Cleveland, D. W., Hwo, S.-Y., and Kirschner, M. W., 1977a, Purification of tau, a microtubule-associated protein that induces assembly of microtubules from purified tubulin, *J. Mol. Biol.* **116:**207–225.
Cleveland, D. W., Hwo, S.-Y., and Kirschner, M. W., 1977b, Physical and chemical properties of purified tau factor and the role in microtubule assembly, *J. Mol. Biol.* **116:**227–247.
Cohen, C., Caspar, D. L. D., Johnson, J. P., Nauss, K., Margossian, S. S., and Parry, D. A. D., 1972, Tropomyosin-troponin assembly, *CSHSQB* **37:**287–297.
Dentler, W. L., Granett, S., and Rosenbaum, J. L., 1975, Ultrastructural localization of the high molecular weight proteins associated with *in vitro* assembled brain microtubules, *J. Cell Biol.* **65:**237–241.
Duerr, A., Pallas, D., and Solomon, F., 1981, Molecular analysis of cytoplasmic microtubules *in situ:* Identification of both widespread and specific proteins, *Cell* **24:**203–211.
Eipper, B. A., 1974, Properties of rat brain tubulin, *J. Biol. Chem.* **249:**1407–1416.
Ellisman, M. H., and Porter, K. R., 1980, Microtrabecular structure of the axoplasmic matrix: Visualization of cross-linking structures and their distribution, *J. Cell Biol.* **87:**464–479.
Fellous, A., Francon, J., Lennon, A. M., and Nunez, J., 1977, Microtubule assembly *in vitro, Eur. J. Biochem.* **78:**167–174.
Gould, R. R., and Borisy, G. G., 1977, The pericentriolar material in Chinese hamster ovary cells nucleates microtubule formation, *J. Cell Biol.* **73:**601–615.
Griffith, L. M., and Pollard, T. D., 1978, Evidence for actin filament-microtubule interaction mediated by microtubule-associated proteins, *J. Cell Biol.* **78:**958–965.
Griffith, L. M., and Pollard, T. D., 1982, The interaction of actin filaments with microtubules and microtubule-associated proteins, *J. Biol. Chem.* **257:**9143–9151.
Herzog, W., and Weber, K., 1978, Fractionation of brain microtubule-associated proteins, *Eur. J. Biochem.* **92:**1–8.
Hirokawa, N., 1982, Cross-linker system between neurofilaments, microtubules, and membranous organelles in frog axons revealed by the quick-freeze, deep-etching method, *J. Cell Biol.* **94:**129–142.
Hodges, R. S., Sodek, J., Smillie, L. B., and Jurasek, L., 1972, Tropomyosin: Amino acid sequence and coiled-coil structure, *CSHSQB* **37:**299–310.
Izant, J. G., and McIntosh, R., 1980, Microtubule-associated proteins: A monoclonal antibody to MAP_2 binds to differentiated neurons, *Proc. Natl. Acad. Sci. USA* **77:**4741–4745.
Jameson, L., and Caplow, M., 1981, Modification of microtubule steady-state dynamics by phosphorylation of the microtubule-associated proteins, *Proc. Natl. Acad. Sci. USA* **78:**3413–3417.
Jameson, L., Frey, T., Zeeberg, B., Dalldorf, F., and Caplow, M., 1980, Inhibition of microtubule assembly by phosphorylation of microtubule-associated proteins, *Biochemistry* **19:**2472–2479.
Kim, H., Binder, L., and Rosenbaum, J. L., 1979, The periodic association of MAP_2 with brain microtubules *in vitro, J. Cell Biol.* **80:**266–276.
Kuznetsov, S. A., Rodionov, V. I., Bershadsky, A. D., Gelfand, V. I., and Rosenblat, V. A., 1980, High molecular weight protein MAP_2 promoting microtubule assembly *in vitro* is associated with microtubules in cells, *Cell Biol. Int. Rep.* **4:**1017–1024.
Leterrier, J.-F., Liem, R. K. H., and Shelanski, M. C., 1982, Interactions between neurofilaments and microtubule-associated proteins: A possible mechanism for intraorganellar bridging, *J. Cell Biol.* **95:**982–986.
Ludueña, R. F., Fellous, A., Francon, J., Nunez, J., and McManus, L., 1981, Effect of tau on the vinblastine-induced aggregation of tubulin, *J. Cell Biol.* **89:**680–683.
Margolis, R. L., and Rauch, C. T., 1981, Characterization of rat brain crude extract microtubule assembly: Correlation of cold stability with the phosphorylation state of a microtubule associated 64K protein, *Biochemistry* **20:**4451–4458.
Margolis, R. L., and Wilson, L., 1979, Regulation of the microtubule steady state *in vitro* by ATP, *Cell* **18:**673–679.

Matus, A., Bernhardt, R., and Hugh-Jones, T., 1981, High molecular weight microtubule-associated proteins are preferentially associated with dendritic microtubules in brain, *Proc. Natl. Acad. Sci. USA* **78:**3010–3014.

McIntosh, J. R., 1974, Bridges between microtubules, *J. Cell Biol.* **61:**166–187.

Miller, P., Walter, U., Theurkauf, W. E., Vallee, R. B., and DeCamilli, P., 1982, Frozen tissue sections as an experimental system to reveal specific binding sites for the regulatory subunit of type II cAMP-dependent protein kinase in neurons, *Proc. Natl. Acad. Sci. USA* **79:**5562–5566.

Murphy, D. B., and Borisy, G. G., 1975, Association of high-molecular-weight proteins with microtubules and their role in microtubule assembly *in vitro, Proc. Natl. Acad. Sci. USA* **72:**2696–2700.

Murphy, D. B., Vallee, R. B., and Borisy, G. G., 1977, Identity and polymerization-stimulatory activity of the non-tubulin proteins associated with microtubules, *Biochemistry* **16:**2598–2605.

Nishida, E., Kuwaki, T., and Sakai, H., 1981, Phosphorylation of microtubule-associated proteins (MAPs) and pH of the medium control interaction between MAPs and actin filaments, *J. Biochem.* (*Tokyo*) **90:**575–578.

Olmsted, J. B., and Lyon, H. D., 1981, A microtubule-associated protein specific to differentiated neuroblastoma cells, *J. Biol. Chem.* **256:**3507–3511.

Peloquin, J. G., and Borisy, G. G., 1979, Cell and tissue distribution of the major high molecular weight microtubule-associated protein from brain, *J. Cell Biol.* **83:**338a.

Peters, A., Palay, S. L., and Webster, H., deF., 1976, *The Fine Structure of the Nervous System,* W. B. Saunders, Philadelphia.

Rangel-Aldao, R., Kupiec, J. W., and Rosen, O. M., 1979, Resolution of the phosphorylated and dephosphorylated cAMP-binding proteins of bovine cardiac muscle by affinity labeling and two-dimensional electrophoresis, *J. Biol. Chem.* **254:**2499–2508.

Runge, M. S., El-Maghrabi, M. R., Claus, T. K., Pilkis, S. J., and Williams, R. C., Jr., 1981, A MAP_2 stimulated protein kinase activity associated with neurofilaments, *Biochemistry* **20:**175–180.

Sattilaro, R. F., LeCluyse, E. L., and Dentler, W. L., 1980, Associations between microtubules and coated vesicles *in vitro, J. Cell Biol.* **87:**250a.

Sattilaro, R. F., Dentler, W. L., and LeCluyse, E. L., 1981, Microtubule-associated proteins (MAPs) and the organization of actin filaments *in vitro, J. Cell Biol.* **90:**467–473.

Selden, S. C., and Pollard, T. D., 1982, Phosphorylation of microtubule-associated proteins (MAPs) regulates their interaction with actin filaments, *J. Cell Biol.* **95:**348a.

Sherline, P., Lee, Y. C., and Jacobs, L. S., 1977, Binding of microtubules to pituitary secretory granules and secretory granule membranes, *J. Cell Biol.* **72:**380–389.

Sheterline, P., 1978, Localisation of the major high-molecular-weight protein on microtubules *in vitro* and in cultured cells, *Exp. Cell Res.* **115:**460–464.

Sheterline, P., 1980, Immunological characterisation of the microtubule-associated protein MAP_2, *FEBS Lett.* **111:**167–170.

Sloboda, R. D., and Dickersin, K., 1980, Structure and composition of the cytoskeleton of nucleated erythrocytes I. The presence or microtubule associated protein 2 in the marginal band, *J. Cell Biol.* **87:**170–179.

Sloboda, R. D., and Rosenbaum, J. L., 1979, Decoration and stabilization of intact, smooth-walled microtubules with microtubule-associated protein, *Biochemistry,* **18:**48–55.

Sloboda, R. D., Rudolph, S. A., Rosenbaum, J. L., and Greengard, P., 1975, Cyclic AMP-dependent endogenous phosphorylation of a microtubule-associated protein, *Proc. Natl. Acad. Sci. USA* **72:**177–181.

Smith, D. S., Jarlfors, U., and Cameron, B. G., 1975, Morphological evidence for the participation of microtubules in axonal transport, *Ann. N.Y. Acad. Sci.* **253:**472–506.

Theurkauf, W. E., and Vallee, R. B., 1982, Molecular characterization of the cAMP-dependent protein kinase bound to microtubule-associated protein 2, *J. Biol. Chem.* **257:**3284–3290.

Theurkauf, W. E., and Vallee, R. B., 1983, Extensive cAMP dependent and cAMP independent phosphorylation of microtubule-associated protein 2, *J. Biol. Chem.* **258:**7883–7886.

Valdivia, M. M., Avila, J., Coll, J., Colaco, C., and Sandoval, I., 1982, Quantitation and characterization of the microtubule associated MAP_2 in procine tissues and its isolation from porcine (PK15) and human (HeLa) cell lines, *Biochem. Biophys. Res. Commun.* **105:**1241–1249.

Vallee, R. B., 1980a, Structure and phosphorylation of microtubule-associated protein 2 (MAP_2), *Proc. Natl. Acad. Sci. USA* **77:**3206–3210.

Vallee, R. B., 1980b, Structure and phosphorylation of MAP_2, in: *Microtubules and Microtubule Inhibitors* (M. DeBrabander and J. DeMey, eds.), pp. 201–211, Elsevier/North Holland Biomedical Press, Amsterdam.

Vallee, R. B., 1982, A taxol-dependent procedure for the isolation of microtubules and microtubule-associated proteins (MAPs), *J. Cell Biol.* **92:**435–442.

Vallee, R. B., and Borisy, G. G., 1977, Removal of the projections from cytoplasmic microtubules *in vitro* by digestion with trypsin, *J. Biol. Chem.* **252:**377–382.

Vallee, R. B., and Borisy, G. G., 1978, The non-tubulin component of microtubule protein oligomers, *J. Biol. Chem.* **253:**2834–2845.

Vallee, R. B., and Davis, S. E., 1983, Low molecular weight microtubule associated proteins are light chains of microtubule-associated protein 1 (MAP_1), *Proc. Natl. Acad. Sci. USA* **80:**1342–1346.

Vallee, R. B., DiBartolomeis, M. J., and Theurkauf, W. E., 1981, A protein kinase bound to the projection portion of MAP_2 (microtubule-associated protein 2), *J. Cell Biol.* **90:**568–576.

Voter, W. A., and Erickson, H. P., 1982, Electron microscopy of MAP_2 (microtubule-associated protein 2), *J. Ultrastruct. Res.* **80:**374–382.

Weatherbee, J. A., Luftig, R. B., and Weihing, R. R., 1980, Purification and reconstitution of HeLa cell microtubules, *Biochemistry* **19:**4116–4123.

Weatherbee, J. A., Sherline, P., Mascardo, R. N., Izant, J. G., Luftig, R. B., and Weihing, R. R., 1982, Microtubule-associated proteins of HeLa cells: Heat stability of the 200,000 molecular weight HeLa MAPs and detection of the presence of MAP_2 in HeLa cell extracts and cycled microtubules, *J. Cell Biol.* **92:**155–163.

Weingarten, M. D., Lockwood, A. H., Hwo, S.-Y., and Kirschner, M. W., 1975, A protein factor essential for microtubule assembly, *Proc. Natl. Acad. Sci. USA* **72:**1858–1862.

Weisenberg, R. C., 1972, Microtubule formation *in vitro* in solutions containing low calcium concentrations, *Science* **177:**1104–1105.

Wolosewick, J. J., and Porter, K. R., 1976, Stereo high-voltage electron microscopy of whole cells of human diploid line, W1-38, *Am. J. Anat.* **147:**303–324.

Zingsheim, H.-P., Herzog, W., and Weber, K., 1979, Differences in surface morphology of microtubules reconstituted from pure brain tubulin using two different microtubule-associated proteins: The high molecular weight MAP_2 proteins and tau proteins, *Eur. J. Cell Biol.* **19:**175–183.

9

Genetic Dissection of the Assembly of Microtubules and Their Role in Mitosis

Fernando Cabral

1. Introduction

Mitosis is a phenomenon which has intrigued biologists ever since Flemming first observed dividing cells under the microscope in 1879. Since then, a good deal has been learned about the events in mitosis and the structures which are involved. For example, we know that in mammalian cells, mitosis generally begins with a change in the morphology of the cell to a more rounded configuration. At the same time, spindle microtubules start to form at the spindle poles and the chromosomes begin to condense. As the nuclear membrane disappears, the chromosomes become attached to the spindle poles through their kinetochore to pole microtubules. Then, in rapid succession, chromosomes align on the metaphase plate, sister chromatids on each chromosome migrate to opposite poles of the spindle, and the spindle itself begins to elongate. Next, a cleavage furrow, created by contractile actin–myosin filaments, forms between the spindle poles and constricts the cytoplasm capturing the interpolar microtubules in a midbody and creating two daughter cells. As the cells exit mitosis and re-enter the G1 portion of the cell cycle, the nuclear membrane reforms, the chromosomes decondense, and cytoplasmic microtubules reform.

Mitosis is an incredible process. It is complicated and very carefully orchestrated. Although much has been learned about the sequence of events

---

Fernando Cabral • Departments of Medicine and of Biochemistry and Molecular Biology, University of Texas Medical School at Houston, Houston, Texas 77025.

and the morphological structures involved, relatively little is known about the molecular details of the process or its regulation. It will not be my aim to review these events in detail or outline the problems which we face in understanding mitosis. For this, the reader is referred to several excellent reviews (Inoue, 1981; Wilson, 1925; Schrader, 1953; McIntosh, 1979). In this review, I will focus on how microtubules might be involved in this complicated process and how genetics can help us to probe this involvement. The discussion will be further restricted to work which has been done on somatic cells. The reader is reminded, however, that the genetic approach has been exquisitely utilized to study the control of cell division in fungi (Hartwell, 1978; Morris, 1980) and *Chlamydomonas* (Warr and Gibbons, 1974; Flavin and Slaughter, 1974; Sato, 1976).

1.1. Microtubules in the Cell Cycle

Cells in interphase have their microtubules arranged in a cytoplasmic microtubule complex (CMTC; Fuller *et al.*, 1975; Weber *et al.*, 1975). These microtubules originate in or near the centrosome, a perinuclear organelle consisting of a centriole pair and associated pericentriolar material (Peterson and Berns, 1980). They radiate out toward the cell periphery, often ending at the plasma membrane (Weber and Osborn, 1979). Biochemically, microtubules consist primarily of α- and β-tubulin, two related proteins of approximately 55,000 daltons each. In addition, they are believed to contain microtubule associated proteins (MAPs), a heterogeneous class of proteins which appear to bind to microtubules and influence their assembly (for a review see Vallee, this volume).

When the G2 interphase cell is ready to progress into mitosis, a dramatic rearrangement of the microtubules occurs. The centriole pair in the G1 centrosome is duplicated during S phase so that the G2 centrosome consists of two centriole pairs. As the cell enters mitosis, the CMTC depolymerizes increasing the pool of free tubulin dimers, and the centrosomes migrate to opposite sides of the nucleus. Here, the centrosomes act as the spindle poles and nucleate the assembly of spindle microtubules by re-utilizing the tubulin from the depolymerized CMTC (Bibring and Baxandall, 1977; Fulton and Simpson, 1979). These spindle microtubules comprise two classes: those which run from one pole toward the opposite pole (interpolar microtubules) and those which run from one pole toward a specialized region on the chromosome called the kinetochore (kinetochore-to-pole microtubules; Brinkley and Stubblefield, 1970; Roos, 1973). In addition to their organization, these microtubules may be distinguished by their sensitivity to cold (4°C) and colcemid. The interpolar microtubules are cold and colcemid labile while the kinetochore to pole microtubules are relatively cold and colcemid stable (Brinkley and Cartwright, 1975; Dustin, 1978). Also, it is believed that calmodulin, a ubiquitous calcium-binding protein, binds preferentially if not exclusively to the kinetochore to pole microtubules (Welsh *et al.*, 1978). The molecular basis for these different properties is not well understood.

Also, it is not clear whether or not the spindle microtubules are biochemically different from their cytoplasmic counterparts. As mentioned previously, spindle microtubules are believed to assemble from the tubulin subunits produced by the disassembly of the cytoplasmic microtubules (Fulton and Simpson, 1979). Still, it is possible that spindle-specific tubulins exist or that different MAPs participate in the assembly of the spindle. Evidence for the latter possibility has recently come from the observation that serum from certain patients contain antibodies which decorate the spindle but not cytoplasmic microtubules (McCarty *et al.*, 1981; Lydersen and Pettijohn, 1980) and that among various monoclonal antibodies produced against HeLa cell MAPs, some decorate only the spindle (Izant *et al.*, 1982). It will be important to identify the antigens in these experiments and to determine if they function directly in spindle assembly or if they are only indirectly associated with the spindle apparatus. Biochemical evidence for differences in the composition of spindle vs. cytoplasmic microtubules has also appeared (Zieve and Solomon, 1982) but confirmation of these results is lacking.

1.2. The Role of Microtubules in Mitosis

The role of the microtubules in mitotic events is a subject of considerable debate. There is ample evidence that they are a major component of the mitotic spindle apparatus (Inoue, 1981). However, it cannot be stated with any certainty whether or not they participate directly in chromosome movement, cell elongation, or cytokinesis. Chromosome movements range from erratic prometaphase motions to a highly ordered poleward migration of chromatids during anaphase. A number of theories have been proposed to explain these movements including actin–myosin contraction, microtubule sliding, microtubule assembly–disassembly, and microtubule treadmilling (Inoue, 1981). Data supporting each of these theories has been reported; but in every case, conflicting experiments also exist (Pickett-Heaps *et al.*, 1982). A similar conclusion can be drawn concerning cell elongation and cytokinesis, i.e., conclusive evidence showing microtubule involvement in these processes is lacking. In the case of cytokinesis, for example, it has long been observed that the cleavage furrow forms along a plane perpendicular to and bisecting the spindle axis (Conrad and Rappaport, 1981). Experiments in which the spindle is mechanically displaced with a microneedle reinforce the idea that the spindle is able to specify the site of cleavage (Conrad and Rappaport, 1981), and suggest that microtubules may be necessary for the assembly or positioning of the contractile ring. Other experiments, however, indicate that the cleavage furrow will form and cytokinesis will proceed even if the spindle is destroyed or mechanically removed from the cell provided the cell has reached metaphase or anaphase at the time of spindle removal (Hiramoto, 1971; Rappaport, 1971). Thus, the early spindle appears to be necessary for cytokinesis to occur normally but its continued presence beyond metaphase may not be required. The role of microtubules in this process is not clear.

Many of the ambiguities regarding the role of microtubules in mitosis arise from the sheer complexity of the problem. Biochemical and morphological approaches alone may not be sufficient to elucidate the control of mitotic events. Comparative studies of mitosis in different organisms are useful for determining which factors are essential to the process, but they can also be confusing in that different organisms may have evolved different mechanisms for carrying out particular steps. An approach which has proven very powerful in dissecting complex biological processes, e.g., metabolic pathways, is genetics. This approach has been especially powerful when used in combination with biochemical and morphological techniques. The rationale is to isolate mutants with specific lesions affecting a particular process, characterize the nature of the lesion biochemically and morphologically, and then determine the effects of the lesion on the operation of the process. This chapter will be devoted to this approach.

2. A Genetic Approach to Microtubule Function

2.1. Why Use Somatic Cells?

At first glance, somatic cells might seem a poor choice for genetic studies of mitosis. They grow slowly, are expensive to culture, have a diploid set of chromosomes which makes the isolation of recessive mutations difficult if not impossible, and cannot be mated, making genetic manipulations cumbersome. It is true that mutant isolation and subsequent genetic characterization is simpler in the lower eukaryotes, but mammalian cells do offer some advantages for the study of mitosis. First, microtubules from mammalian cells are better characterized biochemically and physiologically. Conditions for isolation and *in vitro* assembly of microtubules are well worked out and a number of MAPs have been described. In addition to their role in spindle assembly, mammalian microtubules have been implicated in a wide variety of cellular functions including cellular morphology, locomotion, saltatory motion, secretion, chemotaxis, and hormone action (Dustin, 1978; Crossin and Carney, 1981). This allows the effects of tubulin alterations to be monitored using a variety of physiological events. Second, mammalian cells have superior morphology for examining the state of microtubule and spindle assembly using immunofluorescence and electron microscopy. Third, karyotypic alterations are more easily followed in mammalian cells. Finally, different organisms carry out mitosis in distinctly different ways (McIntosh, 1979). It is of considerable interest to determine how mitosis in mammalian cells compares with the same process in lower eukaryotes.

For our own studies, we have chosen the Chinese hamster ovary (CHO) cell line (Puck *et al.,* 1958). These are hearty cells which grow rapidly (generation time of approximately 12 hr), can be easily synchronized, have been well characterized, and offer a number of advantages for genetic studies. For example, they have a stable karyotype, can be grown from a single cell, and grow on dishes or in suspension. In addition, markers for genetic manipulations are easily introduced and a number of recessive mutations in the cell line have already been isolated (Siminovitch, 1976).

2.2. *Isolation of Mutants Altered in Mitosis*

2.2.1. *Temperature-Sensitive (ts) Cells Blocked in Mitosis*

Since mitosis is presumably necessary for cell division and viability, it is expected that only mutants with a conditional defect in mitosis can be obtained. One way of obtaining such cells is to look for temperature-sensitive mutants which are blocked in mitosis. This may be done by killing cells which grow at the nonpermissive temperature or by making use of the observation that many cell lines round up during mitosis and become less well attached to the tissue culture dish. In an early paper, Smith and Wigglesworth (1972) selected temperature-sensitive BHK-21 cells by killing the cells able to grow at the nonpermissive temperature with cytosine arabinoside and then shifting the cells back to the permissive temperature. Several of the surviving clones were found to have a defect in cytokinesis. Chinese hamster fibroblasts with a ts block in cytokinesis were also isolated by Hatzfeld and Buttin (1975) using a BUdR suicide selection. Unlike the mutants described by Smith and Wigglesworth, these cells became multinucleated and sometimes anucleate. Even in the absence of cell division at the nonpermissive temperature, the cells continued to replicate chromosomes leading to the accumulation of cells with greatly elevated numbers of chromosomes.

Using similar selections, Wang and co-workers (Wang, 1974; Wang and Yin, 1976) isolated over 50 ts hamster cell mutants and of these, ts-546 was found to block at a prometaphase-like state at the nonpermissive temperature. The reversion frequency of this mutant was low (5×10^{-7}). When mitotic figures were observed at the nonpermissive temperature, scattered chromosomes were seen, suggesting the existence of an aberrant spindle, but this was not directly shown. These cells go on to enter an "interphase-like" state in which nuclear membranes reform around the scattered chromosome aggregates. A second class of mutants with defective prophase progression were later isolated by the same author (Wang, 1976). The chromatin in these cells begins to condense and the nuclear membrane is lost, but discernable chromosomes do not form. The chromatin eventually coalesces into dense-staining clumps and the nuclear membrane fails to reform. These cells would appear to have a defect in chromosome condensation, but again the biochemical lesion was not characterized.

Shiomi and Sato (1976) used a similar selection scheme in a murine leukemic cell line to obtain a mutant, ts2, with properties similar to the ts-546 cell line isolated by Wang. Again, these cells block in mitosis and accumulate abnormal mitotic figures with scattered chromosomes reminiscent of colchicine-treated cells. The cells fail to divide but apparently reenter interphase as large multinucleated cells. Interestingly, cells in G2 at the time of shift up to the nonpermissive temperature progress through mitosis normally but cells in

G1 at the time of shift-up do not. This implies that a temperature-sensitive event in G1 affects mitotic progression or that several hours at the nonpermissive temperature are required for the mitotic defect to be manifested.

A variation on these selections was performed by Thompson and Lindl (1976). Making use of the rounded morphology of cells in mitosis, these authors looked for variants of mutagenized CHO cells which did not attach to tissue culture dishes at the nonpermissive temperature. After several rounds of selection, they isolated a clone, MSI-1, which attached at 34° but was blocked in mitosis at 38.5°. These cells have a higher incidence of polyploidy than do wild-type cells even at the permissive temperature (34°), but the cultures accumulate even higher levels of polyploid cells after a shift to the elevated temperature (38.5–39.5°). These effects were believed to result from a defect in the terminal stages of cytokinesis. Ultrastructurally, microtubules, microfilaments, and other structures appeared to be normal.

This brief summary indicates that selections for cells with ts defects in mitosis can yield mutants with interesting properties. These mutants have not proven to be as useful as was hoped, however, because it has been difficult to demonstrate the biochemical lesion in these cells. For this reason, it has been proposed that selections designed to give mutants with more specific defects, i.e., in known proteins and functions, will be more useful (Stanners, 1978). The following section will describe one way in which such mutants have been obtained.

2.2.2. *Selection for Resistance to Drugs Which Interfere with Spindle Components*

One method for obtaining mutants with specific alterations in mitosis is to pick a known component of mitosis and look for resistance to agents which are known to affect that component. The component with which I will deal in some detail in this review is the microtubule system. As discussed earlier, microtubules are known to be a major constituent of the mitotic spindle apparatus. In addition, a large number of drugs have been found which interfere with microtubule polymerization (Luduena, 1979). These drugs are well characterized structurally and their interactions with tubulin or microtubules have been intensively studied. In some cases, e.g., colchicine, their mechanism of action is reasonably well understood (Margolis and Wilson, 1977). One property which these drugs have in common and which make them useful for mutant selection is cytotoxicity. Implicit in this argument is the assumption that the cytotoxicity is mediated through the interaction of the drugs with microtubules. The rationale for mutant selection, then, is to look for cells resistant to the cytotoxic effects of various microtubule-active drugs. The expectation is that at least some of these resistant cells will have alterations in their microtubules.

This approach has proven to be quite successful in the isolation of tubulin mutants in *Aspergillus nidulans* (Sheir-Neiss *et al.*, 1978; Morris *et al.*, 1979; Oakley, 1981) and in CHO cells (Cabral *et al.*, 1980, 1981, 1983; Cabral, 1983; Ling *et al.*, 1979; Ling, 1981; Warr *et al.*, 1982). In our own laboratory, we use a selection protocol such as the one diagrammed in Fig. 1. Wild-type cells are first tested for their sensitivity to the selecting drug by measuring the plating efficiency of the cells at various drug concentrations. Cells which have been mutagenized by treatment with ethylmethanesulfonate (EMS) or exposure to ultraviolet light and allowed to recover for two days are then plated at a density of approximately 5×10^5 cells per 100 mm tissue culture dish and incubated in complete medium containing twice the concentration of the drug needed to reduce the plating efficiency to below 0.1%. After 7–10 days at 37° in a humidified incubator containing 5% CO_2, surviving colonies growing over a background of dead cells may be seen. These clones are then picked, recloned, and grown for characterization. This kind of single-step isolation procedure generally yields mutants which are 2 to 3-fold more resistant to the selecting drug than are the wild-type cells. Mutant cells with 10-fold or even greater resistance to these drugs may be obtained by reselecting the resistant cell population stepwise to progressively higher drug concentrations and then cloning the cells. This procedure is not usually advisable, however, because stepwise selections lead to the isolation of cells having multiple mutations and thus greatly complicates subsequent interpretations of the mutant cell properties. This problem may be alleviated by characterizing the mutation at each step before proceeding with further selection.

Once resistant cells are isolated, they must be screened for those which have alterations in microtubules. It is clear from the work of Ling and his associates (Ling and Thompson, 1974) as well as our own (Cabral *et al.*, 1980) that the majority of mutants obtained with the drug-resistance phenotype have a permeability defect, i.e., they are more resistant to the drug because they do not take up the drug as readily as do the wild-type cells. A simple way to screen for these permeability mutants was pointed out by Bech-Hansen *et al.*, 1976). These authors found that cells resistant to microtubule-active drugs by virtue of a permeability defect, were also resistant to a variety of other

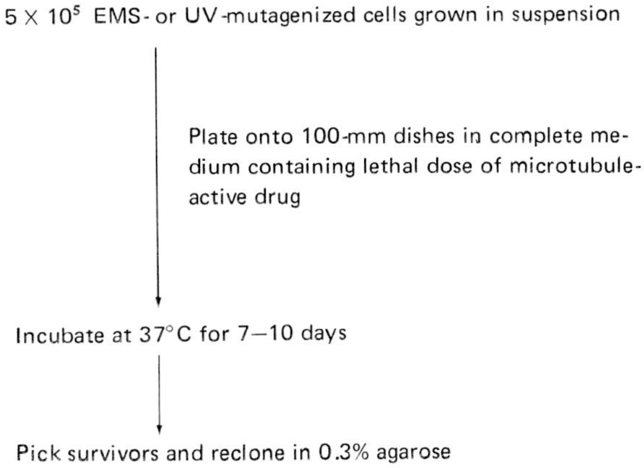

Figure 1. Selection scheme for obtaining drug-resistant mutants in CHO cells. The selection protocol is described in the text.

drugs unrelated to the selecting drug in their mechanism of action but similar in their hydrophobicity. Thus, by testing putative tubulin mutants for their cross-resistance to puromycin, ethidium bromide, actinomycin D, etc., permeability mutants can be readily identified. We were subsequently able to show the converse, i.e., that mutants with clearly defined alterations in tubulin are not cross-resistant to unrelated drugs such as puromycin (Cabral *et al.*, 1980).

Although permeability mutants may be quickly screened out, it is still necessary to identify tubulin mutants among the remainder of the drug-resistant cells. We have approached this problem in a manner similar to the one employed by Morris and his colleagues (Sheir-Neiss *et al.*, 1978; Morris *et al.*, 1979), i.e., to look for electrophoretic shifts in the tubulin proteins using two-dimensional gel electrophoresis. An alternative approach is to isolate tubulin from the putative tubulin mutants and demonstrate an increased resistance to drug inhibition of microtubule polymerization *in vitro*. Each method has certain advantages. The electrophoretic approach is relatively quick and many mutants can be screened in a short time. Furthermore, if an alteration is found, then the identity of the altered protein is known. The main disadvantage is that an alteration in tubulin need not necessarily alter its electrophoretic mobility and thus, some mutants will be lost. Also, without proper controls (see below) one cannot be certain that the electrophoretically altered protein is the result of a structural gene mutation, an altered modifying enzyme or, indeed, that it is even responsible for the drug-resistance phenotype. Still, with the proper controls, this method can unambiguously identify tubulin mutants, and the alteration in tubulin can be shown to be responsible for the drug resistance. Altered drug sensitivity of *in vitro* microtubule assembly provides better linkage between alterations in microtubule proteins and drug resistance but does not identify the altered species. Furthermore, it is time consuming, does not eliminate the possibility of post-translational modification of tubulin as the cause of increased drug resistance, and would miss mutants which are affected in functions not measured by *in vitro* microtubule assembly, e.g., spindle assembly.

3. Mutants with Altered Tubulin

3.1. Survey of Mutants Isolated

As discussed above, we have used two-dimensional gel analysis (Cabral *et al.*, 1980, 1981, 1982) to identify those drug-resistant mutants which carry an altered tubulin. This approach is made feasible by the fact that tubulin is a major cellular constituent and can be readily identified on the gel among the proteins from a whole cell extract. Examples of such gels are shown in Fig. 2. Panel A shows the two-dimensional gel of a wild-type CHO cell. Among the major cellular proteins which can be readily identified are α-tubulin (a), β-tubulin (b), vimentin (c), and actin (d). Most of the drug-resistant cells have an

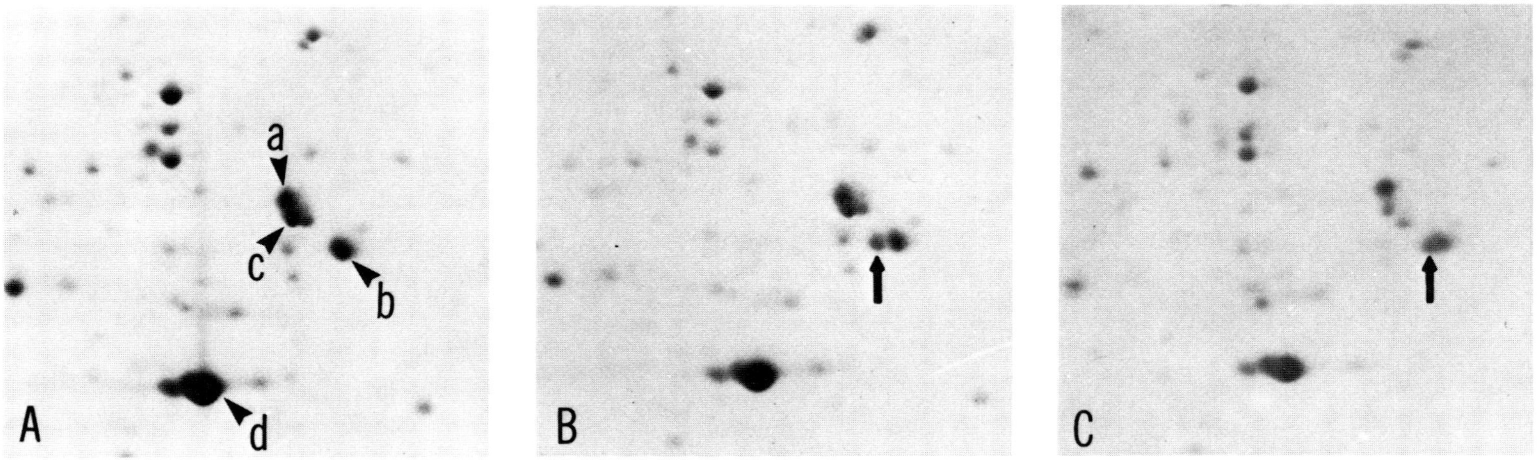

Figure 2. Two-dimensional gel autoradiograms of wild-type and mutant CHO cells. Cells were metabolically labeled for 30 min in [^{35}S]methionine, lysed in hot SDS and run on two-dimensional gels (Cabral *et al.*, 1980). The horizontal dimension represents electrofocusing with the acidic end to the right; the vertical dimension represents SDS gel electrophoresis in 10% acrylamide. Only a portion of the gel is shown. (A) Wild-type cells, (a) α-tubulin, (b) β-tubulin, (c) vimentin, (d) actin. (B) Cmd-4, a colcemid-resistant mutant. Note the presence of an "extra" spot (arrow) to the basic side of β-tubulin. (C) Grs-2, a griseofulvin-resistant mutant. The "extra" spot (arrow) in this mutant has a different electrophoretic mobility from the "extra" spot in Cmd-4.

identical gel pattern. In some cases, however, an "extra" spot is seen. Panel B, for example, shows the gel pattern for a colcemid-resistant mutant that we isolated (Cabral *et al.*, 1980). In this mutant, there is a new spot (arrow) which is not present in the wild-type pattern and which migrates to the basic side of β-tubulin. Tryptic peptide mapping showed that this "new spot" represents an electrophoretic variant of β-tubulin. Notice that the wild-type β-tubulin continues to be expressed. This result is not surprising since these are quasidiploid cells. Similarly, Panel C shows the gel pattern of a griseofulvin-resistant CHO mutant which again exhibits an extra spot representing an electrophoretically variant β-tubulin. The altered β-tubulin of the griseofulvin-resistant mutant has a different electrophoretic mobility than does the altered β-tubulin of the colcemid-resistant mutant.

Using similar methods, we have also identified an α-tubulin mutant among our taxol-resistant cells (Cabral *et al.*, 1981). While this might super-

Figure 3. Two-dimensional gel autoradiogram of Tax-2, a taxol-resistant mutant with an altered β-tubulin. The cells were labeled and run on two-dimensional gels as described in the legend for Fig. 2. Note the presence of an "extra" spot (arrow) not seen in wild-type cells (Fig. 2A) to the basic side of β-tubulin.

ficially suggest that mutation for resistance to taxol requires a different genetic locus than mutation for resistance to other microtubule-active drugs, we have since found a second taxol-resistant mutant which has an alteration in β-tubulin (unpublished work). The two-dimensional gel of this mutant is shown in Fig. 3. Again, there is an "extra spot" (arrow) migrating to the basic side of β-tubulin. These results point out the danger in assuming that demonstration of an altered species in a drug-resistant mutant identifies the drug-binding subunit.

These and several other mutants are summarized in Table 1. As may be seen, various microtubule-active drugs have been used to generate mutants with alterations in both α- and β-tubulin. Many of the mutants listed, however, have not been biochemically characterized to the point where the identity of the mutant species is known. A mutant isolated by Gupta (Gupta, 1981; Gupta *et al.*, 1982) for resistance to podophyllotoxin is reported to carry an alteration in a 66,000-dalton protein. The author suggests that this may be a previously unidentified MAP. While this is a potentially exciting mutant, one should be cautious in linking a protein having an altered mobility on two-dimensional gels with the mutant phenotype. To establish this linkage clearly, reversion experiments are needed to show that the mutant phenotype and the altered protein corevert at high frequency (see below).

Most of the mutants in Table 1 express their drug-resistance phenotype in a codominant manner when probed by somatic cell hybridization experiments. Possible exceptions to this are the taxol-resistant α- and β-tubulin mutants whose dominance has not yet been clearly established. A pattern of codominance would be expected since the mutant strains continue to express wild-type tubulin subunits as shown in Fig. 2. The recessive taxol-resistant mutant would appear to constitute a new class of mutants and will be dealt with in detail below.

3.2. Mechanism of Drug Resistance

An important question regarding the mutants listed in Table 1 concerns their mechanism of drug resistance. The simplest explanation which comes to mind is that these cells contain tubulin with an alteration in its drug-binding site. However, there are several problems with this explanation. First, we have measured drug binding in our mutants and it appears to be normal (unpublished results). Second, at least some microtubule-active drugs such as colchicine are believed to poison microtubules when substoichiometric amounts bind to tubulin dimers. These drug-containing dimers then assemble into the growing microtubule and inhibit the polymerization (Margolis and Wilson, 1977; Sternlicht and Ringel, 1979; Sternlicht *et al.*, 1980). Since we know that both wild-type and mutant subunits are present in the mutant cells, loss of a drug-binding site cannot explain the mutant phenotype, i.e., if part of the tubulin in the cell has lost its ability to bind the drug, the unmodified tubulin in the same cell or in a hybrid cell should still bind the drug and poison microtubule assembly. Third, and most importantly, microtubule-active drugs are believed to bind to different sites on tubulin or, in some cases,

Table 1. Summary of Chinese Hamster Ovary Mutants Resistant to Microtubule-Active Drugs

Strain	Altered protein	Codominant (D) or recessive (R)	Selecting drug	Temperature sensitivity	Relative resistant[c]	Single (S) or multiple (M) step selection	References
Cmd-4	β-Tubulin	D	Colcemid	ts[a]	2–3 X	S	Cabral *et al.*, 1980
Col-2	β-Tubulin	D	Colchicine	ts	2–3 X	S	Cabral *et al.*, 1980
Grs-1	β-Tubulin	D	Griseofulvin	?	2–3 X	S	Cabral *et al.*, 1980
Grs-2	β-Tubulin	D	Griseofulvin	ts	2–3 X	S	Cabral *et al.*, 1980
Tax-1	α-Tubulin	?	Taxol	ts	2–3 X	S	Cabral *et al.*, 1981
Tax-2	β-Tubulin	?	Taxol	?	2–3 X	S	
Tax-18	?	R	Taxol	nts[b]	2–3 X	S	Cabral, 1983; Cabral *et al.*, 1983
CM^R114	?	D	Colcemid	?	6 X	M	Ling *et al.*, 1979
CM^R117	?	D	Colcemid	?	6 X	M	Ling *et al.*, 1979
CM^R761	?	D	Colcemid	?	6–7 X	M	Ling *et al.*, 1979
CM^R795	α-Tubulin	D	Colcemid	?	8–9 X	M	Ling *et al.*, 1979
$Pod^{RII}6$	66,000-D protein	D	Podophyllotoxin	?	8–10 X	M	Gupta, 1981; Gupta *et al.*, 1982
Tax_1	?	?	Taxol	?	6 X	S	Warr *et al.*, 1982
BEN_1	?	?	Benzimidazole carbamate	?	2 X	S	Warr *et al.*, 1982

[a]ts, Cells are temperature sensitive.
[b]nts, Cells are not temperature sensitive.
[c]Ratio of drug concentrations required to reduce plating efficiencies of mutant and wild-type cells to less than 0.1%.

to microtubule-associated proteins (Luduena, 1979; Roobol *et al.*, 1977). Yet, many of these mutants are cross resistant to a variety of microtubule-active drugs (Cabral *et al.*, 1980; Ling *et al.*, 1979; Warr *et al.*, 1982). How can an alteration in the binding site for one drug render the cell resistant to drugs which bind at distinctly different sites? *Ad hoc* explanations for each of these arguments can be made, but the weight of evidence indicates that our mutants do not have an alteration in drug-binding sites. Rather, these mutants are envisioned to carry altered tubulin subunits which may add on to the ends of drug-poisoned microtubules and restore microtubule elongation. This mechanism is consistent with the codominant character and drug cross resistance of these mutants. Further evidence for this type of a mechanism comes from reversion studies discussed below.

There is one mutant, CM^R795, listed in Table 1 for which tubulin with lowered drug-binding capacity has been reported (Ling *et al.*, 1979). It is not yet clear how this alteration leads to codominant expression of drug resistance.

Another interesting feature of the mutants in Table 1, resides in the fact that single-step selections yield cells which are only 2 to 3-fold resistant to the selecting drugs. This can perhaps be explained by the fact that tubulin is an essential protein in mitosis (see below) and therefore only minor changes in structure will be nonlethal to the cell. The nature of these structural changes could be predicted from the following hypothetical mechanism of drug action.

Assume microtubule assembly can be described by the following simple equilibrium:

$$n(\alpha\beta) \underset{kd}{\overset{ka}{\rightleftharpoons}} (\alpha\beta)_{T\text{-}n}$$

where $(\alpha\beta)$ represents a tubulin dimer, T represents the total number of tubulin dimers (free plus polymerized), n represents the number of free tubulin dimers, ka represents the rate constant for assembly of an $\alpha\beta$ dimer, and kd represents a dissociation rate-constant of a dimer in the microtubule. Drugs such as colchicine bind to free $\alpha\beta$ dimers (Shelanski and Taylor, 1967) and would thereby lower the ka for assembly, perhaps by causing a conformational change. Evidence for conformational changes in tubulin upon drug binding has in fact appeared (Garland, 1978) and a mechanism such as this has been reported (Sternlicht and Ringel, 1979; Sternlicht *et al.*, 1980). Colchicine-resistant mutants might then produce an altered $\alpha\beta$ dimer having an increased affinity for the microtubule which could compensate for the lowered affinity produced by drug binding. Such mutants would be expected to produce hyperstable microtubules in the absence of the drug and this hyperstability would be removed by addition of the drug. Still further addition of the drug would again shift the equilibrium toward the dimer. Mutants with these characteristics have been reported by Oakley and Morris (1981) in *Aspergillus*. Such mutants would be expected to exhibit only low levels of resistance to the drugs. Assuming that the equilibrium of free and polymerized tubulin is tightly regulated in the cell, changes in *ka* which could lead to higher levels of drug resistance, might perturb the equilibrium to such an extent that it would be lethal. In an alternative mechanism, the mutant αβ dimer might only show increased affinity when the drug is bound. In this case, the microtubules would not be hyperstable in the absence of the drug.

Taxol has recently been shown to bind to and stabilize polymerized microtubules (Manfredi *et al.*, 1981). It would therefore be expected to have the opposite effect of drugs such as colchicine, i.e., it would lower *kd,* thus shifting the equilibrium to the polymerized form. This might imply that at least some mutants resistant to drugs such as colchicine and which as explained above could have hyperstable microtubules, should be supersensitive to taxol. This has indeed been recently reported to be the case for CM^R795 as well as for BEN1 (Warr *et al.*, 1982). Similarly, mutants selected for resistance to taxol might be expected to have a lower *ka* (or an increased *kd*) resulting in more labile microtubules. These microtubules would be expected in some cases to be supersensitive to microtubule depolymerizing drugs such as colchicine. This has been found to be true for Tax-18 (Cabral, 1983).

It should be emphasized that mutation to hyperstability or hyperlability of microtubules could occur through alterations in tubulin itself, or through alterations in microtubule associated proteins. These MAPs are believed to bind to microtubules and lower the concentration of tubulin necessary for assembly (see chapter by Vallee). Thus, they shift the equilibrium toward the polymerized state. Mutations which increase the affinity of MAPs for microtubules would result in hyperstabilization of microtubules, while mutations which lower the affinity would result in hyperlability. At present, it is difficult to analyze mutants for alterations in microtubule associated proteins. MAPs from cultured cells have not been well characterized, probably occur in lower abundance relative to tubulin, and are difficult to purify. Once putative MAPs have been identified for a given cell line, however, a mutant analysis should provide strong evidence for whether or not a particular MAP plays an *in vivo* role in microtubule assembly, and should help to elucidate the mechanism by which MAPs influence microtubule assembly.

3.3. Temperature Sensitivity

If microtubules are necessary for cell viability, then mutations affecting essential microtubule functions should only be recovered in conditional mutants. This has in fact been our experience. Of the mutants in which we have been able to show clear alterations in tubulin, virtually all have proven to be temperature sensitive. One additional mutant, Tax-18, in which we have not yet found the biochemical lesion but in which microtubule assembly is clearly affected, is drug dependent for growth (see below). There is no *a priori* reason, however, that all drug resistant mutants should have defects in microtubule function, and in fact, conditionality of the other mutants listed in Table 1 has not been demonstrated. Our results, however, suggest that tubulin is necessary for cell viability and that the vital process or processes in which tubulin is involved may be identified by studying mutant cell physiology at the nonpermissive temperature, or in the absence of drug in the case of Tax-18. Such studies, described below, have allowed us to more precisely define the role of microtubules in vital cellular events.

3.4. Isolation of Revertants

Conditional mutants are also important to obtain in drug-resistance selections to allow the selection of revertants. The demonstration of the biochemical lesion in a mutant cell line is important for understanding the mutant phenotype. As mentioned above, ts mitosis mutants have not proven as useful as anticipated because of a lack of this critical information. At the same time, finding a biochemical lesion in a mutant cell line is not sufficient to be able to argue that the altered protein is responsible for the mutant phenotype. To support this argument, it is necessary to perform *in vitro* reconstitution studies, gene transfer experiments, or reversion studies. Of these, the last is the most straightforward. The argument is made that if two events, e.g., altered protein and altered function, are controlled by a single gene, then the two should corevert at a frequency consistent with a single mutational event.

We have taken this approach to show that the altered tubulin in our mutants is responsible for their drug resistance and temperature sensitivity. Since it is difficult to select for drug-sensitive revertants, we have employed an approach that was used by Morris and his colleagues with *Aspergillus* (Morris *et al.*, 1979). If it is assumed that drug resistance, temperature sensitivity, and altered tubulin are controlled by a single gene, then reversion of any one should lead to coreversion of the other two. We thus selected cells able to grow at the nonpermissive temperature and screened those "temperature-resistant" revertants for altered tubulin by two-dimensional gel electrophoresis and for drug resistance (Cabral *et al.*, 1982). The results of this approach for mutant Cmd-4 are summarized in Fig. 4.

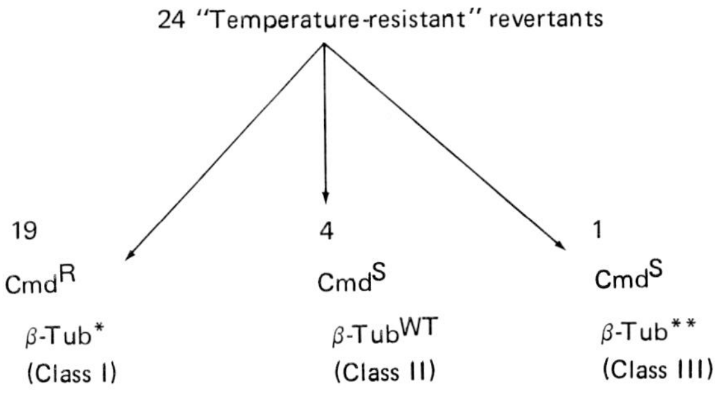

Figure 4. Summary of classes of revertants isolated from mutant Cmd-4. Mutant Cmd-4, a colcemid-resistant and temperature-sensitive mutant with an altered β-tubulin, was plated at a temperature which is nonpermissive for growth of these cells (40.5°C) but which permits the growth of wild-type cells (Cabral *et al.*, 1982). The revertants obtained fell into three distinct classes as described in the text.

Of the 24 revertants selected for loss of their temperature-sensitive phenotype, 19 retained their drug-resistance and their altered β-tubulin. This class of revertants is still largely uncharacterized, but may contain further alterations in the mutant β-tubulin which are electrophoretically silent and which restore stability to the microtubules without removing the ability of the altered β-tubulin to confer drug resistance. A more exciting possibility is that some of these revertants have alterations in other proteins, e.g., MAPs, which can interact with and stabilize the microtubules. If true, this would provide a means of identifying proteins which interact to form microtubules *in vivo.* The second most abundant class of revertants (4/24) are those which have lost their drug resistance and their altered β-tubulin. This is the class of revertants which was predicted and which strongly supports our conclusion that the altered β-tubulin, the drug resistance, and the temperature sensitivity of the mutant result from a mutation in a single gene. This conclusion is reinforced by the single revertant which constitutes the third class. In this case, there is a further alteration in the mutant β-tubulin itself which results in loss of drug resistance and temperature sensitivity. The mechanism of this reversion is discussed in the next section.

4. Use of Mutants to Study Microtubule Assembly and Tubulin Regulation

Elegant studies employing a genetic approach to study flagellar assembly and function in *Chlamydomonas* have been reported (Huang *et al.*, 1981, 1982; Witman *et al.*, 1978). The work in mammalian cells has not yet progressed to this level of sophistication, but preliminary work from our laboratory and in *Drosophila* from Raff and her colleagues (Kemphues *et al.*, 1979, 1980) suggests that a similar approach is feasible in higher organisms.

As mentioned above, the isolation of revertants is important to show that a demonstrated biochemical alteration in a mutant is responsible for the mutant phenotype; but revertants can also be intrinsically interesting. A good example of this is the class III revertant (Fig. 4), which we have named, Cmd-4-A5. Shown in Fig. 5 are two-dimensional gels of mutant Cmd-4 (Figure 5A) and revertant Cmd-4-A5 (Fig. 5B) as well as crude microtubules prepared from these cells (Figs. 5C and D). These crude microtubules were made by lysing the cells in a microtubule stabilizing buffer (Pipeleers *et al.*, 1977) and then centrifuging them at low speed (Cabral *et al.*, 1980). The pellet fractions are shown in Figs. 5C and D. Note that the altered tubulin in revertant Cmd-4-A5 (β**; Fig. 5B arrow) has a different mobility than the altered tubulin in mutant Cmd-4 (β*; Fig. 5A, arrow). When the cellular microtubules are assayed for the presence of these altered tubulins, we find that β* appears in the pellet from mutant Cmd-4, i.e., it assembles into microtubules (Fig. 5C), but β** does not appear in the pellet from revertant Cmd-4-A5 (Fig. 5D). Thus, it appears that Cmd-4-A5 reverted from mutant Cmd-4 by acquiring a second alteration in β* (to produce β**) which prevents it from

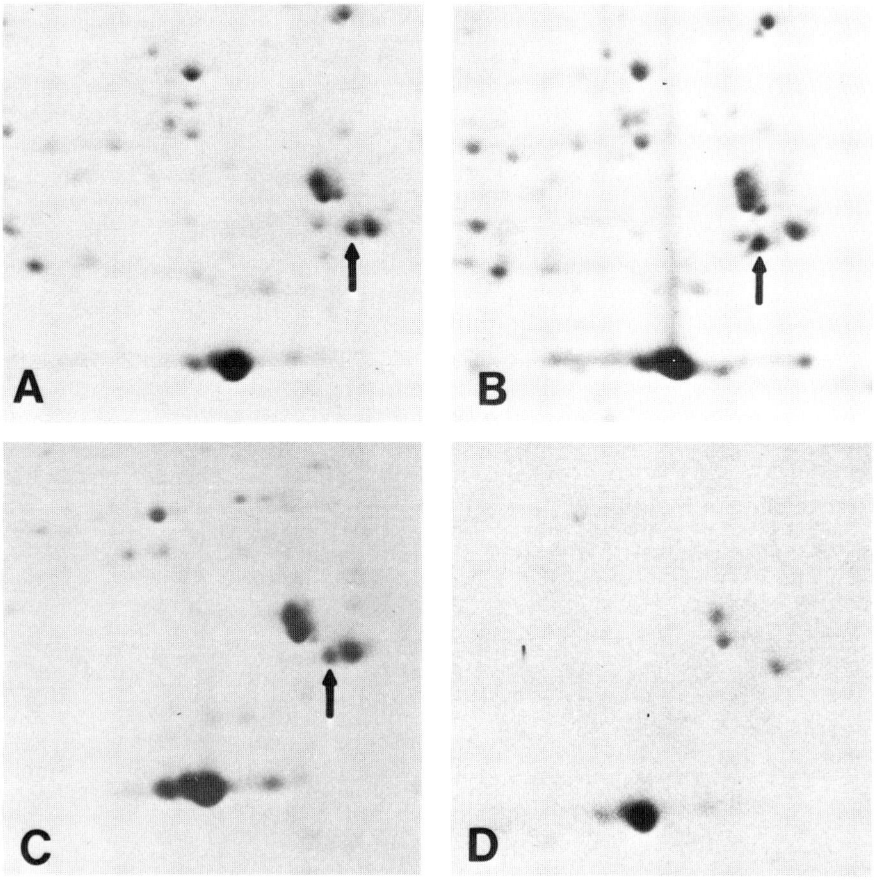

Figure 5. Two-dimensional gel autoradiograms of Cmd-4 and one of its revertants, Cmd-4-A5. Cells were labeled and run on two-dimensional gels as described in the legend for Fig. 2 except that in (C) and (D), the cells were first extracted with a microtubule stabilizing buffer as described in the text. (A) Cmd-4, whole cell lysate. The arrow indicates the presence of an altered β-tubulin. (B) Cmd-4-A5, whole cell lysate. The arrow indicates the presence of an altered β-tubulin with a different electrophoretic mobility than the altered β-tubulin in Cmd-4. (C) A crude microtubule preparation from Cmd-4. Note the continued presence of the altered β-tubulin (arrow). (D) A crude microtubule preparation from Cmd-4-A5. Note the absence of the altered β-tubulin.

assembling into microtubules. Since the microtubules in the revertant contain only wild-type subunits, the strain is no longer temperature sensitive or drug resistant. This result implies that altered β-tubulin must assemble in order to confer drug resistance. This conclusion is reinforced by some recent experiments in which revertant Cmd-4-A5 was used to again select for colcemid-resistant mutants. Among the mutants which were obtained, was a strain having an alteration in α-tubulin (α*; unpublished work). Apparently, this mutant has an altered α-tubulin that can interact with β** and cause it to reassemble into microtubules. Thus, the strain is again drug resistant and temperature sensitive. This further implies that β** cannot assemble because its site of interaction with α-tubulin is altered. Mapping the alterations in β** from Cmd-4-A5 and α* from its second-site mutant may allow us to define the regions in α- and β-tubulin which interact in the formation of microtubules.

Another feature of revertant Cmd-4-A5 which is evident from comparing Figs. 5A and B is the increased expression of the altered β-tubulin in the revertant. It should be emphasized here that Fig. 5 represents a gel autoradiogram of pulse labeled cells. Long labeling times or Coomassie blue staining of the gel show a vastly decreased presence if not a total absence of β** in Cmd-4-A5. This suggests that β** is synthesized at a high rate but is rapidly degraded. The enhanced degradation of β** presumably results from an inability to interact with α-tubulin, causing it to be recognized by intracellular proteases. Recently, work from at least two laboratories (Ben-Ze'ev *et al.*, 1979; Cleveland *et al.*, 1981) suggests that the pool of free tubulin in the cell can somehow regulate its own synthesis, i.e., when the pool of free tubulin is high, e.g., in colchicine treated cells, the synthesis rate for tubulin is depressed. If, on the other hand, the pool size is low, e.g., in vinblastine or taxol-treated cells, the synthesis rate for tubulin is high. We hypothesize that in revertant Cmd-4-A5, β** does not interact with α-tubulin to form tubulin dimers. As a result, it is recognized by intracellular proteases which rapidly degrade it. Since it is unable to accumulate in the cytoplasm, it cannot inhibit its own synthesis and so its synthesis rate is near maximal. Experiments to test this hypothesis are in progress.

It should also be pointed out that the experiments by Penman's and Kirschner's groups suggest coordinate control of α- and β-tubulin synthesis by this feedback mechanism. However, the use of drugs ensures the simultaneous change in pool size of all the tubulin subunits. In revertant strain Cmd-4-A5, only β** is unable to accumulate and only its synthesis rate appears to be affected. Thus, I suggest that this feedback mechanism for control of tubulin expression, if it indeed exists, may not be a coordinate control but may in fact be gene or gene product specific.

These results, while still preliminary, serve to show that somatic cell genetics may provide a good approach for studying tubulin assembly and expression in mammalian cells.

5. Use of Mutants to Define Microtubule Function

Microtubules are believed to play a role in many of the motile events of the cell including mitosis, cell locomotion, determination of cellular morphology, saltatory motion, and mobility of cell-surface proteins. Much of the evidence for microtubule involvement in these processes has been obtained using drugs which interfere with microtubule assembly. While these drugs have provided us a powerful tool to study microtubules, great care should be exercised in interpreting results obtained with them. For example, it should be kept in mind that these drugs have effects on the cell unrelated to their action on microtubules and that these drugs are cytotoxic to cells. Thus, one cannot always be certain that the results obtained with these drugs result from microtubule malfunction or from a more generalized cytotoxic condition induced in the cells by the drug. One reason for isolating mutants is to remove some of the ambiguities, and to define more precisely the mechanisms of microtubule involvement in cellular events.

Ambiguities concerning drug specificity may be eliminated using mutants that are drug resistant by virtue of an alteration in tubulin or MAPs. If more drug is required to inhibit a process in the mutant than is required in wild-type cells, one may be reasonably confident that the drug is inhibiting the process by interacting with microtubules. This paradigm has been used to show that microtubules are involved in nuclear movement in *Aspergillus nidulans* (Oakley and Morris, 1980) and in the capping phenomenon in CHO cells (Aubin *et al.*, 1980).

In addition to being drug resistant, mutants with alterations in microtubules might be expected to exhibit altered microtubule function. If the microtubule mediated process is essential for cell growth or viability, however, it should only be expressed under a particular set of conditions. This has been found to be the case for every mutant we have isolated in which a clear alteration in tubulin has been demonstrated. As summarized in Table 1, all our tubulin mutants (Tax-18 is an exception, but see below) are ts for growth, i.e., an elevated temperature (usually around 40.5°) can be found at which wild-type cells grow but mutant cells do not. This suggests that the alterations in tubulin make the microtubules less able to function at the elevated temperature and further demonstrate that microtubules are vital for normal cellular growth.

The ts phenotype of these mutants has allowed us to study the role of microtubules in cell growth. Examination of the cells at the nonpermissive temperature indicates that spindle microtubules are preferentially affected (Abraham, Cabral, and Gottemsan, in preparation). The cytoplasmic microtubules were found to persist at the elevated temperature and at least one cytoplasmic microtubule function, saltatory motion, does not appear to be inhibited. Spindle assembly, however, is clearly affected.

Similar results were found with mutant Tax-18. Although not ts for growth, this cell line requires the continuous presence of taxol for cell division (Cabral, 1983; Cabral *et al.*, 1983). In the presence of taxol, the cells grow normally with a doubling time of 12 hr during logarithmic growth. In the absence of the drug, however, the cells become larger, rounder, flatter, and exhibit micronucleation. Furthermore, their cell number does not increase during four days in culture.

Since taxol is a microtubule stabilizing drug (Schiff and Horwitz, 1980; Schiff *et al.*, 1979), we reasoned that the cells might require the drug for microtubule assembly. Examination of wild-type and mutant cells by indirect immunofluorescence with antibodies to tubulin quickly showed that this assumption was only partially correct. Again, cytoplasmic microtubules are clearly able to assemble in the absence of the drug and these microtubules are functional as assayed by the persistance of saltatory motion. The mitotic spindle apparatus, however, is not able to assemble (Cabral *et al.*, 1983). A summary of these results is shown in Figs. 6 and 7. Wild-type cells without taxol (Fig. 6A), wild-type cells with taxol (Fig. 6B), Tax-18 cells without taxol (Fig. 6C) and Tax-18 cells with taxol (Fig. 6D) are all competent to assemble cytoplasmic microtubules. The presence of taxol (0.2 μg/ml) in the wild-type culture (Fig. 6B) produces bundling of the cytoplasmic microtubules. A similar effect is not found in the mutant cell culture (Fig. 6D). This may reflect the increased resistance to taxol in the mutant cells.

When the mitotic cells in the population are examined, mutant cells

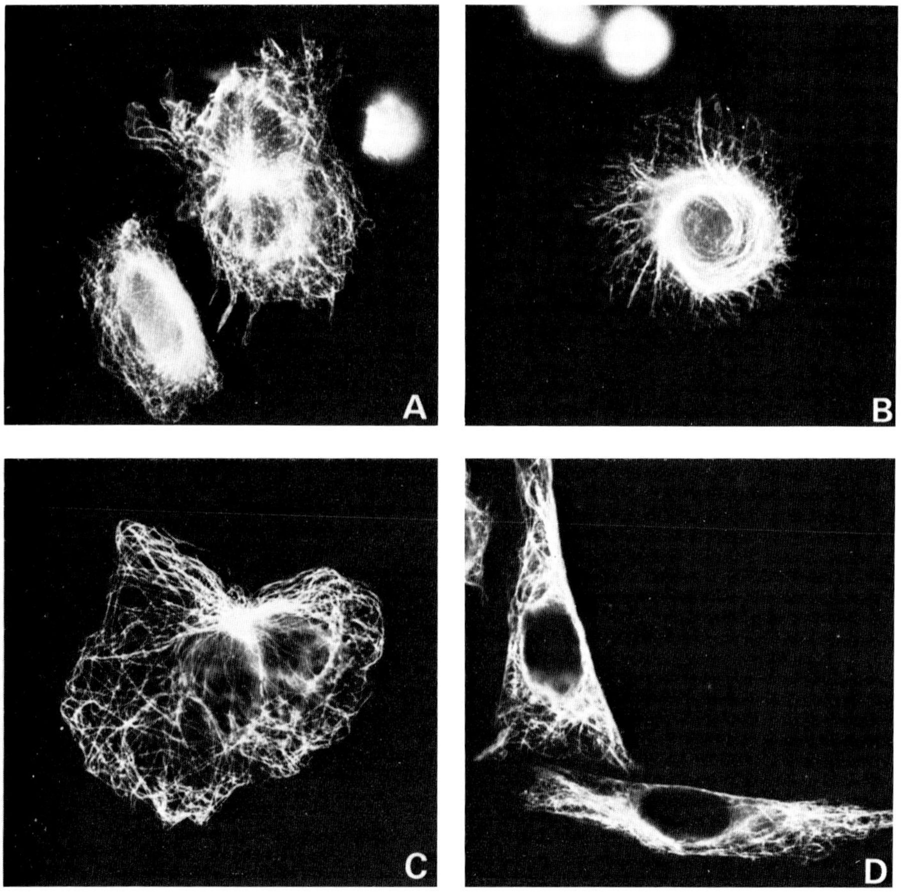

Figure 6. Indirect immunofluorescence of wild-type (A,B) and Tax-18 cells (C,D) using antibodies to tubulin (Brinkley *et al.*, 1980). (A) Wild-type cells not treated with taxol. (B) Treated for 2 days in 0.2 μg/ml taxol. Note the presence of highly bundled perinuclear microtubules in (B). (C) Tax-18 cells deprived of taxol for 2 days still demonstrate cytoplasmic microtubules. (D) The mutant cells were continuously cultured in 0.2 μg/ml taxol. Note the absence of microtubule bundles (compare to B). The cells were photographed using a Leitz orthoplan microscope with a Leitz 50X oil objective.

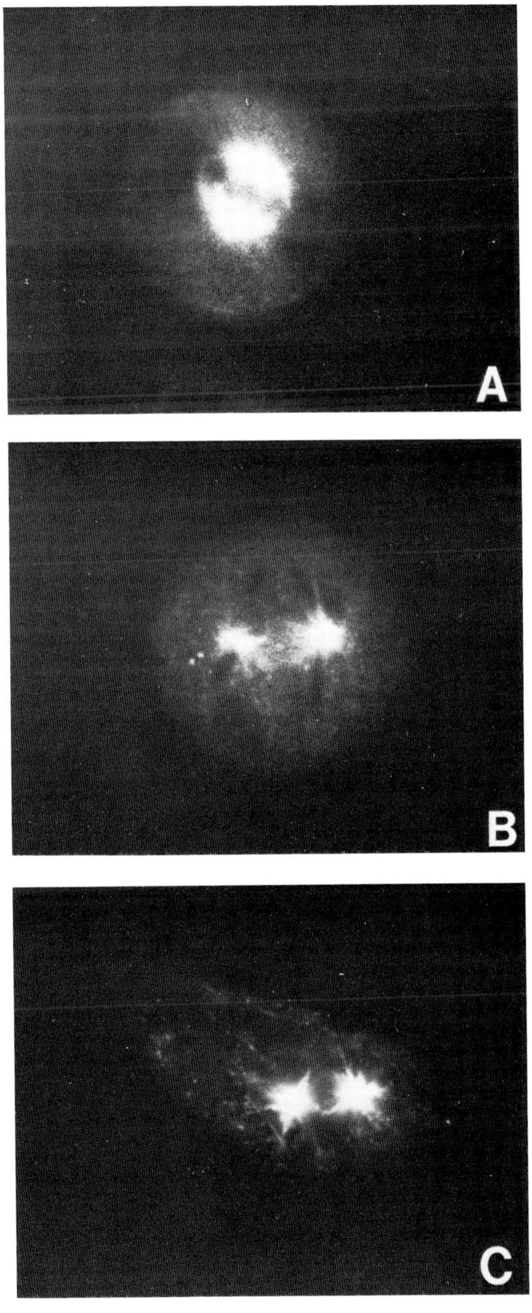

Figure 7. Spindle microtubule immunofluorescence of Tax-18 cells using tubulin antibodies. (A) The cells were cultured continuously in 0.2 μg/ml taxol. A metaphase spindle is shown. (B and C) The cells were deprived of taxol for 2 days. Only prometaphase-like spindles are apparent. Metaphase and anaphase spindles are not seen.

grown in taxol are found to exhibit all stages of mitosis. A Tax-18 metaphase cell is shown in Fig. 7A. The mutant spindle always has a more delicate appearance than the wild-type spindle, indicating that spindle assembly is not completely normal in Tax-18 even in the presence of taxol. Mutant cells deprived of taxol for 1 or 2 days have very atypical spindles (Figs. 7B and C). These rather primitive structures resemble the spindles found in prometaphase cells. They do not, however, go on to form metaphase, anaphase, or telophase structures. Cells deprived of taxol also fail to exhibit midbodies. This is consistant with a lack of cell division in the mutant. I wish to emphasize that the rudimentary spindle found in the taxol-deprived cells is different from the spindle of a colchicine-treated cell. Unlike the latter cells, taxol-starved Tax-18 cells have clearly separated spindle poles and prominent microtubule asters at the poles. The defect in Tax-18, then, appears to block the progression of spindle assembly from the prometaphase to the metaphase stage of mitosis. Assembly of cytoplasmic microtubules is not blocked.

Electron microscopic examination of the spindle in Tax-18 cells indicates that centriole and kinetochore structures are normal, suggesting but not proving, that the defect in these cells is not one of microtubule nucleation at the organizing centers (Cabral *et al.*, 1983). The presence of kinetochore-to-pole microtubules can be demonstrated, but evidence for interpolar microtubules has not yet been obtained, and we tentatively suggest that the mutant cells may be unable to assemble this class of microtubules.

The biochemical lesion which blocks spindle assembly has not yet been identified. Analysis of seven mutants with the Tax-18 phenotype by two-dimensional gel electrophoresis has thus far failed to reveal any alterations in either the amount or the electrophoretic mobility of tubulin. Furthermore, the mutation in Tax-18 behaves recessively in somatic cell hybridization experiments suggesting that the altered gene might exist as a single copy. Our tubulin mutants on the other hand clearly demonstrate the existence of multiple tubulin genes in CHO cells. This conclusion has been confirmed by direct demonstration of multiple α- and β-tubulin genes by DNA hybridization of Southern blots of CHO DNA gels (Abraham and Gottesman, personal communication). These observations lead us to suggest that the lesion in Tax-18 may not reside in tubulin itself but rather in a MAP which is required for spindle microtubule assembly. We further suggest that this MAP is missing or defective in Tax-18 and that taxol performs a function similar to the function supplied by this protein. Our laboratory is currently attempting to isolate spindles from wild-type and mutant cells in order to directly test some of these hypotheses.

Whatever the biochemical defect in Tax-18, spindle assembly is clearly blocked and so we can begin to ask what happens to a cell which is unable to assemble a spindle. First, and not surprisingly, chromosome organization during mitosis is disrupted. A prometaphase arrangement of chromosomes is achieved presumably because of the persistence of the kinetochore-to-pole microtubules. Furthermore, prometaphase chromosome movements are still present. However, chromosome alignment on the metaphase plate and the orderly segregation of sister chromatids to opposite spindle poles do not occur. As the cell exits mitosis, chromosomes are tightly organized around the spindle poles and restitution nuclear membranes form around scattered groupings of chromosomes leading to the formation of micronuclei.

Time-lapse observation indicates that the taxol-deprived mutant cells require 2–5 times longer to go through mitosis and that the cells accumulate at the prometaphase stage. The mitotic cell eventually goes into a telophase-like stage which is very prolonged compared to normal and in which violent membrane blebbing occurs. At the end of this period, the cell again becomes quiescent and spreads out as a single large cell, i.e., cell division fails to occur. Thus, defective spindle assembly not only affects chromosome organization and delays mitosis, but it also results in an inability of the mitotic cell to divide. The reason for this defect in cytokinesis is not yet apparent. Two possibilities are (1) that the contractile ring is not able to assemble, or (2) that the contractile ring does assemble but cannot completely "pinch off" the cytoplasm. We are currently testing these alternatives. The second explanation seems most attractive to us. If one assumes that a stable midbody requires the interaction of the contractile ring with the microtubules spanning the spindle poles, then an absence of interpolar microtubules, which as mentioned above may be the case in this mutant, would be sufficient to explain the lack of cell division.

It is interesting that in spite of the inability of these cells to assemble a spindle, segregate chromosomes, and divide, they continue to progress through the cell cycle. This is evident from the large size which the cells attain, their continued incorporation of precursors into DNA and protein, their continued chromosome replication, and their continued centriole duplication (Cabral, 1983; Cabral *et al.*, 1983). Thus, the cells become giant cells with octaploid and greater DNA contents. This observation defines the role of microtubules in cell viability more precisely. Spindle microtubules are not directly necessary for cell viability, at least in the short term. It is the lack of chromosome segregation and cell division which eventually leads to the cells' destruction. After 3–4 days without taxol, the cells become very large and as mentioned above, the "telophase-like" state becomes very violent, possibly resulting in the loss of membrane and occasional chromosomes from the cell. Also, after 3–4 days, many chromosomal abnormalities appear (experiments in progress), probably leading to a gross imbalance of gene dosage in the cell. At this stage, extensive cell death becomes apparent.

The study of this single mutant has taught us much about the role of microtubules in mitosis. Many of the arguments presented above are still speculative and much work remains to be done especially to characterize the biochemical lesion in this cell line. It is noteworthy, however, that we are finding much the same results in the temperature-sensitive cell lines, in which the biochemical lesion is clearly defined as an alteration in α- or β-tubulin (Abraham, Cabral, and Gottesman, in preparation). This suggests that α- and β-tubulin and the defective protein in Tax-18 may form a critical complex in microtubule assembly which is tightly regulated and which may serve as a major control point in mitosis. It is tempting to speculate that such a control point could be used by living organisms to produce giant cells in the body such as megakaryocytes and osteoclasts.

6. Future Prospects

It is my hope that this review has served to summarize the current status of genetic approaches to microtubule function in mammalian cells. I have attempted to review the kinds of approaches that may be used, the problems encountered in these approaches, the types of mutants which have been isolated and, most importantly, a glimpse of the kinds of information regarding microtubule assembly, function, and regulation which can be obtained. I believe that genetics provides a very powerful tool for studying microtubules and compliments the exciting work being done in other laboratories using biochemical, morphological, immunological, and micromanipulative techniques and the new techniques of molecular biology. Integration of these various methodologies in the study of mutant phenotypes should greatly expand our understanding of microtubules.

For the future, a number of problems remain. First, only a handful of mutants have been isolated. If we are to understand a process as complicated as mitosis, many more mutants need to be isolated, and, in particular, selections must be devised which will yield other classes of mutants not yet isolated. For example, mutants blocked in the entry into mitosis, in centriole separation, in anaphase progression, and in cytokinesis would provide us with tools to study the regulation of processes not easily studied by other approaches. While microtubules may be involved in some of these events, it is clear that a study of microtubules alone will not be sufficient to understand these phenomena.

Among the mutants which have been isolated, much remains to be done. In many of the mutants, the biochemical lesion is not yet known. Even in those cases where a defect in α- or β-tubulin has been found, the mechanism by which this altered tubulin confers drug resistance is still speculative. Furthermore, for those mutants in which defective spindle assembly has been found, the molecular details of the assembly block need to be explored and the relationship between a defective spindle and the absence of cytokinesis needs to be more precisely defined.

Finally, I believe that mutants will prove to be crucial for defining what proteins are necessary for spindle assembly and for determining how these proteins interact. First, however, the biochemical groundwork for spindle isolation and for identifying putative spindle MAPs must be laid out. Once this is accomplished, I predict that the isolation of second site suppressor mutations will progress with great rapidity providing us with unambiguous proof for the involvement of these spindle MAPs in spindle microtubule assembly.

ACKNOWLEDGMENTS. I wish to thank the many excellent people with whom I have had the privilege to collaborate in the development and characterization of some of the mutants described in this review. Among these people are Michael Gottesman and Irene Abraham of the NIH and Bill R. Brinkley and Linda Wible of Baylor College of Medicine. I also wish to thank M. Gottesman and I. Abraham for communication of data in advance of publication, L. Wible and B. R. Brinkley for the data used to prepare Figs. 6 and 7, and Sandi Jackson for typing the manuscript. This work was supported by grants GM29955 from the NIH and CD154 from the American Cancer Society. The author is the recipient of a Junior Faculty Research Award from the ACS.

References

Aubin, J. E., Tolson, N., and Ling, V., 1980, The redistribution of fluoresceinated concanavalin A in Chinese hamster ovary cells and in their colecmid-resistant mutants, *Exp. Cell Res.* **126:**75.

Bech-Hansen, N. T., Till, J. E., and Ling, V., 1976, Pleiotropic phenotype of colchicine-resistant CHO cells: Cross-resistance and collateral sensitivity, *J. Cell. Physiol.* **88:**23.

Ben-Ze'ev, A., Farmer, S. R., and Penman, S., 1979, Mechanisms of regulating tubulin synthesis in cultured mammalian cells, *Cell* **17:**319.

Bibring, T., and Baxandall, J., 1977, Tubulin synthesis in sea urchin embryos: Almost all tubulin of the first cleavage mitotic apparatus derives from the unfertilized egg, *Dev. Biol.* **55:**191.

Brinkley, B. R., and Cartwright, J., 1975, Cold-labile and cold stable microtubules in the mitotic spindle of mammalian cells, *Ann. N.Y. Acad. Sci.* **253:**428.

Brinkley, B. R., and Stubblefield, E., 1970, Ultrastructure and interaction of the kinetochore and centriole in mitosis and meiosis, in: *Advances in Cell Biology*, Vol. 1 (D. M. Prescott and L. E. McConkey, eds.), pp. 119–185, Appleton-Century Crofts, New York.

Brinkley, B. R., Fistel, S. H., Marcum, J. M., and Pardue, R. L., 1980, Microtubules in cultured cells: Indirect immunofluorescent staining with tubulin antibody, *Int. Rev. Cytol.* **63:**59.

Cabral, F., 1983, The isolation of CHO mutants requiring the continuous presence of taxol for cell division, *J. Cell Biol.* **97:**22.

Cabral, F., Sobel, M., and Gottesman, M. M., 1980, CHO mutants resistant to colchicine, colcemid, or griseofulvin have an altered β-tubulin, *Cell* **20:**29.

Cabral, F., Abraham, I., and Gottesman, M. M., 1981, Isolation of a taxol-resistant Chinese hamster ovary cell mutant that has an alteration in α-tubulin, *Proc. Natl. Acad. Sci. USA* **78:**4388.

Cabral, F., Abraham, I., and Gottesman, M. M., 1982, Revertants of a CHO mutant with an altered β-tubulin: Evidence that the altered tubulin confers both colcemid resistance and temperature sensitivity on the cell, *Mol. Cell Biol.* **2:**720.

Cabral, F., Wible, L., Brenner, S., and Brinkley, B. R., 1983, A taxol requiring mutant of CHO cells with impaired mitotic spindle assembly, *J. Cell Biol.* **97:**30.

Cleveland, D. W., Lopata, M. A., Sherline, P., and Kirschner, M. W., 1981, Unpolymerized tubulin modulates the level of tubulin mRNAs, *Cell* **25:**537.

Conrad, G. W., and Rappaport, R., 1981, Mechanisms of cytokinesis in animal cells, in: *Mitosis/Cytokinesis* (A. Zimmerman and A. Forer, eds.), pp. 365–396, Academic Press, New York.

Crossin, K. L., and Carney, D. H., 1981, Evidence that microtubule depolymerization early in the cell cycle is sufficient to initiate DNA synthesis, *Cell* **23:**61.

Dustin, P., 1978, *Microtubules*, Springer-Verlag, Berlin.

Flavin, M., and Slaughter, C., 1974, Microtubule assembly and function in *Chlamydomonas:* Inhibition of growth and flagellar regeneration by antitubulins and other drugs and isolation of resistant mutants, *J. Bacteriol.* **118:**59.
Fulton, C., and Simpson, P. A., 1979, Tubulin pools, synthesis, and utilization, in: *Microtubules* (K. Roberts and J. S. Hyams, eds.), pp. 117–174, Academic Press, New York.
Fuller, G. M., Brinkley, B. R., and Boughter, J. M., 1975, Immunofluorescence of mitotic spindles by using monospecific antibody against bovine brain tubulin, *Science* **187:**948.
Garland, D. L., 1978, Kinetics and mechanism of colchicine binding to tubulin: Evidence for ligand-induced conformational change, *Biochemistry* **17:**4266.
Gupta, R. S., 1981, Resistance to the microtubule inhibitor podophyllotoxin: Selection and partial characterization of mutants in CHO cells, *Somat. Cell Genet.* **7:**59.
Gupta, R. S., Ho, T. K. W., Moffat, M. R. K., and Gupta, R., 1982, Podophyllotoxin-resistant mutants of Chinese hamster ovary cells, *J. Biol. Chem.* **257:**1071.
Hartwell, L. H., 1978, Cell division from a genetic perspective, *J. Cell Biol.* **77:**627.
Hatzfeld, J., and Buttin, G., 1975, Temperature-sensitive cell cycle mutants: A Chinese hamster cell line with a reversible block in cytokinesis, *Cell* **5:**123.
Hiramoto, Y., 1971, Analysis of cleavage stimulus by means of micromanipulation of sea urchin eggs, *Exp. Cell Res.* **68:**291.
Huang, B., Piperno, G., Ramanis, Z., and Luck, D. J. L., 1981, Radial spokes of *Chlamydomonas* flagella: Genetic analysis of assembly and function, *J. Cell Biol.* **88:**80.
Huang, B., Ramanis, Z., and Luck, D. J. L., 1982, Suppressor mutations in *Chlamydomonas* reveal a regulatory mechanisms for flagellar function, *Cell* **28:**115.
Inoue, S., 1981, Cell division and the mitotic spindle, *J. Cell Biol.* **91:**131s.
Izant, J. G., Weatherbee, J. A., and McIntosh, J. R., 1982, A microtubule-associated protein in the mitotic spindle and the interphase nucleus, *Nature (London)* **295:**248.
Kemphues, K. J., Raff, R. A., Kaufman, T. C., and Raff, E. C., 1979, Mutation in a structural gene for a β-tubulin specific to testes in *Drosophila melanogaster, Proc. Natl.. Acad. Sci. USA* **76:**3991.
Kemphues, K. J., Raff, E. C., Raff, R. A., and Kaufman, T. C., 1980, Mutation in a testis-specific β-tubulin in *Drosophila:* Analysis of its effects on meiosis and map location of the gene, *Cell* **21:**445.
Ling, V., 1981, Mutations as an investigative tool in mammalian cells, in: *Mitosis/Cytokinesis* (A. M. Zimmerman, and A. Forer, eds.), pp. 197–209, Academic Press, New York.
Ling, V., and Thompson, L. H., 1974, Reduced permeability in CHO cells as a mechanism of resistance to colchicine, *J. Cell Physiol.* **83:**103.
Ling, V., Aubin, J. E., Chase, A., and Sarangi, F., 1979, Mutants of Chinese hamster ovary (CHO) cells with altered colcemid-binding affinity, *Cell* **18:**423.
Luduena, R. F., 1979, Biochemistry of tubulin, in: *Microtubules* (K. Roberts and J. S. Hyams, eds.), pp. 65–116, Academic Press, New York.
Lydersen, B. K., and Pettijohn, D. E., 1980, Human-specific nuclear protein that associates with the polar region of the mitotic apparatus: Distribution in a human/hamster hybrid cell, *Cell* **22:**489.
Manfredi, J. J., Parness, J., and Horwitz, S. B., 1981, Taxol binds to cellular microtubules, *J. Cell Biol.* **94:**688.
Margolis, R. L., and Wilson, L., 1977, Addition of colchicine-tubulin complex to microtubule ends: The mechanism of substoichiometric colchicine poisoning, *Proc. Natl. Acad. Sci. USA* **74:**3466.
McCarty, G. A., Valencia, D. W., Fritzler, M. J., and Barada, F. A., 1981, A unique antinuclear antibody staining only the mitotic-spindle apparatus, *N. Eng. J. Med.* **305:**703.
McIntosh, J. R., 1979, Cell division, in: *Microtubules,* (K. Roberts and J. S. Hyams, eds.), pp. 381–441, Academic Press, New York.
Morris, N. R., 1980, Chromosome structure and the molecular biology of mitosis in eukaryotic micro-organisms, in: *The Eukaryotic Microbial Cell* (G. W. Gooday, D. Loyd, and A. P. J. Trinci, eds.), pp. 41–76, Cambridge University Press, New York.

Morris, N. R., Lai, M. H., and Oakley, C. E., 1979, Identification of a gene for α-tubulin in *Aspergillus nidulans, Cell* **16:**437.

Oakley, B. R., 1981, Mitotic mutants, in: *Mitosis/Cytokinesis* (A. M. Zimmerman and A. Forer, eds.), pp. 181–196, Academic Press, New York.

Oakley, B. R., and Morris, N. R., 1980, Nuclear movement is β-tubulin dependent in *Aspergillus nidulans, Cell* **19:**255.

Oakley, B. R., and Morris, N. R., 1981, A β-tubulin mutation in *Aspergillus nidulans* that blocks microtubule function without blocking assembly, *Cell* **24:**837.

Peterson, S. P., and Berns, M. W., 1980, The centriolar complex, *Int. Rev. Cytol.* **64:**81.

Pickett-Heaps, J. D., Tippit, D. H., and Porter, K. R., 1982, Rethinking mitosis, *Cell* **29:**729.

Pipeleers, D. G., Pipeleers-Marichal, M. A., Sherline, P., and Kipnis, D. M., 1977, A sensitive method for measuring polymerized and depolymerized forms of tubulin in tissues, *J. Cell Biol.* **74:**341.

Puck, T. T., Ciecuira, S. J., and Robinson, A., 1958, Genetics of somatic mammalian cells III. long-term cultivation of euploid cells from human and animal subjects, *J. Exp. Med.* **108:**945.

Rappaport, R., 1971, Cytokinesis in animal cells, *Int. Rev. Cytol.* **31:**169.

Roobol, A., Gull, K., and Pogson, C. I., 1977, Evidence that griseofulvin binds to a microtubule associated protein, *FEBS Lett.* **75:**149.

Roos, U. P., 1973, Light and electron microscopy of PtK_2 cells in mitosis. II. kinetochore structure and function, *Chromosoma* **41:**195.

Sato, C. H., 1976, A conditional cell division mutant of *Chlamydomonas reinhardii* having an increased level of colchicine resistance, *Exp. Cell Res.* **101:**251.

Schiff, P. B., and Horwitz, S. B., 1980, Taxol stabilizes microtubules in mouse fibroblastic cells, *Proc. Natl. Acad. Sci. USA* **77:**1561.

Schiff, P. B., Fant, J., and Horwitz, S. B., 1979, Promotion of microtubule assembly *in vitro* by taxol, *Nature (London)* **277:**665.

Schrader, F., 1953, *Mitosis. The Movements of Chromosomes in Cell Division,* Columbia University Press, New York.

Sheir-Neiss, G., Lai, M. H., and Morris, N. R., 1978, Identification of a gene for β-tubulin in *Aspergillus nidulans, Cell* **15:**639.

Shelanski, M. L., and Taylor, E. W., 1967, Isolation of a protein subunit from microtubules, *J. Cell Biol.* **34:**549.

Shiomi, T., and Sato, K., 1976, A temperature-sensitive mutant defective in mitosis and cytokinesis, *Exp. Cell Res.* **100:**297.

Siminovitch, L., 1976, On the nature of hereditable variation in cultured somatic cells, *Cell* **7:**1.

Smith, B. J., and Wigglesworth, N. M., 1972, A cell line which is temperature-sensitive for cytokinesis, *J. Cell. Physiol.* **80:**253.

Stanners, C. P., 1978, Characterization of temperature-sensitive mutants of animal cells, *J. Cell. Physiol.* **95:**407.

Sternlicht, H., and Ringel, I., 1979, Colchicine inhibition of microtubule assembly via copolymer formation, *J. Biol. Chem.* **254:**10540.

Sternlicht, H., Ringel, I., and Szasz, J., 1980, The co-polymerization of tubulin and tubulin-colchicine complex in the absence and presence of associated proteins, *J. Biol. Chem.* **255:**9138.

Thompson, L. H., and Lindl, P. A., 1976, A CHO-cell mutant with a defect in cytokinesis, *Somat. Cell Genet.* **2:**387.

Wang, R. J., 1974, Temperature-sensitive mammalian cell line blocked in mitosis, *Nature (London)* **248:**76.

Wang, R. J., 1976, A novel temperature-sensitive mammalian cell line exhibiting defective prophase progression, *Cell* **8:**257.

Wang, R. J., and Yin, L., 1976, Further studies on a mutant mammalian cell line defective in mitosis, *Exp. Cell Res.* **101:**331.

Warr, J. R., and Gibbons, D., 1974, Further studies on colchicine-resistant mutants of *Chlamydomonas reinhardi, Exp. Cell Res.* **85:**117.

Warr, J. R., Flanagan, D. J., and Anderson, M., 1982, Mutants of Chinese hamster ovary cells with altered sensitivity to taxol and benzimidazole carbamates, *Cell Biol. Int. Rep.* **6:**455.

Weber, K., and Osborn, M., 1979, Intracellular display of microtubular structures revealed in indirect immunofluorescence microscopy, in: *Microtubules* (K. Roberts and J. S. Hyams, eds.), pp. 279–313, Academic Press, New York.

Weber, K., Bibring, Th., and Osborn, M., 1975, Specific visualization of tubulin-containing structures in tissue culture cells by immunofluorescence. Cytoplasmic microtubules, vinblastine-induced paracrystals, and mitotic figures, *Exp. Cell Res.* **95:**111.

Welsh, M. J., Dedman, J. R., Brinkley, B. R., and Means, A. R., 1978, Calcium-dependent regulator protein: Localization in the mitotic apparatus of eukaryotic cells, *Proc. Natl. Acad. Sci. USA* **75:**1867.

Wilson, E. B., 1925, *The Cell in Development and Heredity,* MacMillan, New York.

Witman, G. B., Plummer, J., and Sander, G., 1978, *Chlamydomonas* flagellar mutants lacking radial spokes and central tubules. Structure, composition, and function of specific axonemal components, *J. Cell Biol.* **76:**729.

Zieve, G., and Solomon, F., 1982, Proteins specifically associated with the microtubules of the mammalian mitotic spindle, *Cell* **28:**233.

10

Cytoskeleton in Platelet Function

Jon C. Lewis

1. Structure and Function of the Circulating Platelet

1.1. Origin and Turnover

The circulating platelet is a discoid shaped cellular fragment which originates in the bone marrow through the orderly demarcation of megakaryocyte cytoplasm (Odell and Jackson, 1969; Behnke, 1968a; Marsh *et al.*, 1955; for review see Ebbe, 1976). The number of platelets in circulation varies widely among species, and among individuals within a species the concentrations in blood typically range from 200,000–500,000 per microliter. On the average, the circulation life is 8–9 days. However, with constant activation to participate in the hemostatic process and contribute to maintenance of blood vessel integrity (Henry, 1977), the turnover times range from 6–11 days (Harker, 1979). In man, it has been estimated that platelet consumption is approximately 35,000 platelets/μl of blood per day (Harker, 1979). Although the number of platelets arising from a single megakaryocyte has been estimated to be as high as 12,000 (Ebbe, 1976; Tavassoli, 1980; Kaufman *et al.*, 1965) and probably varies with species and ploidy of the megakaryocyte (Kaufman *et al.*, 1965), it is clear from ultrastructural studies of both circulating and bone marrow cells that megakaryocyte demarcation (Behnke, 1968a; Marsh *et al.*, 1955) proceeds in a well-controlled fashion to ensure structural and functional integrity of the platelet (Kaufman *et al.*, 1965; White and Gerrard, 1976; White, 1979; Lewis *et al.*, 1980a; Lewis and Bowie, 1978; Barnhart, 1978).

Jon C. Lewis • Department of Pathology, Bowman Gray School of Medicine, Wake Forest University, Winston-Salem, North Carolina 27103.

1.2. Ultrastructure vs. Function

1.2.1. The Granulomere

In general, three distinct organizational zones can be identified in the platelet (Fig. 1). The center of the discoid cell, the granulomere, as recognized (Wolpero and Ruska, 1939; Bessis and Burstein, 1948a) and described in some detail (Bessis and Bricka, 1948) by early microscopists, contains mitochondria, granules, and elements of the dense tubular system, a specialized complex of smooth endoplasmic reticulum (Shultz *et al.*, 1958) associated with peroxidase activity (Breton-Gorius and Guichard, 1972). In addition, a network of membrane-limited channels, the open canalicular system (Aleksandrowicz *et al.*, 1957), having lumina which are in continuity with the extracellular medium are typically found. The granule population is heterogeneous and is comprised of several distinct organelles including acid phosphatase positive lysosomes (Jean and Racine, 1962; Bentfield and Bainton, 1976; Lewis *et al.*, 1976a), catalase positive peroxisomes (Breton-Gorius and Guichard, 1972), specific or α-granules (Shultz *et al.*, 1958), and electron dense bodies (the delta granulomere of Shultz *et al.*, 1958) known to contain serotonin, adenine nucleotides, and calcium (Lewis *et al.*, 1976b; Da Prada and Pletscher, 1974; White, 1969). No specific arrangement of granulomere constituents has been described, and the various organelles appear to be randomly distributed in an ill-defined matrix or cytoplasmic ground substance, which has been described as having a delicately granular consistency (White and Gerrard, 1976; Bessis and Bricka, 1948; Shultz *et al.*, 1958; De Robertis *et al.*, 1953; Hutter, 1957; Bessis and Breton-Gorius, 1965; Behnke, 1965, 1966; Lewis *et al.*, 1980b, 1982a; Nachmias, 1980).

Neither the lack of matrix definition nor the apparent randomness of organelle distribution in the resting cell imparts a functional haphazardness, for when stimulated to participate in a hemostatic event, the granulomere undergoes rapid reorganization culminating in the controlled, sequential release of dense-body and α-granule constituents (for reviews see MacIntyre, 1976; Kaplan *et al.*, 1979).

1.2.2. The Sol-Gel Zone

Surrounding and providing the demarcation for the granulomere is a region that has been described by numerous authors (White, 1979; Barnhart, 1978) as the sol-gel zone, a name which implies a dynamic state involving polymerization and depolymerization of the ground substance. The most prominant and distinctive feature of this zone is the band of microtubules which encircles the granulomere (Bessis and Breton-Gorius, 1965; Behnke, 1965; for review see White and Gerrard, 1979). The cytoplasmic matrix adjacent to the marginal band of microtubules consistently appears less electron dense than that of surrounding areas (Lewis *et al.*, 1980a,b; Bessis and Breton-Gorius, 1965; Behnke, 1965; White and Gerrard, 1979) and delicate filaments

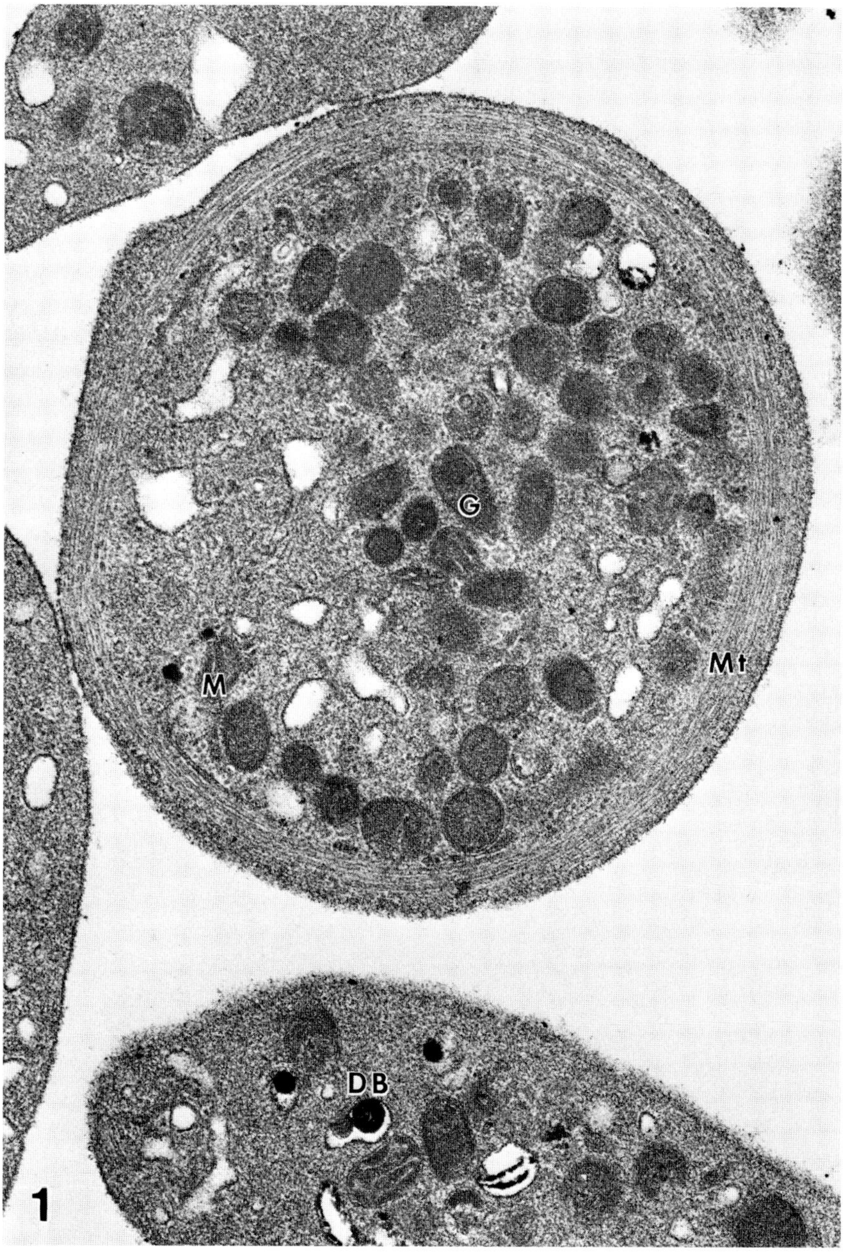

Figure 1. Representative mammalian platelets illustrating the three zones of organization. The granulomere contains α-granules (G), mitochondria (M), and dense bodies (DB) for serotonin storage. Note the prominant band of microtubules (Mt) enmeshed in an ill-defined matrix, the sol-gel zone, in the outermost region of the cells, the peripheral zone. Elements of the open canalicular system (arrows) are closely apposed to the plasma membrane, ×34,000. Reproduced from Lewis *et al.* (1980a).

have been described associated with or interconnecting the tubules (Bessis and Breton-Gorius, 1965; Lewis *et al.*, 1980b, 1982a; Nachmias, 1980; White and Gerrard, 1979; Phillips, 1980). The relationship of the filamentous proteins in this area of the platelet to the rest of the ground substance is unknown, but it has been estimated that the proteins which constitute the sol-gel zone account for 55% of the total cell protein (White, 1979) and are responsible for numerous activities including the granule-release reaction mentioned above (MacIntyre, 1976; Kaplan *et al.*, 1979).

1.2.3. The Peripheral Zone

The final zone is the peripheral zone (White, 1979; Barnhart, 1978) which appears outside of the marginal band of microtubules in platelets sectioned in the equitorial plane (White, 1979; Lewis *et al.*, 1976b, 1980a; Bessis and Breton-Gorius; 1965; Behnke, 1965; see Fig. 1).

The peripheral region, also evidenced as a dense band in whole-mounted platelets (Lewis *et al.*, 1980b; Nachmias, 1980) or as a condensed region in metal replicas of adherent platelets (Bessis and Bricka, 1948), contains submembranous (cortical) proteins in close association with the outer plasma membrane and membranes of the surface-connected open canalicular system. As has been described for other cell types, the platelet plasma membrane is an integrated macromolecular complex of proteins (57%), neutral and phospholipid (33%), glycolipid (about 10% of total lipid), and glycoproteins (carbohydrates comprises 8%) which provides receptors and transducers to mediate and modulate the response of the platelet to its environment (Lewis *et al.*, 1976a; Phillips, 1980; Mills and MacFarlane, 1976; Behnke, 1968b; Crawford and Taylor, 1977; see Table 1).

2. *Cytoplasmic Changes with Activation*

2.1. *Shape Change*

The discoid shape, reflecting a quiescent state, is largely conserved during circulation, and morphometric analyses of platelets from a variety of species including man indicate that between 50–90% of the cells are of this morphology (Lewis and Bowie, 1978; Barnhart, 1978; Lewis *et al.*, 1980b; Barnhart *et al.*, 1972). The remainder of the cells fall into several morphologic categories including those with one pseudopod (10–25% of the circulating cells), those having a spherical shape and numerous pseudopods (10–25% of cells), and those whose morphology cannot be clearly classified. Although considerable variability can be found in the literature both with respect to the nomenclature for describing nondiscoid cells and in the absolute percentage of nondiscoid forms (Lewis and Bowie, 1978; Lewis *et al.*, 1980b; Rebuck *et al.*, 1960; Jain, 1975; for a review of this topic see Frojmovic and Milton, 1982), it is clear than in nonpathological situations (Walsh *et al.*, 1975), the discoid

Table 1. Enzymatic and Receptor Functions of Platelet Plasma Membranes[a]

Enzymes	Receptors
Phosphodiesterase	Thrombin
Acid phosphatase	ADP
Mg^{2+}-dependent ATPase	Collagen
Ca^{2+}-dependent ATPase	Fibrinogen
cAMP-dependent protein kinase	5-Hydroxytryptamine
Nucleosidediphosphate kinase	Vassopressin
Cholinesterase	Aggregated IgG
	Factor VIII—vWf
	Fibronectin
	Immune complexes

[a]Partial listing shown as representative example.

shape predominates. The distribution of platelet forms changes rapidly after stimulation of the cells, for within a short time following challenge (5–15 sec), platelets become spherical in shape and extend numerous pseudopods ranging in length from 0.5–3.0 μm (see Fig. 2). As is the case with pseudopod

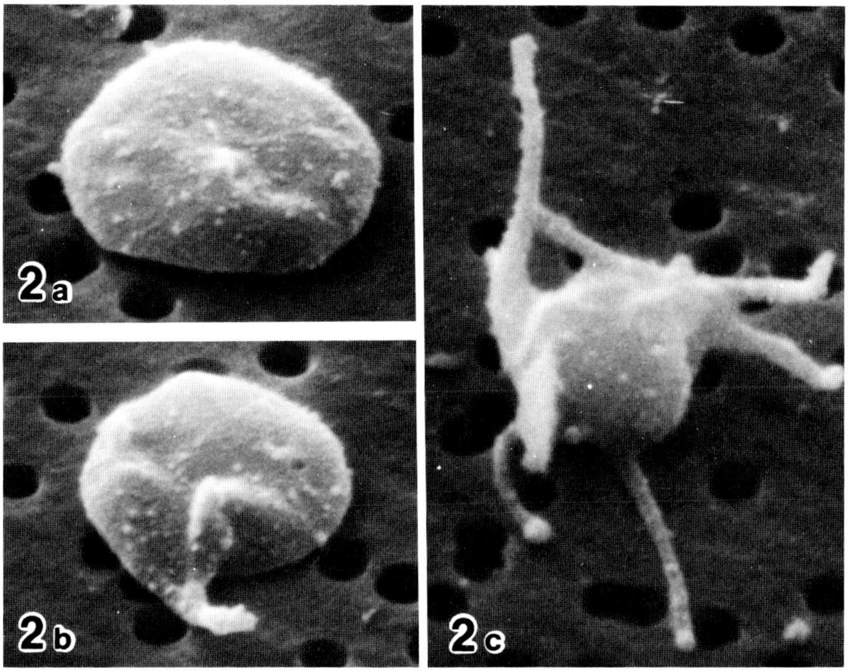

Figure 2. Platelets illustrating the initial stages of cell activation. Discocytes (a) when stimulated rapidly extend pseudopods (b) and ultimately become spiney spheres (c). The transition time from a–c has been estimated at less than 10 min, ×15,000. Reproduced from Lewis *et al.* (1980b).

length, the number of pseudopods per cell is variable with 3–6 being typical (Frojmovic and Milton, 1982) and as many as 20 being observed in some cases.

2.2. Sol-Gel Reorganization with Shape Change

Recognition of platelet shape change upon stimulation is not a new observation, for the extension of pseudopods and the development of a delicate hyalomere were among the first events studied at the ultrastructural level (Bessis and Burstein, 1948a,b; Bessis and Bricka, 1948). In what should be regarded as a classic paper, Bessis and Burstein (1948a) described the activation process as an orderly sequence involving transition of the circulating discocyte through a dendritic stage to a form characterized as having an extensive hyalomere (Fig. 3). This early work of Bessis and Burstein (1948a,b) is also significant since it highlighted the filamentous nature of the hyaloplasm in the activated cells and demonstrated distinct organizational zones. Included in the description were the centrally located granulomere, a circumferentially located filamentous zone, and a condensed periphery (Fig. 4). In a parallel study, Bessis and Bricka (1948) focused attention upon the cytoplasm of the activated human platelets and correlated the hyaline appearance as observed in whole-mount preparations, with the fibrillar nature of cytoplasm as revealed in metal-shadow replicas. The authors summarized their observations by stating, "that the protoplasm of the spread thrombocyte is composed of a network of small fibers spreading in two directions, some radiating, and others circular or curved; the periphery seems to be formed by the prefixation of many small curved fibers." These early observations of Bessis and associates have been confirmed in recent years by several investigators (Lewis *et al.*, 1980b, 1982a, 1983; Nachmias, 1980; Barnhart *et al.*, 1972; Rebuck *et al.*, 1960; Albrecht and Lewis, 1982), and it is evident that the fibrillar component reported in this pioneering work corresponds to part of what is now recognized as the platelet cytoskeleton.

3. Cytoskeletal Structures and Platelet Function

3.1. Cytoplasmic Contraction and Hemostasis

The relationship between the hemostatic process and platelet function has intrigued hematopathologists for nearly a century, since the description by Hayem (1878) of platelet aggregation in whole blood that was placed between two pieces of glass. This process, named "viscous metamorphous" by Elberth and Schimmelbusch (1885) and described in some detail by Bizzozero (1882), was extensively studied by Wright and Minot who pointed out that during the process, platelets undergo a series of morphologic changes beginning with an initial spreading and culminating with contraction (Wright and Minot, 1917). The experiments of Wright and Minot (1917) were extended by Lüscher (1956a) who correlated platelet viscous metamorphosis with clot retraction. In addition to demonstrating that clot retraction was a platelet-de-

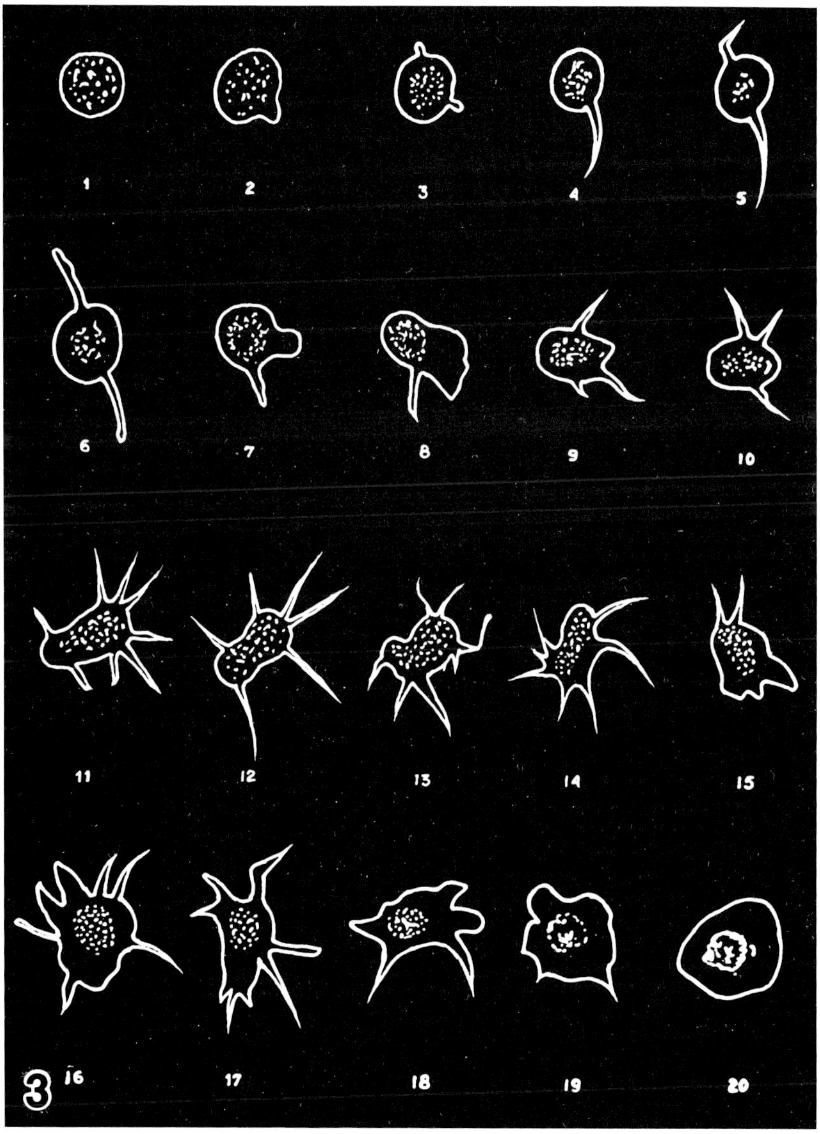

Figure 3. Diagrammatic interpretation by Bessis of the transition stages in platelet activation. The figure is extrapolated from early electron microscope studies. As shown, the circulating discocyte (1) is activated through a series of pseudopodial forms (2–12). The platelets then have pseudopods extending from an enlarged base (13–18) which ultimately gives rise to the fully spread cell (19–20). Reproduced with permission from Bessis and Burstein (1948a,b).

pendent process and proportional to the concentration of platelets, Lüscher documented a calcium requirement and suggested that proteins released from platelets during the initial stage of viscous metamorphosis were responsible for the contractile event. Interestingly, although Lüscher suggested that platelet proteins provided the mechanism for contraction, it was clear from

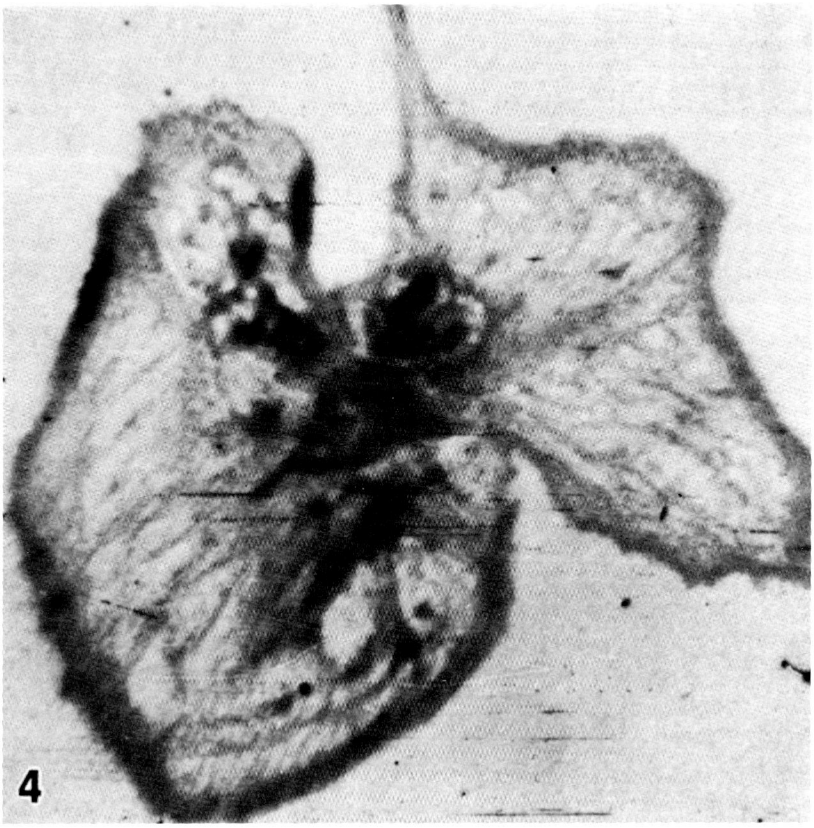

Figure 4. Electron micrograph from the 1948 works of Bessis and Burstein documenting the filamentous character of the spread platelet. This classic micrograph also illustrates three distinct organizational zones. The central dense granulomere is surrounded by a more open filamentous zone, and at the cell periphery a condensed ring of cortical filaments is shown. Reproduced with permission from Bessis and Burstein (1948b).

the studies that platelet extracts could not be used to drive the retraction of fibrin clots; rather an intact platelet mechanism was required. These biochemical studies paralleled the early electron microscopic observations of fibrin-platelet thrombi in which platelets were observed entwined within the fibrin mass by extension of delicate cytoplasmic processes (Wolpers and Ruska, 1939).

3.2. *Identification of Contractile Filaments in Activated Platelets*

Indirect evidence in support of a platelet contractile mechanism similar to that of muscle was substantial by the late 1950's and included the identification of glucose as a factor which improved retraction (Lüscher, 1956b), the recognition of a constant relationship between an active glycolytic system and maximum retraction capacity (Bettex-Galland and Lüscher, 1959), and the observation that fresh platelets which have maximal retraction capacity have a high ATP content (Bettex-Galland and Lüscher, 1959; Born, 1956) that drops dramatically during clot formation (Bettex-Galland and Lüscher, 1959; Born, 1956). Definitive support for the presence of a contractile system within the platelet was provided by Bettex-Galland and Lüscher (1959), who were successful in isolating an actomyosin-like protein from human platelets. Using techniques developed for extraction of muscle actomyosin, these investigators were able to prepare a platelet extract that would, at low ionic strength, undergo contraction in the presence of Mg^{2+} and ATP. The relationship between this reversible contractile process and viscous metamorphosis was immediately apparent, and with subsequent study, its metabolic association to clot retraction was demonstrated (Bettex-Galland and Lüscher, 1963).

4. Characterization of the Platelet Cytoskeleton

4.1. Microtubules: The First Cytoskeletal Elements

The cytoskeletal concept was introduced to the platelet literature by Behnke (1965), who when describing the marginal band of microtubules in rat and human platelets, suggested a supportive role for these organelles in maintaining the discoid shape (Behnke, 1965). This suggestion was strengthened by observations of cells which were in the early stages of activation and, therefore, had lost the discoid shape while extending numerous pseudopods. Under such activation conditions as reported by Behnke (1965) "the marginal bundle was disorganized and the microtubules were found to be oriented more or less at random within the cytoplasm." Microtubules, as further noted in the paper, were not in the pseudopods. This latter observation was significant, since it dispelled any association between microtubules and the dynamic process of cytoplasmic extension while relegating a role of structural support. This distinction in function has been a point of considerable interest over the ensuing years as investigators have attempted to elucidate the contractile mechanism and correlate cell structural organization to the force-generating process. (For reviews of this topic see Henry, 1977; White and Gerrard, 1979; Bettex-Galland and Lüscher, 1963; Crawford, 1976)

4.2. Contractile Proteins and the Cytoskeleton

Since filamentous proteins had been associated with contractile events in many cell types and since the actinomyosin-like protein, thrombosthenin, had been isolated from platelets by Bettex-Galland and Lüscher (1959, 1963), the ultrastructural identification of muscle-like proteins was of great interest. Such identification appeared in a pair of papers published in 1965 and 1966. The first of these was a report by Bessis and Breton-Gorius (1965) in which organization of the platelet hyalomere was described for cells adherent to carbon films. In addition to describing the marginal band of microtubules, these authors drew attention to the presence throughout the cytoplasm of a delicate fibrillar network comprised of 60–80 Å filaments. As noted by Bessis and Breton-Gorius, the filaments appeared in the cells during adhesion; and within the cytoplasm, the filamentous network extended from the cell center, the granulomere, to the periphery and into dendritic extensions (pseudopods). The potential contractile nature of the filaments, as noted by Bessis and Breton-Gorius (1965), was emphasized by Behnke (1966) in the second of the two papers. In the studies reported by Behnke, the cytoplasm of platelets in experimental thrombi contained randomly-oriented microtubules and a floccular material "which in places had the appearance of filaments, approximately 50 Å in diameter. . . ." In the discussion of the paper, Behnke placed his observations into context with the amassing literature on thrombosthenin biochemistry and compared the 50 Å filaments to those reported the year before by Bessis and Breton-Gorius. Behnke further suggested the 50 Å filaments and microtubules as a possible "morphological substrate" of contractile proteins. The presence in platelets of filaments having diameters of approximately 50 Å was again noted in a subsequent paper by Behnke and Zelander (1967) where the filaments were associated with microtubules isolated from osmotically lysed platelets. One of the major, although unexpressed, ramifications of identifying the 60–80 Å filaments was an expansion of the cytoskeletal concept to include both microtubules, as most clearly identified in the resting cell, and the smaller filaments which predominated following activation. This expansion was and still is paradoxical, since on the one hand the cytoskeleton was envisioned as consisting of a fairly rigid band of tubular support elements, and on the other as a network of dynamically changing filaments whose role was to facilitate cell reorganization.

4.3. Morphological Heterogeneity of Cytoskeletal Elements

4.3.1. Heterogeneity in Two-Dimensional Preparations

The concept of the cytoskeleton evolved and expanded over the next several years as the early observations of Bessis and Breton-Gorius (1965) and Behnke (1965) were confirmed by several investigators (Sixma and Molenaar, 1966; White, 1967; Zucker-Franklin, 1969, 1970) many of whom noted variability in filament diameters. Most of the filaments reported by these various investigators fell in the size range 40–120 Å, but many had diameters approaching 180 Å. Based upon a detailed study of human platelet filament sizes, Zucker-Franklin (1969) proposed two major filament groups. The first had diameters in the range 50–70 Å, and the second was comprised of filaments 80–120 Å in diameter. This grouping was in accordance with the later identification by heavy meromyosin binding (Fig. 5) of the smaller size filaments as actin and with the suggestion that the larger filaments may be myosinoid in character (Behnke *et al.*, 1971; Zucker-Franklin and Grusky, 1972). The relationship of size to biochemistry was summarized by Behnke *et al.* (1971), who pointed out: "Any filament of a diameter from 40 Å and up to

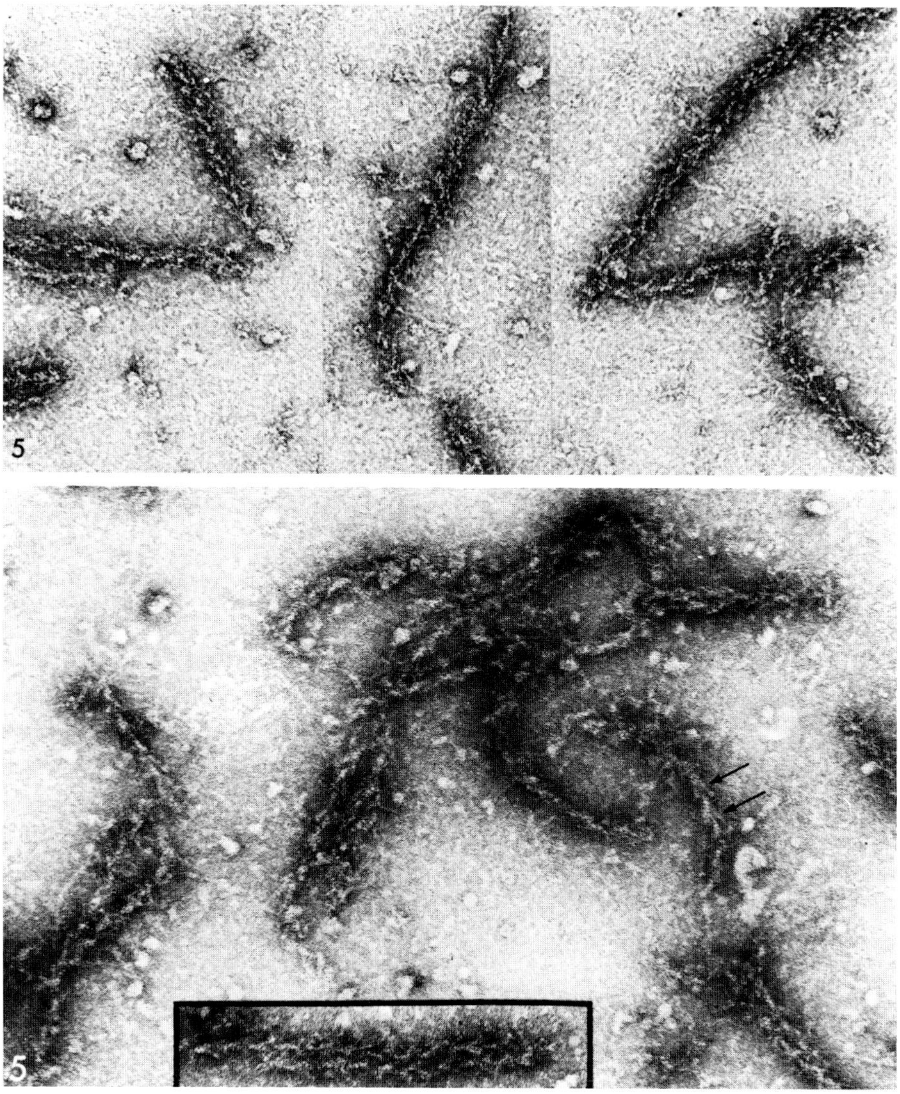

Figure 5. Negative stain preparation of thin filaments extracted from human platelets. Decoration with heavy meromyosin demonstrates the characteristic arrowhead arrays. Reproduced with permission from Zucker-Franklin and Grusky (1972).

180 Å can be myosinoid filament. Filaments measuring more than 70 Å or less than 50 Å are not actinoid, whereas filaments in the 50–60 Å range may be, but are not necessarily, actinoid."

4.3.2. Heterogeneity in Three-Dimensional Preparations

The variability in platelet cytoskeleton filament size has recently been studied in our own laboratories using computer-assisted morphometrics in conjunction with whole-mount, high-voltage electron microscopy of adherent platelets (Lewis *et al.*, 1980b, 1982a, 1983, unpublished). A major difference between our recent studies and the earlier work cited above is in the use of critical-point drying to preserve the three-dimensional integrity of the cytoplasmic matrix in the cells. As illustrated in Fig. 6, the cytoskeleton in such platelets appears as a complex network of filaments similar to those described by the earlier investigators. Although similar in appearance, a significant distinction is noted at the fine structure level where it is apparent that the cytoskeleton is even more heterogenous than previous reported. This heterogeneity is summarized in Fig. 7, where based upon morphometric analysis, three, rather than two, major classes of filaments are identified. The first class, having diameters of 20–50 Å with a mean of 39 Å, comprises about 20% of all filaments and is morphologically comparable to the microtrabecular lattice reported by Wolosowick and Porter (1979) for cells grown in culture. The second major class, containing filaments in the 50–100 Å size range (91 Å mean), encompasses 30% of the platelet cytoskeleton. Filaments within this second class are similar both in size (Lind and Stossel, 1982) and cytoplasmic distribution (Zucker-Franklin and Grusky, 1972; Debus *et al.*, 1981; see Section 5 for description of distribution) to platelet actin. The third major size group which accounts for 28% of the cytoskeleton has filament diameters ranging from 150–180 Å with a mean of 165 Å. This latter measurement is not unlike that reported for platelet myosin (Pollard, 1979), but it also is in the range reported for laterally associated actin (Gonnella and Nachmias, 1981). A fourth group of filaments, those having sizes greater than 200 Å and less than 280 Å, may represent platelet microtubules which were reported by Bessis and Breton-Gorius (1965) and by Behnke (1966) to have sizes varying from 200–230 Å and 200–260 Å, respectively. Identification of these larger organelles as microtubules cannot be made with certainty, however, since the preparative techniques used for whole mounts have been associated with some condensation artifact in the cytoskeleton of platelets (see Section 5) and other cell types (Small and Langanger, 1981). Furthermore, Gonnella and Nachmias (1981) have shown that platelet actin in cells activated with the ionophore A-23187 occurs in bundles of closely associated filaments and that nonfilamentous actin, isolated from tetracaine-treated platelets could, under appropriate *in vitro* conditions, be polymerized to yield similar filament bundles. The bundling of actin is significant with respect to this fourth class of filaments for, as reported by us, large filaments in platelet cytoskeletons often appear to be comprised of smaller, closely entwined filaments that converge and diverge along their major axes (Lewis *et al.*, 1983, and unpublished data). Variability is found not only among the major classes of filaments, but also within each of the three major classes (Fig. 8). The cytoskeleton, in light of this morphologic variability, must be viewed as a highly complex system.

4.4. Cytoskeletal Biochemical Complexity

It is beyond the scope of this chapter to review the biochemistry of the platelet cytoskeleton, and such a task is unnecessary in light of the many

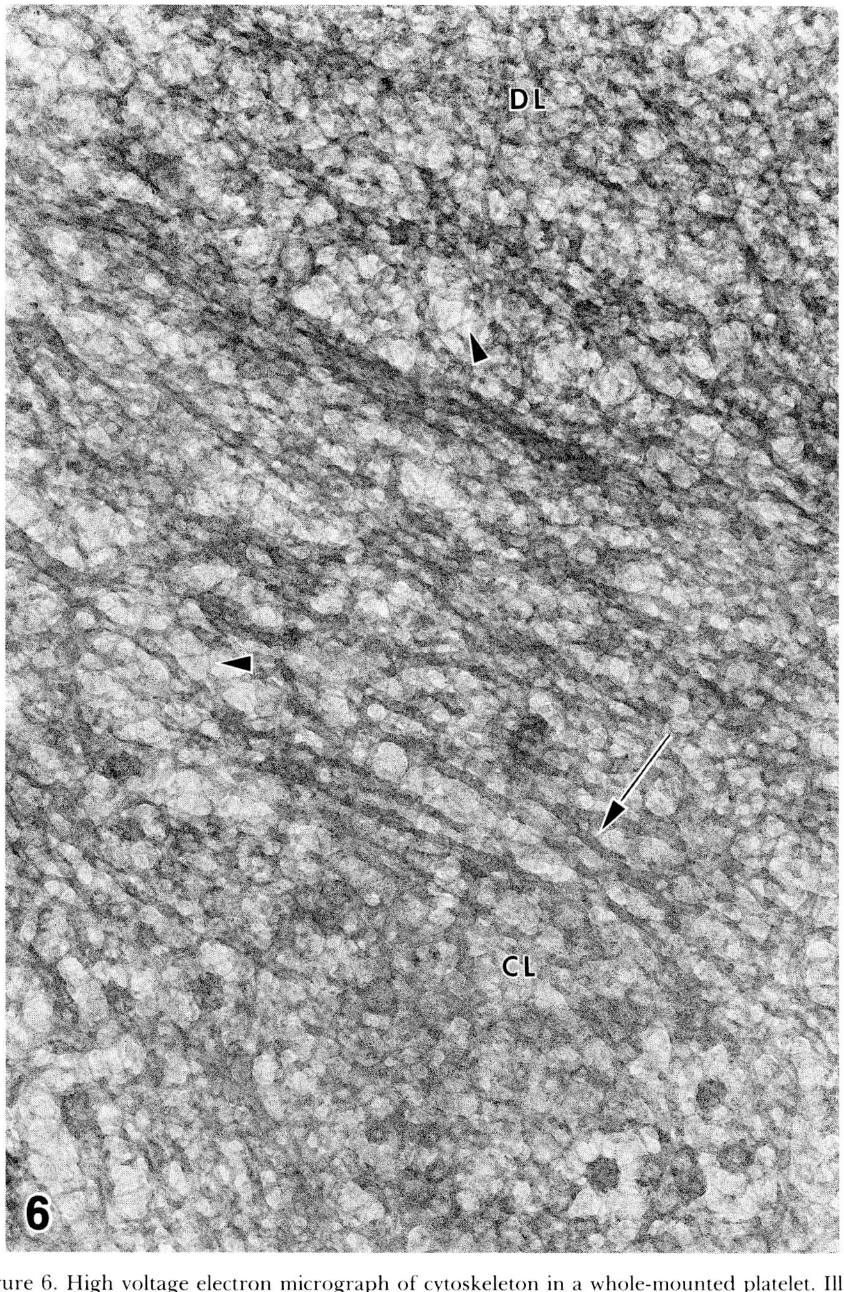

Figure 6. High voltage electron micrograph of cytoskeleton in a whole-mounted platelet. Illustrated is the diversity of filament sizes with diameters ranging from 30–40 Å (arrowheads) to those with diameters approaching 170 Å (long arrow). Coarse lattice (CL) is toward granulomere region whereas the delicate lattice (DL) is near the cell periphery, ×73,000.

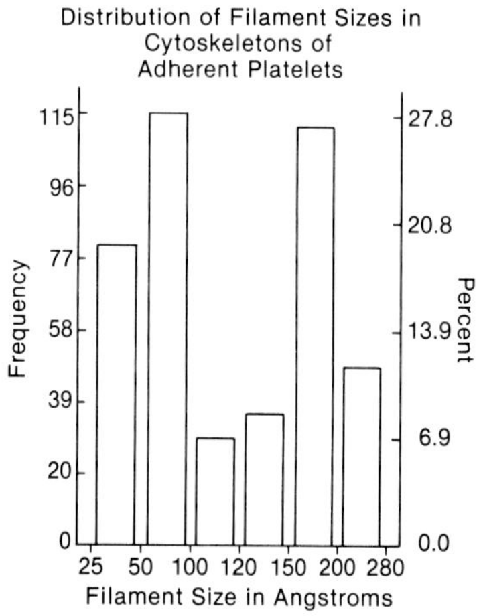

Figure 7. Numerical distribution of filaments in the platelet cytoskeleton. Data are based upon 425 measurements made of whole-mount critical-point dried intact cells. Similar data (not shown) were obtained from post-triton cytoskeleton of spread cells. Note the three major size classes with diameters in the range 30–45 Å, 60–100 Å, and 155–180 Å.

excellent reviews which have been recently published (White and Gerrard, 1979; Crawford, 1976; Cohen, 1979; Lind and Stossel, 1982; Adelstein and Pollard, 1978; Pollard, 1979; Lüscher, 1980; Feinstein and Walenga, 1981; Feinstein, 1982). It is, however, important for the purposes of this chapter to acknowledge, that in addition to the actin, myosin, and microtubules studied extensively by early investigators, the cytoskeletal apparatus of platelets contains a high molecular weight (270,000 daltons) actin-binding protein (Lucas *et al.*, 1976; Rosenberg *et al.*, 1981a), a 105,000 dalton α-actinin-like protein which cross reacts with antibodies to cardiac muscle α-actinin (Rosenberg *et al.*, 1981b), and a host of regulatory molecules including tropomyosin (Cohen and Cohen, 1972; Cote and Smillie, 1981), profilin (Carlsson *et al.*, 1977; Markey *et al.*, 1981) which binds to actin monomers, and gelsolin (Lind *et al.*,

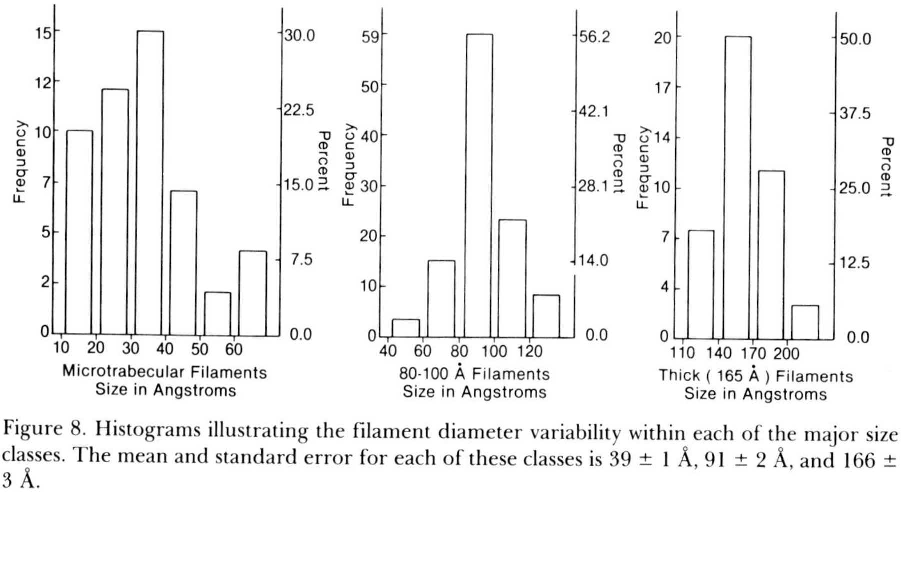

Figure 8. Histograms illustrating the filament diameter variability within each of the major size classes. The mean and standard error for each of these classes is 39 ± 1 Å, 91 ± 2 Å, and 166 ± 3 Å.

1982; Collier and Wang, 1982), a 90,000 dalton peptide which can bind to actin filaments and prevent further polymerization (see Lind and Stossel, 1982 for a complete discussion of cytoskeletal regulatory proteins). Recently, Collier and Wang have reported isolation of a new high molecular weight protein, P235, which also restricts the length of actin filament (Collier and Wang, 1982). In addition to identification of the major structural and regulatory proteins, recent studies have shown that the contractile event, actin polymerization, and perhaps the development of the three-dimensional cytoskeletal network are calcium-dependent mechanisms (Lüscher, 1980; Feinstein and Walenga, 1981; Feinstein, 1982; Rosenberg *et al.*, 1981b; Wang and Bryan, 1981). Furthermore, many of the individual biochemical events appear to be mediated through calcium-dependent phosphorylation reactions (Hathaway *et al.*, 1981; Hidaka and Nishikawa, 1980; Daniel *et al.*, 1981; Carroll and Gerrard, 1982; Fox and Phillips, 1982).

5. *Organization of the Cytoskeleton following Activation*

5.1. *Actin Polymerization and Formation of the Cytoskeleton*

The cytoplasmic ground substance of resting platelets, as already noted, appears as a slightly granular yet nondescript matrix in which microtubules are the only consistently observed filamentous structures. This ultrastructural appearance parallels biochemical data, as several investigators have found that actin, which accounts for 20–30% of platelet protein is to a large degree in a depolymerized state. Based upon use of the DNase 1 inhibition assay (Fox *et al.*, 1981), estimates of G-actin in the resting cell range from 50–90% (Carlsson *et al.*, 1979; Jennings *et al.*, 1981; Pribluda *et al.*, 1981; Davies and Palek, 1982a). This situation changes rapidly upon activation, for, along with the appearance of filaments, there occur dramatic change in cytoplasmic viscosity and a significant reduction in the amount of G-actin (Jennings *et al.*, 1981; Pribluda *et al.*, 1981; Davies and Palek, 1982a; Rotman and Heldman, 1981). The kinetics of actin polymerization have recently been compared to the rate of shape change, aggregation, and the release reaction in human platelets stimulated by either ADP or thrombin (Pribluda *et al.*, 1981). Although differences were found between the two aggregating agents, in both cases there occurred a rapid phase (10–15 sec) of actin mobilization following challenge. This rapid initial phase compares favorably to the rate of shape change and pseudopod extension (Frojmovic and Milton, 1982), and in the studies of Pribluda *et al.* (1981) the actin mobilization preceeded either aggregation or the release reaction.

5.1.2. *Nonactin Protein*

Although actin polymerization has received extensive attention in recent years, it is important to point out that the cytoskeletons which develop following platelet stimulation also contain many of the proteins enumerated in Section 4.4. Jennings *et al.* (1981), while studying the post-triton cytoskeletons of thrombin activated cells, found that with a 15 sec activation time, 65% of the actin was recovered in the cytoskeletal pellet. In addition, these residues contained 14% of the cell myosin and 6% of the actin-binding protein. When the activation period was extended to 30 sec the pellet myosin increased to 90% and the actin-binding protein to 20% of the cell total. In separate studies, Gonnella and Nachmias (1981) and Markey *et al.* (1981) have also documented increases in the precipitable proteins following activation. In addition to actin, myosin, and actin-binding protein described by Jennings *et al.* (1981), it is known that peptides having molecular weights of 100,000, 95,000, 90,000, 50,000–65,000, and 30,000–35,000 either increase in quantity or become associated with the developing fibrillar complex (Markey *et al.*, 1981; Gonnella and Nachmias, 1981).

5.2. Structural Organization of the Cytoskeleton in Adhesion-Activation

5.2.1. *Stages of Adhesion*

Identification of the biochemical events associated with activation and recognition of the biochemical and temporal complexities of the polymerization reaction have rekindled interest in the associated structural events (Nachmias, 1980; Lewis *et al.*, 1982a,b, 1983; Albrecht and Lewis, 1982, Nachmias *et al.*, 1979a,b; Nachmias and Sullender, 1978). This interest has been compounded by the realization that platelet activation is not a single all-or-none event, but rather it is a controlled process involving the orderly transition of the cell through several states. The orderliness has been elegantly demonstrated in the recent studies of Allen *et al.* (1979), who used high extinction differential interference contrast microscopy to describe eight stages in the transformation of a platelet from an early speroidal to a fully-spread form (Fig. 9). Although the stages described by Allen are not particularly different from those illustrated by Bessis and Burstein (1948a,b), the work is highly significant since it describes the stages as they occurred in a single cell as that cell was observed over a time span of 13 min. Whereas earlier workers were able to surmise the transition from a sequence of static photographs of different cells, Allen and his co-workers were able to directly observe and record the events. If it can be assumed that most surface-activated platelets undergo such a transitional sequence (recent studies in our own laboratory have confirmed and extended the observations of Allen to platelets from nonhuman primates; see Lewis *et al.*, 1983), then it is possible to ask whether changes of a consistent nature can be found in the cytoplasm of cells at different stages of activation. To explore this question we (Lewis *et al.*, 1982a) and others (Albrecht and Lewis, 1982; Steele and Porter, personal communication) have been using high voltage and stereo (3-D) microscopy in conjunction with whole-mount preparations of adherent platelets (Fig. 6).

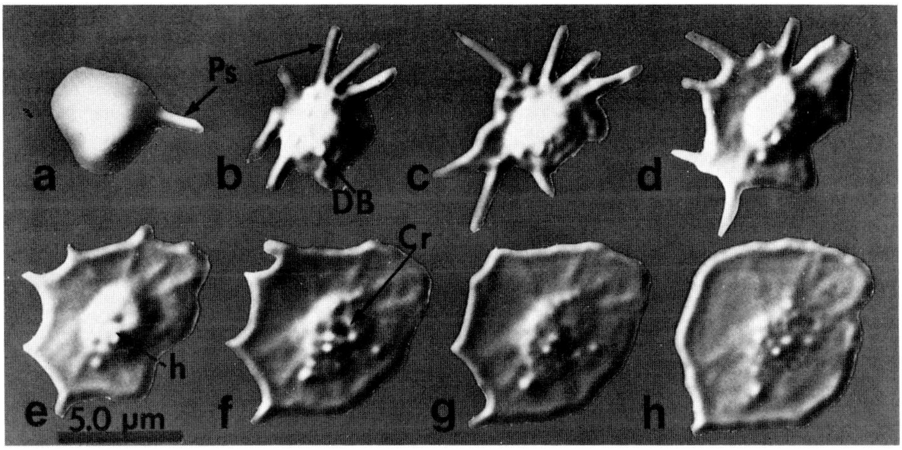

Figure 9. High extinction DIC micrographs of the same platelet at various stages in its transformation. The time from the initial spheroid cell at (a) to the final spread state is 13 min. This figure should be compared to Fig. 3. The time at which the photographs were taken are 0, 1, 2, 3, 4, 7, 8, and 13 min (a–h), respectively. Reproduced with permission from Allen *et al.* (1979).

The results of several of these studies are summarized below to illustrate the changes which can be found in platelets as they undergo activation.

5.2.2. *Cytoskeletal Correlates of Adhesion Stages*

Prior to adhesion (as has been reported by others and noted in Section 1) the cytoplasm of platelets was primarily granular in nature and in the electron microscope appeared as a nondescript moderately electron dense matrix. Following activation and during a pseudopodial stage which corresponds to the 1–2 min time described by Allen *et al.* (1979), a filamentous lattice comprised of the major filament classes described in Fig. 7 was identified (Fig. 10). When observed at high magnification in 3-D, no consistent organizational pattern was evident in the developing lattice at these early times, but as a general rule, the larger filaments were restricted to a thin band of cytoplasm circumferential to the granulomere. In this band, the larger filaments, 80–100 Å and 165 Å, were generally oriented with their major axis concentric to the plasma membrane. The smallest elements, 30–50 Å, extended between and interlaced the large filaments (Fig. 11). During the more advanced stages of adhesion (Fig. 12 illustrates a cell at about 3–4 min and Fig. 13 a cell at about 10–12 min), the organizational pattern became much more pronounced, and three distinct zones were identified in the cells. The first of the zones, which we (Lewis *et al.*, 1982b) have named the central matrix, corresponds to the granulomere reported by Bessis and Burstein (1948a) and has a cytoplasmic ground consistency similar to that reported for the unactivated cell. A similar observation has been reported by Nachmias (1980) and Mattson

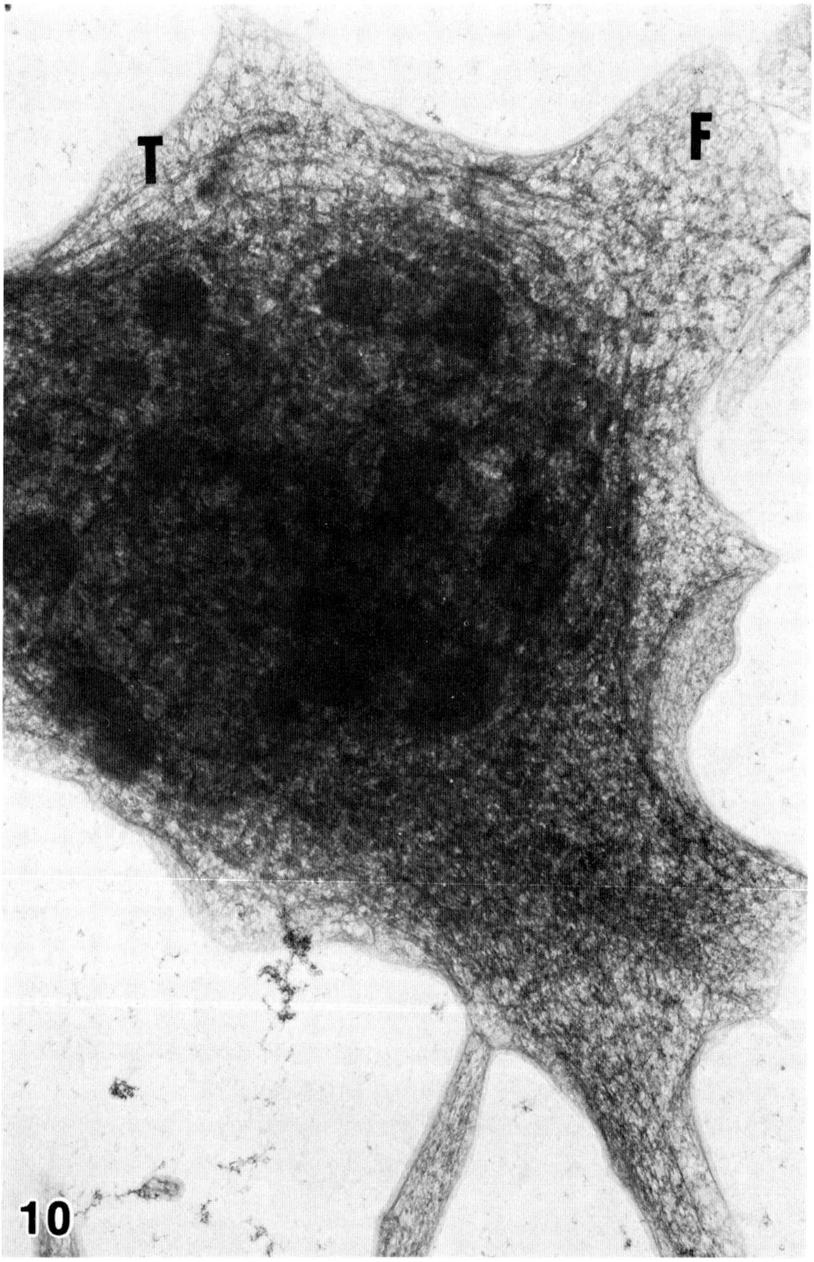

Figure 10. High voltage electron micrograph of a platelet in the early stage of activation. The cytoplasmic ground substance in this cell has been transformed from the amorphous resting state to an intricate filamentous network. Note the complex array of tubules (T) and thick filaments (F) in the peripheral cytoplasm. This cell corresponds to the 1–2 min time frame as defined in Fig. 9, ×30,000.

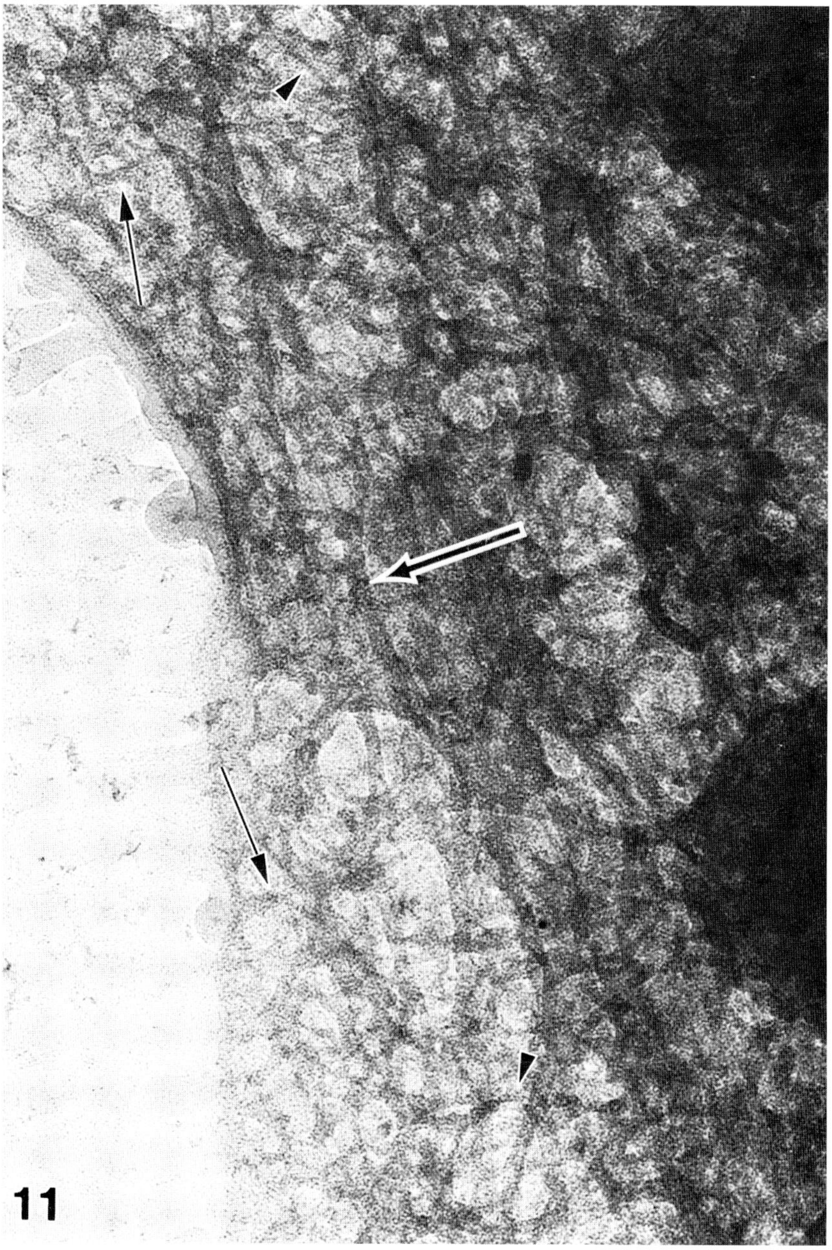

Figure 11. Micrograph of select region from the periphery of a cell similar to that shown in Fig. 10. The lattice consists of 165 Å (large arrow), 80–100 Å (small arrows) interlaced by delicate 30–50 Å microtrabulae (arrowheads). Note the tortuous paths of the larger filaments which are oriented concentric to the plasma membrane, ×210,000.

Mattson and Zuiches (1981). Surrounding the central matrix is a second zone, the trabecular zone, which is characterized as being a course, open lattice with spaces often several thousand angstroms in diameter. It is noteworthy that the major structural elements in this zone are the 165 Å filaments and lesser amounts of 80–100 Å filaments. The 30–50 Å filaments, microtrabeculae, are also found in this region and are organized as described in Fig. 9 for the early

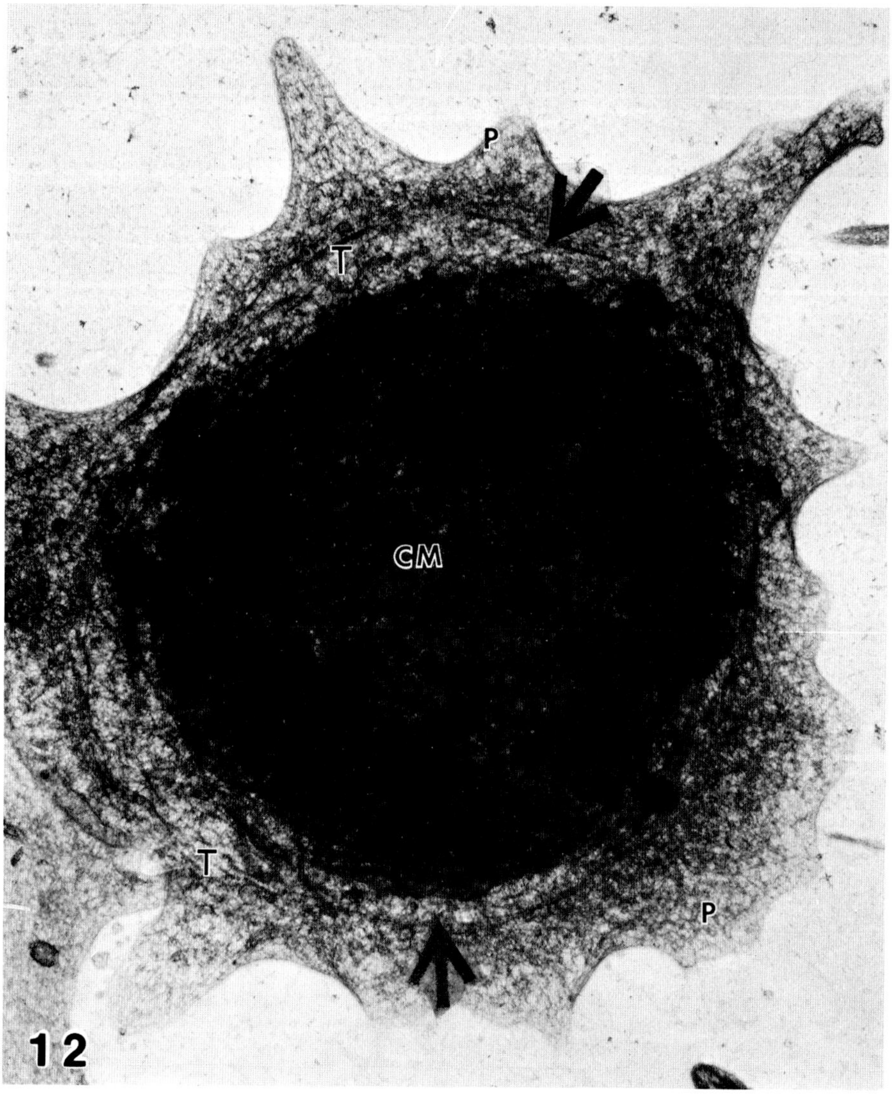

Figure 12. High voltage electron micrograph of a platelet at a 3–4 min adhesion state. Three distinct zones, the central matrix (CM), the trabecular zone (T), and the peripheral web (P) are shown. The larger filaments (arrows) in the trabecular zone have diameters in the range 155–170 Å, ×25,000.

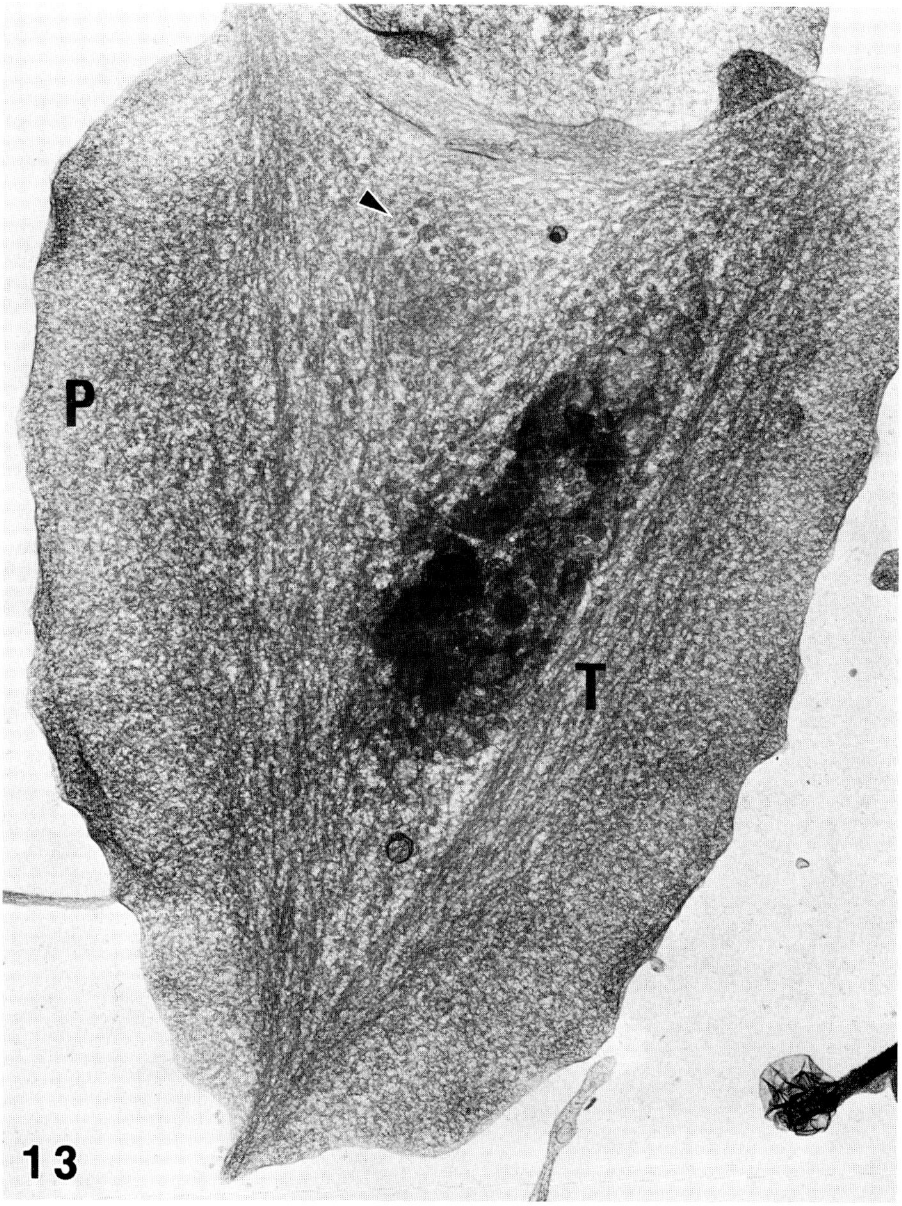

Figure 13. Fully spread adherent platelet in which the three organizational zones are distinctly defined. As illustrated in this micrograph, spreading from an intermediate stage such as in Fig. 12 results in an expansion of both the trabecular zone (T) and the peripheral web (P) region. Compare the number of thick filaments (165 Å) to that in Fig. 12. Careful study of the cell reveals a gradual transition between the coarse lattice of the trabecular zone and the honeycomb lattice of the peripheral web. A few microtubules are identified in the spread hyaloplasm (arrowhead), ×25,000.

stages of activation. Located at the cell margin is the third major organizational zone, the peripheral web, which is comprised of 70 Å filaments of short length arranged in a tight honeycomb configuration. Although these three zones often appear to be distinct, upon close observation orderly transition can generally be identified between adjacent regions. The organization of the two outer zones is graphically illustrated in Fig. 14, which also gives an interpretation of the transition between the two. It should be emphasized that although we (Lewis *et al.*, 1982b), as a matter of convenience, have labeled the three organizational areas in platelets from African green monkeys, the observation of the three zones is unique neither to this species nor to our laboratory. Mattson and Zuiches (1981), Steele and Porter (personal communication), and Albrecht and Lewis (1982) have all observed similar organization in whole-mount preparations of adherent human platelets (Fig. 15).

5.2.3. *Cytoskeleton in Surface-Bound Aggregates*

The studies of Mattson and Zuiches (1981) are particularly noteworthy, for in addition to describing the intact whole-mount cells, these investigators demonstrated a similar organization in post-triton cytoskeletons prepared from adherent cells (Fig. 16). This observation has been confirmed both by us (Lewis *et al.*, 1982b) and by Albrecht and Lewis (1982). It is possible, but unconfirmed, that the post-triton cytoskeletons of adherent cells are composed of contractile proteins. Such a composition would parallel that reported for cells activated while in suspension, a condition that would result in aggregation (Fox *et al.*, 1981; Carlsson *et al.*, 1979; Jennings *et al.*, 1981; Pribluda *et al.*, 1981; Davies and Palek, 1982a,b). Comparisons between aggregation cytoskeletons and adhesion cytoskeletons have not been reported, but Steele and Porter (personal communication) have observed nascent aggregates which formed subsequent to adhesion. As shown in Fig. 17, the cells located basally within the aggregates have the three zones described for singly-adherent platelets. However, the association of the cells apparently influences the overall geometry, for the granulomere region in each of the constituent platelets is placed eccentrically toward the center of the aggregate near the region of platelet confluence. The hyalomere regions, containing the trabecular zone and the peripheral web in each cell, are on the outer side of the aggregate creating an overall radial symmetry.

5.2.4. *Microtubules in the Activation-Cytoskeleton*

It is obvious from our own recent studies (Lewis *et al.*, 1982a) and those of Mattson and Zuiches (1981) that microtubules play a minor, if any, role in the development of the platelet cytoskeleton following adhesion activation. This observation is in dramatic contrast to function ascribed to these organelles in the circulating cell (Behnke, 1965; Behnke and Zelander, 1967; White and Gerrard, 1979). Recently, Nachmias (1980) has used 3-D microscopy in conjunction with negative staining to describe the microtubule band in the cells

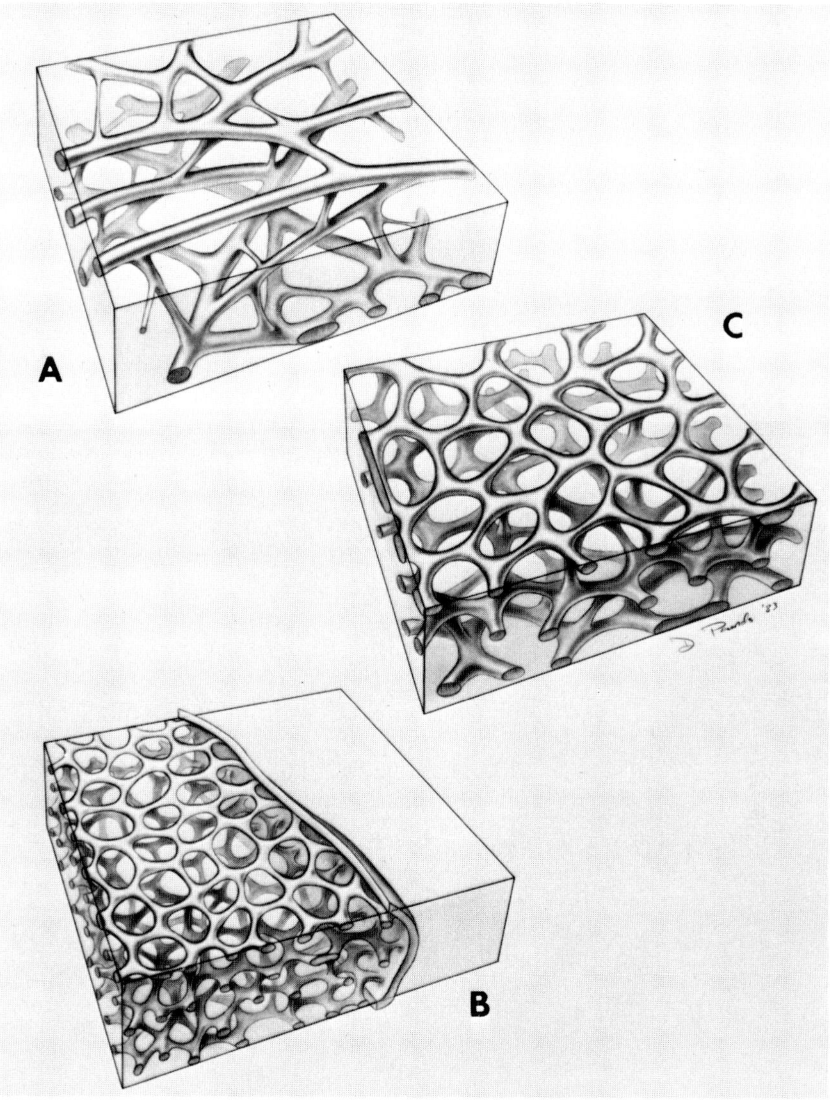

Figure 14. Artist's interpretation of the lattice structure in the trabecular zone (A) and the peripheral web (B) regions of the platelet. As shown (C), the region between the two zones is a transition lattice. The three images are drawn on the same scale so comparisons can be made of the filament sizes (diameter and length), their orientation, and the amount of intralattice space. Interpretation and drawings by Mr. David Pounds, Division of Audio Visual Resources, Bowman Gray School of Medicine.

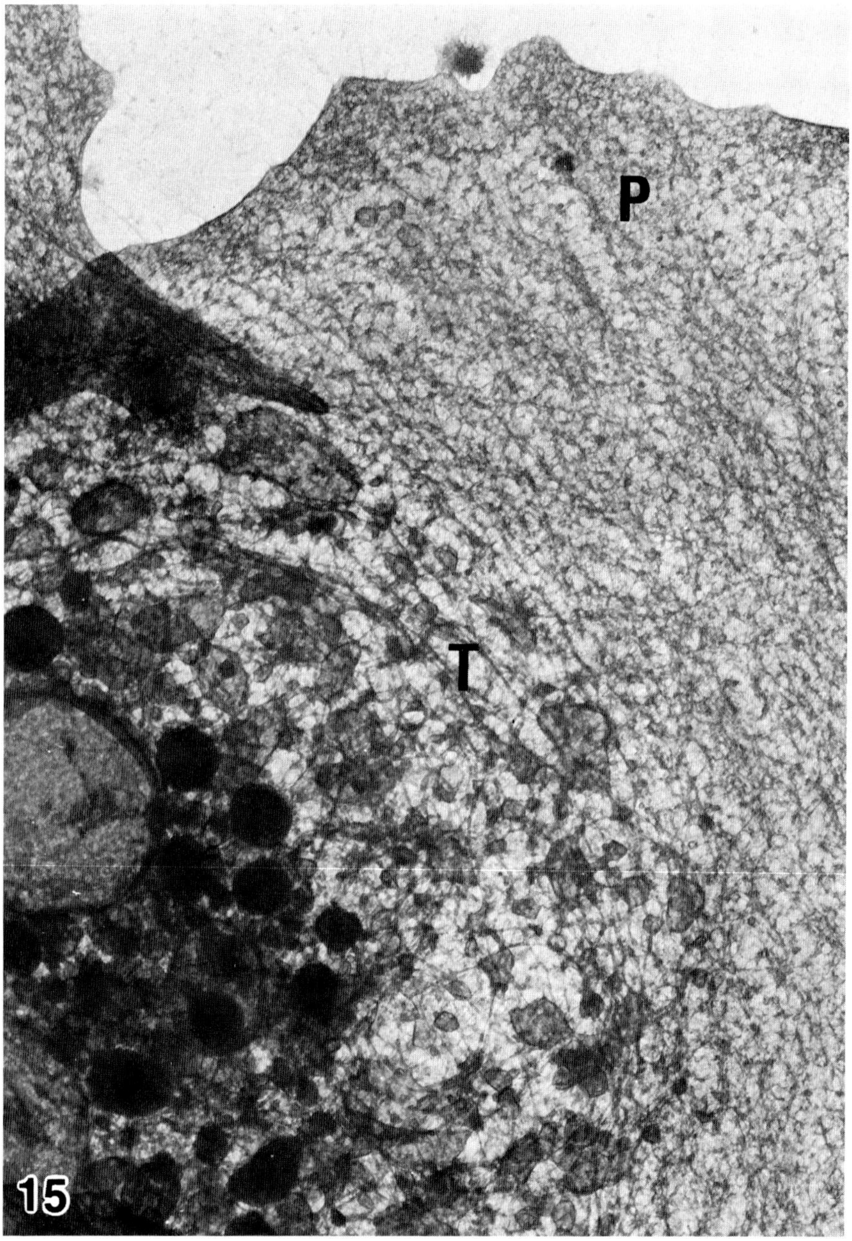

Figure 15. High voltage electron micrograph of a select region from an adherent spread, human platelet. The cytoplasm in this cell compares favorably to that of the African green monkey cells and contain the three zones as described in Figs. 10–12. Trabecular zone (T) and peripheral web (P) are marked, ×35,000. Micrograph kindly provided by Dr. Keith R. Porter and Ms. Robin J. Steele, University of Colorado, Boulder.

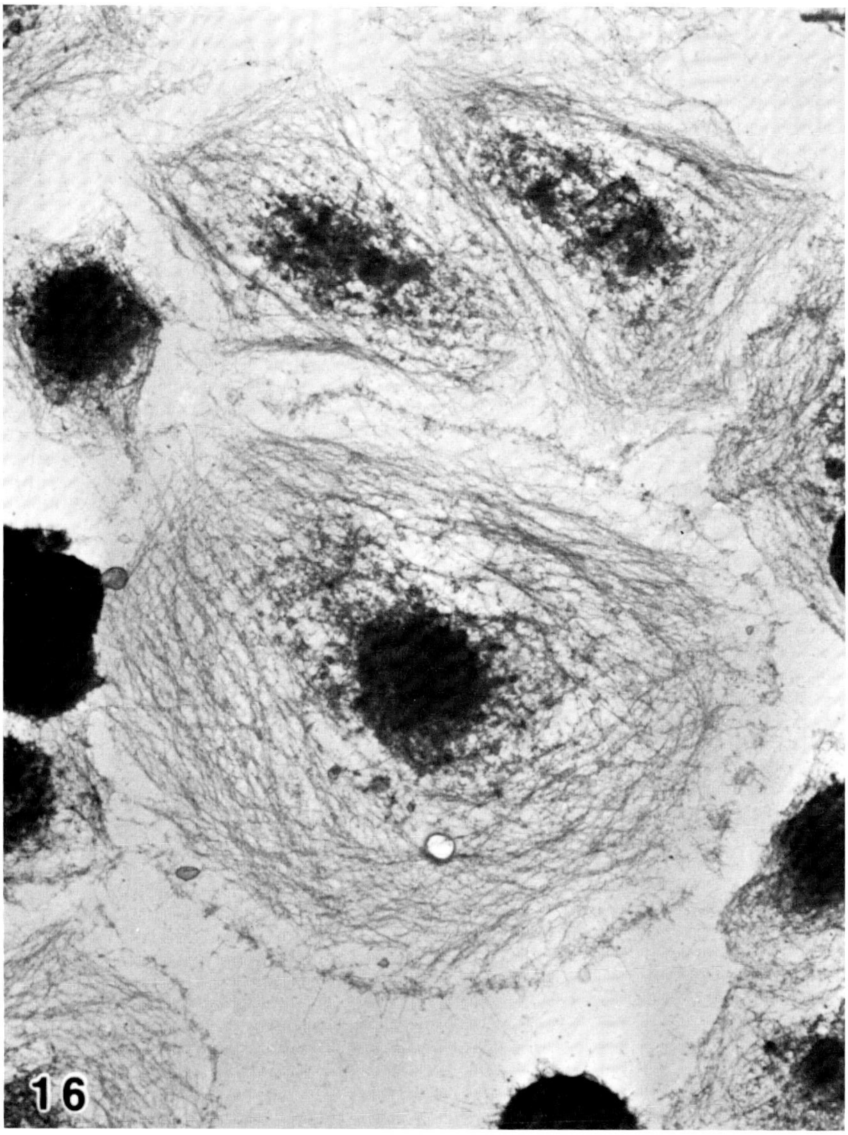

Figure 16. Platelet cytoskeletons subsequent to triton extraction. Adherent cells were rinsed for 1 min in 0.1% triton in the adhesion medium prior to fixation and drying by the critical point method. As evident in this preparation each of the residual cytoskeletons has the three distinct zones as identified in the intact cells, ×10,000.

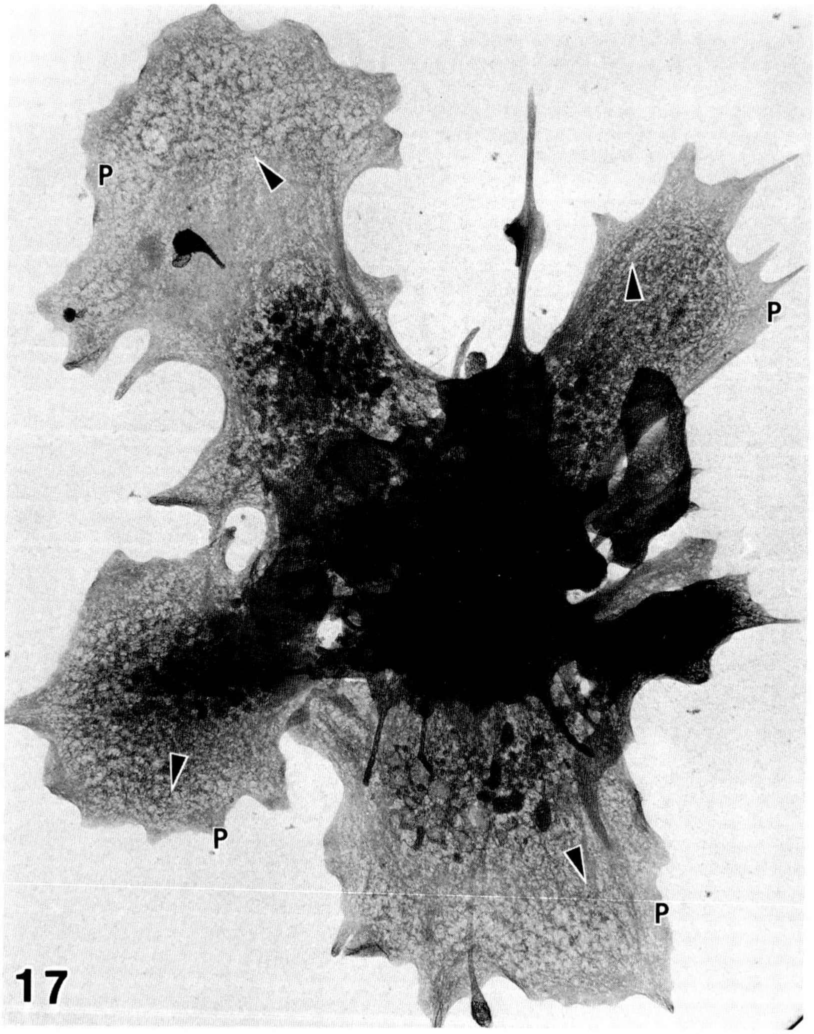

Figure 17. A small aggregate of human platelets in which the basally located cells are organized with the granulomere regions at the center of the aggregate. In each cell, the hyalomere with peripheral web (P) and trabecular zone (arrowheads) has spread outward from the aggregate center resulting in an overall radial configuration, ×10,000. Micrograph kindly provided by Dr. Keith R. Porter and Ms. Robin J. Steele, University of Colorado, Boulder.

following a brief period of spreading on polylysine-coated grids. In cells which had been in contact with the surface for periods of 45 sec or less and had relatively little spread cytoplasm, the microtubule coil remained intact. This observation is consistent with our own observation of platelets from African green monkeys (Lewis *et al.*, 1982b). However, when platelets were allowed to spread more fully, as noted by Nachmias, the ground cytoplasm consisted entirely of a delicate network of interlacing fibrils similar to those described in Section 5.2.2. Microtubules were not consistently observed. In our studies, dissociation of the microtubule coil was evident in cells that had morphologies reflecting 1–2 min contact with the foreign surface. At all morphologic stages suggesting more prolonged surface contact, the microtubules were generally in a configuration which reflected conformity to the overall symmetry of the cell (Lewis *et al.,* 1982b). This conformity appears not to be required for cell shape transition, however, since Mattson and Zuiches (1981) rarely observed microtubules in adherent human platelets. When observed by these investigators the tubules were seen, "coursing in a gently curved fashion through the periphery of the cytoplasm" (Mattson and Zuiches, 1981). Furthermore, treatment with either colchicine or vinblastine sulfate alters neither the spreading of the cells on the surface nor the release reaction which culminates the adhesion process (Lewis *et al.,* 1982b). The lack of microtubule participation in the adhesion events again brings to question the role that these organelles play in platelet structural physiology. This question has also recently been raised for cells undergoing aggregation. In a series of experiments reported by White (1983), treatment of platelets with taxol in the cold resulted in dissociation of the microtubule coil, and the reassociation of the tubulin as single straight elements or in bundles of straight elements which projected radially throughout the platelet cytoplasm. Warming the cells to 37°C did not reform the circumferential band, but the cells responded normally with respect to aggregation and release. The implication of these studies is that neither the microtubule coil nor some other specific microtubule–skeletal arrangement is required for normal function of these cells. Whether microtubules are necessary at all is a question that requires further attention, for as noted by White (1983) and as recently shown by Menche *et al.* (1980), internal reorganization, secondary aggregation, and secretion are inhibited by antimitotic agents. This issue becomes even more complex in light of Coller's (1981) recent studies suggesting that cytoskeletal filaments, particularly microtubules, may in fact play a role in mediating the cAMP-related inhibition of platelets treated with the prostaglandins PGI_2 and PGE_1.

5.2.5. Cytoskeletal Variability among Platelets

Although observations from several laboratories report similar three-dimensional structure in adhesion-activated platelets, variation on the common pattern does occur. This variability, which can be expressed either as a generalized flocculent lattice or as a highly condensed lattice (Fig. 18), can be found in all cell preparations, but the number of cells having such aberrant appearance is relatively small. In spite of the fact that the proportion of such cells is not great, their presence is important and questions the "trueness" of the three-dimensional studies. The question becomes even more important in light of the recent report by Small and Langanger (1981) on the effects of osmium tetroxide treatment and dehydration on the ultrastructure of actin meshworks in chick fibroblasts. To better understand the source of variability in platelet cytoskeleton, a series of experiments involving the use of 0.1% and

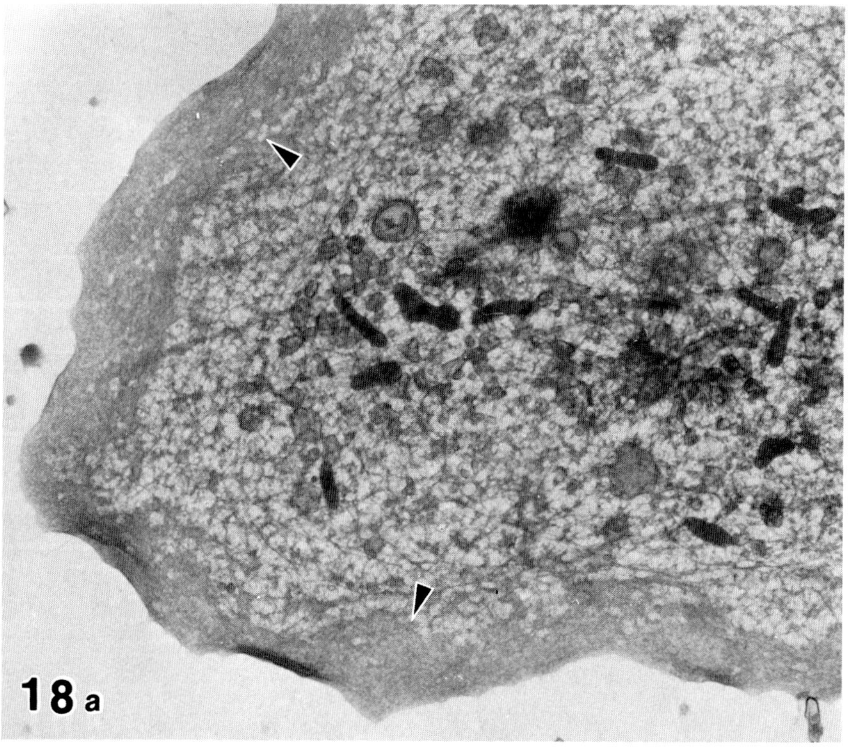
18 a

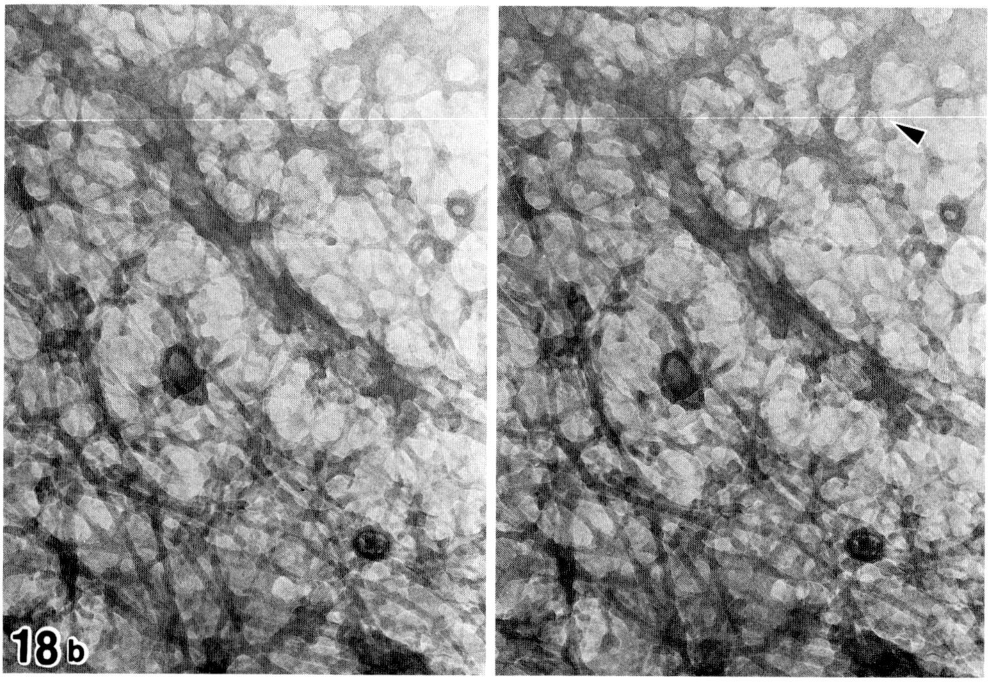
18 b

Table 2. Osmium Tetroxide Effects on the Platelet Cytoskeleton[a]

Osmium (%)	Filament[b]	Microtubule[c] (diameter in angstroms)	Lattice space[b]
0.0	60	200	250
0.1	40	200	270
1.0	50	200	370

[a]Date based upon duplicate experiments conducted by Ms. Gwen Campbell, Salem College, Winston-Salem. Each number is the mean of 80 individual measurements. In each of the experiments, washed platelets in Puck's balanced salts were allowed to adhere to formver coated grids for 45 min at 37°C prior to fixation for 10 min in cacodylate buffered glutaraldehyde. Post fixation in osmium at the concentrations shown was for 5 min. Observations were made of critical-point dried, intact whole-mount platelets.
[b]Filament sizes and lattice spacing are from peripheral web region of the cells.
[c]Microtubules were from various locations since no precise distribution is found in adherent platelets. The diameters shown, although on the lower end of those reported for platelets, have been confirmed by measuring cross-sectional area of tubules in tannic acid-glutaraldehyde fixed cells (data not shown).

no osmium treatment has been conducted. These studies have revealed a subtle shift in the morphology of the cells with a more pronounced coarse fibrillar appearance in cells treated with the higher concentrations of osmium. Morphometric analysis of the osmium-treated cells revealed that the coarse fibrillar appearance was due both to an increase in the size of the intratrabecular spaces (open cytoplasmic areas) and a decrease in average filament diameter (Table 2). In an attempt to better understand the effect of osmium on structure of the lattice, we have also treated cells with osmium vapors. No differences of a consistent nature were found, however, when cells treated in this manner were compared to the immersion fixed samples. It is important to point out that, although there is a subtle effect of osmium on the cytoskeleton of platelets, the three major organizational zones described in Section 5.2.2 were consistently observed.

Porter has suggested that the degree of cell hydration at the time of fixation will influence the degree of cytoskeletal condensation, and as water is removed from the cell, the lattice will condense similar to that shown in Fig. 18. Since this configuration, in our experience, is the most commonly observed aberration, we conducted a series of experiments to determine the effects of osmolality on the cytoskeleton of platelets. When cells are maintained at 150–170 milliosmoles prior to fixation, the lattice has a more deli-

Figure 18. Micrographs illustrating the condensed appearance of the cytoskeletal lattice which can occur experimentally with increasing osmolality of the adhesion medium or naturally if the cell's capability to regulate water has been compromised. (a) Condensed region shown at the cell periphery (arrows). ×20,000. (b) High magnification stereo pair micrographs of a select area in a cell such as shown in (a), The thickened lattice at the edge of the condensed zone appears to be derived from more delicate filaments in the surrounding cytoplasm, ×80,000. Negatives kindly provided by Dr. Keith R. Porter and Ms. Robin J. Steele, University of Colorado, Boulder.

cate appearance as compared to the fibrillar controls. In contrast, increasing the osmolality to 600 milliosmoles condensed the lattice in a manner similar to that shown in Fig. 18. Since platelets constitute a heterogenous population with regard to age and functional capacity, it is conceivable that the deviations normally observed are a reflection of varying capacity to regulate cell water (Porter and Tucker, 1981).

6. Membrane–Cytoskeletal Interaction

6.1. Effects of Surface-Active Agents on Cytoskeleton and Cell Function

The ability of membrane-active agents such as the local anaesthetics, tetracaine and lidocaine, to inhibit platelet functions such as aggregation and secretion has been well documented in the literature (Feinstein *et al.*, 1975; Prowse *et al.*, 1982). The mechanism for the inhibitory effect has not been clearly established, but notations have been made that the effects of these compounds correlate both with alterations in cellular calcium movements (Prowse *et al.*, 1982) and with differential intercollation into the inner and outer leaflets of the plasma membrane (Nachmias *et al.*, 1979b). Irrespective of the immediate mode of action, a relationship between cytoskeletal disruption and platelet inhibition by surface agents has clearly been shown (Nachmias *et al.*, 1977, 1979b). When human platelets were treated with lidocaine in the concentration range 1–60 mM, a dose-dependent suppression of pseudopod extension was observed, and at 30 mM the suppression was evident in 70% of the cells (Nachmias *et al.*, 1977). Most applicable to this review is the observation that cytoskeletons from the lidocaine-treated cells had a dramatic reduction in the number of 70–80 Å filaments. Furthermore, the normal net-like arrangement associated with these filaments was replaced by a somewhat coarse lattice and a granular matrix. In a subsequent work (Nachmias *et al.*, 1979b), lidocaine treatment at 30 mM was found to cause retraction of pseudopods in 90% of the treated cells, and tetracaine at a concentration of 2–4 mM caused retraction in 100% of the platelets. The tetracaine treatment is significant because SDS gel electrophoresis of post-treatment cell lysate revealed the proteolysis of two high molecular weight (250,000 and 230,000 dalton) peptides. Although the identity of the peptides was not established by Nachmias *et al.* (1979b), their size similarity to actin-binding protein was mentioned.

The correlative effects local anaesthetics have on platelet function and cytoskeletal composition, both structural and biochemical, are representative of the expanding literature in the area of membrane–cytoskeletal interaction. Since the ultrastructural demonstration by Zucker-Franklin (1969) of actin filaments as submembranous proteins, numerous investigators have attempted to elucidate the role played at this location. Evidence in varying degrees of substantiation has been put forth that cytoskeletal proteins may constitute integral membrane proteins (Gerrard *et al.*, 1979; Booyse *et al.*,

1971), or may be exposed at the extracellular surface following cell activation (Booyse and Rafelson, 1972; George *et al.*, 1980; Puszkin *et al.*, 1977; Bouvier *et al.*, 1977; Diggle *et al.*, 1979) to facilitate cell–cell interaction or cell–surface reaction. Although questions have been raised about some of these suggestions (Gogstad *et al.*, 1981; Langer *et al.*, 1982; Bennett *et al.*, 1981), it is clear that the association of the cytoskeletal complex with the plasma membrane in platelets is intimate, and that cytoskeletal organization upon platelet stimulation is accompanied by predictable changes in the cell surface (Davies and Palek, 1982b). Furthermore, it is becoming increasingly more evident that a two-way street exists for regulation within the platelet. Events at the cell surface can regulate the cytoskeleton (Nachmias *et al.*, 1977, 1979b), or alternatively, the cytoskeleton may regulate events at the cell surface.

6.2. Cell-Surface Regulation of Cytoskeletal Action

The relationship of cell function to events occurring at the plasma membrane is not a new concept to platelet cell biology, for it has long been recognized that molecular defects in the plasma membrane result in impaired hemostasis (Phillips *et al.*, 1980). This impairment takes on a new dimension in light of recent cytoskeletal studies of platelets from patients having Glanzmann's thrombasthenia, a congenital disorder characterized by abnormal platelet aggregation and clot retraction. The major molecular deficiency in Glanzmann's patients is a reduction in the plasma membrane glycoprotein IIb-III complex, which among other roles, appears to be a fibrinogen receptor and seems to facilitate platelet–platelet binding during aggregation. This latter point has been substantiated by Phillips *et al.* (1980), who found with thrombin aggregation, the cytoskeletal structure from platelets contained 26% of glycoprotein IIb and 14% of glycoprotein III. In platelets stimulated by concanavalin A, Painter *et al.* (1982) found 80–90% of these two glycoproteins in the post-triton cytoskeletons. Additionally, Painter *et al.* (1982) reported a doubling of the sedimentable actin with Con A, which suggests a stabilization of the cytoskeleton in the complex. The identification of the surface membrane glycoprotein IIb-III complex in cytoskeletons from normal individuals is in dramatic contrast to cytoskeletons from patients having Glanzmann's thrombasthenia, for as noted by Tuszynski, both platelets and their corresponding cytoskeletons from patients bound less than 17% of the fibrinogen found with control cells, and interestingly, an antibody raised against triton insoluble cytoskeletons inhibited both ADP and thrombin-induced aggregation (Tuszynski *et al.*, 1982a). The deficient cytoskeleton-surface membrane association described by Tuszynski is consistent with earlier reports showing an α-actinin deficiency (Gerrard *et al.*, 1979) and an impaired association of actin with the platelet membrane following thrombin-induced secretion (George and Morgan, 1981). This latter point may explain the abnormal clot retraction observed with Glanzmann's, for Cohen *et al.* (1982) have recently shown that thrombasthenic platelets failed to develop tension or to allign pseudopods with fibrin strands, processes felt to be required for generation of the contractile force. Improper association of the cytoskeleton with the membrane in Glanzmann's thrombasthenia has also been linked to impaired cell adhesion and spreading. In a morphometric study, comparing the adhesion of platelets from thrombasthenic patients to normal controls, Rosenstein *et al.* (1981) found about 30% of the patients' platelets reached a fully spread state, whereas under identical conditions, more than 90% of the control cells were in the advanced state.

6.3. Cytoskeletal Control of Membrane Receptor Organization

Cytoskeletal–membrane associations appear to be involved with events other than those of a contractile nature, for triton-insoluble thrombin-induced cytoskeletal preparations have recently been found to potentiate the factor Xa-catalyzed activation of prothrombin (Tuszynski *et al.*, 1982b) and control distribution of fibrinogen receptors (Albrecht and Lewis, 1982). In a series of recent reports, Tuszynski and co-workers (1982b,c) found factor Va in triton-insoluble cytoskeletons from thrombin-activated human platelets. The cytoskeletal activity was inhibited by a monoclonal antibody to factor V, and when treated with the antibody, the cytoskeleton's ability to potentiate the factor Xa-catalyzed activation of prothrombin was reduced. In addition, the ability of the normal cytoskeleton to correct the clotting time of factor V-deficient plasma was eliminated if the cytoskeletons were first treated with the antibody. Since the cytoskeletons also bound factor Xa, the observations of Tuszynski *et al.* suggest that orientation of the prothrombinase complex, Xa-Va, on the platelet surface may be under cytoskeletal control.

A similar relationship of surface organization and cytoskeletal configuration has been reported by Albrecht and Lewis (1982), who used 5 nM and 18 nM gold beads to localize fibrinogen and the glycoprotein IIb-III complex on whole-mount adherent platelets. The observations of Albrecht are interesting because they not only show colocalization of the ligand and its receptors, but they also demonstrate a redistribution of the receptors during cell adhesion. The pattern of surface redistribution was indicative of cytoskeletal control, for in the advanced stages, the bound markers were exclusively located over the region described by us (Lewis *et al.*, 1982b) as the trabecular zone. Whether or not the cytoskeletal reorganizations associated with trabecular zone development has determined fibrinogen binding site location is unknown, but the experiments of Tuszynski *et al.* (1982a,b) and those of Albrecht and Lewis (1982) open the door on a relatively unexplored area of platelet cell biology.

References

Adelstein, R. S., and Pollard, T. D., 1978, Platelet contractile proteins, in: *Progress in Hemostasis and Thrombosis 4* (T. H. Spaet, ed.), pp. 37–58, Grune & Stratton, New York.

Albrecht, R. M., and Lewis, J. C., 1982, Examination of platelet activation by HVEM and SEM: Cytoskeleton receptors sites and dense tubular system, *J. Cell Biol.* **95:**466a.

Aleksandrowicz, J., Blicharski, J., and Feltynowski, A., 1957, Mikroskopia elektronowa krwinek badanych metoda ultracienkich skrawow, *Haematol. Clin. Med. Intern. Acad. Med. Cracoviensis* **1**:3.

Allen, R. D., Zacharski, L. R., Widersky, S. T., Rosenstein, R., Zaitlin, L. M., and Burgess, D. R., 1979, Transformation and motility of human platelets: Details of the shape change and release reaction observed by optical and electron microscopy, *J. Cell Biol.* **83**:126–142.

Barnhart, M. I., 1978, Platelet responses in health and disease, *Mol. Cell Biochem.* **22**:113–137.

Barnhart, M. I., Walsh, R. T., and Robinson, J. A., 1972, A 3-D view of platelet response to chemical stimuli, *Ann. N.Y. Acad. Sci.* **201**:360–391.

Behnke, O., 1965, Further studies on microtubules. A marginal bundle in human and rat thrombocytes, *J. Ultrastruct. Res.* **13**:469.

Behnke, O,, 1966, Morphological changes in the hyalomere of rat blood platelets in experimental venous thrombi, *Scand. J. Haemat.* **3**:136–148.

Behnke, O., 1968a, An electron microscope study of the megakaryocyte of the rat bone marrow. I. Development of the demarcation membrane system and the platelet surface coat, *J. Ultrastruct. Res.* **24**:412–433.

Behnke, O., 1968b, Electron microscopical observation on the surface coating of human platelets, *J. Ultrastruct. Res.* **24**:51–69.

Behnke, O., and Zelander, T., 1967, Filamentous substructure of microtubules of the marginal bundle of mammalian blood platelets, *J. Ultrastruct. Res.* **19**:147–165.

Behnke, O., Kristensen, B. I., and Engdahl-Nielson, L., 1971, Electron microscopical observations on actinoid and myosinoid filaments in blood platelets, *J. Ultrastruct. Res.* **37**:351–369.

Bennett, J. S., Vilaire, G., Coleman, R. F., and Colman, R. W., 1981, Localization of human platelet membrane-associated actomyosin using the affinity label 5-p-fluorosulfonylbenzoyl adenosine, *J. Biol. Chem.* **256**:1185–1190.

Bentfield, M. E., and Bainton, D. F., 1976, Primary lysosomes of rat megakaryocytes and platelets, *J. Clin, Invest.* **56**:1635–1639.

Bessis, M., and Breton-Gorius, J., 1965, Les microtubules et les fibrilles dans les plaquettes etalees, *Nouvelle Revue Francaise d'Hematologie* **5**:657–662.

Bessis, M., and Bricka, M., 1948, Etude sur L'ultra-structure du protoplasm des thrombocytes au microscope electronique, *Biochim. Biophys. Acta* **2**:339–349.

Bessis, M., and Burstein, M., 1948a, Etudes sur les thrombocytes au microscope electronique, *Revue D'Hematologie* **3**:48–68.

Bessis, M., and Burstein, M., 1948b, Une technique pour examiner les plaquettes au microscope électronique, *Comptes Rendus Société de Biologie* **142**:27–28.

Bettex-Galland, M., and Lüscher, E. F., 1959, Extraction of an actomyosin-like protein from human thrombocytes, *Nature* **184**:276–277.

Bettex-Galland, M., and Lüscher, E. F., 1963, Studies on the metabolism of human blood platelets in relation to clot retraction, *Thrombosis et Diathesis Haemorrhagica* **4**:178–195.

Bizzozero, J., 1882, Ubër einen neuen formbestandtil des blutes and dessen rolle bie de thrombose and der blutgerinnung, *Virchows Arch. Pathol. Anat.* **90**:261.

Booyse, F. M., and Rafelson, M. E., 1972, Regulation and mechanism of platelet aggregation, *Ann. N.Y. Acad. Sci.* **301**:37–60.

Booyse, F. M., Sternberger, L. A., Zschocke, D., and Rafelson, M. E., 1971, Ultrastructural localization of contractile protein (thrombobosthen) in human platelets using an unlabeled antibody-peroxidase staining technique, *J. Histochem. Cytochem.* **19**:540–550.

Born, G. V. R., 1956, The break-down of adenine triphosphate in blood platelets during clotting, *J. Physiol.* **133**:61–62.

Bouvier, C. A., Gabbiani, G., Badonnel, G. B., Manjo, G., and Luscher, E. G., 1977, Binding of anti-actin autoantibodies to platelets, *Thrombosis and Haemostasis* **37**:321–328.

Breton-Gorius, J., and Guichard, J., 1972, Ultrastructural localization of peroxidase activity in human platelets and megakaryocytes, *Am. J. Pathol.* **66**:277–294.

Carlsson, L., Nyström, L. E., Sundkvist, I., Markey, F., and Lindberg, U., 1977, Actin polymerization is influenced by profilin, a low molecular weight protein in nonmuscle cells, *J. Mol. Biol.* **115**:465–483.

Carlsson, L., Markey, F., Blikstad, I., Persson, T., and Lindberg, U., 1979, Reorganization of actin in platelets stimulated by thrombin as measured by the DNase I inhibition assay, *Proc. Natl. Acad. Sci. USA* **76:**6376–6380.

Carroll, R. C., and Gerrard, J. M., 1982, Phophorylation of platelet actin-binding protein during platelet activation, *Blood* **59:**466–471.

Cohen, I., 1979, The contractile system of blood platelets and its function, *Meth. Achiev. Exp. Pathol.* **9:**40–86.

Cohen, I., and Cohen, C., 1972, A tropomyosin-like protein from human platelets, *J. Mol. Biol.* **68:**383–387.

Cohen, I., Gerrard, J. M., and White, J. G., 1982, Ultrastructure of clots during isometric contraction, *J. Cell Biol.* **93:**775–787.

Coller, B. S., 1981, Inhibition of von Willebrand factor-dependent platelet function by increased platelet cyclic AMP and its prevention by cytoskeleton-disrupting agents, *Blood* **57:**846–855.

Collier, N. C., and Wang, K., 1982, Purification and properties of human platelet P235, *J. Biol. Chem.* **257:**6937–6943.

Cote, G. P., and Smillie, L. B., 1981, The interaction of equine platelet thropomyosin with skeletal muscle actin, *J. Biol. Chem.* **256:**7257–7261.

Crawford, N., 1976, Platelet microfilaments and microtubules, in: *Platelets in Biology and Pathology* (J. I. Gordon, ed.), pp. 121–153, North Holland, New York.

Crawford, N., and Taylor, D. G., 1977, Biochemical aspects of platelet behavior associated with surface membrane reactivity, *Br. Med. Bull.* **33:**199–206.

Daniel, J. L., Molish, I. R., and Holmsen, H., 1981, Myosin Phosphorylation in intact platelets, *J. Biol. Chem.* **256:**7510–7514.

Da Prada, M., and Pletscher, A., 1974, Mechanisms of 5-hydroxy tryptamine storage in subcellular organelles of blood platelets, *Adv. Biochem. Psychopharmacol.* **10:**311–320.

Davies, G. E., and Palek, J., 1982a, The state of actin polymerization in tetracaine treated platelets, *Thrombosis Haemostasis* **48:**153–155.

Davies, G. E., and Palek, J., 1982b, Platelet protein organization: Analysis by treatment with membrane-permeable cross-linking reagents, *Blood* **59:**502–513.

Debus, E., Weber, K., and Osborn, M., 1981, The cytoskeleton of blood platelets viewed by immunofluorescence microscopy, *Eur. J. Cell Biol.* **24:**45–52.

De Robertis, E., Paseyro, P., and Ressig, M., 1953, Electron microscopic studies of the actin of thrombin on blood platelets, *Blood* **8:**587–597.

Diggle, T. A., Toh, B. H., Firkin, B. G., and Pfueller, S. L., 1979, Human platelet actin surface expression after platelet activation, *Thrombosis Haemostasis* **42:**799–802.

Ebbe, S., 1976, Biology of megakaryocytes, in: *Progress in Hemostasis and Thrombosis 3* (T. H. Spaet, ed.), pp. 211–230, Grune & Stratton, New York.

Eberth, J. C., and Schimmelbusch, C., 1885, Die blutplätchen and die blutgerinnung, *Virchows Arch. Pathol. Anat.* **101:**201.

Feinstein, M. B., 1982, The role of calcium in hemostasis, in: *Progress in Hemostasis and Thrombosis* (T. H. Spaet, ed.), pp. 25–63, Grune & Stratton, New York.

Feinstein, M. B., and Walenga, R., 1981, The role of calcium in platelet adhesion, in: *Biochemistry of the Acute Allergic Reactions* (J. Golden, ed.), pp. 279–293, Alan R. Liss, New York.

Feinstein, M. B., Fiekers, J., and Fraser, C., 1975, An analysis of the mechanism of local anaesthetic inhibition of platelet aggregation and secretion, *J. Pharmacol. Exp. Ther.* **197:**215–228.

Fox, J. E. B., and Phillips, D. R., 1982, Role of phosphorylation in mediating the association of myosin with the cytoskeletal structures of human platelets, *J. Biol. Chem.* **257:**4120–4126.

Fox, J. E. B., Dockter, M. E., and Phillips, D. R., 1981, An improved method by determining the actin filament content of nonmuscle cells by the DNase I inhibition assay, *Anal. Biochem.* **117:**170–177.

Frojmovic, M. M., and Milton, J. G., 1982, Human platelet size, shape and related functions in health and disease, *Physiol. Rev.* **62:**185–261.

George, J. N., and Morgan, R. K., 1981, Glanzmann's thrombasthenia: Deficient association of actin with the platelet membrane following thrombin-induced secretion, *Thrombosis Res.* **22:**503–506.

George, N. N., Lyons, R. M., and Morgan, R. K., 1980, Membrane changes associated with platelet activation, *J. Clin. Invest.* **66:**1–9.

Gerrard, J. M., Schollmeyer, J. V., Phillips, D. R., and White, J. G., 1979, α-actinin deficiency in thrombasthenia, *Am. J. Pathol.* **94:**509–522.

Gogstad, G. O., Solum, N. O., and Hagen, I., 1981, Platelet glycoprotein III and α-actinin are different proteins, *Thrombosis Res.* **24:**157–162.

Gonnella, P. A., and Nachmias, J. T., 1981, Platelet activation and microfilament bundling, *J. Cell Biol.* **89:**146–151.

Harker, L., 1979, Platelet survival game: Its measurement and use, in: *Progress in Hemostasis and Thrombosis 4* (T. H. Spaet, ed.), pp. 321–348, Grune & Stratton, New York.

Hathaway, D. R., Eaton, C. R., and Adelstein, R. S., 1981, Regulation of human platelet myosin light chain kinase by the catalytic subunit of cyclic AMP-dependent protein kinase, *Nature* **291:**252–254.

Hayem, G., 1878, Recherches sur l'évolution des Hématies dans le sang de l'homme et des vertébrés, *Arch. Physiol. Norm. Pathol.* **2:**692–733.

Henry, R. L., 1977, Platelet function, *Seminars Thrombosis Hemostasis* **5:**93–122.

Hidaka, H., and Nishikawa, M., 1980, Platelet aggregation and protein phosphorylation, *Acta Haematol. Japonica* **43:**1124–1129.

Hutter, R. V. P., 1957, Electron microscopic observations on platelets from human blood, *Am. J. Clin. Pathol.* **28:**447–460.

Jain, N. C., 1975, A scanning electron microscopy study of platelets of certain animal species, *Thrombosis Diath. Haemorrhage* **33:**501–507.

Jean, G., and Racine, L., 1962, *Proceedings of the 5th International Congress on Electron Microscopy, Philadelphia Vol. 2,* pp. 1–8, Academic Press, New York.

Jennings, L. K., Fox, J. E. B., Edwards, H. H., and Phillips, D. R., 1981, Changes in the cytoskeletal structure of human platelets following thrombin activation, *J. Biol. Chem.* **256:**6927–6932.

Kaplan, K. L., Broekman, M. J., and Chernoff, A., 1979, Platelet α-granule protein studies on release and subcellular localization, *Blood* **53:**605–618.

Kaufman, R., Airo, R., Pollack, S., and Crosby, W., 1965, Circulating megakaryocytes and platelet release in the lung, *Blood* **26:**720–729.

Langer, B. G., Leung, L. L., Gonnella, P. A., Nachmias, V. T., Nachman, R. L., and Pepe, F. A., 1982, α-Actinin and membrane glycoprotein IIIa are different proteins in human blood platelets, *Proc. Natl. Acad. Sci. USA* **79:**432–435.

Lewis, J. C., and Bowie, E. J. W., 1978, Ultrastructural studies of platelets of von Willebrand and normal swine, *Mayo Clin. Proc.* **53:**179–183.

Lewis, J. C., Maldonado, J. E., and Mann, K. G., 1976a, Phagocytosis in human platelets: Localization of acid phosphatase positive phagosomes following latex uptake, *Blood* **47:**833–840.

Lewis, J. C., Maldonado, J. E., Mann, K. G., and Moertel, C. G., 1979b, Ultrastructural cytochemistry of platelets and megakaryocytes in the carcinoid syndrome, *Mayo Clin. Proc.* **51:**585–593.

Lewis, J. C., Cowley, L. H., Taylor, R. G., and Clarkson, T. B., 1980a, Ultrastructural analysis of platelets in nonhuman primates. I. Comparative morphometrics on six species, *Exp. Mol. Pathol.* **32:**175–187.

Lewis, J. C., Prater, T., Taylor, R. G., and White, M. S., 1980b, The use of correlative SEM and TEM to study thrombocyte and platelet adhesion to artificial surfaces, *Scan. Elect. Microsc.* **3:**189–202.

Lewis, J. C., White, M. S., Prater, T., Taylor, R. G., and Davis, K. S., 1982a, Ultrastructural analysis of platelets in nonhuman primates. III. Stereo microscopy of microtubules during platelet adhesion and the release reaction, *Exp. Mol. Pathol.* **37:**370–381.

Lewis, J. C., White, M. S., Prater, T., Hartle, F. A., Campbell, G., and Wray, G., 1982b, Cytoskeletal 3-dimensional organization during platelet adhesion and release, in: *40th Annual Proceedings of the Electron Microscopy Society of America* (G. E. Bailey, ed.), pp. 12–13, Claitor's, Baton Rouge.

Lewis, J. C., White, M. S., Prater, T., Porter, K. R., and Steele, R. J., 1983, Cytoskeletal changes during adhesion and release: Observations of human and nonhuman primate platelets, *Scan. Elect. Microsc.*, in press.

Lind, S. E., and Stossel, T. P., 1982, The microfilament network of the platelet, in: *Progress in Hemostasis and Thrombosis 6* (T. H. Spaet, ed.), pp. 63–78, Grune & Stratton, New York.
Lind, S. E., Yin, H. L., and Sotssel, T. P., 1982, Human platelets contain gelsolin: A regulator of actin filament length, *J. Clin. Invest.* **69:**1384–1387.
Lucas, R. C., Detwiler, T. C., and Stracher, A., 1976, The identification and isolation of a high molecular weight (270,000 dalton) actin-binding-protein from human platelets, *J. Cell Biol.* **70:**259a.
Lüscher, E. F., 1956a, Viscous metamorphosis of blood platelets and clot retraction, *Vox Sanquinis* **1:**133–156.
Lüscher, E. F., 1956b, Glukose als cofactor bei retraktion des blutgerinnsels, *Experientia* **12:**294.
Lüscher, E. F., 1980, Regulation of the contractile system of blood platelets, *Eur. J. Cancer* **16:**5–6.
MacIntyre, D. E., 1976, The platelet release reaction: Association with adhesion and aggregation and comparison with secretory responses in other cells, in: *Platelets in Biology and Pathology* (J. I. Gordon, ed.), pp. 61–85, Elsevier/North Holland, Amsterdam.
Markey, F., Persson, T., and Lindberg, U., 1981, Characterization of platelet extracts before and after stimulation with respect to the possible role of profilactin as microfilament precursor, *Cell* **23:**145–153.
Marsh, Q. B., Kautz, J., Motulsky, A. G., 1955, An electron microscope study of sectioned platelets and megakaryocytes, *J. Clin. Invest.* **34:**929–930.
Mattson, J. C., and Zuiches, C. A., 1981, Elucidation of the platelet cytoskeleton, *Ann. N.Y. Acad. Sci.* **370:**11–21.
Menche, D., Israel, A., and Karpatkid, S., 1980, Platelets and microtubules, *J. Clin. Invest.* **66:**284–291.
Mills, D. C. B., and MacFarlane, 1976, Platelet receptors, in: *Platelets in Biology and Pathology* (J. I. Gordon, ed.), Elsevier/North Holland, Amsterdam.
Nachmias, V. T., 1980, Cytoskeleton of human platelets at rest and after spreading, *J. Cell Biol.* **86:**795–802.
Nachmias, V. T., and Sullender, J. S., 1978, The cytoskeleton of human platelets at rest and after spreading: Whole mounts viewed at 200 KV correlated with negatively stained specimens examined at 50 KV, *Ninth International Congress on Electron Microscopy* **II:**458–459.
Nachmias, V., Sullender, J., and Asch, A., 1977, Shape and cytoplasmic filaments in control and lidocaine-treated platelets, *Blood* **50:**39–53.
Nachmias, V. T., Sullender, J., Fallon, J., and Asch, A., 1979a, Observation on the "cytoskeleton" of human platelets, *Thrombosis and Haemostasis* **42:**1661–1665.
Nachmias, V. T., Sullender, J. S., and Fallon, J. R., 1979b, Effects of local anesthetics on human platelets: Filopodial suppression and endogenous proteolysis, *Blood* **53:**63–72.
Odell, T. T., and Jackson, C. W., 1969, Megakaryocytopoiesis, in: *Symposium on Hemapoietic Cellular Proliferation* (F. Stohlman, ed.), pp. 278, Grune & Stratton, New York.
Painter, R. G., Ginsberg, M., and Jaques, B., 1982, Concanavalin A induces interactions between surface glycoproteins and the platelet cytoskeleton, *J. Cell Biol.* **92:**565–573.
Phillips, D. R., 1980, An evaluation of membrane glycoproteins in platelet adhesion and aggregation, in: *Progress in Hemostasis and Thrombosis* (T. H. Spaet, ed.), pp. 81–109, Grune & Stratton, New York.
Phillips, D. R., Jennings, L. K., and Edwards, H. H., 1980, Identification of membrane proteins mediating the interaction of human platelets, *J. Cell Biol.* **86:**77086.
Pollard, T. D., 1979, Platelet contractile proteins, *Thrombosis Haemostasis* **42:**1634–1637.
Porter, K. R., and Tucker, J. B., 1981, The ground substance of the lining cell, *Sci. Am.* **244:**57–67.
Pribluda, V., Laub, F., and Rotman, A., 1981, The state of actin in activated platelets, *Eur. J. Biochem.* **116:**293–296.
Prowse, C., Pepper, D., and Dawes, J., 1982, Prevention of the platelet actin-granule release reaction by membrane-active drugs, *Thrombosis Res.* **25:**218–227.
Puszkin, E. G., Maldonado, R., Spaet, T. H., and Zucker, M. B., 1977, Platelet myosin. Localization of the rod myosin fragment and effect of its antibodies and platelet function, *J. Cell Biol.* **252:**4371–4378.

Rebuck, J. W., Riddle, J. M., Johnson, S. A., Monto, R. W., and Spurrock, R. M., 1960, Contributions of electron microscopy to the study of platelets, *Henry Ford Hosp. Med. Bull.* **8:**273–292.

Rosenberg, S., Stracher, A., and Lucas, R. C., 1981a, Isolation and characterization of actin and actin-binding protein from human platelets, *J. Cell Biol.* **91:**201–211.

Rosenberg, S., Stracher, A., and Burridae, K., 1981b, Isolation and characterization of a calcium-sensitive α-actinin-like protein from human platelet cytoskeletons, *J. Biol. Chem.* **256:**-12986–12991.

Rosenstein, R., Zacharski, L. R., and Allen, R. D., 1981, Quantitation of human plaetlet transformation on siliconized glass: A comparison of "normal" and "abnormal" platelets, *Thrombosis Haemostasis* **46:**521–524.

Rotman, A., and Heldman, J., 1981, Intracellular viscosity changes during activation of blood platelets: Studies by fluorescence polarization, *Biochemistry* **20:**5995–5999.

Shultz, H., Jurgens, R., and Hiepler, E., 1958, Die ultrastruktur dur thrombozyten bei der konstitutionellen thrombopathie (v. Willebrand-Jurgens) mit einem beitrag zur submikroskopischen orthologie der thrombozyten, *Thrombosis Diath. Haemorrhage* **2:**319–323.

Sixma, J. J., and Molenaar, I., 1966, Microtubules and microfilaments in human platelets, *Thrombosis et Diathesis Haemorrhagica* **16:**153–162.

Small, J. V., and Langanger, G., 1981, Organization of actin in the leading edge of cultured cells: Influence of oxmium tetroxide and dehydration on the ultrastructure of actin meshworks, *J. Cell Biol.* **91:**695–705.

Tavassoli, M., 1980, Megakaryocyte-platelet axis and the process of platelet formation and release, *Blood* **55:**537–545.

Tuszynski, G. P., Kornecki, E., Niewiarowski, S., Knight, L., and Srivastava, S., 1982a, Platelet cytoskeletons contain receptors for fibrinogen, *J. Cell Biol.* **95:**3a.

Tuszynski, G. P., Walsh, P. N., Schick, P., and Koshy, A., 1982b, Platelet cytoskeletons possess platelet factor 3 activity, *Circulation* **66** (Suppl. II)**:**176.

Tuszynski, G. P., Walsh, P. N., Piperno, J. R., and Koshy, A., 1982c, Association of coagulation factor V with the platelet cytoskeleton, *J. Biol. Chem.* **247:**4557–4563.

Walsh, R. T., Bauer, R. B., and Barnhart, M. I., 1975, Platelet function in transient ischemia and cerebrovascular disease, in: *Platelets, Recent Advances in Basic Research and Clinical Aspects*, pp. 367–377, Excerpta Medica, Amsterdam.

Wang, L. L., and Bryan, J., 1981, Isolation of calcium-dependent platelet proteins that interact with actin, *Cell* **25:**637–649.

White, J. G., 1967, The submembrane filaments of blood platelets, *Am. J. Pathol.* **56:**267–277.

White, J. G., 1969, The dense bodies of human platelets: Inherent electron opacity of serotonin storage particles, *Blood* **33:**598–606.

White, J. G., 1979, Current concepts of platelet structure, *Am. J. Clin. Pathol.* **71:**363–378.

White, J. G., 1983, Ultrastructural physiology of platelets with randomly dispersed rather than circumferential band microtubules, *Am. J. Pathol.* **110:**55–63.

White, J. G., and Gerrard, J. M., 1976, Ultrastructural features of abnormal blood platelets, *Am. J. Pathol.* **83:**590–614.

White, J. G., and Gerrard, J. M., 1979, Interaction of microtubules and micorfilaments in platelet contractile physiology, *Meth. Achiev. Exp. Pathol.* **9:**1–39.

Wolosowick, J. J., and Porter, K. R., 1979, Microtrabecular lattice of the cytoplasmic ground substance. Artifact or reality, *J. Cell Biol.* **82:**114–139.

Wolpero, C., and Ruska, H., 1939, Strukturuntersuchungen zur blutgerinnung, *Klinische Wochenschrift* **19:**1078–1083.

Wright, J. H., and Minot, G. R., 1917, The viscous metamorphosis of the blood platelets, *J. Exp. Med.* **26:**395–409.

Zucker-Franklin, D., 1969, Microfibrils of blood platelets: Their relationship to microtubules and the contractile protein, *J. Clin. Invest.* **48:**165–171.

Zucker-Franklin, D., 1970, The submembranous fibrils of human blood platelets, *J. Cell Biol.* **47:**295.

Zucker-Franklin, D., and Grusky, G., 1972, The actin and myosin filaments of human and bovine blood platelets, *J. Clin. Invest.* **51:**419–429.

11

Monoclonal Antibodies to Intermediate Filament Proteins

USE IN DIAGNOSTIC SURGICAL PATHOLOGY

Arthur M. Vogel and Allen M. Gown

1. Introduction

Surgical pathologists base diagnoses upon characteristic features of neoplasms such as gland formation, papillary structures, and the spindle shape of tumor cells. Tumors without obvious differentiated features pose diagnostic problems because they lack the features that allow an accurate identification of tumor type. In such situations, one depends upon ultrastructural analysis by electron microscopy or relatively nonspecific histochemical stains to gain a clue as to the identity of the neoplasm. Clearly, what has been lacking in diagnostic pathology is a set of well-characterized tissue-specific reagents capable of distinguishing among different cells. Recently, anti-intermediate filament protein antibodies have been employed to distinguish cell type in poorly differentiated neoplasms.

Intermediate filaments are unique components of the cytoskeleton, differing from the other cytoskeleton arrays, microfilaments, and microtubules, by filament diameter and protein composition. The actin-containing microfilaments and tubulin-containing microtubules are composed of similar proteins in different cell types, but the intermediate filaments of different cells are composed of a family of immunochemically distinct proteins. Thus, the different intermediate filament proteins can serve as tissue-specific markers. Five such classes of intermediate filament proteins have been described

---

Arthur M. Vogel and *Allen M. Gown* • Department of Pathology SM-30, University of Washington, Seattle, Washington 98195.

(Table 1; Lazarides, 1980, 1982). Vimentin (decamin) is the intermediate filament protein found in mesenchymal tissue such as endothelium, lymphocytes, fibroblasts, and some vascular smooth muscle cells (Starger and Goldman, 1977; Hynes and Destree, 1978; Franke *et al.*, 1979). Desmin, a lower molecular weight protein than vimentin, is found only in muscle cells including skeletal, cardiac, and smooth muscle (Gard *et al.*, 1979; Tuszynski *et al.*, 1979). Vimentin and desmin exist together in some cells, and these two proteins can form copolymers *in vitro* (Gard *et al.*, 1979; Steinert *et al.*, 1981). Sequence analysis clearly demonstrates that they are nonidentical proteins (Geisler and Weber, 1981). The keratins (prekeratin, cytokeratin) are a group of 15–20 proteins ranging in size from 40–70-kd, found mostly in epithelial cells (Fuchs and Green, 1978; Sun *et al.*, 1979; Franke *et al.*, 1981). Squamous epithelium (epidermis, esophagus, vagina) contains the greatest number of different keratin molecules, and squamous epithelium and nonsquamous epithelium contain different keratin molecules (Gown and Vogel, 1982; Moll *et al.*, 1982; Tseng *et al.*, 1982; Wu *et al.*, 1982). Glial fibrillary acidic protein (GFAP) is localized to astroglia and ependymal cells, and has a lower molecular weight than both desmin and vimentin (Dahl and Bignami, 1973; Eng and Kosek, 1974). Vimentin and GFAP often coexist in glial cells (Dahl *et al.*, 1981; Yen and Fields, 1981). Neurofilament proteins are found only in neurons. Three proteins of molecular weight 68-kd, 160-kd, and 200-kd interact to form neurofilaments (Liem *et al.*, 1978).

Because of these differences, an antibody capable of recognizing only one of these molecules could function as a tissue-specific marker. Antibodies to vimentin stain only mesenchymal cells, while antidesmin antibodies are muscle-specific (Lazarides and Hubbard, 1976; Franke *et al.*, 1981a). Similarly, GFAP antibodies decorate only glial and ependymal cells, antineurofilament antibodies are neuron-specific, and antibodies made to keratin are specific for epithelial cells (Franke *et al.*, 1981b; Sun *et al., 1979; Schlegel et al.*, 1980a; Dahl *et al.*, 1981; Shaw *et al.*, 1981; Yen and Fields, 1981). In general, tumors

Table 1. Classes of Intermediate Filament Proteins

Type	Molecular weight (kd)	Location
Vimentin (decamin)	55,000–58,000	Mesenchymal cells Fibroblasts, endothelium Muscle cells
Desmin (skeletin)	55,000	Muscle cells: smooth, skeletal, cardiac
Keratin (cytokeratin, prekeratin)	Multiple proteins 40,000–70,000	Epithelium Mesothelial cells
Glial fibrillary acidic protein (GFAP)	52,000	Glial cells: astrocytes, ependymal cells
Neurofilament proteins	200,000 160,000 68,000	Neurons of CNS and PNS

possess the same intermediate filament proteins present in normal tissue, so that tumors interact with these antibodies in a fashion comparable to the tissues of origin. Antikeratin antibodies mainly stain carcinomas, while tumors of mesenchymal origin (sarcomas and lymphomas) react with antivimentin antibodies (Bannasch *et al.*, 1980; Schlegel *et al.*, 1980b; Altmannsberger *et al.*, 1981; Gabbiani *et al.*, 1981; Ramaekers *et al.*, 1981). Rhabdomyosarcomas and leiomyosarcomas, tumors originating in skeletal and smooth muscle, respectively, are decorated by antidesmin antibodies, while glial tumors react with anti-GFAP (Bignami *et al.*, 1980; Lehtonen *et al.*, 1982; Denk *et al.*, 1983). Therefore, these anti-intermediate filament antibodies are useful tissue-specific reagents.

We have isolated hybridoma antibodies to various intermediate filament proteins to obtain antibodies with greater specificity (Gown and Vogel, 1982). In this review, we describe the reactivity of these antibodies on normal and neoplastic tissue, and show that a battery of these anti-intermediate filament protein antibodies can be used as tissue-specific reagents in the practice of surgical pathology.

2. *Description of Monoclonal Antibodies*

Hybridoma antibodies were generated against a variety of intermediate filament proteins isolated from different sources. Cytoskeleton was isolated from human fibroblasts and a human hepatocellular carcinoma cell line. Stratum corneum was solubilized in SDS and β-mercaptoethanol, and purified 200-kd rat neurofilament protein was kindly provided by Dr. F. T. Chiu, Department of Neurology, Albert Einstein College of Medicine. The different preparations are displayed on an 8% polyacrylamide gel in Fig. 1. Human fibroblasts provide a source of vimentin (58-kd), the mesenchymal-specific protein (lane a). The hepatoma cells contain a 54-kd protein, not found in any of the other intermediate filament preparations except A431 cells (lanes b and c). A small amount of 58-kd material is present in the Hep3B cells, but two-dimensional peptide maps indicate that this molecule and vimentin from fibroblasts are different (Gown and Vogel, 1982). Stratum corneum yielded proteins in the molecular weight range of 40–70-kd, with prominent bands at 66-kd (doublet), 57-kd, 51-kd, and 49-kd (land d). The 200-kd neurofilament preparation contains only this protein (lane f).

Monoclonal antibodies were generated against each of these four preparations (Fig. 1, lanes a, b, d, and f) and assayed initially on tissue-culture cells by immunofluorescence. Table 2 lists the different antibodies and the immunogens against which they were made. Two antivimentin antibodies were isolated, 17 βG3 and 43 βE8, both of which manifest similar reactivities. Antibody 35βH11 arose from the fusion with Hep3B cytoskeleton, and specifically recognizes the 54-kd protein present in these cells. Antibodies 34βE12 and 34βB4 resulted from fusions using solubilized stratum corneum as immunogen. These two antibodies differ with respect to their staining of

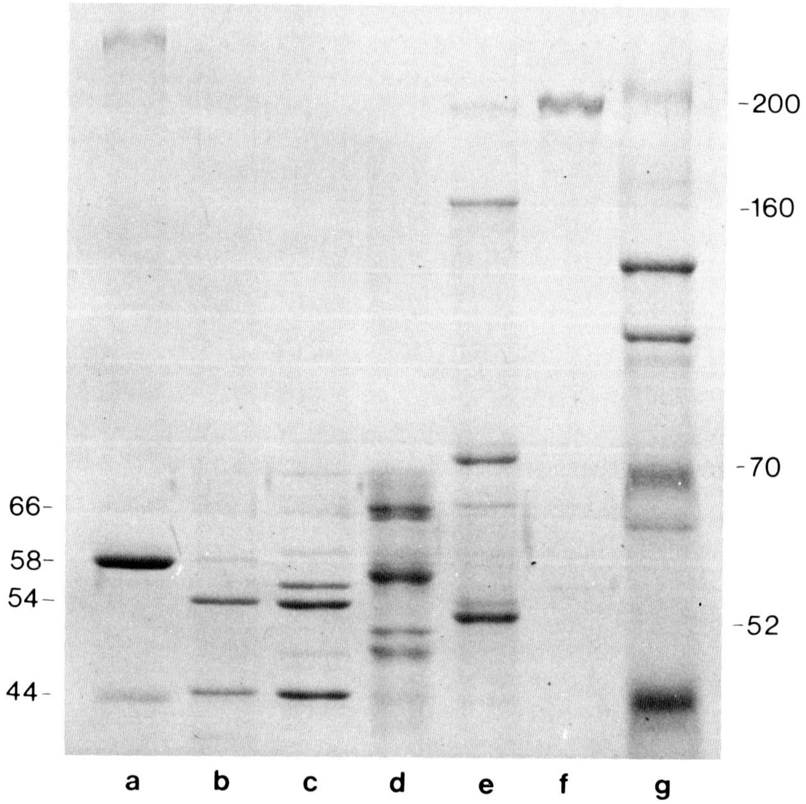

Figure 1. SDS polyacrylamide gel (8%) of cytoskeletal preparations from different sources. Cell lines were extracted in Triton high-salt, and the insoluble material suspended in SDS sample buffer (Gown and Vogel, 1982). Keratin from stratum corneum was solubilized in 5% SDS, 1% β-mercaptoethanol, as described by Sun and Green (1978). (a) Fibroblasts, (b) Hep3B (hepatocellular carcinoma), (c) A431 (epidermoid carcinoma), (d) keratin from stratum corneum, (e) rat neurofilament preparation, (f) rat 200-kd neurofilament protein, (g) molecular weight markers of 200-kd, 116-kd, 94-kd, 68-kd; and 43-kd. Approximately 10–20 μg of protein was added per well. Reproduced from *Journal of Cell Biology*, 1982, Vol. 95, p. 414, with copyright permission from Rockefeller University Press.

Table 2. Hybridoma Antibodies to Intermediate Filament Proteins

Antibody name	Immunogenic preparation
17βG3	Cytoskeleton from fibroblasts
43βE8	
35βH11	Cytoskeleton from hepatocellular carcinoma cells
34βE12	Keratin from stratum corneum
34βB4	
31γA11	Rat 200-kd neurofilament protein
31αF3	

Table 3. Reaction of Hybridoma Antibodies on Tissue-Culture Cells[a]

		Reactivity		
Antibody	Antibody specificity	Fibroblasts	Hep3B	A431
17βG3	Vimentin	+	−	−
43βE8		+	−	−
35βH11	54-kd protein of Hep3B cells	−	+	+
34βE12	Keratin from stratum corneum	−	−	+
34βB4		−	−	+
31γA11	Rat 200-kd neurofilament protein	−	−	−
31αF3				

[a]Acetone-fixed cells were assayed by immunofluorescence using ascites fluid containing the different antibodies.

tissue (see below). Fusions involving the rat 200-kd neurofilament protein yielded two antibodies, 31γA11 and 31αF3.

The antivimentin antibodies 17βG3 and 43βE8 stain the intermediate filaments of fibroblasts, but not hepatoma cells or A431 cells (epidermoid carcinoma cell line; Table 3). By western-blot analysis, these antibodies bind to the 58-kd protein of human fibroblasts, but don't consistently recognize proteins in Hep3B or A431 cells (Fig. 2A). These antibodies also crossreact with GFAP (Gown and Vogel, 1982; Table 4).

Antibody 35βH11, made to hepatoma cytoskeleton, reacts with hepatoma cells and other epithelial cells in culture, but fails to decorate the intermediate filaments of fibroblasts (Table 3). Immunoblot experiments show that 35βH11 recognizes the 54-kd protein present in Hep3B and other epithelial cells (Fig. 2B). This antibody does not recognize any other intermediate filament protein, including those isolated from stratum corneum (Table 4). To

Table 4. Reactivity of Different Monoclonal Antibodies with Different Intermediate Filament Proteins by Immunoblot Method

			Keratin from stratum corneum					Neurofilament proteins		
Antibody	Vimentin (58-kd)	Hepatoma cells 54-kd protein	66	57	51	49	GFAP	200	160	68
17βG3	+	−	+	+	−	−	+	−	−	−
43βE8	+	−	N.D.[a]	N.D.	N.D.	N.D.	+	−	−	−
35βH11	−	+	−	−	−	−	−	−	−	−
34βE12	−	−	+	+	+	+	−	−	−	−
34βB4	−	−	+	+	+	+	−	−	−	−
31γA11	+	−	−	−	−	−	−	+	+	+
31αF3	−	−	−	−	−	−	−	+	+	−

[a]N.D. = not done.

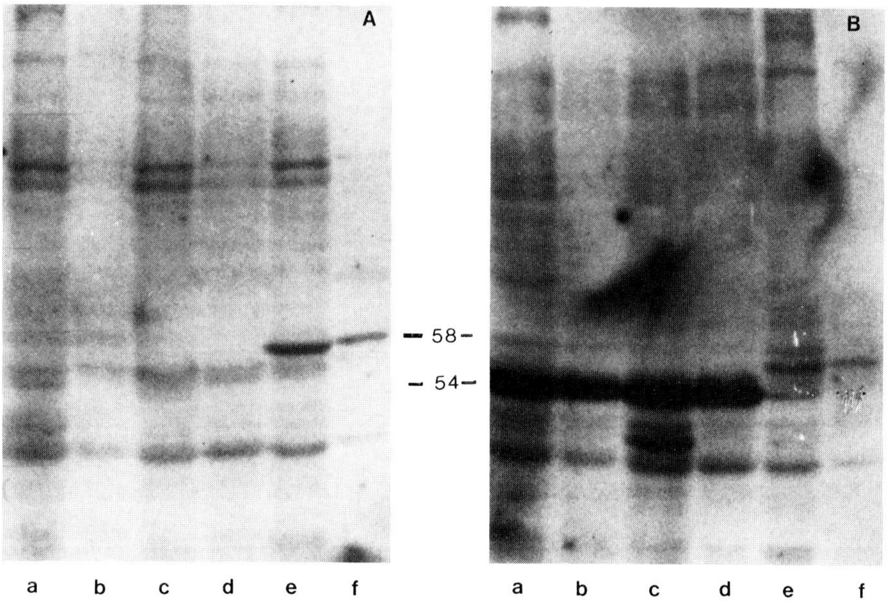

Figure 2. Western-blot identification of antigens by hybridoma antibodies. Confluent cultures of cells were solubilized in SDS sample buffer, electrophoresed on an 8% SDS polyacrylamide gel, electrophoretically transferred onto nitrocellulose paper, and incubated with the different antibodies (Gown and Vogel, 1982). (A) 17βG3, (B) 35βH11, (C) 34βE12, (D) 34βB4, (E) 31γA11, (F) 31αF3. (A and B) Lane a, A431 (epidermoid carcinoma); lane b, Hs-0700T (adenocarcinoma); lane c, HepG2 (hepatocellular carcinoma); lane d, Hep3B (hepatocellular carcinoma); lane e Hs-0578T (carcinosarcoma of breast); lane f, human fibroblasts. (C and D) Lane a, A431 (epidermoid carcinoma); lane b, keratin from stratum corneum. (E and F) Lane a, human fibroblasts; lane b, keratin from stratum corneum; lane c, rat neurofilament preparation; lane d, rat 200-kd neurofilament protein.

date, in analyzing approximately 10–12 human epithelial tumor cell lines, we find an exact correlation of reactivity with 35βH11 and the presence of the 54-kd protein.

Antibodies 34βE12 and 34βB4 arose from fusions using solubilized stratum corneum. Both antibodies stain the filaments of only A431 cells, an epidermoid carcinoma cell line (Table 3), with 34βE12 consistently resulting in brighter fluorescence than 34βB4. In western-blot experiments, both antibodies recognize proteins of 66-kd and 57-kd in stratum corneum (Figs. 2C and D), and cross react with no other intermediate filament protein, including the 54-kd protein in hepatoma cells (Table 4).

Antineurofilament antibodies 31γA11 and 31αF3 both fail to react with any of a number of selected *in vitro* cell lines (Table 3). In immunoblot experiments, 31γA11 recognizes the three neurofilament proteins, 200-kd, 160-kd, 68-kd, while 31αF3 binds to only the 200-kd and 160-kd proteins (Figs. 2E and F). 31γA11 can recognize vimentin in the western blots, but does not stain fibroblasts in immunofluorescence assays. We cannot explain this apparent

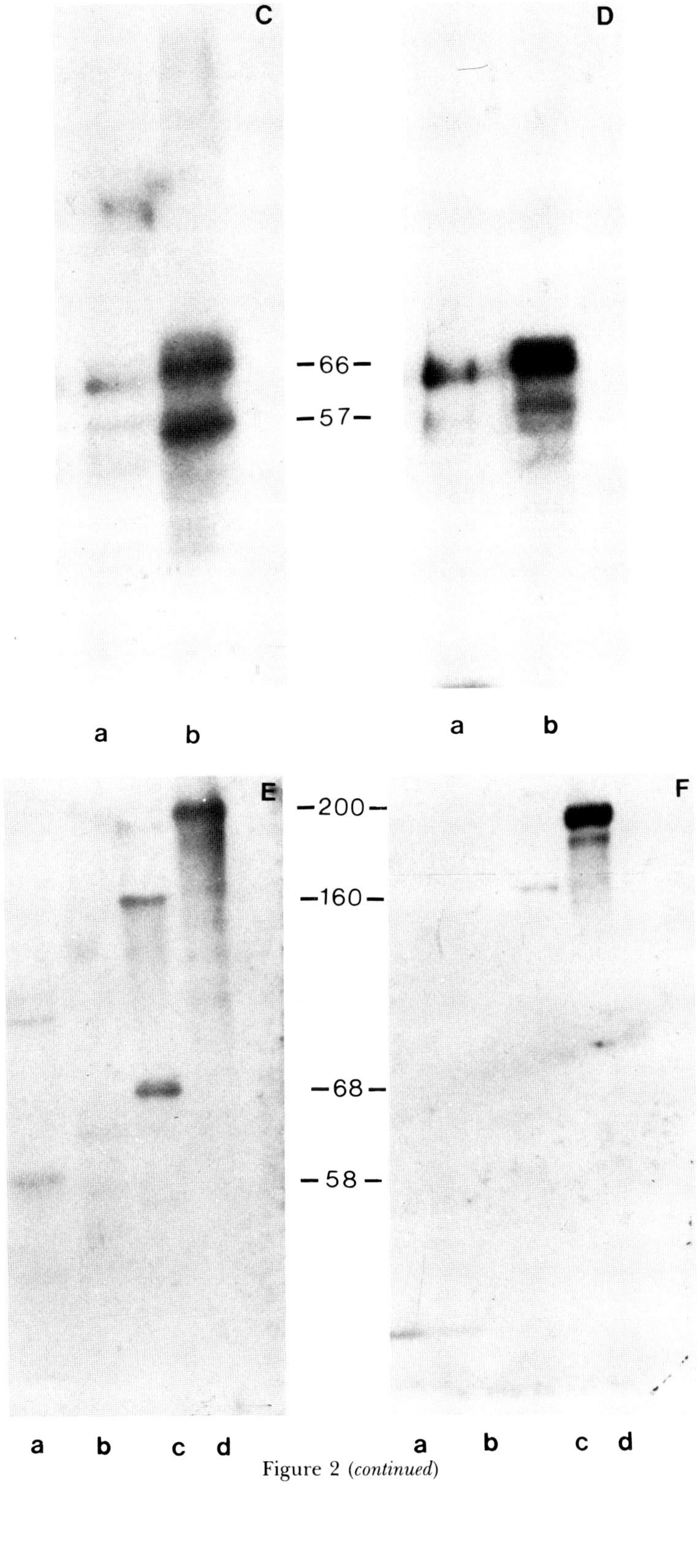

Figure 2 (*continued*)

discrepancy. Therefore, based upon reactivity on tissue culture cells and antigen identification by western-blot experiments, we have generated a series of specific antibodies capable of distinguishing different cell types.

3. Reactivity on Fixed Embedded Tissue

We have assayed these antibodies on unfixed frozen sections and paraffin sections of fixed tissue and found both to be satisfactory. We report here the results on fixed, paraffin-embedded tissue because one can counterstain the sections to identify the structures containing the positive material. No differences in reactivity have been noted when the antibodies are assayed on frozen sections or fixed paraffin-embedded sections.

Formalin fixation reduces the reactivity of some of our antibodies, so we routinely fix tissue in methanol Carnoy's fixative (Methacarn: 60% methanol, 30% chloroform, 10% acetic acid). Reactivity is demonstrated using biotinylated antimouse antibodies, avidin, and biotinylated peroxidase (ABC method; Hsu *et al.*, 1981). This procedure yields very low background staining and easily identified reaction products.

3.1. Antivimentin Antibody 43βE8

Antibody 43βE8 reacts more clearly and reproducibly on fixed tissue than 17βG3; consequently, only this hybridoma will be discussed. It decorates blood vessels, histiocytes, and stromal fibroblasts in all organs, and melanocytes within the epidermis (Figs. 3C and 4C). The endothelium of blood vessels is heavily stained, while the smooth muscle cells of vessels show variable staining with medial cells of small vessels exhibiting greater staining than those of large vessels (Fig. 12C). Cells within glomeruli are stained by 43βE8, and we believe these are the capillary endothelial cells and epithelial cells (Fig. 6C). Curiously, lymphocytes do not react with 43βE8, even though lymphoid cells contain vimentin (Gabbiani *et al.*, 1981). We do not understand the reason for this, but one possible explanation is that lymphocyte vimentin may differ slightly from vimentin in other tissues. As expected, this antibody is nonreactive on neurons, skeletal muscle, smooth muscle of the gastrointestinal tract, cardiac muscle, and squamous and nonsquamous epithelium.

3.2. 35βH11, Anti-54-kd Antibody

This antibody stains all nonsquamous epithelium, but fails to react with squamous epithelium (Fig. 3A). It decorates the pneumocytes that line the pulmonary alveoli, and also stains bronchial epithelium. In the liver, both hepatocytes and bile ducts are positive (Fig. 4A). Similarly, pancreatic acinar cells, ducts, and islet cells react, the latter cells being somewhat less reactive than ducts and acinar cells (Fig. 5A). The mucosa of the gastrointestinal tract,

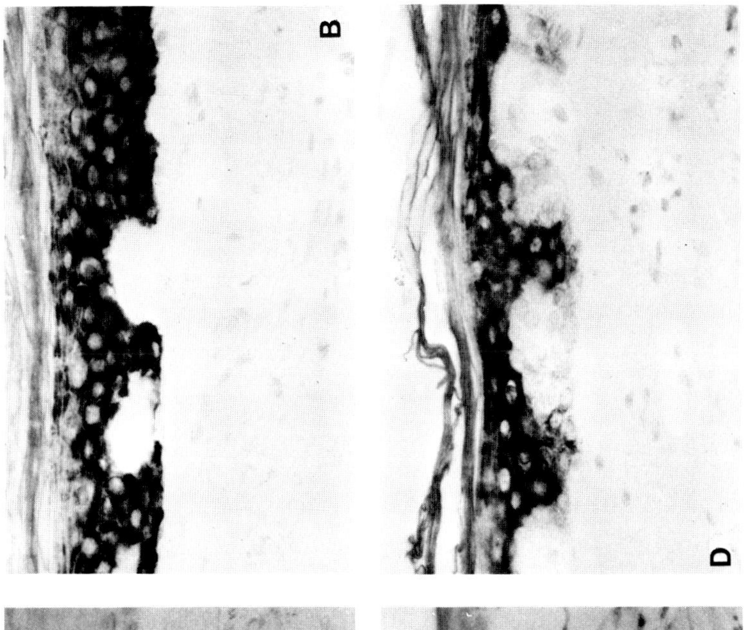

Figure 3. Skin reacted with anti-intermediate filament antibodies. (A) 35βH11, (B) 34βE12, (C) 43βE8, (D) 34βB4. Sections were counterstained with methyl green. 34βE12 stains the full thickness of the epidermis. 34βB4 spares the basal layer of the epidermis. 43βE8 stains only the melanocytes at the dermal–epidermal junction and dermal blood vessels.

from stomach to large bowel, is markedly positive. All of the renal tubules, proximal, distal, and collecting ducts, are recognized, as are the cells lining Bowman's capsule (Fig. 6A). Cells within the glomerular tuft are negative. Transitional epithelium of the urinary tract is uniformly positive, and the epithelial components of prostate and breast react with 35βH11 (Fig. 7A). Finally, endocrine organs such as thyroid, parathyroid, and pituitary are also stained. The antibody fails to stain the squamous epithelium of skin, esopha-

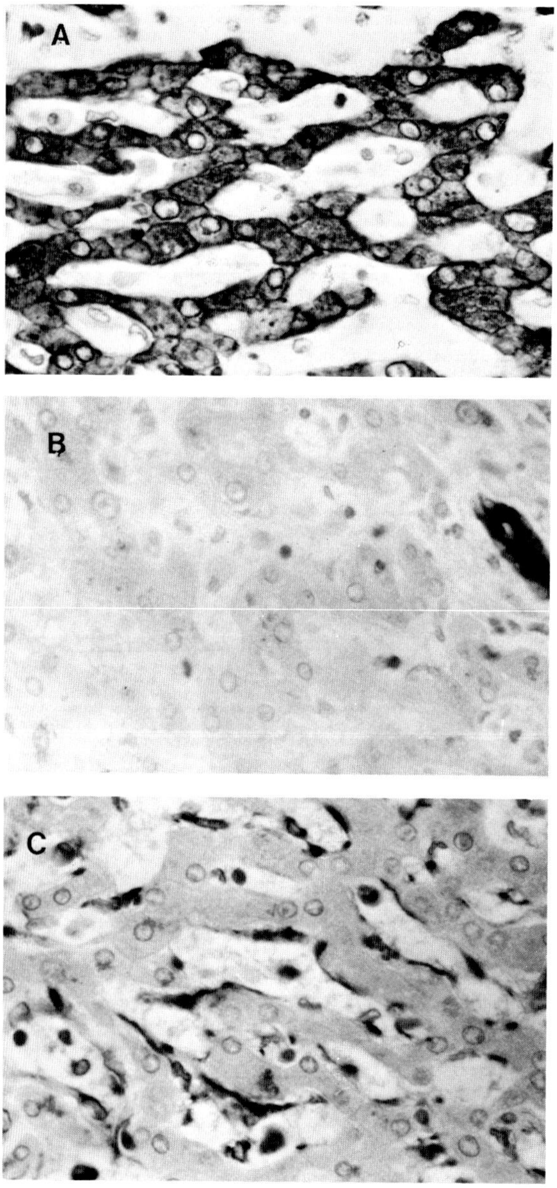

Figure 4. Liver. (A) 35βH11, (B) 34βE12, (C) 43βE8. 35βH11 stains hepatocytes and spares Kupffer's cells. 34βE12 only decorates a bile duct. 43βE8 identifies Kupffer's cells and blood vessels.

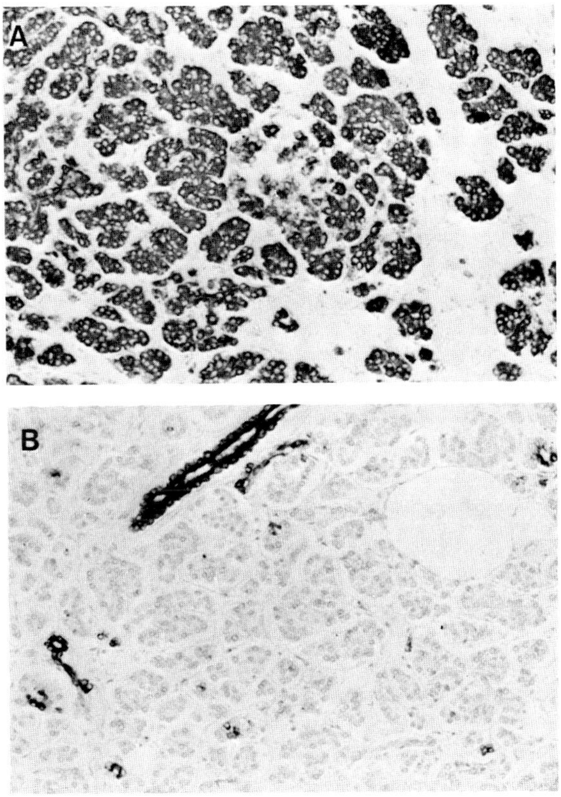

Figure 5. Pancreas. (A) 35βH11, (B) 34βE12. 35βH11 stains acinar cells, an islet, and ducts. 34βE12 is positive only on the pancreatic ducts.

gus, and vagina, and is nonreactive with nerve, all types of muscle, blood vessels, lymphocytes, and stromal fibroblasts.

Western blots performed on tissue solubilized in SDS and β-mercaptoethanol demonstrate that 35βH11 recognizes a 54-kd molecule in epithelial tissue, and that this molecule is absent from nonepithelial tissues (Gown and Vogel, manuscript submitted). The unique feature of 35βH11 is its ability to distinguish squamous from nonsquamous epithelium, and this phenomenon is probably explained by the presence of the 54-kd protein in nonsquamous epithelium and its absence from squamous epithelium. Thus, it appears that the 54-kd molecule is a cytokeratin unique to nonsquamous epithelium.

3.3. Anti-Stratum Corneum Antibody 34βE12

Antibody 34βE12 stains the full thickness of the squamous epithelium of the skin, esophagus, and vagina (Fig. 3B). It also stains nonsquamous epithelium such as sweat glands, bile ducts, pancreatic ducts, and mammary ducts and lobules (Figs. 4B and 5B). The mucosal lining of the gastrointestinal

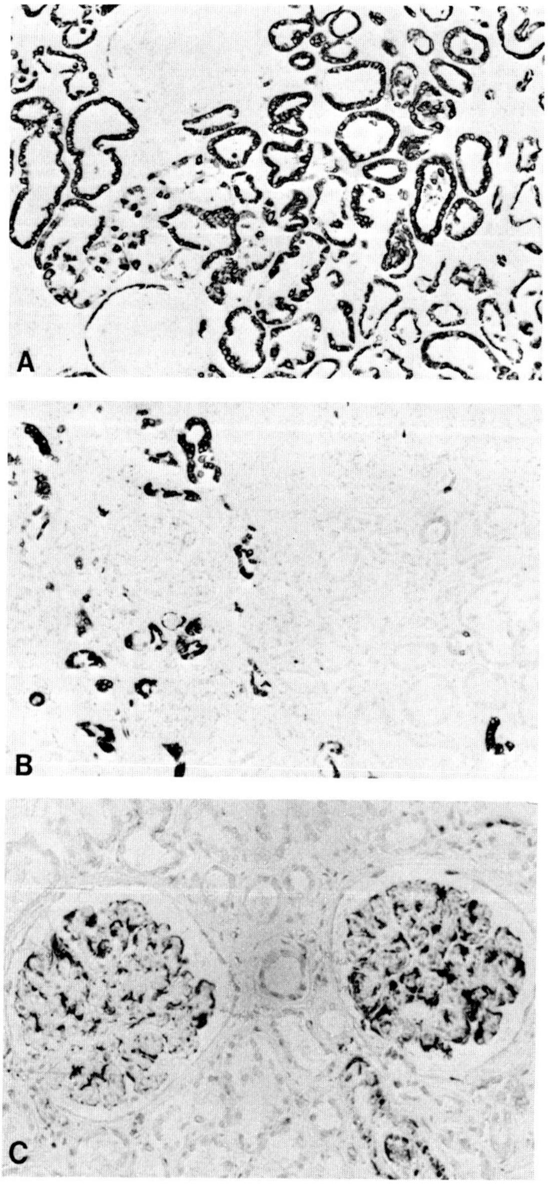

Figure 6. Kidney. (A) 35βH11, (B) 34βE12, (C) 43βE8. 35βH11 stains all the tubules and Bowman's capsule; glomeruli are negative. 34βE12 decorates only Bowman's capsule and collecting ducts. 43βE8 stains only the glomeruli and vessels.

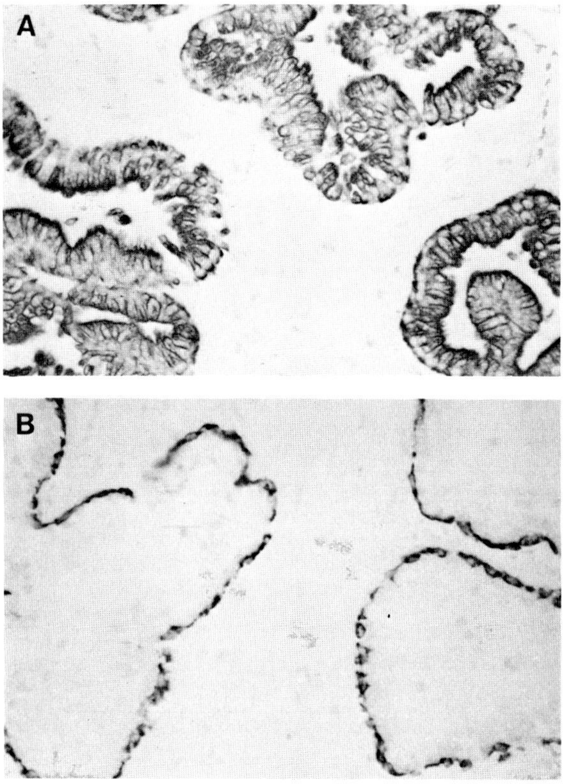

Figure 7. Prostate. (A) 35βH11, (B) 34βE12. 35βH11 stains all the epithelial cells, while 34βE12 stains only the basal layer of epithelial cells.

tract is positive in the basal portions of the glands, with the more superficial cells of the gland staining less strongly. A similar situation is observed in prostate where the epithelial cells at the base of the glands stain more strongly than the rest of the glandular epithelium (Fig.7B). Endocrine cells of the thyroid, parathyroid, and pituitary are largely unreactive with 34βE12. In the kidney, 34βE12 decorates only the distal collecting tubules and the lining of Bowman's capsule (Fig. 6B). The antibody does not react with hepatocytes, pancreatic acinar cells, proximal tubules of the kidney, and endometrial glands, all of which react with 35βH11 (Figs. 4B, 5B, and 6B). Additionally, no reactivity is observed on nerve, muscle, blood vessels, or lymphoid cells. The selective reactivity of 34βE12 distinguishes if from 35βH11 and the other anti-stratum corneum antibody, 34βB4.

3.4. Anti-Stratum Corneum Antibody 34βB4

This antibody stains only the suprabasal layers of squamous epithelium, failing to react with the basal layer (Fig. 3D). It reacts with no other tissues, including epithelial and nonepithelial tissues. Antibodies 34βB4 and 34βE12 both identify 66-kd and 57-kd proteins in stratum corneum, yet exhibit different staining capacities on epithelial tissue. We do not understand the reason for this observation.

3.5. *Antineurofilament Antibodies*

Both antibodies stain only the axons of neurons of the central and peripheral nervous system (Figs. 8A and B). The neuronal cell bodies, glial cells, ependymal cells, meninges, and blood vessels are negative, as are cells derived from the neural crest such as pancreatic islets, melanocytes, and adrenal medullary cells. All other tissues, both mesenchymal and epithelial, are also nonreactive.

4. *Reactivity of Neoplasms*

For these antibodies to be useful as diagnostic reagents, tumors must mimic the reactivity of the normal tissues in which they originate. To date, this has been true.

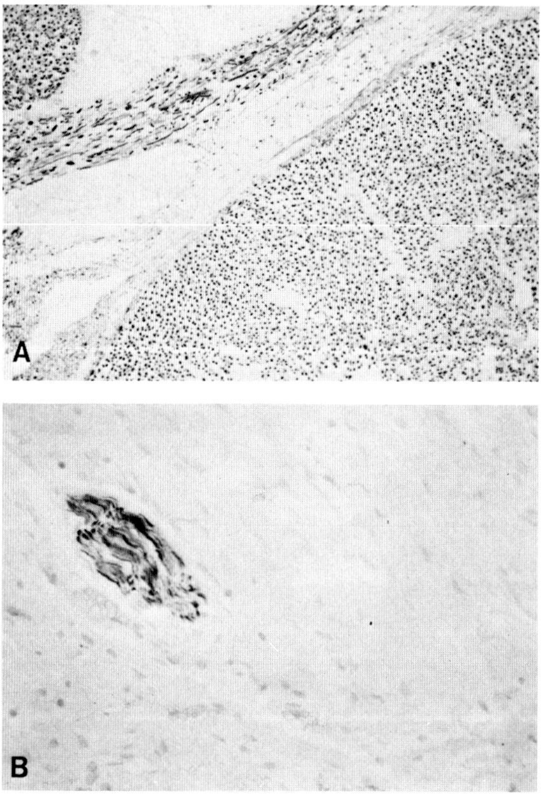

Figure 8. Neural tissue. (A) Spinal cord, 31αA11; (B) peripheral nerve, 31αF3.

4.1. 43βE8, Antivimentin

This antibody is a marker for mesenchymal tissue, therefore, mesenchymal tumors should stain while epithelial tumors should not. Sarcomas, such as fibrosarcomas and malignant fibrous histiocytomas are decorated by 43βE8 (Fig. 9A). Vascular tumors, such as hemangiomas, are quite positive, a finding consistent with the extensive staining of endothelial cells by 43βE8. Meningiomas and giant-cell tumors of bone also react (Fig. 10A). We also have preliminary evidence that malignant melanoma cells react with 43βE8. As predicted by the lack of staining of lymphoid cells, lymphomas do not react with 43βE8. Carcinomas (epithelial tumors) rarely stain with 43βE8 (Figs. 12C and 13C), and we do not know if this antibody will recognize tumors of muscle such as rhabdomyosarcomas or leiomyosarcomas. We conclude that 43βE8 can be used as a marker for at least some mesenchymal tumors.

4.2. Anticytokeratin Antibodies 35βH11, 35βE12, and 34βB4

All carcinomas tested react with at least one of these antibodies, while lymphomas, sarcomas, gliomas, and a single melanoma do not. Carcinomas

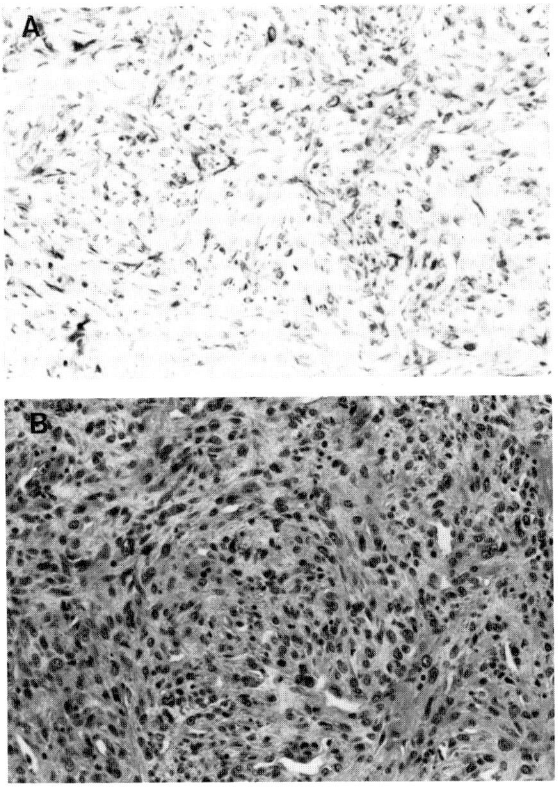

Figure 9. Lung sarcoma. (A) 43βE8, (B) hematoxylin-eosin stained section. The tumor is an undifferentiated sarcoma reacting only with 43βE8.

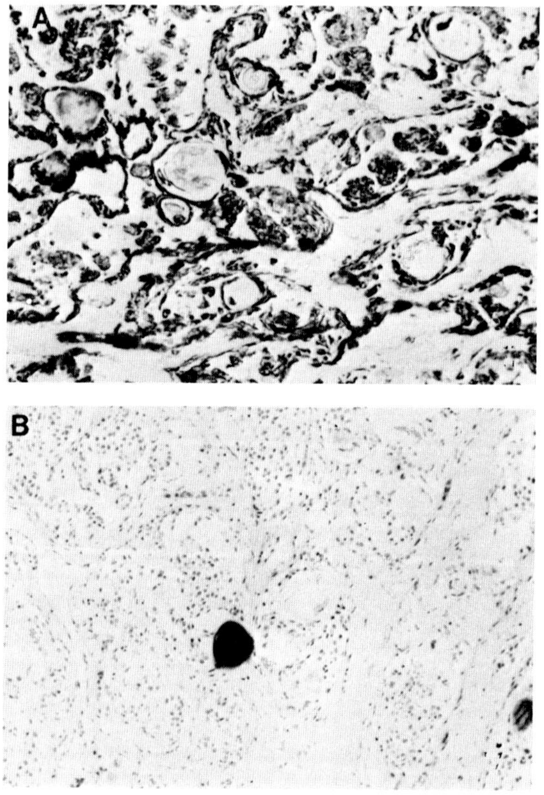

Figure 10. Meningioma. (A) 43βE8, (B) hematoxylin-eosin stained section.

behave like their parental normal tissue, so that one can distinguish different carcinomas based upon reactivity with these three antibodies.

Squamous carcinomas react with 34βE11 and 34βB4, but fail to react with 34βH11 (Figs. 11A–D). 34βE12 is uniformly positive on squamous cell carcinomas, irrespective of the degree of differentiation of the tumor, while 34βB4 stains better differentiated (more keratinized) tumors, appearing to recognize areas of keratinization within these more differentiated neoplasms (Fig. 11C). Squamous tumors are the only carcinomas not recognized by 35βH11 (Table 5). All other epithelial tumors tested react with this antibody.

The ability of 34βE12 to decorate nonsquamous carcinomas is determined by the tissue of origin of the tumor. Neoplasms arising in ductular structures (breast carcinomas, pancreatic carcinomas, cholangiocarcinomas) are uniformly positive (Fig. 12B). Gastrointestinal tumors and prostatic tumors are decorated, but not all of the cells are stained, and these tumors react more strongly with 35βH11 than 34βE12. Hepatocellular carcinomas and endometrial carcinomas do not react with 34βE12, but stain heavily with 35βH11 (Figs. 13 and 14). One can therefore distinguish hepatomas from cholangiocarcinomas by reaction with these two anticytokeratin antibodies.

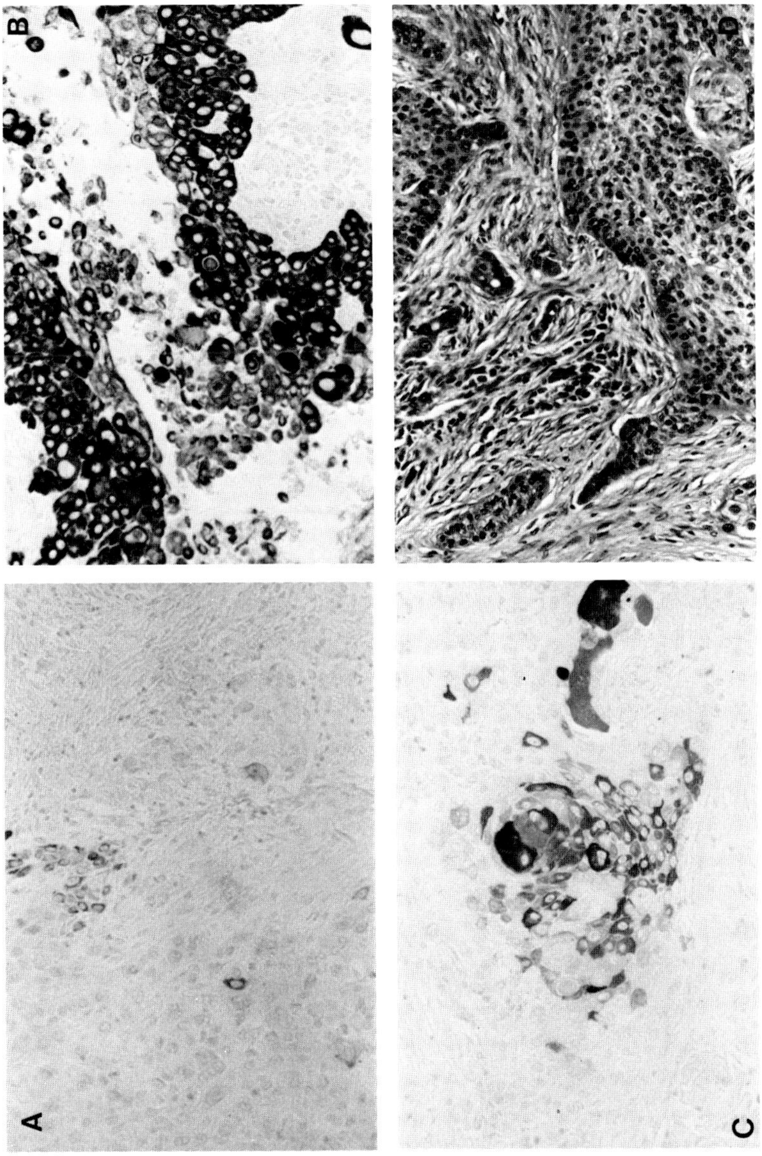

Figure 11. Squamous-cell carcinoma. (A) 35βH11, (B) 34βE12, (C) 34βB4, (D) hematoxylin-eosin stained section. 35βH11 reacts weakly with a few cells, while 34βE12 heavily stains each tumor cell. 34βB4 decorates only keratinizing areas.

Table 5. Reactivity of Anticytokeratin Antibodies on Carcinomas

Origin of neoplasm	Antibodies		
	35βH11	34βE12	34βB4
Squamous	−	+++	+
Mesothelium	+++	+++	−
Breast	+++	+++	−
Thymus	−	+++	N.D.[a]
GI tract	+++	+	−
Hepatoma	+++	−	−
Bile ducts (cholangiocarcinoma)	+++	+++	−
Endometrium	+++	−	−
Thyroid	+++	+	−

[a]N.D. = not done.

Hepatomas are 35βH11-positive and 34βE12-negative (Fig. 13), while cholangiocarcinomas are positive with both antibodies (Fig. 12).

Both 35βH11 and 34βE12 stain mesotheliomas, tumors of the mesothelial lining of the thorax and abdomen. Additionally, 34βE12 stains the keratinizing cells within thymomas, while 35βH11 does not. A summary of the staining of the different carcinomas is presented in Table 5. It is clear that these anticytokeratin antibodies are capable of distinguishing among different carcinomas.

4.3. Antineurofilament Antibodies

Both antibodies fail to react with many types of neoplasms including pheochromocytomas, carcinoid tumors, three neuroblastomas, glial tumors, and a single melanoma. It is not yet clear if these antibodies will be useful in tumor diagnosis.

5. Discussion

We have generated tissue-specific reagents by isolating hybridoma antibodies to the different intermediate filament proteins. These antibodies are more useful than conventional polyclonal antisera because they can be produced in large quantities, recognize a single epitope, and do not vary in activity. They can be used to characterize biochemical differentiation in morphologically undifferentiated neoplasms. Usually, the anatomic location and histologic features of the tumor are sufficient for accurate diagnoses. In a small percentage of cases, however, neoplasms are so undifferentiated that it is difficult to determine the cell type by standard histology. The differential diagnosis of undifferentiated tumors might include carcinoma, lymphoma, sarcoma, and melanoma. The monoclonal antibodies described here are use-

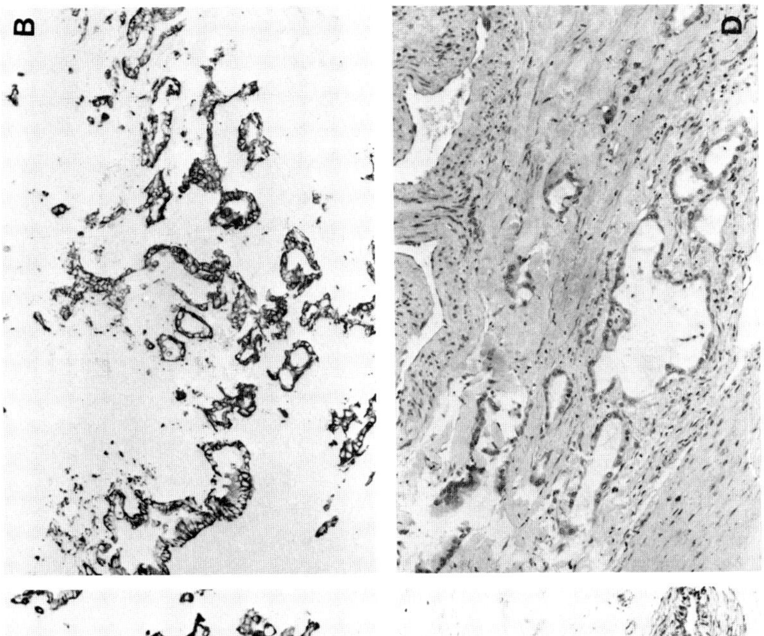

Figure 12. Cholangiocarcinoma. (A) 35βH11, (B) 34βE12, (C) 43βE8, (D) hematoxylin-eosin stained section. 35βH11 and 34βE12 each react strongly with tumor cells. 43βE8 decorates only the blood vessels, failing to stain the tumor cells.

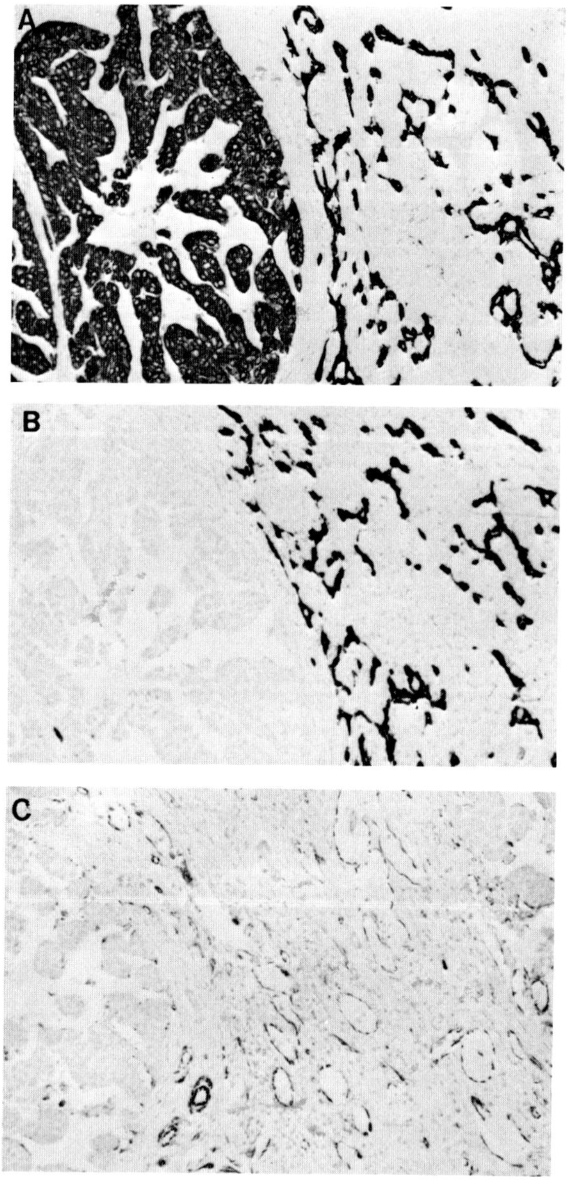

Figure 13. Hepatocellular carcinoma. (A) 35βH11, (B) 34βE12, (C) 43βE8. 35βH11 stains the nest of malignant hepatocytes and isolated benign bile ducts. 34βE12 recognizes only the bile ducts, not the tumor, and 43βE8 recognizes only blood vessels.

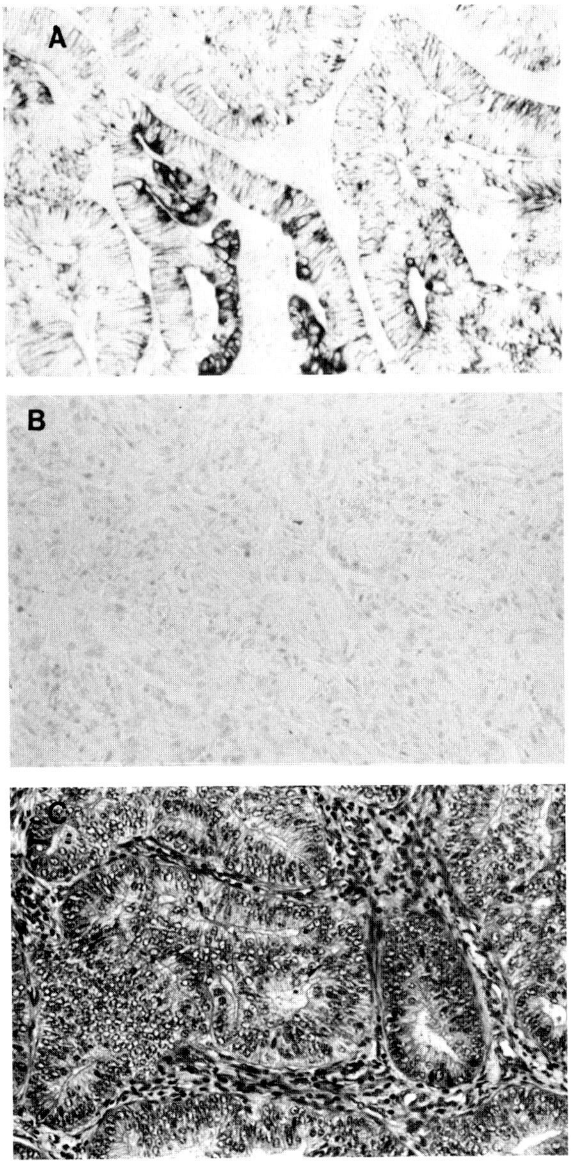

Figure 14. Endometrial carcinoma. (A) 35βH11, (B) 34βE12, (C) hematoxylin-eosin stained section. 35βH11 stains the malignant glands, while 34βE12 fails to recognize the tumor.

ful in identifying these different types of tumors. Any neoplasm stained by an anticytokeratin antibody is a carcinoma, while tumors decorated by 43βE8, the antivimentin antibody, are likely sarcomas. Different kinds of carcinomas can be distinguished based on reactivity with the three different cytokeratin antibodies. Lymphomas can be diagnosed by reaction with various nonfilament lymphoid-specific markers. Thus, useful panels of antibodies are being developed as diagnostic aids in surgical pathology.

This work clearly shows that antibodies to different cytoskeletal components can be used as tissue-specific markers. It may be possible to generate antibodies with greater specificity by immunizing with other cell-specific cytoskeletal components. For example, there are many different cytokeratins, some of which are found in only one or two types of cells (Moll *et al.*, 1982; Wu *et al.*, 1982). Antibodies to these proteins might be useful as specific markers for a single type of epithelial cell. Additionally, the vimentin and desmin of different cells may be slightly different (as suggested by the restricted reactivity of antibody 43βE8), and monoclonal antibodies capable of detecting these differences could be highly specific reagents. Finally, antibodies capable of distinguishing the three forms of actin (α, β, and γ) might also function as tissue-specific markers. An α-actin-specific antibody might selectively stain skeletal or cardiac muscle and be useful in detecting tumors arising in these tissues. The cytoskeleton may thus provide a rich source of cell-specific molecules which may be used to generate tissue-specific markers. Such markers will allow greater accuracy in the diagnosis of human tumors, which in turn should facilitate more effective treatment modalities.

ACKNOWLEDGMENTS. The excellent technical assistance of Marina Ferguson, Elaine Yamanaka, Helen Wan, and Jane Caughlan is gratefully acknowledged. Hepatoma lines Hep3B and HepG2 were obtained from Drs. David Aden and Barbara Knowles, Wistar Institute, Philadelphia, Pennsylvania. Dr. Helene Smith, Peralta Institute, Oakland, California, provided Hs-0578T and Hs-0700T. These latter two cell lines were produced with support from the National Cancer Institute, Biological Carcinogenesis Branch, Division of Cancer Cause and Prevention, under the auspices of the Office of Naval Research and Regents of the University of California. Dr. F. T. Chiu, Department of Neurology, Albert Einstein College of Medicine, provided the rat 200-kd neurofilament protein. This work was supported by grants HL-03174 and CA-28238 from the National Institutes of Health, and American Cancer Society Institutional Research Grant IN-26 administered through the Division of Oncology, University of Washington, via the American Cancer Society Institutional Cancer Grant Committee, chaired by Dr. Alexander Fefer.

References

Altmannsberger, M., Osborn, M., Schauer, A., and Weber, K., 1981, Antibodies to different intermediate filament proteins: Cell type specific markers in paraffin-embedded human tissue, *Lab. Invest.* **45**:427–434.

Bannasch, P., Zerban, H., Schmid, E., and Franke, W., 1980, Liver tumors distinguished by immunofluorescence microscopy with antibodies to proteins of intermediate-sized filaments, *Proc. Natl. Acad. Sci. USA* **77:**4948–4952.

Bignami, A., Dahl, D., and Rueger, D., 1980, Glial fibrillary acidic protein (GFA) in normal neural cells and in pathological conditions, *Adv. Cell Neurobiol.* **1:**285–310.

Dahl, D., and Bignami, A., 1973, Immunochemical and immunofluorescence studies of the glial fibrillary acidic protein in vertebrates, *Brain Res.* **61:**279–293.

Dahl, D., Bignami, A., Weber, K., and Osborn, M., 1981, Filament proteins in rat optic nerves undergoing Wallerian degeneration. II. Localization of vimentin, the fibroblastic 100 Å filament protein in normal and ractive astrocytes, *Exp. Neurol.* **73:**496–506.

Denk, H., Krepler, R., Artlieb, U., Gabbiani, G., and Franke, W., 1983, Immunocytochemical and biochemical approach to the classification of soft tissue tumors, *Cancer*, in press.

Eng, L., and Kosek, J., 1974, Electron microscopic localization of the glial fibrillary acidic protein and S-100 protein by immunoenzymatic techniques, *Trans. Am. Soc. Neurochem.* **5:**160–175.

Franke, W., Schmid, E., Osborn, M., and Weber, K., 1978a, Different intermediate sized filaments distinguished by immunofluorescence microscopy, *Proc. Natl. Acad. Sci. USA* **75:**5034–5038.

Franke, W., Weber, K., Osborn, M., Schmid, E., and Freudenstein, C., 1978b, Antibody to prekeratin: Decoration of tonofilament-like arrays in various cells of epithelial character, *Exp. Cell Res.* **116:**429–445.

Franke, W., Schmid, E., Winter, S., Osborn, M., and Weber, K., 1979, Widespread occurrence of intermediate sized filaments of the vimentin-type in cultured cells from diverse vertebrates, *Exp. Cell Res.* **123:**25–46.

Franke, W., Schiller, D., Moll, R., Winter, S., Schmid, E., Engelbrecht, I., Denk, H., Krepler, R., and Platzer, B., 1981, Diversity of cytokeratins: Differentiation specific expression of cytokeratin polypeptides in epithelial cells and tissues, *J. Mol. Biol.* **153:**933–959.

Fuchs, E., and Green, M., 1978, The expression of keratin genes in epidermis and cultured epidermal cells, *Cell* **15:**887–897.

Gabbiani, G., Kapanci, Y., Barazzone, P., and Franke, W., 1981, Immunochemical identification of intermediate-sized filaments in human neoplastic cells: A diagnostic aid for the surgical pathologist, *Am. J. Pathol.* **104:**206–216.

Gard, D., Bell, P., and Lazarides, E., 1979, Coexistence of desmin and the fibroblast intermediate filament subunit in muscle and normal cells: Identification and comparative peptide analysis, *Proc. Natl. Acad. Sci. USA* **76:**3894–3898.

Geisler, N., and Weber, K., 1981, Comparison of the proteins of two immunologically distinct intermediate-sized filaments by amino acid sequence analysis: Desmin and vimentin, *Proc. Natl. Acad. Sci. USA* **78:**4120–4123.

Gown, A., and Vogel, A., 1982, Monoclonal antibodies to intermediate filament proteins of human cells: Unique and cross-reacting antibodies, *J. Cell Biol.* **95:**414–424.

Hus, S., Raine, L., and Fanger, H., 1981, Use of avidin-biotin-peroxidase complex (ABC) in immunoperoxidase techniques: A comparison between ABC and unlabeled antibody (PAP) procedures, *J. Histochem. Cytochem.* **29:**577–580.

Hynes, R., and Destree, A., 1978, 10 nm filaments in normal and transformed cells, *Cell* **13:**151–163.

Lazarides, E., 1980, Intermediate filaments as mechanical integrators of cellular space, *Nature (London)* **283:**249–256.

Lazarides, E., 1982, Intermediate filaments: A chemically heterogeneous, developmentally regulated class of proteins, *Annu. Rev. Biochem.* **51:**219–250.

Lazarides, E., and Hubbard, B., 1976, Immunological characterization of the subunit of the 100 Å filaments from muscle cells, *Proc. Natl. Acad. Sci. USA* **73:**4344–4348.

Lehtonen, K., Agikainen, J., and Badley, R., 1982, Rhabdomyoma: Ultrastructural features and distribution of desmin, muscle type of intermediate filament protein, *Acta Pathol. Microbiol. Immunol. Scand.* **90:**125–129.

Liem, R., Yen, S., Salomon, G., and Shelanski, M., 1978, Intermediate filaments in nervous tissue, *J. Cell Biol.* **79:**637–645.

Moll, R., Franke, W., Schiller, D., Geiger, B., and Krepla, R., 1982, The catalog of human cytokeratins: Patterns of expression in normal epithelia, tumors and cultured cells, *Cell* **31:**11–24.

Ramaekers, F., Puts, J., Kant, A., Moesker, O., Jap, P., and Vooiss, G., 1981, Use of antibodies to intermediate filaments in the characterization of human tumors, *Cold Spring Harbor Symp. Quant. Biol.* **46:**331–339.

Schlegel, R., Banks-Schlegel, S., and Pinkus, G., 1980a, Immunohistochemical localization of keratin in normal human tissues, *Lab. Invest.* **42:**91–96.

Schlegel, R., Banks-Schlegel, S., McLeod, J., and Pinkus, G., 1980b, Immunoperoxidase localization of keratin in human neoplasms: A preliminary study, *Am. J. Pathol.* **101:**41–49.

Shaw, G., Osborn, M., and Weber, K., 1981, An immunofluorescence microscopical study of the neurofilament triplet proteins, vimentin and glial fibrillary acidic protein with the adult rat brain, *Eur. J. Cell Biol.* **24:**20–27.

Starger, J., and Goldman, R., 1977, Isolation and preliminary characterization of 10 nm filaments from baby hamster kidney (BHK-21) cells, *Proc. Natl. Acad. Sci. USA* **74:**2422–2426.

Steinert, P., Idler, W., Cabral, F., Gottesman, M., and Goldman, R., 1981, In vitro assembly of homopolymers and copolymers filaments from intermediate filament subunits of muscle and fibroblastic cells, *Proc. Natl. Acad. Sci. USA* **78:**3692–3696.

Sun, T., and Green, H., 1978, Keratin filaments of cultured human epidermal cells. Formation of intermolecular disulfide bonds during terminal differentiation, *J. Biol. Chem.* **253:**-2053–2060.

Sun, T., Shih, C., and Green, H., 1979, Keratin cytoskeletons in epithelial cells of internal organs, *Proc. Natl. Acad. Sci. USA* **76:**2813–2817.

Tseng, S., Jarvinen, M., Nelson, W., Huang, J., Woodcock-Mitchell, J., and Sun, T., 1982, Correlation of specific keratins with different types of epithelial differentiation: Monoclonal antibody studies, *Cell* **30:**361–372.

Tuszynski, G., Frank, E., Damsky, C., Buck, C., and Warren, L., 1979, The detection of smooth muscle desmin-like protein in BHK21/C13 fibroblasts, *J. Biol. Chem.* **254:**6138–6143.

Wu, Y., Parker, L., Binder, N., Beckett, M., Sinard, J., Griffiths, C., and Rheinwald, J., 1982, The mesothelial keratins: A new family of cytoskeletal proteins identified in cultured mesothelial cells and nonkeratinizing epithelia, *Cell* **31:**693–703.

Yen, S., and Fields, K., 1981, Antibodies to neurofilament, glial filament, and fibroblast intermediate filament proteins bind to different cell types of the nervous system, *J. Cell Biol.* **88:**115–126.

Addendum to Chapter 1

Manfred Schliwa

In this addendum, I will briefly discuss some papers, pertaining to the subject of intracellular organelle transport, that appeared while this review was written and in press. This additional material is organized according to the subject headings used in the text.

2.1. Foreign particles such as latex beads microinjected into crab axons will move anterogradely in a manner indistinguishable from that of endogenous organelles (Adams and Bray, 1983). This observation can be interpreted in terms of a transport system capable of carrying almost any particle with suitable surface properties.

2.2. Frixione (1983) described the existence of structural associations between pigment granules and microtubules in crayfish retinula cells which appear strong enough to survive cell disruption.

3.1.1. Possible mechanisms of force generation for fast axoplasmic transport have been considered in two theoretical papers (Weiss and Gross, 1983; Gross and Weiss, 1983) which expand on the microstream hypothesis for axoplasmic transport that was formulated previously. Current concepts for force-generating mechanisms are contrasted with a list of the most important properties of axoplasmic transport. The authors have concluded that the available experimental data best fit the microstream concept which holds that an essentially nonspecific shear force is generated near microtubules and thus creates low viscosity domains of streaming cytoplasm.

Hayden *et al.* (1983) have formulated a similar concept for particle movement in cultured cells which also proposes the existence of microtubule-associated low viscosity channels. In addition, this model incorporates Ochs's transport filament hypothesis. In this model, microtubules would define low viscosity channels in which the actin filaments (with their attached load) move by interacting with soluble myosin molecules; the direction of movement depends on the actin filament polarity. This model is derived from the finding that particles move bidirectionally along linear structures believed to be single microtubules. Microtubules would have no part in the force-generating mechanism.

3.1.2. Beckerle and Porter (1983) have analyzed the relative importance of microtubules and of an actin-based contractile system in *Holocentrus* erythrophores. As in other pigment cell types, some residual and undirected movement of granules can be elicited in the complete absence of microtubules, even though their presence is essential for organized, radial transport. Microinjected phalloidin, DNase I, and NEM-HMM have no effect on pigment granule motility.

3.1.3. The existence of an ATPase activity intimately associated with neuronal microtubules has been questioned (Murphy *et al.*, 1983a). The ATPase activity previously discovered in preparations of brain microtubule protein is greater than 90% particulate in nature and is probably derived from contaminating membraneous material. It seems to be identical with the F1-ATPase present in mitochondria (Murphy *et al.*, 1983b). The well-known proteins associated with neuronal microtubules (MAP_1, MAP_2, and tau factors) do not contain significant amounts of ATPase activity.

The effect of vanadate on saltatory organelle movements has been studied in a cell system which possesses an internal control. Buckley and Stewart (1983) microinjected vanadate into ciliated epithelial cells and observed an immediate cessation of ciliary beating, while saltatory organelle movements in the cytoplasm of the same cell proceeded undisturbed. This pronounced differential inhibitory effect on the movement of cilia and organelles suggests that organelle movements are not brought about by a dynein ATPase similar to that of cilia, and that the vanadate sensitivity of cytoplasmic dynein, if proven to exist, is vastly different from that of ciliary dynein. The results of this study vary with previous observations of the effect of vanadate on saltatory organelle movements.

3.1.7. Microinjection of antibodies against keratin filaments into cultured PtK_2 cells will not interfere with saltatory movement of intracellular particles, even though it will induce a complete disruption of the keratin filament bundle network (Klymkowsky *et al.*, 1983). Thus, it appears that the collapse of intermediate filaments, whether of the vimentin or the keratin type, does not affect saltatory particle motion.

In their study of the intracellular movement of Concanavalin A-containing vesicles, Herman and Albertini (1982) came to a different conclusion. These endocytic vesicles are proposed to associate initially with intermediate filaments, which are believed to be held in an extended configuration by cytoplasmic microtubules. Microtubule disruption promotes centripetal displacement of the vesicle-intermediate filament complex.

3.2.3. Shimmen and Tazawa (1983) have improved their permeabilization procedure of internodal cells by using a pretreatment with an EGTA-containing medium, followed by plasmolysis at low temperature. No detergent was used. Streaming at almost normal rates continues for more than 30 min, but is inhibited by magnesium depletion or by cytochalasin B. Calcium at concentrations above 0.5 μM completely and reversibly inhibits streaming (Tominaga *et al.*, 1983).

3.2.4. Sheetz and Spudich (1983) have developed an assay for myosin-dependent movement based on the interaction of HMM-coated fluorescent beads with actin filament bundles in dissected *Nitella* internodal cells. Myosin-coated beads are observed to move unidirectionally along actin cables at rates comparable to those in muscle and other cells, but at rates far lower than streaming in intact *Nitella* cells. The experiments demonstrate that myosin molecules can walk along actin filament bundles, and that this actin-myosin interaction is capable of translocating a particle. Whether streaming in intact cells is based on this principle remains to be demonstrated.

References

Adams, R. J., and Bray, D., 1983, Rapid transport of foreign particles microinjected into crab axons, *Nature* **303:**718–720.

Beckerle, M. C., and Porter, K. R., 1983, Analysis of the role of microtubules and actin in erythrophore intracellular motility, *J. Cell Biol.* **96:**354–362.

Buckley, I., and Stewart, M., 1983, Ciliary but not saltatory movements are inhibited by vanadate microinjected into living cultured cells, *Cell Motility* **3:**167–184.

Frixione, E., 1983, Firm structural associations between migratory pigment granules and microtubules in crayfish retinula cells, *J. Cell Biol.* **96:**1258–1267.

Gross, G. W., and Weiss, D. G., 1983, Intracellular transport in axonal microtubular domains. II. Velocity profile and energetics of circumtubular flow, *Protoplasma* **114:**198–209.

Hayden, J. H., Allen, R. D., and Goldman, R. D., 1983, Cytoplasmic transport in keratocytes: direct visuazization of particle translocation along microtubules, *Cell Motility* **3:**1–20.

Herman, B., and Albertini, D. F., 1982, The intracellular movement of endocytic vesicles in cultured granulosa cells, *Cell Motility* **2:**583–598.

Klymkowsky, M. W., Miller, R. H., and Lane, E. B., 1983, Morphology, behavior, and interaction of cultured epithelial cells after an antibody-induced disruption of keratin filament organization, *J. Cell Biol.* **96:**494–509.

Murphy, D. B., Hiebsch, R. R., and Wallis, K. T., 1983a, Identity and origin of the ATPase activity associated with neuronal microtubules. I. The ATPase activity is associated with membrane vesicles, *J. Cell Biol.* **96:**1298–1305.

Murphy, D. B., Wallis, K. T., and Hiebsch, R. R., 1983b, Identity and origin of the ATPase activity associated with neuronal microtubules. II. Identification of a 50,000-dalton polypeptide with ATPase activity similar to F-1 ATPase from mitochondria, *J. Cell Biol.* **96:**1306–1315.

Sheetz, M. P., and Spudich, J. A., 1983, Movement of myosin-coated fluorescent beads along actin cables *in vitro*, *Nature* **303:**31–35.

Shimmen, T., and Tazawa, M., 1983, Control of cytoplasmic streaming by ATP, Mg^{2+}, and cytochalasin B, *Protoplasma* **115:**18–24.
Tominaga, Y., Shimmen, T., and Tazawa, M., 1983, Control of cytoplasmic streaming by extracellular Ca^{2+} in permeabilized *Nitella* cells, *Protoplasma* **116:**75–77.
Weiss, D. G., and Gross, G. W., 1983, Intracellular transport in axonal microtubular domains. I. Theoretical considerations on the essential properties of a force generating mechanism, *Protoplasma* **114:**179–197.

Index